Rapid Eye Movement Sleep

B.N. Mallick ■ S. Inoué (Editors)

RAPID EYE MOVEMENT SLEEP

Marcel Dekker, Inc.
New York • Basel

EDITORS
Birendra Nath Mallick, Ph.D
Associate Professor of Neurobiology
School of Life Sciences
Jawaharlal Nehru University
New Delhi-110 067, INDIA

Shojiro Inoué, D.Sc
Professor and Head, Division of Biocybernetics
Institute for Medical and Dental Engineering
Tokyo Medical and Dental University
Kanda-Surugadai 2–3–10, Chiyoda-ku
Tokyo 101–0062, JAPAN

Exclusive distribution in Europe, North and South America, Australia and Japan by Marcel Dekker, Inc., 270 Madison Avenue, New York, New York 10016

ISBN 0-8247-0322-7 Marcel Dekker, Inc., USA

Printed in India

Foreword

Around the Discovery of REM Sleep in Cats

Michel Jouvet
Claude Bernard University, Lyon, France

It is the privilege of one of the oldest hypnologists to recall how Neurophysiology in the 1950's was very different from now.

During my internship in the Medical School (1951–53) I spent one year as a Resident in Neurosurgery. This is why I became interested in Brain Physiology. At this time, Pavlov was considered as a great neurophysiologist, and I still remember the visit of two pupils of Pavlov, Bykov and Asratian, at the Lyon's Medical School (France) in 1952. I had read Pavlov's books with much interest and I was convinced that the cerebral cortex was necessary for conditioned reflex and even for sleep (through the process of the so-called "internal inhibition").

Later I started to learn about electroencephalography when the first 6-channel EEG polygraph was bought in the Neurology Department. One day, I read the first issue (1949) of *Electroencephalography and clinical neurophysiology* and discovered the historical paper by Moruzzi and Magoun about the reticular formation. To my great surprise, I learnt that there was a waking system in the brainstem, and that the EEG was certainly a better method to study the brain than the drops of saliva from the parotid duct of a conditioned dog. For this reason, thanks to a Fullbright fellowship, I went to work for one year (1954) in Professor Horace W. Magoun's laboratory located in the Veteran's Administration Hospital in Long Beach (California).

I had the privilege to share a small office with Professor Toshihiko Tokizane who was also spending a year in Magoun's laboratory before going to Minneapolis in Pr Gellhorn's laboratory. I remember very well the extraordinary patience of Pr Tokizane who was measuring with a decimeter the intervals between unit recording from muscle spindles afferent. Everyday, he used to draw by hand some histograms from the kilometers of oscilloscope recordings on films that he was receiving from Tokyo. Finally he obtained a two peak histogram showing that there were two kinds of unit activity. He did in six months what a microcomputer could do now in a few minutes.

I became a close friend of Pr Tokizane and we discussed often about the future strategy of developing brain research together in Japan, in Europe and in the

USA. This was the beginning of a long friendship which continued until his death.

Besides Pr Tokizane, there were many brilliant neurophysiologists working in Magoun's lab, like Drs French, Livingstone, Ross Adley, J. Green, T. Sawyer, Terzuolog, Eva King and many others. I explained to Pr Magoun that I was interested in conditioned reflex and that I would try to investigate the putative alterations of auditory evoked potential in the auditory cortex of a cat when a tone became associated with a painful shock to the leg. At this time, everybody was working on acute, curatized or anesthetized preparations, but in order to study conditioning, I had to develop a technique for recording chronically-implanted cats. Finally, I could obtain good recording of auditory evoked potentials. Curiously, they were bigger and also appeared in the sensory-motor cortex when associated with shocks. There was in fact some kind of conditioning or plasticity of electrical activity. My interest in this electrophysiological approach to learning was later increased by two different factors. First, Pr Herbert Jasper came to visit Pr Magoun and gave a lecture on habituation of the arousal. I borrowed from him for a week-end the thesis of Sharpless on this subject. At that time, there was no way to reproduce a thesis in the laboratory and I had to copy it with the help of my girl-friend, during two days and nights (I still have this hand-made copy).

Secondly, Raul Hernandez-Peon came in Magoun's lab. He was from Yucatan in Mexico and he really looked like a Mayan statue... Hernandez-Peon was also much interested in the electrical events which could accompany learning and habituation or attention. We started together to record from the cochlear nucleus of chronically implanted cats and we discovered the so-called "afferent neuronal habituation", i.e., the decrease of amplitude of auditory evoked activity (from a loud-speaker) in the cochlear nucleus when the tone was repeated for a long time (hours). We also noticed that auditory evoked activity would decrease when the cat was attentive to another stimulus (a bird or the odor of a fish). This was really the beginning of the new age of electrophysiology correlated with psychophysiology (Hernandez-Peon et al., 1956, 1957). During this work, Pr Tokizane was very much interested in our results and he gave me the address of many of his students whom I met during my return trip to Europe when I visited Japan. This was the first time I met Pr Mannen in Tokyo and Pr Matsumoto in Osaka. Pr Matsumoto was also much interested in Pavlovian physiology and had acquired a good knowledge of Russian. We decided that if he could get the money, he would come to work in my laboratory. Finally, he came in 1963, when he started his pioneering work on monoamines and sleep. Pr Matsumoto was the first Japanese of a long series of about 30 Japanese scientists who came to my laboratory and played a determinant role in the research of my laboratory.

In Osaka, I also met Pr Yoshii who was interested in a new form of conditioning "the frequency specific response", i.e., if a cat was subjected to a 3-sec bright light accompanied by intermittent auditory stimulations at 8-12-16-20 Hz, after many repetitions, the light alone would induce the same rhythm in the auditory cortex. Then I left Osaka and after a 3-month trip returned to France in November 1955 and decided to work on electrophysiological conditioning (Jouvet, 1956). At this time I had to start a neurophysiological laboratory from nothing. I "borrowed" the EEG polygraph from the operating room and built a very primitive Horsley-Clark apparatus. I got more and more disappointed by the so-called "frequency specific repetitive responses" because the frequency specific response was waning and waning as cortical spindles. In fact, the cat became so drowsy during the association of light and tone that it was impossible to obtain a stable background of fast cortical activity. This kind of sleep was in fact looking like the "internal inhibition of Pavlov". Since, at this time, according to Jasper, the spindle activity would be under the control of intralaminar nuclei of the thalamus, I wrote a paper about the role of non specific thalamus in the "frequency specific conditioning response". This paper (one of the worst in my entire life) did not raise any question from the audience at the International Congress in Brussels (Jouvet and Courjon, 1951). For this reason, at the end of 1957, I decided to study the mechanism of the habituation of cortical arousal in chronic cats. It was very easy to observe it in normal cats. The first presentation of a tone in a sleeping cat would induce a cortical arousal lasting about 2 min whereas the fifth or sixth ones would not arouse the cat any more.

In order to disclose which theory was right—Pavlov and the cortex or Magoun and the reticular formation—we decided, together with Francois Michel, a young intern working with me, to study habituation of arousal in chronic decorticated or chronic mesencephalic or pontine cats (in which we had removed all the brain in front of the transaction, in a true Sherringtonian tradition). Since we could record slow wave activity during sleep (slow wave sleep, the only sleep known at this time) in the thalamus or the ponto-mesencephalic reticular formation together with slow wave activity at the cortical level, we thought it would be easy to recognize sleep in those decorticated or pontine preparations.

If we could still observe the habituation of arousal in a chronic decorticated or mesecephalic cat, then Pavlov's hypothesis was wrong and the reticular formation hypothesis was perhaps right.

It took some time to learn to keep chronic decorticated cats or mesencephalic cats in good condition, and we found that we could not decide which theory was right because it was impossible to record any slow wave from the subcortical structures (thalamus or reticular formation) of these preparations. On the other hand, similar subcortical recordings of partially decorticated cats still exhibited quite well the alternation from slow wave to fast activity (and some capricious habituation of the arousal). We also observed that our totally decorticated cats were sometimes behaviorally asleep even without exhibiting any spindles or slow waves in the brainstem. Since we could not record any spindles or slow waves in the total absence of the neocortex, we thought that the neocortex was responsible for slow-wave activity in the brainstem (Jouvet and Michel, 1958).

As a next step, we decided to record EMG (electromyographic) activity to record objectively some motor reaction which could habituate easily. As any neurophysiologist knows, the easiest placement of an EMG electrode when implanting subcortical electrodes is in the neck muscles because they are very close to the skull. Since we were operating on mesencephalic and pontile cats (from which we had removed all the neural structures in front of the pons), the only places to implant subcortical electrodes were the pons and, of course, the reticular formation. We were afraid that the implantation of electrodes in the medulla would be dangerous for these very fragile preparations.

At first we were very disappointed with our experiment. Even very loud auditory stimuli did not evoke much in the way of behavioral reactions in our chronic mesencephalic or pontile preparations, and it was almost impossible to record the EMG startle reaction because of the increase of muscle tone due to the decerebrate rigidity. However, during long-lasting EEG recordings (at this time only 3 to 6 hours) we were surprised to see, every 30 to 40 min, a periodic appearance of "spindle-like" activity in the pons which coincided with the total disappearance of the EMG of the neck. These curious episodes lasted about 6 min. There were also some movements of the vibrissae, but we did not notice the rapid eye movements which are almost unobservable in mesencephalic cats (at least during the first days after the brainstem transaction). We first thought that the "spindle-like" activity we had noted was an artifact of a central response evoked by the movements of the vibrissae. As good neurophysiologists, we cut the vibrissae! Fortunately, the central phasic EEG phenomena persisted unaltered.

Apparently, then, we were faced with some kind of "hindbrain (rhombencephalic) sleep" which contrasted with slow-wave sleep (SWS). Very quickly we started similar polygraphic recording in normal cats. We were surprised to see that the cortical activity was similar to the cortical activity seen during waking at the time of total disappearance of EMG, but that the threshold for arousal was much increased. This was a paradoxical finding. At this time Dement (1958) had just published his classical paper on "activated sleep". It became evident that this "activated sleep" with rapid eye movements was in fact something very different from SWS, and that it was a different phase or state of sleep. To our surprise, we had to conclude that "dreaming" had to be triggered by a structure located in the lower brainstem (Jouvet and Michel 1959; Jouvet et al, 1959b).

Since "rhombencephalic or paradoxical sleep (PS)" could be recognized readily on behavioral (atonia and rapid eye movements) or EEG (spindle-like activity in the pons) criteria, it was relatively easy to delimit the structures responsible for its triggering by local coagulation of the pontile reticular formation, Next, together with Daniele Mounier, we observed that the lesions destroying the dorsolateral part of the pontine tegmentum could abolish selectively PS. Unfortunately, at this time there was not yet a histochemical map of the brainstem and the locus coeruleus was still a *terra incognita* on most of the stereotaxic atlases. Thus, the caudal part of *N. reticularis pontis oralis* and the rostral part of *N. reticularis pontis caudalis* were thought to be responsible for PS. There was nothing either in the histological aspect of these nucleri or in their projections which could give us some clue for understanding the mechanism of PS or its functions (Jouvet and Mounier, 1960). Apparently this state of sleep was an active one, since we could trigger it by stimulating the pons during SWS. But the results of stimulation were rather capricious and we soon learned that PS could not be induced or lengthened at will, but that some "refractory period" existed after each episode. This led us to speculate about some possible "neurohumoral" mechanism which would "discharge" periodically during sleep. Since atropine had a potent and selective suppressor effect, and eserine a facilitatory effect on PS, we speculated that cholinergic mechanisms were in some way implicated in PS.

It was evident, also, that some logical explanation had to be found concerning the relationship between the two states of sleep. It was tempting to speculate about some evolutionary process. Since PS still existed in pontile cats, it could be described as rhombencephalic sleep, whereas SWS could be described as telencephalic sleep, since it was possible that it was related to the complex organization of the neocortex.

The hypothesis that SWS depends upon the cortex and PS depends upon the rhombencephalic sleep is still valid today. PS is also found in animals without eyes (as the mole) and in birds which do not move their eyes (the owl). This is why, I still believe that the term of REM sleep is probably not the best to describe this strange state of sleep whose function is still unknown.

References

1. Dement, W.C. The occurrence of low voltage, fast, electroencephalographic patterns during behavioral sleep in the cat, Electroenceph. Clin. Neurophysiol., 10 (1958) 291–295.
2. Hernandez-Peon, R., Scherrer, H. and Jouvet, M. Modification of electrical activity in the cochlear nucleus during “attention” in unanesthetized cats, Science, 123 (1956) 331–332.
3. Hernandez-Peon, R., Jouvet, M. and Scherrer, H. Auditory potentials at cochlear nucleus during acoustic habituation, Acta Neurol. latino Amer., 3 (1957) 144–156.
4. Jouvet, M. Analyse électroencéphalographique de quelques sortes de conditionnement chez le chat, Acta Neurol. latino Amer., 2 (1956) 107–115.
5. Jouvet, M. and Courjon, J. Discussion du rapport sur le conditionment et l’apprentissage. 1 er Congrés International des Sciences Neurologiques et IVe Congrès International d’E.E.G. et de Neurophysiologie Clinique. Volume des rapports, 1957, 334–338.
6. Jouvet, M. and Michel, F. Recherches sur l’activité électrique cérébrale au cours du sommeil. C.R. Soc. de Biol., 152 (1958) 1167–1170.
7. Jouvet, M. and Michel, F. Corrélations électromyographiques du sommiel chez le chat décortiqué et mésencéphalique chronique. C.R. Soc. de Biol., 153 (1959) 422–425.
8. Jouvet, M., Michel, F. and Courjon, J. Aspects électroencéphalographiques de deux mécanismes inhibiteurs, télencéphalique et rhombencéphalique entrant en jeu au cours du sommiel., J. de Physiol., 51 (1959a) 490–492.
9. Jouvet, M., Michel, F. and Courjon, J. Sur un stade d’activité électrique cérébrale rapide au cours du sommeil physiologique., C.R. Soc. de Biol., 153 (1959b) 1024–1028.
10. Jouvet, M. and Mounier, D., Effects des lésions de la formation réticulée pontique sur le sommeil du chat. C.R. Soc. de Biol., 154 (1960) 2301–2305.

Preface

The knowledge of Rapid Eye Movement (REM) Sleep has increased significantly and is difficult to be covered in one review. Hence, we thought it was time to bring out an edited book on the subject containing advanced and up-to-date scientific information on this special and exclusive state of sleep. We expect that such a book is likely to get a global circulation among all sections of students, teachers and researchers alike.

Fortunately, in response to our appeal, a number of leading scientists in the field across the globe agreed to contribute to the book. Thus, this book deals with various aspects of REM sleep, from phenomena to molecular processes. This is probably the first comprehensive monograph that has ever been published exclusively on the subject. We are sincerely grateful to all the contributors for keeping the deadline in spite of limitations. We are also thankful to others who could not contribute due to prior commitments but encouraged us to go ahead with this publication. We do hope that this book will serve our original purpose for every reader.

We are particularly proud of carrying in this book the 'Foreword' by Prof. Michel Jouvet, who found REM sleep (paradoxical sleep) in 1957 in cats, and the 'Introduction' by Prof. Eugene Aserinsky, who discovered REM sleep in 1953 in humans. On behalf of the contributors, we would like to express our heart-felt thanks to them.

We have not included the disease aspects in this volume. Although we tried to include the latest knowledge in the field, we do not claim it to be complete and absolute. There are certain limitations which we hope the readers will bear with us and ignore for the present and take advantage of the knowledge contained in it. If possible, an attempt will be made to overcome the limitations in future editions.

We are also thankful to the publisher for being so understanding and helping us to bring out this book to promote the science of REM sleep.

EDITORS

Contents

Rapid Eye Movement Sleep
B.N. Mallick and S. Inoué (Eds)

1. Eyelid Condition at Birth: Relationship to adult mammalian sleep-waking patterns

Eugene Aserinsky[†]

Professor Emeritus, Department of Physiology, School of Medicine, Marshall University, West Virginia

SUMMARY

Twentyeight mammalian species having 24 h sleep data were derived from the literature and divided into two groups—those born with the eyes open (EO) and those born with the eyes closed (EC). The eyelid condition at birth is indicative of an advanced embryological stage of development, and together with the duration of the gestation period can provide an index of the rate of maturation. The EO condition implies relatively rapid brain growth *in utero* while the EC condition signifies relatively slower growth.

The EO species had a mean total sleep duration of 8.8 h and a mean duration of 1.3 h of REM sleep, whereas the EC species had significantly longer sleep durations—14.0 h total sleep and 2.5 h REM sleep. There was no significant difference in REM % (REM time/total sleep time) between the two groups.

Body size, duration of the gestation period and the rate of brain maturation are all positively correlated with each other. Therefore, the EO species tend to be larger and to have longer gestation periods than those in the EC group. But whether a given species is in the EO or in the EC group is dependent upon its relationship to other species in the same family or order. Thus the chinchilla and the guinea pig which are in the EO group are relatively large and have relatively long gestation periods compared to other rodents but are considerably smaller than most other species in the EO category.

Recognition that mammalian species comprise two subpopulations (i.e. EO and EC) requires that statistical treatments be handled accordingly.

Key Words: REM sleep, Sleep, Brain Maturation, Mammals, Eyelids, Birth.

This book which is devoted exclusively to REM sleep is a testament to the scientific progress which has been made in the first fifty years since the discovery of the REM state. Well, almost fifty years. Historically, eye movements in sleep had been mentioned in an ancient Chinese legend and in more recent times by at least three scientists but in each of those instances

[†]To our great sorrow, Dr. Aserinsky was killed on July 22, 1998 in a traffic accident.

the observed eye movements were actually 'slow eye movements' (Aserinsky, 1996) and, at any rate, there was not yet any recognition of the existence of a REM state. The first public report of REMs and the REM state was made in 1953 while the first suspicion that rapid eye movements were not an artifact was made about a year earlier.

The immediate impact of the discovery of REM sleep on scientific reasoning was at first disbelief of the validity of the finding, and then shock. It was with reluctance that Nathaniel Kleitman finally recognized that his universally accepted notion that sleep is a purely passive phenomenon was incontrovertibly wrong; ironically, the work which negated his theory was carried out several meters away from his office at the University of Chicago. And then, of course, there was the inevitable conclusion that Freud's fanciful physiological concepts regarding sleep and dreaming were no longer tenable. There followed a spate of scientific papers on almost every conceivable parameter that could be measured before during and after REM sleep. In 1954, *Psychological Abstracts* published its first reference to REM sleep and just ten references indexed under 'Sleep'. Just 13 years later *Citation Index* included some 1400 titles under the category of 'Sleep' and about 120 titles under 'REM'. In 1996 Citation Index listed 4500 'Sleep' titles and about 380 'REM' titles. Sleep clinics mushroomed across the world, licensure for sleep laboratory technicians became commonplace, and sleep became a recognized medical specialty.

The total impact of the REM discovery is impossible to gauge because numerous scientific publications have been triggered by that discovery but are not indexed under the category of 'REM sleep'. Thus a histological study of the Nucleus Ceruleus would not necessarily be indexed under 'Sleep' although the impetus for that work could be its connection to REM.

Although there are numerous books on sleep which, of course, includes REM sleep, this book may be the first dedicated solely to REM sleep topics. In the succeeding chapters, basic scientific work of the highest caliber is presented by key investigators devoted to the study of REM sleep. Clearly, not all topics could be covered in one volume. Particularly, 'soft' subjects such as the conjectures regarding the purpose of sleep are not addressed. This I will remedy to a very small extent in the remaining portion of this chapter wherein I report on some contentious conclusions based on a survey of the 24-hr sleep data of just about every mammalian species reported in the literature. The suggestion is made that sleep length, especially that of REM sleep, may be primarily dictated by the rate of brain maturation in utero. This will require further research for confirmation but the reward will be a new slant on 'purposefulness'. Thus far the focus has been on utilitarian purposes such as neural rest, memory consolidation, etc. but there may be a more primordial cause for sleep. CO_2 for example has a vital function in controlling respiration, a function that developed over eons. But, prior to its adaptation CO_2 served as the end product of metabolism.

If the caveman pondered the purpose and value of sleep he must surely have recognised that its function was obligatory and that the end result was restful. Basically the same notions persist in our modern age although they are cloaked in the more sophisticated language of metabolism and energy saving. With the discovery of REM sleep (Aserinsky, 1953) many theories sprouted but the 'purpose' of sleep remains as enigmatic as ever. As Hobson (Hobson, 1988) pointed out, the saving of about 10% in energy during sleep as compared to the resting state would hardly justify the evolution of so complicated a condition as sleep. As for the energy requirement of the brain, cerebral oxygen consumption during slow wave sleep in man is reduced by only 2.8% (Mangold et al., 1955). Nevertheless, in accordance with Aristotle's view (Aristotle, 1961) that, "There is a reason behind everything in Nature", the energy-saving purpose of sleep is still being expounded (Zepelin and Rechtschaffen, 1974).

The present discussion has its roots in the observation (Aserinsky, 1977) published in1977 that mammalian species could be characterised as either long or short sleepers depending upon whether they were born with the eyes closed or open. Borrowing from avian terminology, the two groups were designated as 'altricial' (helpless nestlings at birth) and 'precocial' (able to run about soon after hatching). However, in the present paper the two groups are simply termed 'eyes-open' (EO) and 'eyes-closed' (EC), respectively because strictly speaking the eyelid parameter alone was used to distinguish the two groups. Moreover, there is something strange in referring to the helpless, immature human infant as 'precocial', a term that was originally for the birds.

Significance of Eyelids at Birth

While the condition of the eyes at birth seems to be an unusual parameter, it is actually an excellent non-invasive measure of cerebral maturation. According to Witschi (Witschi, 1956), the opening of the eyes *in utero* or after birth is the principal feature characterizing the last stage of fetal development known either as Witschi stage 36 or Fetal stage III. It is relevant that the only animals exhibiting a REM-NREM sleep pattern are the mammals and birds whose brains attain the complexity of Stages 35 and 36 whereas reptiles, for example, when fully mature only reach Stage 34.

The actual opening of the eyes, as observed in newborn rabbits and kittens some days after birth, has not been associated with any significant change in the development of sleep (Shimizu and Himwich, 1968). Yet, as a marker of brain development, the time when the eyes open is a highly significant event. When considered in conjunction with the length of gestation it is possible to make an estimate of the 'relative' rate of brain maturation, comparing one species with another. This measure can be made by dividing the time from fertilization to the eye-opening stage (Witschi stage 36) by the length of the gestation period. For the human being this ratio is ca 0.57 (140 days/247 days) whereas in the rat it is 1.73, the numerator of the ratio being obtained from the sum of the gestation length plus the time from the end of gestation to when the eyes open after birth. The lower the ratio, the faster is the rate of development to the eye-opening stage. Differences between species, including body size, were compensated for by inclusion of the gestation duration in the denominator of the ratio. With the exception where partial correlations were obtained, these ratios were not employed in statistical analyses. Instead, where correlations were required between a sleep parameter and the EO-EC groups, the statistical test employed was the Glass Rank-Biserial Correlation which allows for determining the relationship between a set of ranks and a dichotomy.

According to the conecept being examined in this paper, it is the *rate* of brain maturation which seems to be linked to the development of sleep. The physiological changes concomitant with the eye-opening stage are barely known but they must be important. It is at this stage that the premature human infant has a chance for survival at birth. Moreover, it is at this time (in the guinea pig and other species) that the blood-CSF barrier to ferrocyanide is established (Stern and Peyrot, 1961). This indicates that at eye-opening time the fluid environment surrounding the CNS neurons is being stabilized and thus affords a mechanism for regulating the flow of hormones and neurotransmitters between brain cells and the general circulation.

Sleep and Rate of Brain Development

Becuase the EC group of species sleep longer than the EO group it is not unreasonable as a tentative presumption to consider that the length of sleep in the adult is predetermined *in utero* by the relative rate of brain development. This rate, in turn, is tied to the duration of the gestation period which in turn is a function of the size of the species and the size of the brain. All of these factors must be considered relative to other species in the same family or order. For example, in the cat (Felidae) family, the jaguar simply by virtue of its size might be expected to require a longer gestation period than say a puma or the domestic cat. Since the gestation period is longer one might expect the brain to develop at an accelerated rate. According to Blinkov and Glezer (1968), the larger the animal the faster the growth of the brain in relation to body weight *before birth*. Indeed the accelerated brain growth in the jaguar is indicated by its eye-open status, and it does sleep less than the smaller cats who are in the EC group.

Although it has been asserted (Jouvet-Mounier et al.,1970) that the degree of maturity at birth affects the perinatal sleep pattern there is no ready explanation as to why that should have an affect on the adult. There are, however, known CNS mechanisms which can allow for permanent changes in brain function simply because of temporal factors in development. For example, temporary interference with fetal growth in rats may result in growth defects after birth which are no longer susceptible to remedy (Widdowson, 1974). Becuase of heterochronism for development of various brain sites there are critical periods for the establishment of certain neural connections. Although the EC animals ultimately attain the same level of cerebral maturation as those in the EO group, the slower rate of brain development could have altered the neural pathways, and that in turn could have an impact on the sleep phenomenon.

Current Phylogenetic Sleep Data

A survey of phylogenetic data in the literature was restricted to all mammalian species for which there were 24 h sleep records and for whom the gestation periods were typical and measurable. Of the many hundreds of species theoretically available, only 28 species (including representatives of the primate, rodent, carnivore, insectivore, lagomorph and ungulate groups) met the criteria for selection. A number of species were excluded because of the murky data on gestation. Among the excluded were the monotremes (egg layers), marsupials (in which gestation continues in the pouch for an unknown period), bears and bats (with gestation interrupted during hibernation), other species because of delayed implantation, and moles because their lids are partially grown together, precluding their classification as EO or EC. Two species included for certain measures were the mouse and the rabbit. The reported difficulty in distinguishing Waking from REM reduced the value of the mouse data; similarly the necessity for preconditioning the rabbit for two weeks in order to record REM placed that animal's data in question. The sources for much of the data as well as the scientific names for the species are summarized to a great extent by Zepelin and Rechtschaffen (Zepelin and Rechtschaffen,1974) and by Campbell and Tobler (Campbell and Tobler, 1984).

Total Sleep Time and Eyes at Birth

When the daily sleep quota for each species is plotted against the logarithm of its gestation time (Fig. 1) the relationship does not appear linear because of three outlying points which represent ungulates. Those three points may be explained as the result of the investigators having arbitrarily included 'drowsy time' with 'waking time' thereby severely shortening the total sleep time for those species. However, that matter is not material to the concept under consideration. What is most striking is that all the closed-eye species (filled circles) are clustered at one end of the scattergram while the EO species (open circles) form a group at the other end. This dichotomy of the population suggests that there are two subpopulations. As a consequence, a statistical analysis treating all the species as a single population with respect to any parameter, whether it is sleep duration or metabolic rate, would be fraught with uncertainty. According to Sachs (Sachs, 1984), inhomogeneity of data requires that subgroups be treated separately or else the results could be erroneous or misleading. He cited a dramatic example in which the correlation coefficients for the Hb content of blood and the surface area of RBCs were close to zero when the data for men, women and children were tested separately but when the data for the three subgroups were combined the resultant

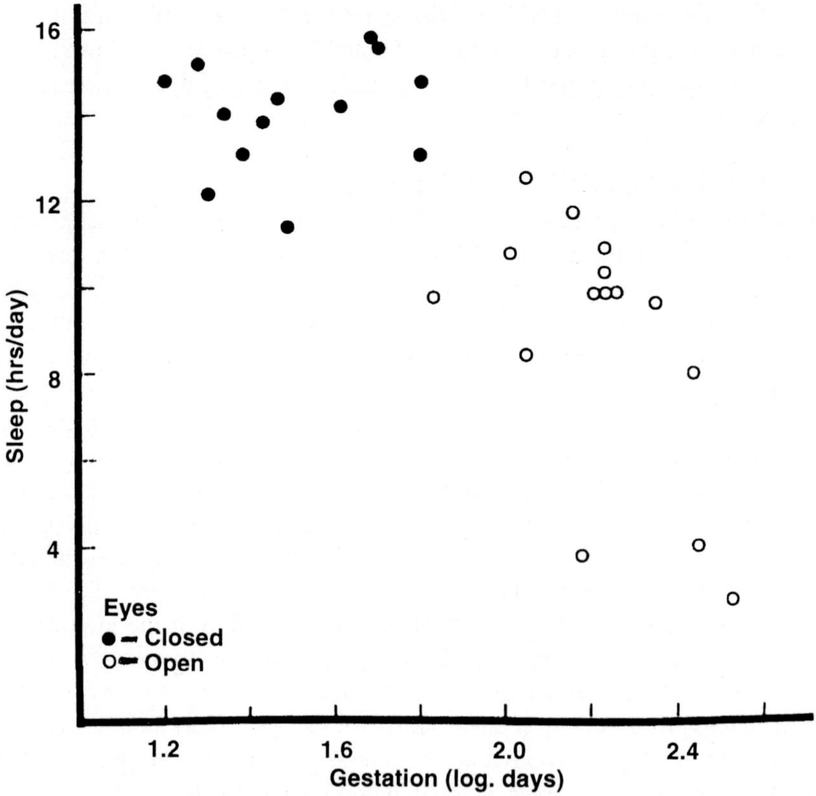

Fig. 1. Relationship of daily sleep duration and the length of the gestation period for 28 mammalian species. Animals born with eyes open and closed are shown by open and closed circles, respectively.

correlation coefficient was a whopping – 0.75—and essentially too obscure for interpretation.

Allowing curiosity to overcome statistical wisdom, the sleep durations for all 28 species were correlated with their gestation times; the resulting Spearman rank coefficient for the entire population was a statistically significant –0.83 but the meaning of this coefficient faded when the correlations were performed separately for the EO and EC groups; for the former group, r_s was 0.18 while for the latter group, r_s was –0.72 ($p < .005$). Interpretation of this result is obviously not clear-cut. To avoid an inhomogeneity correlation, subsequent analyses were performed on each group separately, either using a rank correlation procedure or a two-tailed Mann-Whitney U-test when comparing mean values of the two subgroups. Of course, it is possible that the lumping of different orders and families together have led to a multimodal distribution rather than just a bimodal distribution. Since there is no statistical measure for multimodality, only bimodality is being considered since its appearance is so blatant.

In Table 1, sleep data for the EO group can be compared with those for the

Table 1. Sleep Data for Species Born with Eyes Closed and for Species Born with Eyes Open

Eyes-Open Species	Gestation (days)	Total Sleep (h)	REM Duration (h)	REM (%)	Eyes-Closed Species	Gestation (days)	Total Sleep (h)	REM Duration (h)	REM (%)
Horse	336	2.86	0.78	27.3	Dog	63	13.12	2.80	21.3
Cow	284	3.97	0.75	18.9	Cat	63	14.79	3.25	22.0
Man	274	8.00	1.73	23.1	Tenrec	50	15.60	2.34	15.0
Chimpanzee	225	9.62	1.43	14.9	Tree shrew	48	15.79	2.59	16.4
Capuchin	180	9.85	1.39	14.1	Hedgehog	41	14.16	3.96	28.0
Squirrel monkey	175	9.90	2.70	27.3	Rabbit	31	11.40	1.09	9.6
Pata	170	10.85	0.86	7.9	Mt. Beaver	29	14.40	2.45	17.0
Baboon	168	10.33	1.25	12.1	Squirrel	27	13.80	3.41	24.7
Rhesus	167	9.91	1.36	13.7	Gerbil	24	13.07	2.12	16.2
Sheep	150	3.85	0.57	14.8	Rat	22	14.00	2.13	15.2
Bush baby	145	11.71	1.31	12.6	Mouse	20	12.11	1.44	11.9
Pig	112	8.42	2.19	26.0	Shrew	19	15.22	1.42	11.7
Chinchilla	111	12.53	1.55	12.4	Hamster	16	14.76	3.46	23.4
Jaguar	103	10.80	—	—					
Guinea pig	68	9.71	0.77	7.9					
$\overline{X} \pm SD$	177.9 ±73.9	8.82 ±2.95	1.33 ±0.59	16.6 ±6.7		34.9 ±16.5	14.02 ±1.31	2.50 ±0.89	17.9 ±5.6

EC group. The results reveal that for every parameter measured, except for REM percentage (REM duration/total sleep time), there is not only a statistically significant difference ($p < .002$) between the two groups but the differences are of very large magnitude. The EO group had a mean daily sleep quota of 8.8 h while the EC species obtained 14.0 h sleep. Similarly, the EO group had about half the REM sleep accumulated by the EC species. All of the gestation periods of the EC animals were longer than any of the gestation periods of the other groups. As for body weight, the range was 0.5–521.6 kg for the EO group, and 0.02–40.0 kg for the EC group, with medians of 19.5 and 0.54 kg, respectively.

The differences between the two groups could conceivably be due to the unequal distribution of animal families comprising the two groups. In the EO category there are eight primates and no carnivores whereas in the other group there are six rodents and no primates. In an attempt at equalization of the distribution, the data were averaged so that each family was represented by only one set of figures. Of course, this did not remedy the total absence of a particular family in one group while it was still represented in the other group. Bearing this deficiency in mind, the results obtained with this procedure were quite similar to those obtained with completely unweighted data. The EO families had a mean daily sleep quota of 9.2 h which was still considerably less than the 13.6 h accumulated by the EC group.

Brain Maturation Measured by Metamorphosis

The idea that sleep time is somehow a function of the rate of brain maturation was dependent in this paper on an all-or-none measure. By the method employed, the eyes at birth are considered either open or closed but not in any intermediate state. To ascertain if sleep time varies in graded fashion with the degree of brain maturation, a measure other than the condition of the eyes was used. From tables of biological data (Spector,1961; Altman and Dittmer, 1972; Evans and Sack, 1973; Kenneth and Ritchie, 1953) it was possible to obtain values for the time when embryological metamorphosis was completed for each of eight species in the EO category and also for six of the species in the EC group. This measure was thus available for half of the 28 species under analysis and, except for the insectivores, all the families were represented. The end of metamorphosis is equivalent to the end of Stage 33 and is the time just previous to the formation of the eyelids.

The measure for the rate of embryological maturation was taken as the ratio of the time for the completion of metamorphosis relative to the length of the gestation period. Therefore the smaller the ratio the faster is the development of the embryo.

Application of this measure revealed a mean (± S.D.) ratio of 0.24 ± .09 for the EO species and 0.61 ± .16 for the EC group, confirming that the split of the entire population into two groups based on an embryological measure is valid. In the EO group, metamorphosis was completed in about the first quarter of gestation while in the EC group maturation was much slower, utilizing about 60% of gestation time to reach the same level of development. However, the relationship between sleep duration and the rate of development within each of the two groups is not so simple. For EO group r_S was 0.79 ($p < .025$), indicating that daily sleep time increased as the ratio increased (i.e. as the rate of development decreased). For the EC group the coefficient of correlation (r_S was 0.09) was not significant. If the two groups are combined (as they should not be) r_S was 0.89 but the meaning of this relationship is uninterpretable. One speculation is that as the rate of brain maturation accelerates, a threshold is reached within each family at which time the duration of sleep begins to be progressively reduced as the maturation rate continues to increase. An alternative explanation is that there is no threshold but rather that the relationship between sleep time and maturation is curvilinear with the slope being extremely small for the EC species.

Eye Condition in Several Animal Families

Granted that the EO species sleep less than those born with eyes closed is there any rational basis for the distribution of these species among the various orders? The answer may be deduced from the information in Table 2 where species for which there are sleep records are placed in their appropriate families along with other species for which there are currently no 24-h sleep records. Within each classification (order, family or ungulate grouping) the species are listed in sequence according to the length of the gestation period. For example, among the rodents, the chinchilla and the guinea pig are the only animals born with the eyes open for which sleep data are available whereas the rat and the squirrel represent a sample of several rodents born with the eyes closed for which there are also sleep records. The animals without sleep data were chosen from the literature, primarily from Walker's Mammals of the World (Nowak, 1991), to illustrate the relation between gestation duration, body weight, and the condition of the eyes at birth. In particular, animals were deliberately selected so that each animal grouping (order, family, etc.) would have at least one species born with eyes closed. In Table 1 which lists species with appropriate sleep data, the EC group did not include a single primate but from Table 2 it is evident that there are primates (dwarf and mouse lemurs) which are born with the eyes closed but for which no sleep data were available. Likewise, there is no lagomorph in the EO group in Table 1 but it is apparent from Table 2 that the hare would be a suitable candidate when sleep data are obtained.

It is clear from Table 2 that within each family grouping, as the length of gestation and size of the species generally increase, a point is reached wherein species having a gestation length of a certain duration or longer are born with the eyes open. Among the rodents, species appear to be born with the eyes open when the gestation length is between 30 and 68 days, or longer. It is perhaps pertinent in this regard that some species of spiny mice (not shown

Table 2. Relationship of Gestation Length and Body Weight to Condition of Eyes at Birth

	Eyes at Birth	Gestation (days)	Weight (kg)		Eyes at Birth	Gestation (days)	Weight (kg)
FELIDAE				*PRIMATES*			
Tiger	Open	111	188.00	*Bush baby	Open	145	0.26
Lion	Open	109	190.00	Large lemurs	Open	115	1.73
*Jaguar	Open	103	95.00	Dwarf lemur	Closed	70	0.24
Leopard	Closed	98	48.00	Mouse lemur	Closed	60	0.07
Puma	Closed	93	68.00				
Lynx	Closed	60	8.00	*INSECTIVORA*			
*Domestic cat	Closed	63	3.00	Elephant shrew	Open	60	0.04
				*Tenrec	Closed	50	2.00
RODENTIA				*Tree shrew	Closed	48	0.19
Capybara	Open	130	54.00				
*Chinchilla	Open	111	0.50	*UNGULATES*			
*Guinea pig	Open	68	0.75	Asiatic mouse deer	Open	159	0.75
Giant squirrel	Open	30	1.75	*Sheep	Open	150	55.50
*Squirrel	Closed	27	0.54				
*Rat	Closed	22	0.25	*HYAENIDAE*			
				Spotted hyena	Open	110	63.00
LAGOMORPHA				Brown hyena	Closed	90	42.50
Hare	Open	44	4.00				
*Rabbit	Closed	31	1.20				

*Sleep data for 24 h period are currently available.

Note: The elephant shrew is considered by some mammalogists as comprising a distinct order—Macroscelidea. The tree shrew has been classified as an insectivore or a primate.

in Table 2) are born with eyes open and others with the eyes closed,while their gestation lengths are reported (Nowak, 1991) to range from 35 to 42 days which are just within the aforementioned transition period for eye opening in the rodents. On the other hand, the transition period between eyes-open and eyes-closed animals can be very narrow, as it is among the cats where the EC puma and the EO jaguar have gestation durations which are separated by a mere five days. Inasmuch as the values for the gestation periods are averages one would expect some overlap in their ranges for the two species of large cats. Indeed, it has been reported that the lion (Witschi, 1956), jaguar (Stehlik, 1971) and Bengal tiger (Baudy, 1971) are born with the eyes open but this opening takes place very close to the time of birth. In conrast, most other EO species have their eyes open *in utero* many days before birth. The largest of the cats could very well be on the margin between being born with the eyes open or closed but the smaller puma appears always to be bornwith the eyes closed.

A positive correlation between the length of the gestation period and body weight (Hamilton, 1939) has long been known and this is borne out in the present analysis too. Accordingly, one would expect that there should be some relationship between body size and the eye-opening at birth. Still, gestation duration is the better key to the eye condition at birth than is body weight. An examination of the primates in Table 2 shows that the Bush baby and dwarf lemur have almost identical body weights but the former which is in the EO group has a gestation period which is twice as long as that of the dwarf lemur in the EC class. Similarly, the ungulates are generally large animals and belong to the EO group; but, there is a miniature species of ungulate, the Asiatic mouse deer,which is about the size of a rabbit. This tiny deer is also in the EO group but regardless of its size it has an unusually long gestation period (159 days). Again there appears to be a close connection between eyelid condition at birth and the length of the gestation period.

Other Sleep Theories Revisited

The data supporting the possibility that sleep is fundamentally a function of the rate of brain development (and indirectly of body size) are still too meager to make bold assertions. Nevertheless, the direction for examining the validity of this idea is available. If an intrepid investigator were to obtain 24-h sleep records for the tiger, miniature lemurs, the hare and the elephant shrew, the total daily sleep times for these species should fall into the EC and EO categories suggested by Table 2. Of course, a direct study of brain development for the various EO and EC species might be a superior approach and perhaps more formidable than sleep studies alone. And, obtaining permission of a zoo curator to study the more exotic species might be the most formidable task of all.

In the meantime, if we assume the correctness of the dichotomy of mammalian sleep patterns, we can examine other theories in that light. In general, sleep theories have been predicated on behavioral adaptation. For example, in 1967 Pellet, Beraud, and Paillard (Pellet, 1967) in an abstract entitled, "A controversial problem: insomnia in the guinea pig" concluded that the guinea pig's low sleeping time would be determined by the herbivorous mode of feeding which necessitates the frequent and prolonged taking of nourishment. Would it not be just as reasonable to conclude the opposite—namely, that because the guinea pig's brain development was so fast (relative to other rodents), its sleeping time was reduced as a consequence of the peculiarity of brain maturation? Therefore, because it had more wakefulness 'thrust' upon it, the animal had more time for foraging. In a similar backward rationalization it has been suggested (Allison and Cicchetti, 1976) that large amounts of slow wave sleep are disadvantageous to predators in contrast to the requirements of prey species. According to the mouse, the domestic pussy is as much a predator as the jaguar; yet, the cat and mouse belong to the same sleep category (EC) while the jaguar is in another (EO). It seems more likely that the amount of sleep is not simply linked to predatory behaviour.

Since multiple correlations are readily found when examining interrelated physiological functions it would not be surprising to find that intelligence is somehow related to sleep. One might speculate that since the very large animals within an order not only tend to sleep less than their smaller cousins but have the potential for greater intelligence. Rensch (1967) stated that larger species by virtue of their absolutely larger brains have the advantage of more and larger neurons with the capability of more and stronger dendritic ramifications. This principle was evident even across phylogenetic levels. There is experimental evidence that a giant domestic fowl was capable of completing more tasks than a mouse which is higher on the phylogenetic scale.

Unquestionably, the oldest and still favored theory of sleep is with respect to its function of conserving energy but the data provided in this paper lend no support for that notion. A great boost to the energy-saving idea came from Zepelin and Rechtschaffen (Zepelin and Rechtschaffen, 1974) who analyzed the combined data from 53 mammalian species and then concluded that, "The correlation between sleep time and metabolic rate suggests that sleep has the function of enforcing rest and limited metabolic requirements, although some inconsistent findings are noted." Indeed, the correlation coefficient they found for all the animals combined (sleep time vs metabolic rate) was a highly significant 0.64.

Unfortunately, the "inconsistency" to which they refer was that when correlations were performed for each order separately, the coefficients (not significant at the 5% level) were 0.27 for carnivores, 0.51 for primates, – 0.01 for rodents and – 0.87 for insectivores. Those results strongly suggest that the statistically significant correlation between metabolic rate and sleep time, for all the animals combined, is spurious and yet another example of a correlation of inhomogeneity. Even if the correlation were significant the authors acknowledged that the correlation was no proof of a cause and effect. If metabolic rate were the determinant of sleep time one might expect women to have a daily sleep quota different from men because their basal metabolic rate is lower by about 10%. In one study (Aserinsky, 1969) aimed specifically at determining the maximal sleep capacity of young adults, no significant difference in sleep times was noted for the sexes.

Maturation at Birth: Dichotomy or Continuum

The analysis of sleep in the various species was initiated through the construction of a three-dimensional graph depicting the durations of REM sleep, non-REM sleep and waking for each species. Each species was then charted in sequence according to its waking duration, with the most wakeful

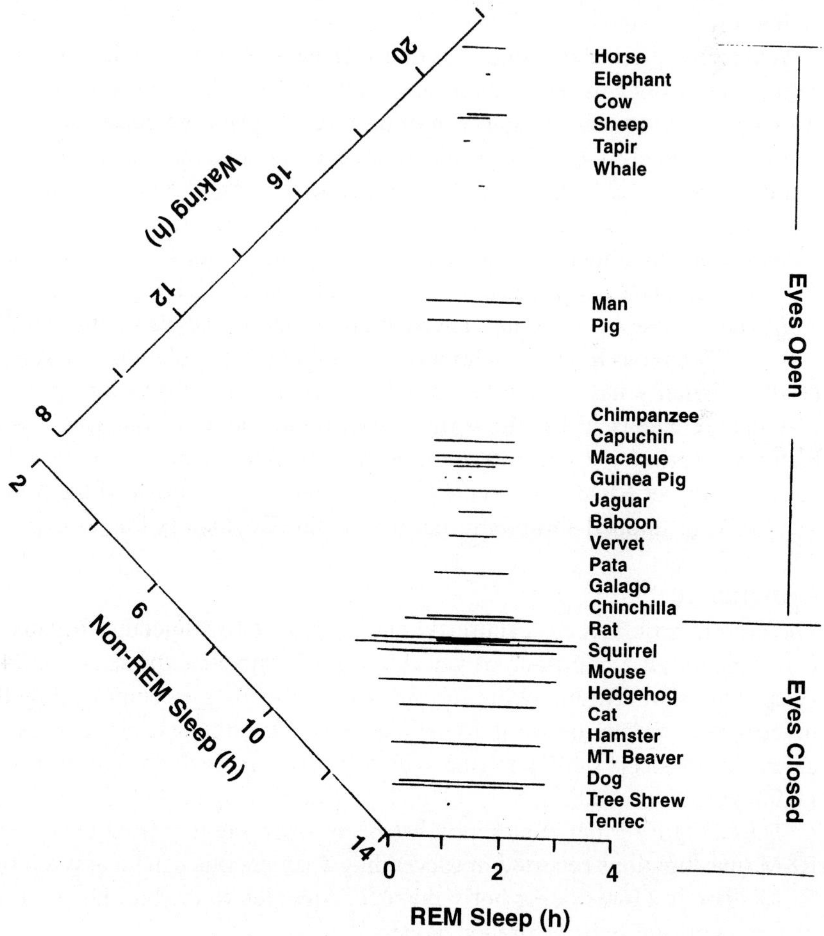

Fig. 2. **Modified three dimensional graph depicting the 24-h sleep-waking pattern of several mammalian species. The bottom point of each bar represents the intersection of the Waking duration and the Non-REM Sleep duration. The length of the bar indicates the duration of REM Sleep. The names of the species are listed in the same sequence as the sleep-waking bars. Eyes Open and Eyes Closed signify the condition of the eyes at birth.**

at the extreme left and the least wakeful at the extreme right. The amount of REM sleep was depicted by a vertical bar. The most obvious feature of the graph was that species to the right of the graph had far more REM sleep than the animals to the left. One glaring feature was that the species closest to man with respect to the sleep-waking pattern was the pig. In searching for a feature common to both species that might explain the similarity of the sleep-waking patterns, factors ranging from social behavior to the nature of the skin were considered but none allowed for a sensible grouping of the remaining species—until the insight that both man and pig were born with the eyes open. Applying the eyelid condition to all the species split the population in two, and subsequently sleep and other parameters seemed to match this dichotomy.

Whether a graded measure of brain rate development would provide a better correlation with sleep duration than the EO-EC division is not known. A graded scale for maturation (not brain rate of development) at birth was applied by Zepelin (Zepelin, 1989) to various mammalian species but the results were marred by the possibility of the occurrence of the 'correlation of inhomogeneity' inasmuch as subgroups were not considered. Furthermore the scale included factors such as hirsuteness and motor behaviour so that the scale cannot be considered as a simple refinement of the EO-EC dichotomy. A graded maturation scale has a certain allure but it may also obfuscate the clarity of a simple EO-EC dichotomy. Examining a population in terms of graded sexuality has certain benefits but a dichotomous division into just male and female is still basic. With respect to dichotomous function it is of historic (Aserinsky, 1996) interest that the research which ultimately led to the discovery of REM was directed at ascertaining the validity of the notion that blinking stopped abruptly rather than gradually upon falling asleep.

Conclusion

The present article treads cautiously on the thin ice of conjecture because it is dependent on a multitude of variables. The number of species with 24-h sleep studies are extremely limited, and the methodologies employed by the investigators are not uniform, to say the least. Despite these difficulties the extension of sleep studies to one critical species in each of five orders or families could substantiate the relation of sleep duration to the EO-EC condition.

The seed for further exploration has been sown. Judging from the diverse REM investigations reported in succeeding chapters one can be assured that REM sleep is a topic being hotly pursued. Any clue to the biological origin of sleep will surely be examined closely.

References

Allison, T. and Cicchetti, D.V. Sleep in mammals: ecological and consitutional variables, Science, 194 (1976) 732–734.

Altman, P.L. and Dittmer, D.S. Biological Data Book, Vol. I, 2nd ed., FASEB, Bethesda, Maryland, 1972.

Aristotle. Cited by R.A. McCance in Characteristics of the newly born. In: G.E. Wolstenholme and M. O'Connor (Eds.), Ciba Foundation Symposium on Somatic Stability in the Newly Born, Little Brown & Co., Boston, 1961.

Aserinsky, E. and Kleitman, N. Eye movements during sleep, Fed. Proc., 12 (1953) 6–7 (Abst.).

Aserinsky, E., The maximal capacity for sleep: Rapid eye movement density as an index of sleep satiety. Recent Adv. Biol. Psychiatr, 1 (1969) 147–159.

Aserinsky, E., Relationship of adult sleep pattern of placental mammals to eyelid condition at birth, Fed. Proc., 36 (1977) 461 (Abst.).

Aserinsky, E., The discovery of REM sleep, J. History Neurosciences, 5 (1996) 213–227.

Baudy, R.E., Notes on breeding felids at the Rare Feline Breeding Center. In: J. Lucas (Ed.), International Zoo Yearbook, The Zoological Society of London, Printed by Aberdeen Univ. Press, Scotland, 1971.

Blinkov, S.M. and Glezer, I.I. The Human Brain in Figures and Tables; A Quantitative Handbook. Translated by Basil Haigh, Basic Books Inc., New York, 1968.

Campbell, S.S. and Tobler, I. Animal sleep: A review of sleep duration across phylogeny, Neurosci Biobehav. Rev., 8 (1984) 269–300.

Case, T.J., On the evolution and adaptive significance of postnatal growth rates in the terrestrial vertebrates, Quart. Rev. Biol., 53 (1978) 243–282.

Evans, H.E. and Sack, W.O. Prenatal Development of Domestic and Laboratory Animals, N.Y. State Veterinary College, Cornell Univ., New York, 1973.

Hamilton, Jr., W.J., American Mammals, McGraw-Hill Co., New York, 1939.

Hobson, J.A., The Dreaming Brain. Basic Books Inc., New York, 1988.

Jouvet-Mounier, D. et al. Ontogenesis of the states of sleep in rat, cat and guinea pig during the first postnatal month, Develop Psychobiol., 2 (1970) 216–239.

Kenneth, J.H. and Ritchie, G.R. Gestation Periods: A Table and Bibliography, Technical Communication No. 5–Commonwealth Agricultural Bureaux, Oliver & Boyd Ltd., Edinburgh, 1953.

Mangold, R. et al. The effects of sleep and lack of sleep on the cerebral circulation and metabolism of young men, J. Clin. Invest., 34 (1955) 1092–1100.

Nowak, R.M., Walker's Mammals of the World, 5th ed., Johns Hopkins Univ. Press, Balto., Md, 1991.

Pellet, J. et al. A controversial problem: insomnia in the guinea pig, Electroencephalogr. Clin. Neurophysiol., 23 (1967) 288 (Abst.).

Rensch, B., The evolution of brain achievements. In: T. Dobzhansky et al. (Eds.). Evolutionary Biology, Vol. I, Apple-Century-Crofts, New York, 1967.

Sachs, L., Applied Statistics: A Handbook of Techniques, 2nd ed., Springer-Verlag, New York, 1984.

Shimizu, A. and Himwich, H.E. The ontogeny of sleep in kittens and young rabbits, Electroencephalogr. Clin. Neurophysiol., 24 (1968) 307–318.

Spector, W.S., Handbook of Biological Data, W.B. Saunders, Philadelphia, 1956, reprinted 1961.

Stehlik, J., Breeding jaguars (Panthera onca) at Ostrava Zoo. In: J. Lucas (Ed.) *International Zoo Yearbook*, Vol. 11, The Zoological Society of London, Printed by Aberdeen Univ. Press, Scotland, 1971.

Stern, L. and Peyrot, R. Cited by D. Richter in The stability of the nervous system. In: G.E. Wolstenholme and M. O'Connor (Eds.), *Ciba Foundation Symposium on Somatic Stability in the Newly Born*, Little Brown & Co., Boston, 1961.

Widdowson, E.M., Immediate and long-term consequences of being large or small at birth:

a comparative approach. In: K. Elliott and J. Knight (Eds.), *Ciba Foundation Symposium 27 (New Series) on Size at Birth*, Amer. Elsevier, New York, 1974.
Witschi, E., Development of Vertebrates. Saunders, Philadelphia, 1956.
Zepelin, H. and Rechtschaffen, A. Mammalian sleep, longevity and energy metabolism, Brain Behav. Evol., 10 (1974) 425–470.
Zepelin, H., Mammalian sleep. In: M.H. Kryger, et al. (Eds.), Principles and Practice of Sleep Medicine, W.B. Saunders, Philadelphia,1989.

Rapid Eye Movement Sleep
B.N. Mallick and S. Inoué (Eds)

2. Phylogeny and Evolution of Rapid Eye Movement (REM) Sleep

Marcos Gabriel Frank

Dept. of Physiology, University of California San Francisco,
513 Parnassus Ave, San Francisco, CA 94143-0444,

SUMMARY

Rest and activity states appear to be ubiquitous in the animal kingdom. Rest states resembling mammalian nonREM sleep have been noted in most vertebrate orders. REM sleep, however, appears to be restricted to birds and mammals. A number of hypotheses have been presented to account for the evolution of REM sleep, but at present, no one hypothesis has isolated the ecological or biological factors that conferred selective advantage to the ancient vertebrates that first expressed REM sleep. The fact that REM sleep is found only in animals with neocortex (or comparable structures) strongly suggests that REM sleep evolved in response to neuronal properties unique to animals with well-developed telencephalons. The observation that REM sleep is found only in animals that express EEG slow wave activity (SWA) during nonREM sleep further suggests that the evolution of synchronized, oscillatory firing in neocortex may have been a pivotal factor in REM sleep evolution.

Key Words: Vertebrate evolution, nonmammalian sleep, natural selection, neocortical development, REM sleep function, comparative physiology.

Introduction

In the years since the first discoveries by Aserinsky, Kleitman and Dement, scientists have made great progress in characterizing the phenomenology and neurobiology of mammalian REM sleep. Neurophysiological studies have revealed a complex neural system that mediates REM sleep expression (Jones, 1991). Combined electrophysiological and behavioral experiments have revealed distinct regulatory mechanisms that control the timing and amounts of REM sleep (Benington and Heller, 1994; Borbely, 1994). Ontogenetic studies have demonstrated dramatic quantitative and qualitative changes in REM sleep during the course of an animal's lifespan (Jouvet-Mounier, Astic et al., 1970). The knowledge gained by these varied approaches, discussed elsewhere in this text, have provided important clues regarding the function of REM sleep. In this chapter, *phylogenetic* studies of REM sleep are addressed. The study of REM sleep in diverse animal species provides important insights into REM sleep physiology and function. In the case of certain animals that have archaic or promitive characteristics, phylogenetic studies provide clues as to the original biological and ecological conditions that may have conferred selective advantage to animals that first expressed REM sleep. Qualitative and quantiative differences in REM sleep expression in extant species may provide information regarding the utility of REM sleep in modern animals. The presence or absence of REM sleep in present-day animals is also a litmus test for current theories of REM sleep function Clearly, any theory of REM sleep function should be able to account for the presence *and* absence of REM sleep in certain animals. In this review, I begin by first discussing some methodological concerns inherent in phylogenetic studies of REM sleep. This is then followed by a review of sleep studies in several taxonomic groups of animals. Several theories regarding the evolution of REM sleep are then reviewed, and a synthesis of primary findings and theories regarding the origin of REM sleep are presented.

Detecting and Defining REM Sleep in the Animal Kingdom

As discussed by Hartse there are a number of considerations that should be kept in mind when evaluating sleep studies in diverse animal species (Hartse, 1994). The first is that only a minute fraction of the over one million animal species known to science have been studied with regard to sleep behavior (Cambell and Tobler, 1984). Consequently, it is difficult to draw strong conclusions about the evolution of REM sleep given the rather small number of studies available to the interested scientist. Secondly, identifying and reproducing those aspects of the natural environment necessary for normal sleep behavior in centain animal species is often difficult. This may be especially true in the case of poikilotherms, like reptiles and amphibians, whose behavior is profoundly affected by changes in ambient temperature. Thirdly, some animals may exhibit heightened vigilance in the unfamiliar laboratory environment, making sleep-like states difficult to detect. A further consideration is that the absence of detailed neuroanatomical and neurochemical studies of nonmammalian nervous systems makes the interpretation of pharmacological studies of nonmammalian sleep especially difficult.

The major impediment in establishing a phylogeny of REM sleepis determining those crieria that best describe REM sleep across diverse animals species. Mammalian REM sleep is characterized by a number of autonomic and neurophysiological events (Zepelin, 1994). Some of these events may not occur during non-mammalian REM sleep-like states. For example, the desynchronized cortical electroencephalogram (EEG) typical of mammalian REM sleep is unlikely to be observed in species that lack well developed (encephalized) brains (e.g. simple invertebrates and insects). Other features of REM sleep, like motor atonia, may be difficult to detect in species that exhibit general motor hypotonia during all stages of rest (as is the case with most poikilotherms). Therefore, any phylogenetic discussion of REM sleep should begin by establishing those fundamental properties of REM sleep that may universally describe this state in the animal kingdom.

As a starting point we can first list criteria that have been established for determining the presence of sleep behavior in non-mammalian species. Flanigan, in a series of classic studies in reptiles, describes a number of characteristics that delineate sleep in any given species (Flanigan, 1973; Flanigan, Wilcox et al., 1973; Flanigan, Knight et al., 1974; Flanigan, 1974):

> "(1) the spontaneous assumption of a stereotypic or species specific posture, (2) the maintenance of behavioral immobility; (3) an elevated behavioral response threshold...(4) repid behavioral reversibility of state with relatively strong stimulation."

Tobler has suggested that a fifth characteristic of sleep, *regulatory capacity*, should be added to the above list (Tobler, 1995). That is, sleep is also defined by the presence of compensatory changes in sleep expression following sleep deprivation. Finally, since sleep and wakefulness reflect different states of neuronal responsiveness and activity, sleep should also be characterized by specific patterns of neuronal activity.

We can add to this list those special criteria that delineate REM sleep from nonREM sleep in mammals. It seems reasonable to suggest that the fundamental, distinguishing property of REM sleep is the episodic expression of wake-like neuronal activation during a state of sleep as defined by the above criteria. Consequently, we can divide the diverse phenomena of mammalian REM sleep into *primary* characteristics (those which best describe this state), *secondary* characteristics (those which are suggestive, but not conclusive evidence for the presence of this state) and *tertiary* characteristics (those which are inconsistently found in mammalian species). The *primary* characteristics of REM sleep-like states are the presence of endogenous neuronal activation combined with heightened arousal thresholds (compared to wakefulness). *Secondary* characteristics are phasic activations of peripheral musculature (myoclonia, REMs), irregularities (and/or) increases in cardiorespiratory output, and the presence of compensatory increases in a REM sleep-like state following its deprivation. *Tertiary* characteristics of REM sleep-like states are the presence of thermoregulatory changes, atonia and specific phasic brain activations (e.g. pontine-geniculate-occipital waves: PGOs). This working definition of REM sleep distinguishes between behavior which may resemble REM sleep (e.g. spontaneous phasic motor activity is common in developing nervous systems but distinct from REM sleep myoclonia) and is not limited to the presence of REMs, which may not occur in species with degenerate ocular systems. or specialized autonomic events that may be unique to endothermic mammals (e.g. changes in thermoregulatory output).

REM Sleep in the Animal Kingdom

Invertebrates and REM sleep

Invertebrate species represent some of the most ancient forms of animal life. Invertebrates first appeared in the Cambrian period of the early Paleozoic era, approximately 570 million years ago (Sarnat and Netsky, 1981). By the Silurian period-about 450 million years ago-terrestrial arthropods (e.g. insects and arachnids) had invaded the land, surviving to this day as perhaps the most successful forms of terrestrial animal life (Sarnat and Netsky, 1981). Despite the fact that invertebrates are the most numerous of all living creatures, there are relatively few studies of rest and activity cycles in invertebrates. There have been reports of sleep-like states in *mollusca* like the seahare *aplysia* (Strumwasser, 1971) and in the octopus (Mather and Mather, 1994), but no evidence for REM sleep was reported. Sleep-like states have also been reported in insects (Andersen, 1968; Kaiser and Kaiser, 1983; Tobler, 1983) and arachnids (Tobler and Stalder, 1988). Rest in these terrestrial arthropods satisfies several of the criteria established by Flanigan and Tobler for the presence of sleep. For example. The honey bee *Apis mellifera* (Kaiser and Kaiser, 1983), moths (Andersen, 1968) the cockroach *Leucophea maderae,* (Tobler, 1983), and the scorpion *Heterometrus pandinus*, (Tobler and Stalder, 1988), all show stereotyped rest-like behaviors associated with increased arousal thresholds. Rest in the honey bee is also associated with decreases in neuronal responsiveness (Kaiser and Kaiser, 1983). Cockroaches and scorpions show compensatory increases in rest states following enforced waking (Tobler, 1983; Tobler and Stalder, 1988). The evidence for REM sleep-like states in terrestrial arthropods, however, is relatively weak. Cockroaches display a state of immobility coupled with movements of antennae of limb twitches, but this state appears to be one of arousal rather than of sleep (Tobler and Stalder, 1988). "Swimming movements" in sleeping moths are elicited when moths are "drastically disturbed", but these movements probably reflect arousal and not phasic motor activations typical of REM sleep-like states (Andersen, 1968). In summary, there are presently no compelling data supporting the presence of REM sleep-like states in invertebrates.

Vertebrates and REM sleep

Fish

Fish represent some of the first vertebrate species to evolve, and currently represent over 70% of all known vertebrate animals (Carrol, 1988). The first archaic fish are observed in the Silurian period of the Paleozoic era, but the great radiation of fishes occurred in the Devonian period of that era, approximately 400 million years ago (Sarnat and Netsky, 1981). As is true for invertebrates, relatively few sleep studies in fish have been conducted, and of these only one taxonomic division of fish (Teleosts) has been examined in any great detail. While there is some consensus regarding the presence of sleep in fish (Melesion-Nolasco and Ayala-Guerrero, 1983; Hartse, 1994), there is less agreement regarding the presence of REM sleep-like states. Tauber and Weitzman reported a variety of eye-movements during behavioral sleep in several species of Bermuda reef fish (Tauber and Weitzman, 1969). However, the eye-movements were not clustered into episodic bursts, as is the case in mammalian REM sleep, nor were they associated with other secondary signs of REM sleep such as irregularities in cardiorespiratory output, or phasic motor activity. Much longer studies of another Teleost, *Tilapsia mossambica,* did not find evidence of REM sleep-like states in fish (Shapiro and Hepburn, 1976). Karmanova and Lazarev reported that the catfish (*Ictalurus nebulosus)* exhibits an unusual state of sleep characterized by EEG spikes and slow waves recorded from the mid- and forebrain, and cardiac arrhythmia (Karmanova and Lazarez, 1980). Karmanova and Lazarev suggest that the EEG spike activity is comparable to phasic brain activations in mammalian REM sleep, but as Hartse points, out similar spiking activity is frequently observed in reptiles and some amphibians, and appears to be a correlate of nonREM sleep (Hartse, 1994). Still, the finding of secondary signs of REM sleep (cardiac arrhythmia) combined with EEG correlates of nonREM sleep is reminiscent of Siegel's findings (Siegel,1995; Siegel. Manager et al., 1996) of a primordial sleep state in primitive egg-laying mammals that displays aspects of both REM and nonREM sleep (see below). In general, however, the evidence for REM sleep-like states in fish is presently unconvincing.

Amphibians

Amphibians first appeared in the fossil record approximately 400 million years ago in the Devonian period, and began their radiation in the Carboniferous period, about 50 million years later (Sarnat and Netsky, 1981; Carrol, 1988). There are presently three orders of extant amphibians including salamanders and newts (*Urodela),* frogs and toads *(Anura*) and limbless amphibians (*Apoda*) (Carrol, 1988). The vast majority of sleep studies have been conducted in *Anurans* (Segura and De Juan, 1966; Hobson, 1967; hobson, 1968; Huntley, Donnely et al., 1969; Karmanova and Lazarez, 1978), although there is at least one study in a *Urodelan,* the salamander Abystoma trigrinum (Lucas, Sterman et al., 1969). With the exception of the bullfrog *Rana catesbiana* (Hobson, 1967), rest in *Anurans*, by most accounts, satisfies several of the criteria established by Flanigan for the presence of sleep. The tree frogs *Hana squirella* and *Hyla cinerea* show stereotyped sleep postures, elevated sensory thresholds and EEG patterns that are specific to behavioral rest (Hobson 1968). The bullfrog, however, has been reported not to sleep at all, maintaining a state of resting vigilance during periods of inactivity in both the laboratory and the field (Hobson, 1967). There is some disagreement over what kinds of EEG activity are associated with sleep-like states in *Anurans*. The specific patterns of EEG activity associated with sleep in tree frogs has been reported as low-amplitude and of fast frequencies (8-30 Hz) (Hobson, 1967; Hobson, 1968). Huntley, however has found "rest/sleep" states in the toad *Bufo boreas* associated with EEG slowing (5-7 Hz) and a reduction in EEG amplitude (Huntley, Donnely et al., 1969). Karmonova and Lazarez also report that sleep in the *Anuran, Rana temporaria* is typified by EEG slow waves (0.5-2.0 Hz) but also find spike-like EEG activity that disappears upon arousal (Karmanova and Lazarez ,1978). In the single study of a *Urodelan,* the salamander *Ambystoma trigrinum,* spectral analyses of the EEG revealed that "quiet" behavioral states were associated with greater power densities in lower EEG frequencies (slow wave activity: SWA) whereas "active" states were associated with greater power densities in higher EEG frequencies (Lucas, Sterman et al., 1969). Whether or not quiet states in a *Urodelan* represent sleep is not yet clear, but the presence of heightened EEG SWA in *Ambystoma trigrinum* during behavioral quiescence and Karmanova and Lazarez's reports of elevated EEG slow wave and spikes during sleep-like states in *Rana temporaria* suggests that amphibians may have a rudimentary form of mammalian nonREM sleep. High levels of EEG slow waves and SWA are typical of mammalian nonREM sleep (Zepelin 1994) and the amphibian spike-like activity may be related to a similar spiking in reptiles associated with nonREM sleep-like states (Hartse, 1994). Amphibians, however, do not appear to have REM sleep-like states. Although, Hobson's findings in *Hyla* and *Rana* of sleep-like states associated with a low-amplitude, fast frequency EEG are suggestive of REM sleep-like states in *Anurans,* in no other study of frogs, toads, or the salamander, was there any evidence for REM sleep-like states.

Reptiles

The first reptiles appeared in the Carboniferous period of the Paleozoic era, approximately 350 million years ago. Reptiles displaced amphibians as the dominant terrestrial vertebrates in the Permian period, about 280 million years ago (Sarnat and Netsky, 1981; Carrol, 1988). There are presently four orders of living reptiles, three of which the *Chelonia* (turtles and tortoises), the *Squamata* (lizards and snakes) and the *Crocodilia* (crocodiles and alligators) having been studied to some extent with respect to sleep and waking states (Carrol, 1988). The *Chelonia* and *Crocodilia* are particularly important reptilian orders because *Chelonians* and *Crocodilians* are believed to be closely related to animals (birds and mammals, respectively) which have REM sleep (Sarnat and Netsky, 1981; Carrol, 1988).

Several studies suggest that *Chelonia, Squamata* and *Crocodilia* have nonREM sleep-like states according to the criteria of Flanigan and Tobler. The red-footed tortoise *Geochelene carbonaria,* and the box turtle *Terrapene carolina,* display rest periods associated with heightened arousal thresholds. EEG patterns distinct from those observed during wakefulness, and compensatory increases in sleep-like states following enforced waking (Flanigan, Knight et al., 1974; Flanigan, 1974). Sleep-like postures in these *Chelonians* are associated with telencephalic EEG spikes and sharp waves, which disappear upon arousal. Sleep-like states associated with EEG spikes are also reported in the turtle tortoise *Gopherus berlandieri* (Ayala-Guerrero, Huitron-Resendiz et al., 1993) and the tortoise *Gopherus flavomarginalis* (Ayala-Guerrero. Calderon et al., 1988). Sleep-like states have also been reported in another tortoise *(Emys orbicularis)* although in this study, no EEG spikes were observed, and slow EEG frequencies were reported instead (Vasilescu, 1970; Vasilescu, 1983) Sleep states comparable to those reported in *Chelonians* have also been reported in several species of *Squamata* and *Crocodilia.* The Iguanid lizards *Ctenosaura pectinata* and *Iguana iguana,* for example, display quiescent periods coupled with eye-closure, heightened arousal thresholds, and distinct EEG patterns, and show compensatory increases in sleep time and EEG spikes following enforced waking (Flanigan, 1973). Similar findings have been reported for the Crocodilian *Caiman sclerops* (Flanigan, Wilcox et al., 1973; Warner and Huggins, 1978). As is true for *Chelonians,* there are contrary findings regarding the type of EEG activity associated with sleep in *Squamata* and *Crocodilia.* Several authors have reported the presence of EEG spikes and/or sharp waves during sleep in Iguanid lizards (Flanigan, 1973; Ayala-Guerrero, Huitron-Resendiz et al., 1994) while others have also reported overall reductions in EEG frequency and amplitude (Tauber, Rojas-Ramirez et al., 1968; Ayala-Guerrero and Huitron-Resendiz, 1991) accompanying sleep. Similar discrepancies are found in reports of the crocodilian *Caiman sclerops,* with Flanigan reporting EEG spikes (Flanigan, Wilcox et al., 1973), and Warner reporting EEG slow waves during sleep (Warner and Huggins, 1987).

While most studies in *Chelonia, Squamata* and *Crocodilia* find evidence of nonREM sleep, there have also been exceptions to these findings. Susic (Susic, 1972), and Walker and Berger (Walker and Berger, 1973), found no evidence for sleep in the sea turtle *Caretta caretta,* or the tortoise *Testudo denticulata,* respectively. Van Twyer also reported that the American alligator, *Alligator mississipiensis* does not sleep (Van Twyer, 1973). Why these authors failed to find evidence of sleep in their animals is not clear. Poor acclimation to the laboratory environment may have contributed to the studies of *Caretta carretta* and *Alligator mississipiensis.* This explanation, however, seems unlikely in the case of the Walker and Berger study which largely duplicated the methodology employed by Flanigan. Therefore any conclusions regarding the presence of nonREM sleep in reptiles must remain tentative pending resolution of these discrepancies in the literature.

There is considerable disagreement in the field regarding the presence of REM sleep in reptiles. Many studies in reptiles do not find evidence for REM sleep-like states (as reviewed in: Monnier, 1980; Melesio-Nolasco, Friedman-Saavedra et al., 1983; Rial, Nicolau et al., 1993; Hartse, 1994). In most reports, slow and saccadic eye movements, or small movements of the head and mouth were sometimes observed during reptilian sleep, but this activity was aperiodic, and not associated with wake-like EEGs (Flanigan, 1973; Flanigan, Wilcox et al., 1973; Flanigan, knight et al., 1974; Flanigan, 1974; Hartse and Rechtschaffen, 1974; Hartse and Rechtschaffen, 1982), nor- in the case of *Testudo denticulata*-with heightened arousal thresholds (Walker and Berger, 1973). The EEG spikes and sharp waves reported during reptilian sleep also appear to be more similar to hippocampal EEG spikes observed in mammalian nonREM sleep than to phasic EEG activity (PGOs) typical of REM sleep-like states (Hartse and Rechtschaffen, 1974; Hartse and Rechtschaffen, 1982).

Several studies, however, have reported very small amounts of REM sleep-like states in *Squamata* (Tauber, Roffwarg et al., 1966; Tauber, Rojas-Ramirez et al., 1968; Huntley, Friedman et al., 1977; Ayala-Guerrero and Huitron-Resendiz, 1991; Ayala-Guerrero, Huitron-Resendiz et al., 1994), and in *Chelonians* (Vasilescu, 1970; Vasilescu, 1983; Ayala-Guerrero, Calderon et al., 1988; Ayala-Guerrero, Huitron-Resendiz et al., 1993). Tauber found "bursts" of REMs lasting 1-3 and 1-7 minutes in sleeping chameleons *Chameleo jacksoni* and *Chameleo melleri,* and in the sleeping iguana *Ctenosaura pectinata* (Tauber, Roffwarg et al., 1966; Tauber, Rojas-Ramirez et al., 1968). Huntley reported "paradoxical sleep" in the iguana *Dipsosaurus dorsalis* characterized by highly irregular cardiorespiratory output, extremely low motor tone coupled with myoclonic bursts, and periodic episodes of EEG theta (5-10 Hz) (Huntley, Friedman et al., 1977). Similar findings have been reported by Ayala-Guerrero who described an "active sleep" state in the lizards *Ctenosaura similis* and *Ctenosaura pectinata* characterized by REMs, "involuntary movements such as head shaking, chewing, blinking, or movements of the body and legs" and slight elevations in heart rate (Ayala-Guerrero and Huitron-Resendiz, 1991; Ayala-Guerrero, Huitron-Resendiz et al., 1994). "Active sleep" as described in *Ctenosaura similis* has also been reported in the desert tortoise *Gopherus flavomarginalis* and the turtle *Gopherus berlandieri* (Ayala-Guerrero, Calderon et al., 1988; Ayala-Guerrero. Huitron-Resendiz et al., 1993). Both *Chelonians* are reported to have REM sleep-like states characterized by eye movements, phasic "motor automatisms", accelerations in heart-rate, and wake-like EEGs (Ayala-Guerrero, Calderon et al., 1988; Ayala-Guerrero, Huitron-Resendiz et al., 1993). Vasilescu 1970 found in another Chelonian, *Emys orbicularis,* sleep characterized by:

> Very marked nuchal hypotonia...sometimes interrupted by isolated or grouped jerks...rapid or slow ocular movements, either isolated or grouped (bursts)...[and] telencephalic EEG arousal (activation). "

Vasilescu has also noted anecdotally the presence of "oneiric" behaviors elicited by optic tract stimulation during sleep in *Emys orbicularis*. The "hallucinatory behaviors" consisted of "swimming or defense reactions, objectively unmotivated" (Vasilescu, 1983) These reptilian REM sleep-like states also appear to be inhibited by pharmacological compounds that reduce REM sleep in mammals. Reserpine (a monoamine depleter) administration reduced the amounts of a REM sleep-like state in *Gopherus berlandieri,* in a manner similar to that reported for mammalian REM sleep (Ayala-Guerrero, Huitron-Resendiz et al., 1993). Similar results were observed in *Ctenosaura pectinata* following PCPA administration (Ayala-Guerrero, Huitron-Resendiz et al., 1994).

These findings while suggestive of REM sleep should be cautiously interpreted. In lizard species where evidence for REM sleep appears to be strong, reports of autonomic and EEG changes during REM sleep states vary considerably. In contrast to Huntley (Huntley, Friedman et al., 1977) or Ayala-Guerrero (Ayala-Guerrero and Huitron-Resendiz, 1991; Ayala-Guerrero, Huitron-Resendiz et al., 1994), Tauber did not observe EEG theta or wake-like EEG patterns associated with REM sleep-like states in chameleons or Iguanid lizards (Tauber, Roffwarg et al., 1996 Tauber, Rojas-Ramirez et al., 1968). Neither Tauber (Tauber, Rojas-Ramirez et al., 1968; Tauber, Roffwarg et al., 1966) nor Ayala-Guerrero (Ayala-Guerrero and Huitron-Resendiz, 1991; Ayala-Guerrero, Huitron-Resendiz et al., 1994) found the "highly irregular" cardiorespiratory changes in their lizards reported for *Dipsosaurus dorsalis* (Huntley, Friedman et al., 1977). Ayala-Guerrero found a slight increase in heart rate during "active sleep" in *Ctenosaura similis* (Ayala-Guerrero and Huitron-Resendiz, 1991), but Tauber found a *decrease* in heart rate during 'REM' sleep in *Ctenosaura pectinata* (Tauber, Rojas-Ramirez et al., 1968). More importantly, in none of the aforementioned studies were presumptive REM sleep-like states adequately distinguished from brief arousals. In several reports, presumptive REM sleep in reptiles was followed by eye-opening and wakefulness, which suggests that the periodic motor bursts and REMs observed in some reptilian species are preparatory activations of motor structures that normally precede arousal. In none of the aforementioned reports were arousal thresholds assessed during presumpive REM sleep periods. Consequently, periods of wake-like EEG during sleep could very well represent wakefulness and not REM sleep. The effects of REM sleep-inhibiting drugs in reptiles, while intriguing, are somewhat difficult to interpret since distinguishing between sleep and wake following the administration of these drugs is not easy (Ayala-Guerrero, Huitron-Resendiz et al., 1993; Ayala-Guerrero, Huitron-Resendiz et al., 1994).

In summary, the presence of REM sleep-like states in reptiles has yet to be convincingly demonstrated. Primary criteria for REM sleep are still lacking since there are presently no studies which demonstrate a sleep-like state characterized by both neuronal activation and heightened arousal thresholds in reptiles. A number of studies in reptiles, however, have reported several secondary signs of REM sleep, such as episodic periods of myoclonia, REMs, and changes in cardiorespiratory output.

Birds

The first archaic birds appeared in the Jurassic period of the Mesozoic era approximately 180 million years ago (Sarnat and Netsky, 1981; Carrol, 1988). Birds are of especial interest to phylogenetic studies of REM sleep because in contrast to other nonmammalian vertebrates, the evidence for avian REM sleep-like states is strong. The presence of avian REM sleep-like states is also of interest because the last vertebrate ancestor shared by birds and mammals prior to the divergence of avian and mammalian lines (a reptile) may not have had REM sleep (Carrol, 1988). This suggests that REM sleep evolved separately in mammals and birds sometime after the segregation of avian and mammalian lines from a common reptilian ancestor.

A wide variety of avian species exhibit rest states that are strikingly similar to mammalian nonREM and REM sleep (Duenas and Ayala-Guerrero, 1983; Amlaner, 1994). Geese, hawks, owls, chickens, and pigeons have all been reported to exhibit resting states that satisfy Flanigan's criteria for the presence of sleep (Cambell and Tobler, 1984). Avian sleep is also associated with EEG patterns comparable to those found in reptiles and mammals (Duenas and Ayala-Guerrero, 1983; Amlaner, 1994). Diphasic EEG spikes, similar to those reported in reptilian sleep, have been reported in certain avian species during sleep (Duenas and Ayala-Guerrero, 1983; Amlaner, 1994). However, unlike amphibians and reptiles, avian sleep is consistently associated with large amplitude, slow waves in the EEG (Duenas and Ayala-Guerrero, 1983; Amlaner, 1994). Birds, therefore, appear to have a form of mammalian-like nonREM sleep.

Birds exhibit REM sleep-like states that satisfy several primary and secondary criteria for mammalian REM sleep (Duenas and Ayala-Guerrero, 1983; Amlaner, 1994). Avian REM sleep-like states are generally observed after long nonREM sleep periods, and are associated with desynchronized EEGs and heightened arousal thresholds(Duenas and Ayala-Guerrero, 1983; Amlaner, 1994). Avian REM sleep-like states are also associated with changes in cardiorespiratory ouput, increases in phasic motor activity, and in most species, REMs (Duenas and Ayala-Guerrero, 1983; Amlaner, 1994).

There are, however, several differences between avian REM sleep-like states and mammalian REM sleep. The motor atonia typical of mammalian REM sleep is infrequently reported in birds. The amounts and durations of avian REM sleep-like states are much smaller than that reported for mammals-roughly 1/4 the amount reported for most placental mammals (Duenas and Ayala-Guerrero, 1983; Amlaner, 1994). Moreover, unlike mammalian REM sleep, avian REM sleep-like states do not rebound following prolonged sleep deprivation-suggesting that REM sleep-like states in birds are not homeostatically regulated (Tobler and Borbely, 1988; Berger and Phillips, 1994). Nevertheless, avian REM sleep-like states do satisfy primary and most secondary criteria for identifying REM sleep (as described in previous sections), and therefore it would appear that birds have REM sleep-like states.

REM sleep in Mammals

Modern mammals are thought to have evolved from a member of the therapsid, or mammal-like reptiles during the Jurassic period of the Mesozoic era, about 215-145 million years ago (Sarnat and Netsky, 1981; Carrol, 1988). Early mammals radiated into modern forms in the Cenozoic era about 65 million years ago (Sarnat and Netsky, 1981; Carrol, 1988). The three extant lines of mammals, the marsupials (pouch-brooders), placentals (e.g.. primates, rodents, felines) and monotremes (egg-laying mammals) have been studied with regard to the presence of REM sleep. REM sleep has been reported in all mammals studied to date with two notable exceptions, a monotremous mammal (the echidna) (Allison, Van Twyer et al., 1972; Siegel, Manger et al., 1996), and a placental Cetacean (the dolphin) Mukhametov, Supin et al., 1977; Mukhametov, 1987). REM sleep in placental mammals has been extensively reviewed elsewhere (Zepelin, 1994), and consequently I will primarily summarize findings from a number of primitive mammals, such as the monotremous platypus and echidna, the marsupial opossum and the placental hedgehog. These latter mammals are considered representative of more ancient marsupial and placental mammalian lines based on fossil records and neocortical organization (Carrol, 1988; Finlay, Innocenti et al., 1991). Cetaceans are also discussed in this section because despite the well-developed neocortex in these species, neuroanatomical and cytoarchitectural studies of the cetacean brain have revealed a number of primitive features which suggest that the cetacean neocortex is more closely related to archaic mammalian brains than to similarly encephalized mammals (Scwerdtfeger, Oelschlager et al., 1984; Morgane, Jacobs et al., 1985).

Marsupials, Placentals and Cetaceans

REM sleep-like states have been reported in a number of archaic mammals like the marsupial opossum, the placental hedgehog, tree shrew, mole and armadillo (Affani and Vacarezza, 1967; Allison and Van Twyer, 1970; Van Twyer and Allison, 1970; Prudom and Klemm, 1973; Allison, Gerber et al., 1977; Monnier, 1980). The opossums, *Didelphis marsupialis* and *Didelphis azurae,* exhibit sleep states characterized by elevated arousal thresholds, desynchronized EEGs, EEG theta (5.0-6.0 Hz) activity, increases in phasic motor activity and decreases in motor tone (Affani and Vacarezza, 1967; Van Twyer and Allison, 1970). Complete motor atonia and REMs during REM sleep in the opossum was inconsistently reported. REM sleep amounts in *Didelphis marsupialis* (29.1% of total sleep time) were comparable to average amounts reported for placental mammals (20 % of total sleep time) (Van Twyer and Allison, 1970). REM sleep also appears to be present in the hedgehog (as reviewed in Monnier, 1980). The hedgehog, *Erinaceus europaeus* displays a REM sleep-like state characterized by phasic motor activity, desynchronized EEG, EEG theta (6.0-8.0 Hz), motor atonia, REMs and heightened arousal thresholds (Monnier, 1980). Similar results have been reported for the shrews,

Suncus murinus, Blarina brevicauda, and *Cryptotis parva* (Allison, Gerber et al., 1977). Sleep states occupying approximately 15-18% of total sleep time characterized by desynchronized EEGs, irregular respiration, increases in phasic motor activity, and generally low levels of motor tone were reported in all three shrew species (Allison, Gerber et al., 1977). Comparable findings are reported in the moles, *Scalopus aquaticus* and *Condylora cristata* (Allison and Van Twyer, 1970). Moles exhibit a sleep state characterized by desynchronized EEGs, increases in phasic motor activity, hippocampal theta (4.0-6.0 Hz) activity, and EEG spindle bursts during the transition from nonREM sleep to the "activated" sleep state (Allison and Van Twyer, 1970). The amount of this sleep state in moles (25% of total sleep time) is similar to REM sleep amounts reported for marsupials, and other placental mammals (Zepelin, 1994). Although arousal thresholds and REMs were not determined during presumptive REM sleep in the mole or shrew, it would appear that REM sleep is present in moles and shrews. The presence of REM sleep in another primitive mammal, the armadillo *Dasypus novemcinctus,* is less clear (Prudom and Klemm, 1973). Prudom reported several sleep states associated with "activated EEG" in *Dasypus novemcinctus* that bore some resemblance to REM sleep in modern placental mammals (Prudom and Klemm, 1973). Sleep states associated with "activated EEG", REMs and atonia constituted 7.5% of total sleep time. However, arousal thresholds were not assessed in the armadillo, and distinguishing quiet wake from REM sleep in these animals was difficult. Nevertheless, the authors concluded that REM sleep was present in the armadillo in amounts that ranged from 7.5%- 24.5% of total sleep time depending on the criteria used to score this state.

There are inconsistent findings regarding the presence of REM sleep in cetaceans. Mukhametov reported that the bottlenose dolphin *(Tursiops truncatus)* and the Amazonian dolphin *(Inia geoffrenis)* have typical patterns of mammalian nonREM sleep with the exception that these species express nonREM sleep uni-hemispherically. Surprisingly, Mukhametov could not find evidence for REM sleep in either the bottlenose or the Amazonian dolphin (Mukhametov, Supin et al., 1977; Mukhametov, 1987). On the other hand, Zepelin cites several studies in dolphin and the killer whale where resting states characterized by REMs, phasic motor activity and penile erections were observed (Zepelin, 1994). In another cetacean, the pilot whale *Globicephala scammoni,* a REM sleep-like state characterized by motor atonia and REMs was reported (Shurley, Serafetinides et al., 1969).

The absence of REM sleep in cetaceans may be related to the constraints of the aquatic environment in which these air-breathing mammals live. Presumably, the atonia, and irregularities in cardiorespiratory output typical of mammalian REM sleep would interfere with surfacing necessary for respiration in these species. Similar arguments have been advanced to explain the uni-hemisphericity of nonREM sleep in cetaceans (Mukhametov, Supin et al., 1977; Mukhametov, 1987). However, as Zepelin points out, other sea-going mammals with unihemispherical sleep and similar ecological constraints have REM sleep (Zepelin, 1994). The peculiarities of the cetacean brain fail to account for the reported absence of REM sleep in these animals. The cetacean brain retains features of more primitive mammals, like the hedgehog, and has neocortical development comparable to primates (Scwerdtfeger, Oelschlager et al., 1984; Morgane, Jacobs et al., 1985); two groups of animals that express abundant amounts of REM sleep. Therefore, as suggested by Siegel, the absence of REM sleep in some cetaceans may be due to poor acclimation to the laboratory setting rather to ecological or neurological factors unique to cetaceans (Siegel, 1995)

Monotremous mammals

The monotremes represent surviving members of an ancient class of mammals known as the prototheria (Kemp, 1982; Carrol, 1988). The prototheria, today represented by the long-beaked echidna, *Zaglossus brujini,* the short-beaked echidna, *Tachyglossus aculeatus,* and the duck-billed platypus, *Ornithorhynchus anatinus* diverged from placental and marsupial mammals lines approximately 130 million years ago during the early Cretaceous period (Kemp, 1982). Monotremes have changed relatively little since their initial divergence from other mammals, and are thought be more representative of the therapsid reptile ancestor common to modern mammals than any other extant mammal (Kemp, 1982). The monotremes, therefore, offer a unique glimpse into the evolution of mammalian REM sleep.

Sleep has been studied in two monotremes, the terrestrial echidna, *Tachyglossus aculeatus,* and the more ancient, amphibious duck-billed platypus, *Ornithorhynchus anatinus*. Two careful studies of the echidna strongly suggest that this monotreme does not have REM sleep (Allison, Van Twyer et al., 1972; Siegel, Manger et al., 1996). Allison et al reported that while the echidna had nonREM sleep, primary criteria for REM sleep were not observed. Presumptive REM sleep-like states were sometimes noted, but these "PS?"[paradoxical sleep?] episodes were not associated with heightened arousal thresholds, and were not homeostatically regulated. Allison et al concluded that the "PS?" episodes were in fact, periods of quiet wakefulness (Allison, Van Twyer et al., 1972). Similar result have been reported by Siegel et al., who found that quiescent periods similar to "PS?" episodes reported by Allison et al generally occurred after active periods and not after nonREM sleep periods. Seigel et al. also demonstrated that the firing rate pattern (FRP) of brainstem neurons in the sleeping echidna was intermediate between FRPs typically observed in mammalian REM and nonREM sleep. The authors concluded that the echidna does not have REM sleep, and instead exhibits a sleep state that combines aspects of both REM and nonREM sleep. According to these authors, echidna sleep represents a primordial mammalian sleep state, preserved in monotremes, that segregated into the distinct states of REM and nonREM sleep during the course of mammalian evolution (Siegel, 1995; Siegel, Manger et al., 1996).

This hypothesis, however, may require revision in light of recent studies in the more ancient duck-billed platypus, *Ornithorhynchus anatinus.* The terrestrial echidna likely evolved from the amphibious platypus (Siegel, Manger et al., 1997), and indeed, the platypus brain appears to be much more primitive, and reptilian-like than the echidna (Bohringer and Rowe, 1977; Divac, Holst et al., 1987). In striking contrast to the echidna, the platypus appears to have an abundance of REM sleep (Siegel, Manger et al., 1997). In the one sleep study of the platypus, more than 60% of sleep time was devoted to a REM sleep-like state characterized by phasic motor activity, REMs, muscle atonia, and heightened arousal thresholds. The EEG of this REM sleep-like state was of moderate to high amplitude, although the frequency of these EEG waves was not reported (Siegel, Manger et al., 1997). It would seem, therefore, that the unusual 'combined' sleep of the echidna may be a unique evolutionary adaptation in the monotreme line, and not necessarily a primordial mammalian sleep as originally proposed.

Summary of Primary Findings

Strong evidence for REM sleep is found only in mammals and birds. REM sleep is not found in invertebrate taxa, and the evidence for REM sleep in fish and amphibians is very weak, or non-existent. Some investigators report REM sleep in certain classes of reptiles, but overall, the presence of reptilian REM sleep has not been convincingly demonstrated. Birds have REM sleep, although there are important differences between avian and mammalian REM sleep. The amount of REM sleep and the duration of individual REM sleep episodes in birds is much smaller than in mammals. In contrast to mammals, sleep deprivation does not elicit rebound increases in avian REM sleep suggesting that REM sleep is not homeostatically regulated in birds. With the exception of the monotremous echidna and possibly the dolphin, all mammals studied to date have REM sleep. Several correlative observations regarding the presence of REM sleep in the animal kingdom can be made from these findings. The first is that REM sleep is not observed in animals that lack or have poorly developed neocortices. For example, insects, fish, amphibians, and reptiles do not have REM sleep and neocortical development in these animals is either non-existent, or extremely rudimentary (Sarnat and Netsky, 1981; Lohman and Smeets, 1991). Mammals and birds, however, have either well-developed neocortex or comparable structures, such as the avian Wulst and dorsal ventricular ridge (Shimizu and Karten, 1991). A second observation is that REM sleep is only found in animals that express EEG SWA during nonREM sleep. Birds and mammals have large amounts of nonREM sleep characterized by large amplitude, slow waves in the EEG. In contrast, nonREM sleep in reptiles and amphibians is usually associated with EEG spikes and/or sharp waves, and only rarely associated with EEG slow waves or SWA. A third observation is that REM sleep is only found in endothermic animals. Poikiloherms such as amphibians and reptiles do not have REM sleep, whereas endothermic birds and mammals do.

These general observations have led to three broad categories of hypotheses regarding the evolution and presumed function of REM sleep. These hypotheses are speculative since data concerning the evolution of neurological structures and comparative neurophysiology are scarce. However, they are representative of current views on how and why REM sleep evolved and are therefore worth considering.

Theories Regarding the Evolution of REM Sleep

Did REM sleep evolve for periodic vigilance?
The 'sentinel' theory of REM sleep

A number of investigators have suggested that REM sleep evolved to provide periodic arousals during nonREM sleep (Drew and Batt, 1972; Vasilescu, 1983). The general hypothesis is that periodic arousals during nonREM sleep allowed sleeping animals opportunities to monitor their environment and thus avoid predation. In the case of animals with consolidated sleep and waking periods, the periodic neuronal excitation of REM sleep may have also prepared the brain for extended waking periods at the end of the rest phase. In support of these hypotheses, these theorists point out that REM sleep is associated with many features typical of wakefulness, namely, neuronal excitation and increased neuronal metabolism. In animals with consolidated sleep/wake cycles, the longest REM sleep episodes occur near the end of the rest period (Zepelin, 1994), and are coupled to circadian mechanisms that provide alerting signals at different times during the subjective day (Edgar, Dement et al., 1993). These findings seem to support the idea that REM sleep prepares the brain for extended waking periods. The 'sentinel' theory of REM sleep, however, seems implausible for several reasons. Several non-mammalian prey species (like frogs and lizards) also have extended rest periods characterized by nonREM sleep-like states and heightened arousal thresholds, but do not have REM sleep. Nor can the sentinel theory of REM sleep adequately explain why REM sleep is accompanied by very high arousal thresholds and motor paralysis. Mammalian REM sleep seems poorly suited to serve a sentinel function since evolution has favored *more* not less disconnection from the external environment during these states.

Did REM sleep evolve in response to endothermy?

The fact that REM sleep is found only in endothermic animals like birds and mammals, suggests that REM sleep evolved as an adaptation to endothermy (as reviewed in Zepelin, 1994). It is possible that the high cost of endothermy in the relatively small ancestors of birds and mammals favored the evolution of sleep states that allowed for energy conservation. The loss of thermoregulatory ability and the cessation of motor output typical of REM sleep may have provided a slight, but significant additional energy savings to these first endothermic animals. Others have argued, however, that the energy savings provided by REM sleep is so small as to be negligible. As pointed out by Zepelin, the net difference in energy expenditure between equal periods of quiet alertness and uninterrupted sleep (nonREM + REM) saves approximately "one frankfurter bun" worth of calories during one rest period for a 200 lb. human (Zepelin, 1994). Considering that REM sleep comprises 1/5 of the total amount of sleep in any given rest period for most mammals (and even less for birds), the net energy savings provided by REM sleep compared to equal amounts of quiet alertness is vanishingly small. It seems a poor evolutionary trade to gain such a small energy savings at the expense of increased risk of predation. In addition, REM sleep amounts in good thermoregulators (modern placental mammals) and poor thermoregulators (archaic placentals like the armadillo) are similar, suggesting that the relation between endothermy and REM sleep amounts, if present at all, is relatively weak.

A different explanation for the correlation between endothermy and REM sleep is provided by Kavanau, who suggests that sleep evolved in order to provide periodic maintenance or "dynamic stabilization" of under-used neuronal connections (Kavanau, 1997). According to Kavanau's hypothesis, the passive reduction in skeletal muscle tone characteristic of wake to nonREM sleep transitions was sufficient to "prevent sleep disruptive movements during reinforcement of synapses in motor circuits" in poikilothermic species. The development of endothermy and the accompanying higher levels of metabolism afforded by endothermy necessitated more powerful means of reducing sleep-disruptive movements (Kavanau, 1997).

"...evolution of more powerful mechanisms for inhibiting muscle contractions would have been favored. In the absence of such mechanisms, skeletal muscle contractions durings DS [dynamic stabilization] of motor circuitry would have severely disrupted sleep... These circumstances apparently led to the adaptive modification of a portion of the primitive sleep state.

"By means of this modification of primitive sleep. DS of infrequently used pathways containing motor components would have become associated largely with REM sleep and DS of infrequently used pathways containing largely non-motor components with nonREM sleep".

Kavanau's hypothesis regarding the evolution of REM sleep seems inconsistent with several phylogenetic and ontogenetic findings. The suggestion that activation of motor components primarily occurs during REM sleep and the activation of non-motor components occurs during nonREM sleep is not supported by studies of neural activity in different brain regions during REM and nonREM sleep. For example, the neuronal excitation characteristic of avian and mammalian REM sleep is not restricted to somatosensory or motor circuits, but is distributed across wide regions of cortical and subcortical structures (Steriade and Hobson, 1976; Amlaner, 1994). If indeed REM sleep evolved in order to block motor commands during DS, then the expression of REM sleep across animal species should be highly correlated with well-developed atonia. There are, however, several mammals and birds that have normal amounts of REM sleep but have no, or poor REM sleep atonia (Ruckebusch, 1972; Amlaner, 1994). According to Kavanau's hypothesis, the abundance of REM sleep in neonatal mammals can be explained if one assumes that neonates "require a high level of non-utilitarian DS" and REM sleep atonia prevents "...full-blown contractions of developing and developed skeletal muscles" (Kavanau, 1997). Neonatal REM sleep, however , is not associated with well-developed atonia. The maturation of REM sleep atonia occurs at relatively late stages of postnatal development (Chase, 1971). It would seem, therefore, that the evolution of REM sleep is not adequately explained by the "dynamic stabilization" hypothesis.

Did REM sleep evolve in response to nonREM sleep SWA?

The correlation between well-developed neocortical structures and the presence of REM sleep suggests that some property of neocortical activity may have produced selection pressure for REM sleep states. The fact that REM sleep is present only in animals with high levels of EEG SWA in nonREM sleep suggests that the evolution of synchronized neocortical oscillations in sleep might have been the critical event that produced selection pressure for REM sleep. There are several findings in mammals that support this hypothesis. First across many mammalian lines, the amount of REM sleep is positively correlated with the amount of nonREM sleep (Benington and Heller, 1994; Zeplin, 1994). Second, nonREM sleep always precedes REM sleep in the sleep cycle, and REM sleep timing is regulated by the amount of prior nonREM sleep (Benington and Heller 1994; Zepelin 1994). Third, REM sleep amounts are conserved as a percentage of total sleep time, and not as a percentage total waking, or total recording time (Benington and Heller, 1994; Zepelin, 1994). These and other findings discussed elsewhere in this text suggest that a need for REM sleep accumulates in prior nonREM sleep.

An accumulation of REM sleep pressure during periods of nonREM sleep may also account for several differences between avian and mammalian REM sleep expression. EEG SWA is not uniformly distributed across the avian telencephalon during nonREM sleep as is the case for mammals. EEG SWA is found primarily in the hyperstriatum accessorium while other regions of the avian brain display reptilian like EEG spikes during nonREM sleep (Amlaner, 1994). Quantitative assessments of EEG SWA during avian nonREM sleep indicate that birds have lower amounts of EEG SWA compared to mammals (Amlaner, 1994) and, in contrast to mammals, EEG SWA in avian nonREM sleep does not intensify following sleep deprivation (Tobler and Borbely, 1988; Berger and Phillips, 1994). Therefore, the amount and intensity of EEG SWA in avian nonREM sleep is considerably less than that found in mammalian nonREM sleep. Consequently, the accumulation of REM sleep pressure during nonREM sleep may take appreciably longer in birds than in mammals. This may explain why birds have much smaller amounts of REM sleep, shorter REM sleep episodes and much longer nonREM sleep periods preceding REM sleep than mammals. Provided that REM sleep need accumulates in response to EEG SWA in nonREM sleep, the absence of EEG SWA intensification following sleep deprivation may account for the absence of REM sleep rebounds in birds.

A more troubling problem for this hypothesis is posed by mammals who have high levels of EEG SWA during nonREM sleep, but do not have REM sleep. While the presence of REM sleep in cetaceans is presently unresolved, it is increasingly clear that the monotremous echidna does not have REM sleep. Echidnas, however, do have high levels of EEG SWA during sleep. What then, are we to make of the echidna? The fact that the echidna combines aspects of both REM and nonREM sleep in one sleep state may prevent a prolonged accumulation of REM sleep pressure, hence obviating the need for discrete episodes of REM sleep. Unfortunately, because so little is known regarding the neural organization of the echidna, the significance of its unusual sleep is not yet clear.

Conclusions and Further Directions

Rest and activity states appear to be ubiquitous in the animal kingdom. Rest states resembling mammalian nonREM sleep have been noted in most vertebrate orders. REM sleep, however, appears to be restricted to birds and mammals. Given that the reptilian ancestor common to birds and mammals may not have had REM sleep, REM sleep may have evolved separately in these two vertebrate lines. A number of hypotheses have been presented to account for the evolution of REM sleep, but at present, no one hypothesis has isolated the ecological or biological factors that conferred selective advantage to the ancient vertebrates that first expressed REM sleep. The fact that REM sleep is found only in animals with neocortex (or comparable structures) strongly suggests that REM sleep evolved in response to neuronal properties unique to animals with well-developed telencephalons. The observation that REM sleep is found only in animals that express EEG SWA during nonREM sleep further suggests that the evolution of synchronized, oscillatory firing in neocortex may have been a pivotal factor in REM sleep evolution.

Strong conclusions regarding the evolution of REM sleep, however, must await the resolution of several vexing mysteries. The issue of reptilian REM sleep has not been adequately resolved, with some investigators reporting REM sleep and others reporting only nonREM sleep states in reptiles. A convincing demonstration of REM sleep in a reptile would force a reassessment of those factors associated with REM sleep evolution, and would place the appearance of REM sleep at an earlier date in vertebrate history. The reported absence of REM sleep in the dolphin and echidna suggest that REM sleep may have disappeared in certain mammals. The absence of REM sleep in the dolphin and echidna is surprising since REM sleep seems necessary for survival in other mammals. If indeed discrete episodes of REM sleep can be dispensed with in certain mammals, then what processes have evolved to take their place? A more complete understanding of the evolution of REM sleep therefore depends on further research in these vertebrate lines.

References

Affani. J.M. and Vacarezza, O.L. Difference in electrical activity during wakefulness and the phase of sleep with muscular twitches as recorded from the cortex of the marsupial *Didelphis Azarae* (South American Opossum), Experientia **23(3)** (1967) 216-217.

Allison, T., Gerber, S.D., Breedlove, S.M. and Dryden, G.L. A behavioral and polygraphic study of sleep in the shrews, Suncus murinos, Blarina brevicauda and Cryptotis parva, Behavioral Biology **20** (1977) 354-366.

Allison, T. and Van Twyer, H. Sleep in the moles, Scalopus aquaticus and Condylora cristata, Experimental Neurology **27** (1970) 564-578.

Allison, T., Van Twyer, H. and Goff, W.R. Electrophysiological studies of the echidna, Tachyglossus aculeatus. I. waking and sleep. Arch. Ital. Biol. **110** (1972) 145-184.

Amlaner, C.J.J., Avian Sleep. In: M.H. Kryger, T. Roth and W.C. Dement Principles and Practice of Sleep Medicine, Philadelphia, W.B. Saunders Company, (1994), 81-94.

Andersen, F.S., Sleep in moths and its dependence on the frequency of stimulation in Anagasta kuehniella, Opuscula Entomological **32** (1968) 15-24.

Ayala-Guerrero, F., Calderon, A. and Perez, M.C. Sleep patterns in a chelonian reptile *Gopherus flavormarginalis,* Physiology & Behavior **44** (1988) 333-337.

Ayala-Guerrero, F. and Huitron-Resendiz, S. Behavioral and electrophsiological patterns of wakefulness-sleep states in a lizard. Bol. Estud. Med. Biol. Mex., 39 (1991) 9-14.

Ayala-Guerrero, F., Huitron-Resendiz, S. and Mexicano, G. Effect of reserpine on sleep patterns in a chelonian reptile *(Gopherus Berlandieri),* Proc. West. Pharmacol. Soc. **36** (1993) 227-231.

Ayala-Guerreo, F., Huitron-Resendiz, S. and Mexicano, G. Effect of parachlorophenylalanine on sleep spikes in the iguanid lizard *Ctenosaura pectinata,* Proc. West. Pharmacol. Soc. **37** (1994) 149-152.

Benington, J.H. and Heller, H.C. Does the function of REM sleep concern non-REM sleep or waking? Progress in Neurobiology **44** (1994) 433-449.

Berger, R.J. and Phillips, N.H. Constant light suppresses sleep and circadian rhythms in pigeons without consequent sleep rebound in darkness, American Journal of Physiology, 267(36) (1994) R945-R952.

Bohringer, R.C. and Rowe, M.J. The organization of the sensory and motor areas of cerebral cortex in the platypus *(Ornithorynchus anatinus),* J. Comp. Neur. **174** (1977) 1-14.

Borbely, A.A., Sleep homeostasis and models of sleep regulation. In: M.H. Kryger, T. Roth and W.C. Dement. Principles and Practice of Sleep Medicine, Philadelphia, W.B. Saunders Company, (1994) 309-320.

Cambell, S.A. and Tobler, I. Animal sleep: A review of sleep duration across phylogeny, Neuroscience and Biobehavioral Reviews, **8** (1984) 269-300.

Carrol, R.L., Vertebrate Paleontology and Evolution. W.H. Freeman and Company, New York, 1988.

Chase, M.H., Brain stem somatic reflex activity in neonatal kittens during sleep and wakefulness, Physiology & Behavior, **7** (1971) 165-172.

Divac, I., Holst, M.C., Nelson, J. and McKenzie, J.S. Afferents of the frontal cortex in the echidna Tachyglossus aculeatus. Indication of an outstandingly large prefrontal area, Brain Behavior and Evolution, **30** (1987) 303-320.

Drew, W.G. and Batt, J. A contribution to the evolutionary theory of dreaming: An hypothesis on the role of the pineal in species and specimen protection, Biological Psychiatry, **4** (2) (1972) 131-146.

Duenas, I.V. and Ayala-Guerrero, F. Filogenia del sueno: Las aves, Bol. Estud. Med. Biol., Mex., Sulemento, **32** (1983) 83-90.

Edgar, D.M., Dement W.C. and Fuller, C.A. Effect of SCN-lesions on sleep in squirrel monkeys: evidence for opponent processes in sleep-wake regulations, Journal of Neuroscience, **13** (1993) 1065-1079.

Finlay, B.L., Innocenti, G. and Scheich, H. The Neocortex. Ontogeny and Phylogeny, Plenum Press, New York and London, 1991.

Flanigan, W.F.J., Sleep and wakefulness in Iguanid lizards, Ctenosaura Pectinata and Iguana Iguana. Brain Behavior and Evolution, **8** (1973) 401-436.

Flanigan, W.F.J., Sleep and wakefulness in chelonian reptiles II. The red footed tortoise *Geochelene carbonaria,* Arch. Ital. Biol., **112** (1974) 253-277.

Flanigan, W.F.J., Knight, C.P., Harste, K.M. and Rechtschaffen. A. Sleep and wakefulness in chelonian reptiles I. The box turtle, *Terropene carolina,* Arch. Ital. Biol., **112** (1974) 227-252.

Flanigan, W.F.J., Wilcox, R.H. and Rechtschaffen. A. The EEG and behavioral continuum of the crocodilian *Caiman sclerops,* Electroencephalography and Clinical Neurophysiology, **34** (1973) 521-538.

Hartse, K.M., Sleep in insects and nonmammalian vertebrates. In: M.H. Kryger, T. Roth and W.C. Dement. Principles and Practice of Sleep Medicine, Philadelphia, W.B. Saunders Company, (1994) 95-104.

Hartse, K.M. and Rechschaffen, A. Effect of atropine on the sleep-related EEG spike activity of the tortoise *Geochelene carbonaria,* Brain Behavior and Evolution, **9** (1974) 81-94.

Hartse, K.M. and Rechtschaffen, A. The effect of amphetamine, nembutal, alpha-methyltyrosine and parachlorophenylalanine on the sleep-related spike activity of the tortoise *Geochelene carbonaria,* and on the cat ventral hippocampal spike, Brain Behavior and Evolution, **9** (1982) 199-222.

Hobson, J.A., Electrographic correlates of behavior in the frog with special reference to sleep. Electroencephalography and Clinical Neurophysiology, **22** (1967) 113-121.

Hobson, J.A., Electrographic correlates of behavior in tree frogs. Nature **220** (1968) 386-387.

Huntley, A.C., Donnely, M. and Cohen, H.B. Sleep in the western toad *Bufo boreas,* Sleep Research, (1969) 141.

Huntley, A.C., Friedman, J.K. and Cohen, H.B. Sleep in an Iguanid lizard *Dipsosaurus dorsalis,* Sleep Research, **6** (1977) 104.

Jones, B., Paradoxical sleep and its chemical/structural substrates in the brain, Neuroscience, **40** (3) (1991) 637-656.

Jouvet-Mounier, D., Astic, L. and Lacote, D. Ontogenesis of the states of sleep in rat, cat and guinea pig during the first postnatal month, Developmental Psychobiology, **2** (4) (1970) 216-239.

Kaiser, W. and Kaiser, J.S. Neuronal correlates of sleep, wakefulness in a diurnal insect, Nature, **301** (1983) 707-709.

Karmanova, I.G. and Lazarez, S.G. Electrophysiological analysis of wakefulness and primary sleep in the frog *rana tempororaria,* Zhurnal Evoliutsionnoi Biokhimii i Fiziologii, **14** (1978) 379-384.

Karmanova, I.G. and Lazarez, S.G. Neurophysiological characteristics of primary sleep in fish and amphibian. In: Sleep 1978, Fourth European Congress on Sleep Research, S. Karger Basel, (1980) 437-442.

Kavanau, J.L., Origin and evolution of sleep: Roles of vision and endothermy, Brain Research Bulletin, **42** (4) (1997) 254-264.

Kemp, T., Mammal-like reptiles and the origin of mammals. Academic Press, London, 1982.

Lohman, A.H.M. and Smeets, W.J.A.J. The dorsal ventricular ridge and cortex of reptiles in historical and phylogenetic perspective. In: B.L. Finlay, G. Innocenti and H. Scheich. *The Neocortex: Ontogeny and Phylogeny,* New York and London, Plenum Press, **200** (1991) 59-74.

Lucas, E., Sterman, M.B. and McGinty, D.J. The salamander EEG: a model of primitive sleep and wakefulness. Psychophysiology, **6** (2) (1969) 230.

Mather, J.A. and Mather, D.L. Skin colours and patterns of juvenile Octopus Vulgaris *(Mollusca cephalopoda)* in Bermuda, Vie et Millieu, **44** (3-4) (1994) 267-272.

Melesio-Nolasco, A. and Ayala-Guerrero, F. Filiogenia del sueno: pisces y anfibos, Bol. Esud. Med. Biol. Mex. Sulemento, **32** (1983) 101-107.

Melesio-Nolasco, S., Friedman-Saavedra, R., Vargas-Reyna, L. and Ayala-Guerreo, F. Filogenia del sueno: Los reptiles, Bol. Estud. Med. Biol., Mex. Sulemento, **32** (1983) 91-99.

Monnier, M., Comparitive electrophysiology of sleep in some vertebrates. Experentia. **36** (1980) 16-19.

Morgane, P.J., Jacobs, M.S. and Galaburda, A. Conservative features of neocortical evolution in dolphin brain, Brain Behavior and Evolution, 26 (1985) 176-184.

Mukhametov, L.M., Unihemispheric slow-wave sleep in the amazonian dolphin, Inia geoffrensis, Neuroscience Letters, **47** (1987) 155-157.

Mukhametov, L.M., Supin, A.Y. and Polyakova, I.G. Interhemispheric asymmetry of the electroencephalographic sleep patterns in dolphins, Brain Research, 134 (1977) 581-584.

Prudom, A.E. and Klemm, W.R. Electrographic correlates of sleep behavior in a primitive mammal, the armadillo *Dasypus novemcinctus,* Physiolgy & Behavior, **10** (1973) 275-282.

Rial, R., Nicolau, M.C., Lopez-Garcia, J.A. and Almirall, H. On the evolution of waking and sleeping, Comp. Biochem. Physiol., **104a**(2) (1993) 198-193.

Ruckebusch, Y., Comparitive aspects of sleep and wakefulness in farm animals. In: M.H. Chase. The Sleeping Brain, Los Angeles. University of California, **1** (1972). 23-29.

Sarnat, H.B. and Netsky, M.B. Evolution of the Nervous System, Oxford University Press, New York, 1981.

Scwerdtfeger, W.K., Oelschlager, H.A. and Stephan, H. Quantitative neuroanatomy of the brain of the La Plata dolphin, *Pontoporia Blainvillei,* Anat. Embryol., **170** (1984) 11-19.

Segura, E.T. and De Juan, A. Electroencephalographic studies in toads, Electroencephalography and Clinical Neurophysiology, **21** (1966) 373-380.

Shapiro, C.M. and Hepburn, H.R. Sleep in a schooling fish *Tilapsia mossambica,* Physiology and Behavior, **16**(5) (1976) 613-615.

Shimizu, T. and Karten, H.J. Multiple origins of neocortex: Contributions of the dorsal ventricular ridge. In: B.L. Finlay, G. Innocenti and H. Scheich. *The Neocortex: Ontogeny and phylogeny,* New York and London, Plenum Press, 200 (1991) 77-85.

Shurley, J.T., Serafetinides, E.A., Brooks, R.E., Elsner, R. and Kenney, D.W. Sleep in cetaceans: I. The pilot whale *Globicephala scammoni,* Psychophysiology, **6**(2) (1969) 230.

Siegel, J.M. Phylogeny and the function of REM sleep, Behavioral Brain Research, **69** (1995) 29-34.

Siegel, J.M., Manger, P.R., Nienhuis, R., Fahringer, H.M. and Pettigrew, J.D. The echidna *Tachyglossus aculeatus* combines REM and non-REM aspects in a single sleep state: implications for the evolution of sleep, Neuroscience, **16**(10) (1996) 3500-3506.

Siegel, J.M., Manger, P.R., Nicnhuis, R., Fahringer, H.M. and Pettigrew, J.D. The platypus has REM sleep, Sleep Research, (1997) 524.

Steriade, M. and Hobson, J.A. Neuronal activity during the sleep-waking cycle, Progress in Neurobiology, **6** (1976) 155-376.

Strumwasser, F., The cellular basis of behavior in Aplysia, Journal of Psychiatric Research, **8** (1971) 237-257.e

Susic, V., Electrographic and Behavioral correlations of the rest-activity cycle in the sea turtle *Caretta caretta,* J. Exp. Mar. Biol. Ecol., **10** (1972) 81-87.

Tauber, E.S., Roffwarg, H.P. and Weitzman, E.D. Eye movements and electroencephalogram activity during sleep in diurnal lizards, nature **212** (1966) 1612-1613.

Tauber, E.S., Rojas-Ramirez, J. and Peon. R.H. Electrophysiological and behavioral correlates of wakefulness and sleep in the lizards *Ctenosaura pectinata,* Electroencephalography and Clinical Neurophysiology, **24** (1968) 424-433.

Tauber, E.S. and Weitzman, E.D. Eye movements during behavioral inactivity in certain Bermuda reef fish, Communications in Behavioral Biology, Part A (3) (1969) 131-135.

Tobler, I., Effect of forced locomotion on the rest-activity cycle of the cockroach, Behavioral Brain Research, **8** (1983) 351-360.

Tobler, I., Is sleep fundamentally different between mammalian species?, Behavioral Brain Research, **69** (1995) 35-41.

Tobler, I. and Bordely, A.A. Sleep and EEG spectra in the pigeon *(Columbia livia)* under baseline conditions and after sleep deprivation, J. Comp. Physiol., **163** (1988) 729-738.

Tobler, I. and Stalder, J. Rest in the scorpion-a sleep-like state?, J. Comp. Physiol. A., **63** (1988) 227-235.

Van Twyer, H., Polygraphic studies of the American alligator, Sleep Research, **2** (1973) 87.

Van Twyer, H. and Allison, T. Sleep in the Opossum Didelphis Marsupials, Electroencephalography and Clinical Neurophysiology, **29** (1970) 181-189.

Vasilescu, E., Sleep and wakefulness in the tortoise Emys Orbicularus. Rev. Roum. Biol. Zoologie, **15**(3) (1970) 177-179.

Vasilescu, E., Phylogenetic and general remarks on sleep, Rev. Roum, Morphol. Embryol. Physiol. Physiologie, **20** (1) (1983) 17-25.

Walker, J.M. and Berger, R.J. A polygraphic study of the tortoise *(Testudo denticulata).* Brain Behavior and Evolution, **8** (1973) 453-467.

Warner, B.F. and Huggins, S.E. An electroencephalographic study of sleep in young caimans in a colony, Comp. Biochem. Physiol. **59a** (1978) 139-144.

Zepelin, H., Mammalian Sleep. In: M.H. Kryger, T. Roth and W.C. Dement. Principles and practice of Sleep Medicine, Philadelphia, W.B. Saunders Company, (1994) 69-80.

Rapid Eye Movement Sleep
B.N. Mallick and S. Inoué (Eds)

3. Ontogenesis of REM Sleep

Masaya Segawa

Segawa Neurological Clinic for Children, 2–8 Surugadai Kanda,
Chiyoda-ku, Tokyo 101 0062, Japan

SUMMARY

REM sleep is a behavioral state with various but state-specific parameters coherently emerged in time. During the developmental course of the brain, these parameters do not develop simultaneously, but each parameter shows a specific time course particular for each. That is, twitch movements or body movements appear early and show periodical occurrence from around 28 gestational weeks, while axial atonia becomes reliable parameters of REM sleep later around 3 months after birth. Rapid eye movements show a similar developmental course as body movements but irregular respiration becomes a measure of REM sleep later taking a similar time course as axial atonia. Periodical occurrence of the irregular heart rate is observed early around 20 gestational weeks, but the periodicity differs from that of body and eye movements. It is around 40 gestational weeks when all parameters synchronize with each other and the state of REM sleep appears.

Each parameter also varies in quality as well as in quantity with weeks of gestation. These processes are observed mainly after birth. In early infancy REM sleep exists in high amount and occurs at shorter intervals. The former decreases rapidly and attains adult levels by 2 yr and the latter prolongs markedly in early childhood and shows adult ultradian rhythm around 10 yr.

Although environmental sensory inputs have influence for its maturation in the postnatal period, most of the developmental course of REM sleep is organized by an endogenous or genetically determined timetable independent of environmental factors. The high amount of REM sleep in the immature brain suggests the roles of this sleep stage for the development of the brain or nervous system. Endogenous activation of the brain by neurons involved in REM sleep show activity dependent development and may influence neuronal growth, synaptic formation and genetic read-out in each area of the brain.

Key Words: Ontogeny of REM sleep, Development particular for each parameter, activity dependent development.

In a normal matured brain, REM sleep is identified by the simultaneous presence of desynchronized cortical EEG, absence of activity in the antigravity muscles, periodic bursts of rapid eye movements, irregularity in respiration, variation of the heart beat rate and phasic muscle activities.

In an immature brain during the developmental course, these components of REM sleep do not appear simultaneously with periodicity. Moreover, each component shows a developmental course particular for each. Phasic muscle twitches show periodical appearance early in the end of the second trimester and muscle atonia accompanies this period later in early infancy. These processes suggest that each parameter of REM sleep is modulated by neurons or the neuronal system particular for each and the developmental variation of these parameters reflects the functional maturation of these neurons, which are located in the brainstem. So by evaluating REM sleep in these age periods we can evaluate the maturational process of the neurons of the brainstem.

Furthermore, the relatively high amount of REM sleep in an immature brain suggests the important roles of REM sleep for the promotion of further brain development.

This article reviews development of the components of REM sleep and discusses the possible roles of REM sleep in the immature brain for the functional development of the brain.

Development of the Components of REM Sleep

In neonates there are two sleep stages, one is the quiet sleep (QS) and the other is the active sleep (AS). QS is defined as a sleep period without eye and body movements and with rhythmic respiration, while AS is defined as a period in which there are eye movements with closed eyes, frequent twitch and gross movements of the trunk, extremities and face, and irreuglar respiration. The former is considered as immature NREM sleep and the latter as REM sleep of an immature brain.

The QS of neonates differs from the slow wave sleep of adults by the presence of sucking and startled response and shows a difference in the pattern of EEG and evoked potentials (Sterman, 1972). This immature state develops rapidly during the first 5 months after birth and has the characteristic features of adult NREM sleep by the end of the 5th month (Sterman, 1972).

On the other hand, neonatal AS shows the characteristics of REM sleep of an adult. That is, in a polysomnogram of an infant of 1 week of age there observed a low amplitude, higher frequency EEG, increase in the respiration rate with irregularity, rapid eye movements and atonia of the mental muscle (Sterman, 1972).

Thus, REM sleep matures to certain levels during the fetal period. But each component of REM sleep shows an identical developmental course, particular for each. Parmelee and Stern (1972) evaluated the usefulness of these parameters as indicators for REM sleep and revealed that body movements become a reliable parameter in the most early weeks of the fetus period and EEG and atonia show their characteristics later around the term of early infancy (Table 1).

Spontaneous movements are observed only at gestational ages of 9–10 weeks. From studies on fetus, Pamelee and Stern (1972) revealed that these spontaneous movements began to show periodicity from around the 5th gestational month. Studies on premature babies showed no periodicity in body movements during sleep in their 24th to 26th gestational weeks (Dreyfus-Brisac, 1968). These body movements began to show waxing and waning with short intervals from arond the 28th to 30th gestational weeks (Dreyfus-Briac, 1967), in which the period of body movements roughly corresponds to AS or REM stage (Roffwarg et al., 1966).

Table 1. A Schematic Organization of the Usefulness of Measures in Defining REM State in the Premature and Young Infant

	Weeks conceptional age				Months post term		
	24	28	32	36	40	3	8
Body movements	±	+	++	+++	++++	++++	++++
Eye movements		+	++	+++	++++	++++	++++
Respiration pattern			±	++	+++	++++	++++
EEG			±	++	+++	++++	++++
Chin MEG				+	+++	++++	++++

With the advancement of gestational age, body movements tend to appear in certain periods of sleep which reflect the increase of the non-movement period. That is, in premature babies of 32 gestational weeks the non-movement period is about 53% of the total sleep period, but it increases to 60% in term infants (Parmelee and Stern, 1972). After birth this period also increases and it is around 80% in infants of 8 months (Parmelee and Stern, 1972).

Rapid eye movements appear in fetus of 24–26 gestational weeks and increase their number with age towards the 32 gestational week (Dreyfus-Brisac, 1967). They are observed in the periods with body movement and show similar developmental course as body movements. In premature babies the period without REMs is 50% at 32 gestational weeks, which increases to 56% in term infants and 70% in infants of 8 months (Dreyfus-Brisac, 1967).

Rhythmic respiration is observed in fetus of 17 gestational weeks but the respiration of premature infants is irregular throghout the sleep stages of 24–26 gestational weeks (Dreyfus-Brisac,1968). It is at 28–30 gestational weeks when the rhythmic respiration begins to be observed in all premature babies (Dreyfus-Brisac, 1967). Parmelee and Stern (1972) evaluated the pattern of respiration in each 20 sec epoch during sleep of premature babies and revealed an increment of the epochs with rhythmic respiration with gestational weeks. That is, they were 6% in premature infants of 32 gestational weeks, 24%at 40 gestational weeks and increased to 54% in infants of 8 months.

Mentalis muscle shows no activities throughout sleep in infants before 32 gestational weeks (Parmelee and Stern, 1972). It is after 40 gestational weeks when reciprocal atonic and tonic phase occurs during sleep and around 3 months of age atonia becomes a reliable parameter of REM sleep (Parmelee and Stern, 1972).

Development of parameters of REM sleep of the fetus are evaluated by the analysis of ultrasound imaging.

Eye movements are observed early at 16–18 gestational weeks, but they are slow eye changes and appear sporadically with a low frequency of less than 5 per min (Prechtl and Nijhuis, 1983, Inoué et al., 1986). At 20 weeks of gestation more rapid eye movements are detectable (Prechtl and Nijhuis, 1983) and at 24–26 gestational weeks, eye movements increase their number and appear in cluster with a higher frequency of more than 20 per min (Inoue et al., 1986). At 35–36 gestational weeks eye movements always occur in cluster and in a higher frequency, when the period with cluster eye movements appear reciprocally with the period of non-eye movement (Inoue et al., 1986).

Birnholz (1981) had also examined fetal eye movements precisely and classified them into four types according to the frequency of occurrence and revealed an increase in frequency with the passage of gestational weeks. According to this study, eye movement with the lowest frequency (< 5 sec) disappeared at 25 gestational weeks and they were replaced by conjugated saccades alternating with the fixation period.

The duration of periods with eye movements and also those without eye movements increased with gestational weeks from the early third trimester until about 37 gestational weeks (Koyanagi et al.,1993). The duration of periods with eye movements increased with two epochs, one at 29–30 gestational weeks and the other at 37–38 weeks. On the other hand the first epoch for increase of the duration of the period without eye movements was delayed at 31–32 weeks but the latter epoch was 37–38 weeks as the period with eye movements. These differences suggest the difference in the neural mechanism for maintaining the periods with eye movement and for the periods without eye movements (Koyanagi et al., 1993). In fetuses of 33–36 and 37–41 gestational weeks two kinds of eye movements with different duration or velocities were observed (Horimoto et al., 1990); one with a shorter duration of 0.07 to 0.6–0.8 sec and the other with a longer duration of 0.6–0.8 to 4–5 sec. When comparing the proportionate amount of time spent for periods of shorter and longer eye movements between these two gestational weeks, that of the former increased in the later gestational weeks.

Marked variation of eye movements between 33 and 41 gestational weeks were shown by Prechtl and Nijhuis (1983). Rapid eye movements (eye movement with shorter duration) and slow eye movements (eye movements with longer duration) are organized in episodes of activity divided by episodes of quiescence which are at first unrelated to cyclic changes in other variables such as body movements, respiratory movements and irregular heart rate. At about 36–38 gestational weeks, periods with eye movements become linked with other variables as coherent behavioral states.

Penile tumescence is closely related to REM sleep in neonates as well as in adults (Karacan, 1966). In adults, penile tumescence was observed in 80–95% of REM sleep while in fetus, it occurred in 78% of the eye movement period and in neonates in 84% (Karacan, 1966). In studies on fetuses of 36–41 gestational weeks, Koyanagi et al. (1991) revealed significantly longer duration of penile tumescence in the eye movement period than in the non-eye movement period. This penile tumescence of fetus could be found in non-eye movement periods, but it was observed rather selectively in the transitional phase from the eye-movement period to the non-eye-movement period or vice versa and only a few were found in the mid-stage or non-eye movement period (Koyanagi et al., 1991).

During NREM sleep of premature babies born at 35–36 gestational weeks or later, mouthing movements were observed with constant intervals (Dreyfus-Brisac, 1970, Prechtl, 1974). In fetuses of 28–31 gestational weeks these mouthing movements appeared with random intervals without any bias of distribution against the stages (Horimoto et al., 1989). However, the movements began to cluster with regular intervals in non-eye movement periods with the gestational weeks. At 35 gestational weeks the movements were observed in non-eye movement periods with high incidence and with regular intervals of 0.3–0.6 sec while in eye movement periods smaller number of the movements occurred at random intervals (Horimoto et al., 1989).

In human fetus, micturition was observed more frequently during the 'active' phase than during the "quiet" phase of fetal heart rate variation. Micturition began to occur in periodicity by around 20 gestational weeks and the interval became longer with the weeks; that is 34 min at 33–36 gestational weeks and 39 min at 37–41 gestational weeks (Visser et al., 1981, Koyanagi et al., 1992). These intervals were different from those between eye-movement and non-eye-movement period in these weeks, that is, 29 and 56 min, respectively and at term, micturition began to occur concomitantly with the onset of eye movement period (Koyanagi et al., 1992).

Each parameter of REM sleep also varies in quality as well as in quantity with the passage of its development. Petre-Quadens (1974) evaluated the intervals of REMs in premature and postterm babies in 40 sec epochs during AS and revealed two modal curves in their development. That is, the frequencies of rapid eye movements with shorter intervals became smaller in premature babies of 37–39 weeks compared to those of 31–36 weeks, but the frequencies began to increase again from the ages of 40–41 gestational weeks and showed a dramatic increase during the period of 2–24 weeks after birth attaining adult levels.

The frequencies of REMs appearing with shorter intervals than 1 sec, that is, in burst, did not show any difference among REMs periods in premature babies but in babies of 40–41 gestational weeks, the frequencies of REMs occurring with burst began to increase in later REM periods during sleep (Petre-Quadens et al., 1971).

Kohyama et al. (1993, 1997) examined the age related changes of phasic muscle activities during REM sleep in 38 normal infants of 34–84 postconceptional weeks and classified the phasic muscle activities into 2 types, one was twitch movement (TMs) lasting 0.5 sec or less and the other was localized movements (LMs) lasting more than 0.5 sec. They evaluated the age related variations of these two phasic activities by assessing the number of TMs against the total number of both phasic activities during REM sleep. They also examined the numbers of both phasic activities occurring during the periods of burst occurrence of REMs and calculated the rate of these phasic activities against the total phasic activities observed during REM sleep.

Kohyama et al. (1993) revealed marked age related variation in these two measures. During infancy and childhood LMs were dissociated into TMs and the number of TMs against the total number of TMs and LMs increased with postconceptional weeks. They called this the dissociation index (DI) and showed its age variation as demonstrated in Fig. 1a.

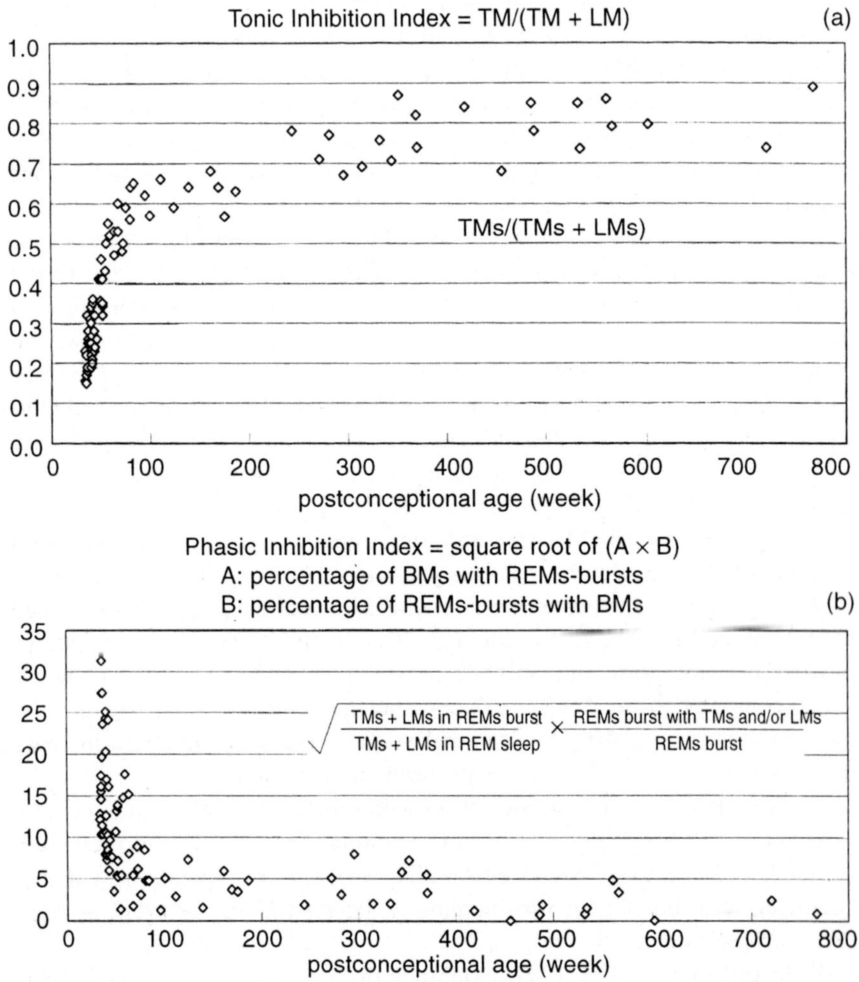

Fig. 1. Age related changes of: (a) tonic inhibition index (calculated by the formula shown above) for normal infants and children, (b) phasic inhibition index (calculated by the formula shown above) for normal infants and children. Abscissae are the age of subjects shown with postconceptional age. Ordinates are the number of the index.

The number of TMs during the periods of REMs decreased with age. They assessed this with the formula shown in Fig. 1b and called it the percent body movements during REMs-burst or phasic inhibition index. This showed exponential decline with conceptional weeks as observed in Fig. 1b.

Maturation of REM Sleep as a Sleep Stage

As shown previously, premature babies of 24–26 gestational weeks show no periodic occurrence of any parameters of sleep. Babies move frequently with atonia of the mental muscle and irregularity in respiration but do not show any eye movements (Dreyfus-Brisac, 1968), or it can be said there is no behavioral state. This stage is defined as undetermined sleep stage (Dreyfus-Brisac, 1968). At 28–30 gestational weeks body movements and rapid eye movements begin to appear periodically in the same period and generate a stage of AS as a prototype of REM sleep. In these weeks, however, the mental muscle is atonic throughout the sleep and there is no period with rhythmic respiration (Dreyfus-Brisac, 1967). Around 36 gestational week, periods with rhythmic respiration appear and irregular respiration is observed periodically coincident with the stage with body movements and rapid eye movement. It is around this week when EEG shows identical pattern in AS or REM sleep. This also implies simultaneous development of QS or the prototype of NREM sleep from 36 gestational weeks, the stage without body movements and REMs and with regular respiration (Pamelee Wenner et al., 1967; see Parmelee and Stern, 1972) and mouthing movements (Horimoto et al., 1989). It is around 40 gestational weeks when atonia of the mentalis muscle begins to occur in the period of AS or REM sleep. But in this period, there is a state which could not be defined as REM sleep or NREM sleep, called the undetermined sleep state or non-state. But by 3 months postterm age all parameters of REM sleep appear selectively in REM sleep and the sleep matures as a behavioral state. Along with the increase of the state of QS or NREM sleep and development of the state of wakefulness, undetermined sleep or no-state disappears.

In neonates REM sleep occupies more than 50% of total sleep. In neonates the percentage of REM is higher in day sleep than in night sleep but this situation reverses around the 2nd postnatal weeks (Aserinski and Klertman, 1955; see Parmelee and Stern, 1972). The total amount of REM sleep decreases markedly in the first two years and at 2 years of age it becomes 20–25% of the sleep period. This level continues throughout life until senior ages when it declines again.

Development of the Ultradian Rhythm

Sterman (1972) reviewed the studies on ultradian or rest-activity cycle of various ages and revealed age related changes of its length. It was around 40 min in premature babies and increased to 50 min in postterm infants from 0 to 8 months. It became dramatically longer during childhood, that is, it was around 70 min at 2–5 yr and attained adult levels of 90 min at 5–10 yr.

Investigation on fetuses revealed the presence of two ultradian rhythm in the prenatal period. That is, the ultradian rhythm for eye and body movements and that for irregular heart rate and micturition which synchronized around the term (Koyanagi et al, 1992).

Roles of REM Sleep for the Maturation of the Brain

Emergence of the behavioral state is a function of brain maturation and each behavioral state may serve an important role for the development of the brain specific for each (Mirmiran, 1995). Presence of a high amount of REM sleep in preterm and early postterm infants and its decline with age suggests importance of the roles of this sleep stage in the early developmental stage and for brain maturation.

From studies on premature babies and fetuses, all components appeared in preterm periods but each component showed a particular developmental course for each.

So it is necessary to consider the neurons and neuronal systems involved in the generation of each component of REM sleep.

Almost all components of REM sleep are generated or executed by cholinergic neurons (see elsewhere in this monograph). The numbers of twitch movements and rapid eye movements are modulated by the nigrostriatal dopamine neurons, at least from early childhood, in the matter of exaggeration and suppression, respectively (Segawa et al., 1987).

Selective occurrence of these parameters in REM sleep is modulated by noradrenergic and serotonergic neurons of the brainstem which are activated during NREM sleep (see elsewhere in this monograph).

So difference in the developmental courses among parameters suggest that neurons which execute and modulate the body and eye movements differ from those for respiration and atonia and neurons that execute body movements differ from those for eye movements but they have connections, probably reciprocal connections to each other. Moreover, the neuron(s) which are involved in the ultradian rhythm for body and eye movements are different from those for heart rate and micturition.

As there were no differences in the development of eye movement activities during REM sleep between pups reared in constant darkness and those exposed to diurnal light-dark alterations (Van Somerson et al., 1990; see Mirmiran, 1995), Mirmiran (1995) suggested that the development of the eye movement in REM sleep is a part of an endogenous timetable of REM sleep for brain maturation, rather than due to an increase of visual stimulation or developmental oculomotor system.

It might be true for all parameters concerning REM sleep which develop during the fetal or preterm period, but the parameters which show maturational or developmental variation after birth might be influenced by the environmental factors, particular 'sensory' input for each.

Nocturnal variation of the REMs periods and the amount of phasic components particularly the numbers of twitch movements are matters which appear after the development of the circadian sleep-wake cycle.

Thus evaluation of each parameter of REM sleep or active sleep during the developmental course is useful for the evaluation of the maturational process of the neuron(s) involved in each parameter and also detect the abnormalities of the function or development of each neuron(s). The critical period for the appearance and maturation of each parameter and for the synchronization or cooccurrence of certain parameters are not only the critical periods for the maturation of the neurons modulating or regulating these parameters and also the vulnerable periods of these neurons. So evaluation of these parameters is useful for clarifying the pathophysiologies of disorders occurring in early infancy (Horimoto et al., 1993; Kohyama et al., 1993; Kohyama, 1998; Segawa, 1997).

The importance of REM sleep in the immature brain has been studied experimentally by deprivation of the stage.

In animal (rats) experimets, deprivation of REMs by an antidepressant, clomipramine, or an antihypertensive drug, clonidine, injected in postnatal day 8–21; the second and third weeks of life, the period with rapid rat brain maturation and with a high amount and intensity of REM sleep, showed in adulthood increased anxiety, (Nijhuis et al., 1992; see Mirmiran, 1995) reduced sexual activity (Mirmiran et al., 1981; and Mirmiran et al., 1983; see Mirmiran, 1995) and a disturbed sleep-wake pattern (Mirmiran, 1995) and showed shorter REM latency, frequent sleep onset REM periods and an abnormal temporal course of REM rebound after REM sleep deprivation (Vogel et al., 1990). The sleep also showed a higher amount of REM sleep with a higher intensity of phasic activity (Mirmiran, 1995). These animals had a significant reduction in the size of the cerebral cortex and the brainstem with proportional reduction of protein (Mirmiran et al., 1983) and lack the growth response to environmental stimulation later in life (Mirmiran et al., 1983). Neonatally clonidine treated rats showed supersensitivity of the pyramidal cells of the hippocampus to noradrenaline (Gorter et al., 1990; see Mirmiran, 1995). In this period it should be taken into account that regulation of noradrenaline receptor sensitivity is thought to be a function of REM sleep in adulthood (Siegal and Rogarski, 1988). These animals also showed reduction of spontaneous firing of the raphe neurons in adulthood (Yavari et al., 1993).

Mirmiran (1995) stresses the important function of fetal/neonatal REM sleep. He suggests that the function is fulfilled by the endogenous activation of the brain and that it influences neuronal growth, synaptic plastically, learning and unlearning, genetic read-out and individual differences and the cardiorespiratory and thermoregulatory centers.

The relatively high amount of REM sleep in early life when the maturation of the nervous system is proceeding at a high rate led Roffwarg and his colleagues (1966) to first postulate that the neural activity controlled by REM state mechanisms may be developmentally functional. His group hypothesized an ontogenic function of REM sleep as that the widespread control of neuronal activity extended by specific REM sleep process helps to direct brain maturation through activity dependent developmental mechanisms (Marks et al., 1995). They examined effects of REM sleep on the development of feline visual system which is known to incorporate information derived from visual experience to establish neural connectivity. By instrumental REM sleep deprivation on mono-ocular deprived feline, they estimated the ratio of the cell-size of the A1 segment against the A segment of the lateral geniculate nucleus (LGN) and revealed enhancement of the reduction caused by monocular deprivation. They also examined the cell size in the LGN monocular segment contralateral to the unoccluded and pathed or occluded eye and found that REM sleep deprivation caused more pronounced decrease of the size of cells caused by eye occlusion. They also revealed reduction of the size in the segment contralateral to the unocculuded eye.

By instrumental deprivation of REM sleep on neonatal cats in which the occurrence of PGO waves was blocked by bilateral lesions of the ponto-mesencephalic isthmus, they found significant reduction of A1/A ratio in context with sham lesion group. This reduction of LGN cell size is considered to be due to the suppression of PGO wave activity and not due to the postsynaptic degeneration caused by the brainstem lesion.

Examination of the neural activity associated with PGO waves in the LGN reveals a distribution of facilitatory influence markedly different from that induced by visual experience. So the source of influence of REM sleep deprivation and block of PGO waves on the visual system appears not to be derived from the environment but rather stems from an intrinsic source of genetic origin.

These data indicate that propagation of phasic influences during REM sleep interact with other processes subserving neural development and Marks et al. (1995) conclude that REM sleep directs the course of brain maturation in early life through the control of neural activity.

Atonia in REM sleep is considered to have roles for suppressing all the reflex activities including motor and autonomic functions. Leakage of atonia into NREM sleep could cause sudden death during sleep by suppressing the wakening response or activation of respiratory movement to the reduction of the levels of O_2 saturation. On the contrary, REM sleep without atonia causes REM sleep behavior disorder, in which dream appears as real behavior. So REM sleep with atonia makes it possible to activate each neuronal system or pathway independently and may adjust or restore the function. During the developmental course, prolonged atonia allows functional and morphological (probably synaptic connection) maturation of each neuronal system or pathway corresponding without any influence on each other.

With appearance of the state without atonia, that is, QS or NREM stage, functional relation of each neuronal system or pathway may develop. But this process becomes more active in the postnatal period particularly after the appearance of the state of wakefulness. Parameters of REM sleep also show 'maturation' in the post-term period in the pattern of co-occurrence and reciprocity among parameters. This process might be due to the functional matruation of the aminergic neurons of the brainstem and their connection for which environmental sensory input may have important roles.

References

Birnholz, J.C. The development of human fetal eye movement patterns, Science, 213 (1981) 679–681.

Dreyfus-Brisac, C., Ontogénése du sommeil chez le préfmaturé humain: etude polygraphique. In: A. Minkowski (Ed.), *Regional Development of the Brain in Early Life,* Blackwell, Oxford, 1967, pp. 437–457.

Dreyfus-Brisac, C. Sleep ontogenesis in early human prematurity from 24 to 27 weeks of conceptional age, Develop. Phychbiol., (1968): 162–169.

Dreyfus-Brisac, C. Ontogenesis of sleep in human prematures after 32 weeks of conceptional age, Developmental Psychobiology, 32(1970) 91–121.

Horimoto, N., Koyanagi, T., Nagata,S., Nakahara, H. and Nakano, H. Concurrence of mouthing movement and rapid eye movement/non-rapid eye movement phases with advance in gestation of the human fetus, Am. J. of Obstet. and Gynecol.,161 (1989) 344–351.

Horimoto, N., Koyanagi, T., Satoh, S., Yoshizato, T. and Nakano, H. Fetal eye movement assessed with real-time ultrasonography; are there rapid and slow eye movements? Am. J. of Obstet. and Gynecol., 163 (1990) 1480–1484.

Horimoto, N., Koyanagi, T., Maeda, H., Satoh, S., Takashima, T., Minami, T. and Nakano, H. Can brain impairment be detected by in utero behavioral patterns?, Archieves of Disease in Childhood, 69 (1993) 3–8.

Inoue, M., Koyanagi, T., Nakahara, H., Hara, K., Hori, E. and Nakano, H. Functional development of human eye movement in utero assessed quantitatively with real-time ultrasound, Am. J. of Obstet. and Gynecol., 155 (1986) 170–174.

Karacan, I., The developmental aspect and the effect of certain clinical conditions upon penile erection during sleep, Excerpta Medica, ICS 150 (1966) 2356–2357.

Kohyama, J., Shimohira, M., Itoh, M., Fukumizu, M. and Iwakawa, Y. Phasic muscle activity during REM sleep in infancy—normal maturation and contrastve abnormality in SIDS/ALTE and West syndrome, J. Sleep Res., 2 (1993) 241–249.

Kohyama, J., Shimohira, M. and Iwakawa, Y. Maturation of motility and motor inhibition in rapid-eye-movement sleep, J. Pediatr., 130 (1997) 117–122.

Kohyama, J. Sleep as a window on the developing brain, Current Problems in Pediatrics, 28 (1998) 69–100.

Koyanagi, T., Horimoto, N. and Nakano, H. REM Sleep determined using in utero penile tumescence in human fetus at term, Biology of the Neonate, 60 (suppl 1) (1991) 30–35.

Koyanagi, T., Horimoto, N., Satoh, S., Inoue, M. and Nakano, H. The temporal relationship between the onset of rapid eye movement period and the first micturition thereafter in the human fetus with advance in gestation. Early Human Development, 30 (1992) 11–19.

Koyanagi, T. Horimoto, N., Takahashi, T., et al. Ontogenesis of ultradian rhythm in the human festus observed through the alternation of eye movement and no eye movement periods, J. of Reproductive and Infant Psychology, 11 (1993) 129–134.

Marks, G.A., Shaffery, J.P., Oksenberg, A., Speciale, S.G. and Roffwarg, H.P. A functional role for REM sleep in brain maturation, Behavioural Brain Research, 69 (1995) 1–11.

Mirmiran, M. The function of fetal/neonatal rapid eye movement sleep, Behavioural Brain Research, 69 (1995) 13–22.

Mirmiran, M., Uylings, H.B.M. and Corner, M.A. Effects of postweaning environmental enrichment on regional brain development in the pre-weaning REM (active) sleep-deprived rats, Dev. Brain Res., 7 (1983) 102–105.

Nijhuis, J.G., Prechtl, H.F.R., Martin, C.B., Jr. and Bots, R.S.G.M. Are there behavioural states in the human fetus? Early human Development, 6 (1982) 177–195.

Parmelee, A.H., Jr. and Stern, E. Development of states in infants. In: C.D. Clement, D.P. Purpura, F.E. Mayer (Eds.), *Sleep and the Maturing Nervous System*, Academic Press, New York, 1972, pp. 199–228.

Petre-Quadens, O., De Lee, C. and Remy M. Eye movement density during sleep and brain maturation, Brain Research, 26 (1971) 49–56.

Petre-Quadens, O., Sleep in human newborn. In: O Petre-Quadens and J.D. Schlag (Eds.), *Basic Sleep Mechanism*, Academic Press, New York and London, 1974, pp. 355–380.

Prechtl, H.F.R. and Nijhuis, J.G., Eye movement in the human fetus and newborn, Behavioural Brain Research, 10 (1983) 119–124.

Roffwarg, H.F., Muzio, J. and Dement W.C. Ontogenetic development of the human sleep-wakefulness cycle, Science, 152 (1966) 604-619.

Segawa, M., Nomura, Y., Hikosaka, O., Soda, M., Usui, S. and Kase, M. Roles of the basal ganglia and related structures in symptoms of dystonia. In: M.B. Carpenter, and A., Jayaraman, (Eds.), *The Basal Ganglia II*, Plenum Publishing Corporation, New York, 1987, pp.489–504.

Segawa, M. Pathophysiology of Rett syndrome from the standpoint of early catecholamine disturbance, European Child & Adolescent Psychiatry, 6 (1997) 56–60 (Suppl. 1).

Siegel, J.M. and Rogarski, M.A. A function for REM sleep: regulation of noradrenergic receptor sensitivity, Brain Res. Rev. 13 (1988) 213–233.

Sterman, M.B., The basic rest-activity cycle and sleep: developmental consideration in man and cats. In: Clement, D.P. Purpura, F.E. Mayer (Eds.), *Sleep and the Maturing Nervous System*, Academic Press, New York, 1972, pp. 175–197.

Vertes, R.P. A life-sustaining function for REM sleep: a theory, Neurosci. Biohehav. Rev., 10 (1986) 371–376.

Visser, G.H.A., Goodman, J.D.S., Levine, D.H. and Dawes, G.S. Micturition and the heart period cycle in the human fetus, British Journal of Obstetrics and Gynaecology, 88 (1981) 803–805.

Vogel, G., Neill, D., Hagler, M., Kors, D. and Hartley P. A new animal model of endogenous depression: a summary of present findings, Neurosci. Biobehav. Rev., 14 (1990) 85–91.

Yavari, P., Vogel, G.W. and Neill, D.B. Decreased raphe unit activity n a rat model of endogenous depression, Brain Res., 611 (1993) 31–36.

Rapid Eye Movement Sleep
B.N. Mallick and S. Inoué (Eds)

4. Initiation of Rapid Eye Movement Sleep: Beyond the brainstem

Adrian R. Morrison[1,2], Larry D. Sanford[1] and Richard J. Ross[2,3]

[1]Laboratory for Study of the Brain in Sleep, [2]Department of Animal Biology, School of Veterinary Medicine; [2]Department of Psychiatry, School of Medicine, University of Pennsylvania; [3]Research Service, Philadelphia Veterans Affairs Medical Center, Philadelphia, PA, USA

SUMMARY

Studies of REM mechanisms have focused on the caudal brainstem for nearly forty years. This focus stems from the demonstration that cats decerebrated at the caudal midbrain level still exhibit many signs of REM, such as atonia, rapid eye movements and the pontine component of PGO waves. In this chapter we challenge the assumption that these results prove that REM is solely initiated by bulbar mechanisms in **intact** animals. We suggest that forebrain reorganization, particularly in the hypothalamus, must occur some time preceding REM onset to permit the relatively non-homeostatic state of REM to occur. One structure with very strong interconnections with the pontine region deemed critical for REM is the amygdala. Because the amygdala charges sensory stimuli with emotional significance and the brain in REM has so many features of the brain during orienting(behavior seen during REM without atonia dramatically demonstrates this), we reasoned that manipulations of this region would dramatically affect sleep. This chapter reviews a variety of manipulations of the amygdala that have demonstrated a role for this structure in the regulation of sleep. Serotonin placed in the central nucleus of the amygdala (CNA) induced a rapid change of state when injected in REM. Injections in NREM resulted in significantly longer latencies to state change. Methysergide in CNA released PGO waves into NREM. Unlike the effect in cats, injection of carbachol into CNA suppressed REM. Prolactin decreased total NREM but did not change REM. This effect is in contrast with the results obtained in the hypothalamus, for injections there increased REM when given diurnally and decreased it with nocturnal injections. Electrical stimulation of CNA increased the amplitudes of sound-induced PGO waves, which are signs of alerting. These studies have revealed the value of looking beyond the caudal brainstem in studying REM mechanisms.

Key Words: Amygdala, Central Nucleus, REM without Atonia, Serotonin, Carbachol, Prolactin.

Introduction

This essay presents a view of REM organization contrary to that which has held sway for more than three decades: that the caudal brainstem is all-important in the generation of REM. We shall present arguments and accumulating supporting data for **a new emphasis** in the study of REM mechanisms: **the forebrain**. First, we shall describe the aspects of REM most relevant to this proposal and, then, review the time-honored view of the manner in which REM is initiated. We then present our hypothesis that the activity of the brain in REM resembles in most respects that during orienting (OR) in wakefulness (W). We shall next argue why the forebrain must be factored into any theory of normal REM control and why the **amygdala** begs to be a focus of investigation. The essay concludes with recent findings consonant with our call for a new focus.

Key Features of REM

REM is a fascinating brain and behavioral state that confounded classical notions of the separate natures of sleep and W when recognized in people and animals (Aserinsky and Keitman, 1953; Dement, 1958). Instead of the distinctly different electroencephalographic (EEG) pattern associated with "classical" sleep (high-amplitude, low-frequency waves), research revealed that REM had an EEG pattern with low-amplitude, high-frequency waves, resembling those of W. Another striking feature of REM is the large-amplitude, macropotential spike called the PGO wave. These are spontaneous waves (usually recorded in the lateral geniculate body in cats and in the pons in rats) preceding REM by up to 1 min and then occuring singly or in clusters associated with rapid eye movements. Chronic, single-unit recordings then revealed that firing rates were high in both W and REM in many brain regions; while during "classical" sleep, now termed NREM or synchronized sleep, activity was lower in many areas (Steriade and McCarley, 1990). And other remarkable features emerged. Vigorous twitches of skeletal muscles merely interrupted an ongoing muscle atonia, which we now know is primarily the result of postsynaptic inhibition of motor neurons (Chase and Morales, 1985). Furthermore, research demonstrated the tightly regulated homeostasis characteristic of homeotherms to be greatly suppressed in REM (Parmeggiani and Morrison, 1991).

Current Concepts of Cellular Mechanisms Regulating REM

At the cellular level, the focus has been on the caudal brainstem and, most particularly, on the pontine tegmentum. Several rostropontine neuronal groups have been demonstrated to pay key roles in REM generation and maintenance (Steriade and McCarley, 1990): cholinergic (ACh) neurons in the laterodorsal tegmental (LDT) and pedunculopontine tegmental (PPT) nuclei; serotonergic (5-HT) neurons of the dorsal raphe nucleus (DRN); and noradrenergic (NA) cells of the locus coeruleus (LC). It is assumed that ACh neurons are active in REM. 5-HT and NA neurons, on the other hand, gradually diminish their firing rates as an animal passes from W to NREM and then remain largely silent throughout an episode of REM. The current view is that the activity of ACh neurons and the silencing of 5-HT and NA neurons as REM approaches underlie the generation of REM (Steriade and McCarley, 1990).

Results obtained by the introduction of substances mimicking or countering the actions of these neurotransmitters reinforce this story. For example, carbachol, an ACh agoinst induces long-lasting REM periods when infused in to the pontine tegmentum where ACh neurons project (Steriade and McCarley, 1990). Isoproterenol, a beta-adrenergic agonist, will inhibit REM when infused in the pons (Tononi, Pompeiano, Gianni and Pompeiano, 1988). Infusion of 5-HT in LDT in rats inhibits REM (Horner, Sanford, Annis, Pack and Morrison, 1997); whereas suppression of DRN activity by microdialysis of the 5-HT_{1A} agonist, 8-OH-DPAT, within DRN increases REM in cats (Portas, Thakker, Rainnie and McCarley, 1996). These findings are consistent with the theorized role of 5-HT in REM control; however, one should note that DRN cells are reactivated in REM without atonia (REM-A) (Trulson, Jacobs and Morrison, 1981), which is created by placement of bilateral, thermal lesions in the pontine tegmentum (Morrison, 1988). This suggests that silence of DRN 5-HT cells is not necessary for REM occurrence and that they play a role in the suppression of motor activity in normal REM.

REM is a State of the Brain Much Like Alert Wakefulness

Morrison (1979) proposed that the linkage of REM's most obvious features, activated EEG and muscle atonia, could be explained in other than teleological terms, i.e., the usefulness of an individual being paralyzed during a brain state characterized by vivid dreaming. In this proposal, the inhibition of motor neurons is regarded as secondary to the reticular activation of REM. Atonia may then be seen as an exaggeration of the suppression of movement to a novel or unexpected stimulus characterizing OR in W. In other words, the mechanisms responsible for OR in W, which links reticular activation with suppression of movement, are likely to be employed in REM when the same linkage occurs, albeit in exaggerated form. The latter would be a splendid example of nature's parsimony. Two observations stimulated this hypothesis:

(1) Skeletal muscle atonia and absence of overt behavior are not necessary features of REM; for in REM-A one observes behaviors ranging from simple head raising to quadripedal locomotion, the head moving in ways that resemble searching or OR in cats (Morrison, 1988). Rats show many similar behaviors (Mirmiran, 1983). As with normal REM, REM-A episodes follow NREM. During W, lesioned cats have significantly increased exploratory locomotion (Morrison, 1988). Thus, the link between reticular activation associated with OR and inhibition of movement is broken in W as well, suggesting that suppression of locomotor drive in addition to motor neuronal inhibition is a feature of normal REM.

(A brief aside on the subject of REM-A is necessary here. The early decision to refer to this state as "paradoxical sleep (REM) without atonia" (Morrison, 1988) has led to some confusion, suggesting that the clumsier "REM with elaborate behavior" may have been a better choice. For example, a recent, extensive review of the cholinergic pedunculopontine nucleus (Rye, 1997) emphasized that destruction of the ACh neurons of that nucleus by a cytotoxin had induced REM-A when, in fact, only nuchal muscle tone had been released (Webster and Jones, 1988). This is not REM-A as various workers (Morrison, 1988) have described it. Rather, elaborate behavior requires that a complex of local neurons and fibers of passage stemming from a number of brain areas be destroyed. Indeed, the site and size of the lesions dictate which behaviors will emerge (Morrison, 1988). We emphasize this point because as workers search for pathological changes associated with various motor disorders of sleep, such as REM behavior disorder, they should not be limited by a misunderstanding of the nature of REM-A).

(2) A wave analogous to spontaneous PGO waves can be elicited by tones in REM, NREM and W in cats and rats (Sanford, Morrison and Ross, 1994). Firing of the cholinergic neurons at the mesopontine border that project to the lateral geniculate body (LGB) induce PGO waves in REM (Steriade and McCarley, 1990), and auditory stimulation activates these cells as well (Steriade, Pare, Datta, Oakson and Curro Dossi, 1990). The largest elicited waves in W are associated with OR (Fig. 1), and those elicited in REM match in amplitude the highest-amplitude spontaneous PGO waves (Sanford, Morrison, Ball, Ross and Mann, 1993). These observations support the hypothesis (Morrison, 1979) that PGO waves signal the spontaneous "internal" alerting that occurs during REM, a sleep state that has most of the characteristics of very alert W when OR occurs (e.g., similar EEG pattern, hippocampal theta rhythm, and increased brain temperature). The characteristic "OR" behavior of REM-A serves to reinforce this hypothesis.

The virtually indistinguishable cellular activity during W and REM in much of the brain, particularly thalamocortical regions, has recently led Llinás and Paré (Paré and Llinás, 1995) to essentially the same conclusion. In their conception, W and REM "are fundamentally identical states, with the provision that the handling of sensory information is altered in REM sleep." They note that sensory responsiveness, as indicated by evoked potentials, is more alike in W and REM than either is to NREM, though higher intensity stimuli are required for awakening from REM. *It is clear, though, that mechanisms other than skeletal muscleparalysis are required to alter the impact of endogenously and exogenously generated "alerting" in REM so that state is maintained.* We think that those mechanisms likely involve forebrain structures.

Reasons to Explore Elsewhere

The emphasis on the caudal brainstem as the central site of REM control over more than three decades stems, of course, from the demonstration that cats decerebrated at the level of the caudal midbrain still exhibit the peripheral signs of REM, i.e., atonia, muscle twitches and rapid eye movements. Also, one can record the pontine component of PGO waves. Much has been learned about the regulation of REM as a consequence of this emphasis; but it has been translated, inappropriately in our opinion (Morrison and Reiner, 1985), into the belief that mechanisms in the caudal brainstem are sufficient for the expression of REM in normal animals.

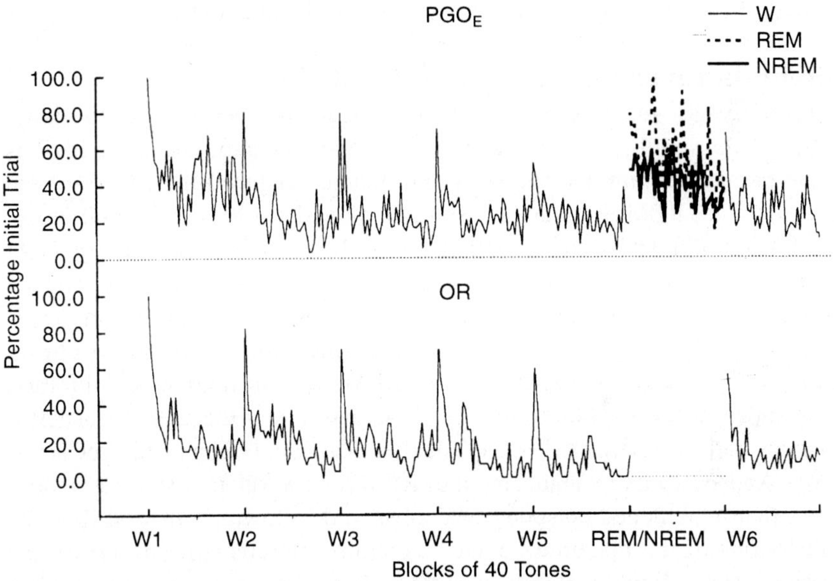

Fig. 1. **High-amplitude elicited PGO waves in waking were closely associated with the highest OR score in W. The OR score was calculated by subtracting the pre-tone score from the post-tone score. Scores were derived from assigned values to body posture and head and ear movements. The highest values were obtained with the averaged responses to the first tone or white-noise burst (90–100 dB) in a block of 40, each block being separated by 20 min in W and 20 min after testing in NREM or REM. Reprinted with permission of Elsvier Scientific Publisher from Sanford et al., 1993.**

There is no direct proof that this is so

Three observations suggest that one should factar in the rostral brain as well: (1) The decerebrate is not homeostatically regulated in the way that an intact animal is; (2) As noted earlier, homeostasis is greatly diminished in REM as well, a consequence of greatly suppressed hypothalamic control (Parmeggiani and Morrison, 1991); and (3) Finally, decerebration abnormally predisposes the cat to enter the REM-like state. A number of stimuli not normally sleep-inducing, such as passing a stomach tube, inserting a rectal probe, and passively flexing and extending the joints, will cause the decerebrate cat to enter REM (Jouvet, 1964). The brainstem, lacking modulation from rostral structures, must be in an unstable, supersensitive state, allowing it to be switched into the REM mode. It is likely, therefore, that forebrain structures have a role in the orderly onset of normal REM as the brain reorganizes to enter this "unusual" state of diminished homeostatic control. We have proposed (Morrison and Reiner,1985) that the midbrain transection creating the decerebrate condition serves as a substitute for the reorganization that must occur prior to the switch in to REM. That reorganization reduces or eliminates, among other things, the homeostatic control exerted by forebrain structures. It must also alter the cognitive significance of the barrage of internally generated signals arriving from the pons (PGO waves) that in W could well signal danger.

Forebrain Involvement in REM Regulation

The past several years have brought forth results in support of our thinking. For example, simultaneous infusion of carbachol into cholinoceptive regions of the basal forebrain and the pons significantly reduced the ability of the latter to induce REM (Baghdoyan, Spotts and Snyder, 1993). Further, ibotenic acid lesions of the posterior hypothalamus induced a 300% increase in REM during the first post-injection day and released the same stimulus-induced REM episodes seen after decerebration (Sakai, Mansari, Lin, Zhang and Vanni-Meercier, 1990). Microdialysis of adenosine into the basal forebrain dramatically decreased W and increased REM, both the number and duration of episodes; while its addition to LDT increased only the number of REM episodes and also NREM (Portas, Thakker, Rainnie, Greene and McCarley, 1997). And a selective enhancement of REM for several days (more episodes rather than lengthened episodes) was observed following bilateral ibotenic acid lesions of the thalamic nucleus, centralis lateralis (Marini, Gritti and Mancia, 1990). Finally, recent work in humans has demonstrated through imaging increased activity in the amygdala during REM (Maquet, Peters, Aerts, et al.,1966).

The Amygdala as an Integral Component of Normal REM Mechanisms

Certain features of the amygdala make it an obvious candidate for a significant role in control, or at least modulation, of caudal brainstem activities as well: (1) *The amygdala is intimately connected with the key pontine nuclei and other critical forebrain structures.* It receives inputs from a wide variety of structures (Amaral, Price, Pitkänen and Carmichael, 1992). Its lateral nucleus (LA) is the main amygdalar recipient of afferents from sensory processing areas of the thalamus and cortex (Amaral, Price, Pitkänen and Carmichael, 1992). The basolateral nucleus (BLA), which transmits information from LA to the amygdala's main output nucleus to the brainstem, the central nucleus (CNA) (Savander, Go, LeDoux and Pitkänen, 1995), is heavily interconnected with the basal forebrain (Amaral, Price, Pitkänen and Carmichael, 1992). The lateral hypothalamus receives amygdalar afferents primarily from CNA (Amaral, Price, Pitkänen and Carmichael, 1992). Efferents from CNA are split between two major amygdalar output tracts, the stria terminalis and the ventral amygdalofugal pathway (VAF), the caudal part of which enters the brainstem (Price, Russchen and Amaral, 1987). CNA has direct projections to LC, DRN, LDT as well as oher pontine regions (e.g., LCα, and subcoeruleus). Also, there are reciprocal projections from many of these regions back to CNA. (Takeuchi, Megirian and Hopkins, 1982) (2) *The amygdala plays a role in investing sensory events with emotional significance.* It evaluates the emotional significance of stimuli and initiates appropriate neurobehavioral responses (LeDoux, 1992). Cells in the amygdala, particularly CNA and BLA, respond to novel and significant stimuli in different sensory modalities (Bordi, LeDoux, Clugnet and Pavlides, 1993).

Thus, the amygdala is anatomically and functionally suited to modulate pontine regions involved in vigilance in waking based on stimulus relevance and on emotional expectations of stimulus importance and to perform in similar fashion regarding endogenously generated activity in these regions during REM. Preventing behavioral arousal during REM may be a significant amygdalar role. Of course, there are other pathways involved in amygdalar modulation of vigilance and behavioral state, such as efferents to the basal forebrain, hypothalamus or midbrain regions (Amaral, Price, Pitkänen and Carmichael, 1992) that have been implicated in arousal and attention (basal forebrain) and the control of REM (Steriade and McCarley, 1990). It would seem that this complex, including the amygdala, must undergo reorganization some time before full entrance into REM.

Modulation of various aspects of sleep by the amygdala

There is now ample experimetal evidence that the amygdala has a role in modulating sleep mechanisms. For example, microinjections of the cholinergic agonist, carbachol, into the region of CNA of rats suppress REM. Rats (N = 6) were implanted with EEG and EMG electrodes and with cannulae aimed at CNA bilaterally. Four rats received a single microinjection of saline (0.2 μl) and were monitored polygraphically for 6 h on two consecutive days. Two days later they received a single microinjection of carbachol (3.0 μl/0.2 ml) and were monitored for 6 h on two consecutive days and for 6 h on the 5th post-injection day. Two rats were recorded on one saline and one carbachol day only. Total REM (Fig. 2), mean REM episode length and number of episodes were suppressed on the first carbachol recording day in 4 rats with cannulae placements in CNA, and all three measures returned to control levels by the second carbachol recording day. Injections outside of CNA (N = 2) were not effective. These results contrast with those obtained in cats, where microinjections of carbachol into CNA produced a prolonged enhancement of REM and NREM with PGO waves for up to 5 days (Calvo, Simón-Arceo and Fernández-Mas, 1996). The suppression of REM episode duration on a single day by carbachol microinjections into the amygdala in rats could, of course, be due to a species difference in behavioral state control, or perhaps to a greater number of cells being affected by the drug in the much smaller rat CNA. But given that the major ACh input into the amygdala is from the basal forebrain (Woolf and Butcher, 1982), which impedes REM induced by pontine carbachol injection when also activated by carbachol (Baghdoyan, Spotts and Snyder, 1993), the results we obtain are easier to understand than those obtained with cats.

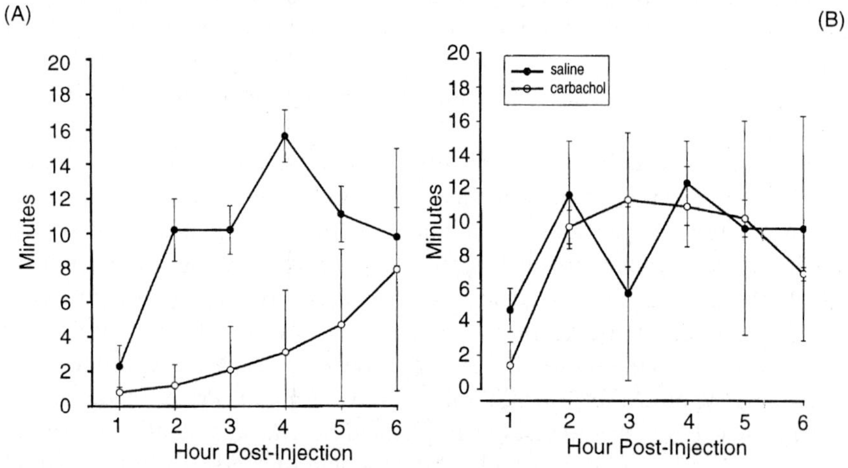

Fig. 2. (A) Total REM/hour after saline or carbachol in CNA (N = 4), (B): Total REM/hour after saline or cabachol with injection sites outside CNA (N = 2). Vertical bars equal standard deviation.

Prolactin, administered subcutaneously, intracerebroventricularly (ICV), or locally into the dorsolateral hypothalamus, an area containing prolactin-immunoreactive neurons, increased REM when given diurnally and decreased REM when given nocturnally (Roky, Valatx, Paut-Pagano and Jouvet,1994). Because high concentrations of prolactin-immunoreactive fibers are found in CNA (Siaud, Manzoni, Balmefrezol, Barbanel, Assenmacher and Alonso, 1989), we questioned whether prolactin would have similar effects on sleep when injected into the amygdala. Eight rats were studied with microinjections (10.0, 100.0, 250.0 ng prolactin/0.2 μl saline) administered in quiet waking during the lights-on period. Prolactin microinjections were counterbalanced across dosages and were preceded and followed by microinjections of saline alone. Microinjections of prolactin into CNA significantly decreased total NREM during the first three hours post-injection. REM was not significantly affected by any dose of prolactin; however, all doses of prolactin marginally increased the duration of transition episodes compared to saline (Fig. 3). Our conflicting results with the earlier reports may have been due to differences in record scoring. We distinguished between transition, which was increased, and REM, which according to our scoring criteria was either reduced or remained the same. We also found that prolactin in CNA could decrease NREM, which was not affected in other studies (Roky, Valatx, Paut-Pagano and Jouvet, 1994).

Evidence is beginning to accumulate linking activity of CNA to the generation of PGO waves of REM, which are also a neural marker of orienting (Sanford, Morrison and Ross, 1994). Electrical stimulation of CNA during REM increases PGO wave frequency in cats (Calvo, Badillo, Morales-Ramirez and Palacios-Salas, 1987) and PGO wave amplitude in rats (DeBoer, Sanford, Ross and Morrison, 1998). Further, electrical stimulation of CNA increases the amplitude of PGO waves elicited by auditory stimuli (DeBoer, Sanford, Ross and Morrison, 1998) (Fig. 4). Single unit activity in CNA and the lateral nucleus of the amygdala of rats (Bordi, LeDoux, Clugnet and Pavlides, 1993) increases during REM. This increase in unit activity may be related to the role of the amygdala in PGO waves (Calvo, Badillo, Morales-Ramirez and Palacios-Salas, 1987).

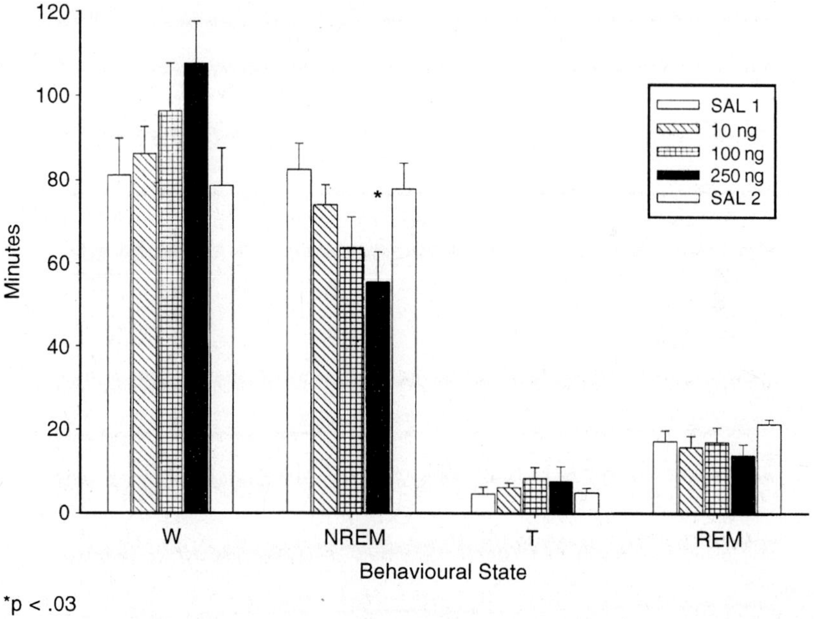

Fig. 3. Total number of minutes spent in W, NREM, Transition (T) and REM for the first 3 hours post-injection of prolactin into CNA. SAL 1: Saline 1; SAL 2: Saline 2. Reprinted with permission of Websciences from Sanford et al., Sleep Research Online 1 (1998): 109–113.

Electrical stimulation of CNA can produce "alerting" behaviors coupled with EEG desynchronization (Kapp, Supple and Whalen, 1994). A recent study (Silvestri and Kapp, 1998) examined the influence of stimulating CNA on putative cholinergic neurons in PPT in rabbits. One category of neuron exhibited "burst" firing in response to acoustic stimuli similar to that observed in cells with firing that precedes PGO waves in cats (Steriade, Pare, Datta, Oakson and Curro Dossi, 1990). Stimulation of CNA produced activation in 43% of these putative PGO-related neurons, further suggesting a functional

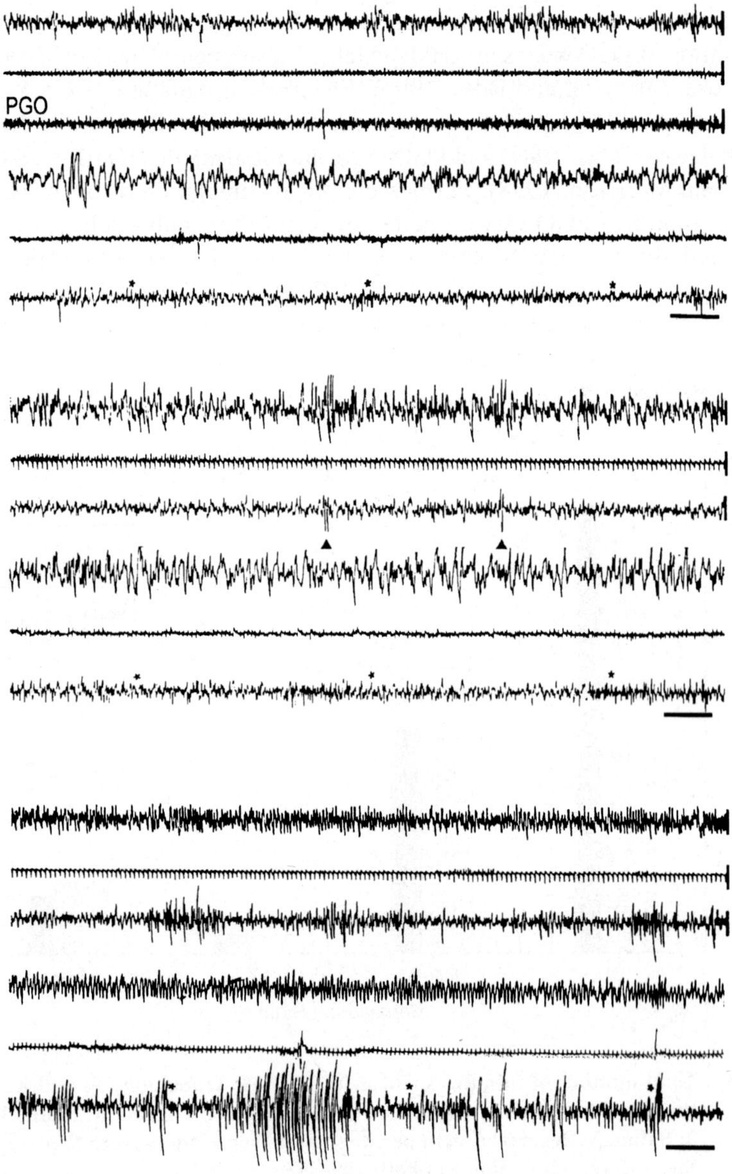

Fig. 4. Polygraph excerpts from one rat illustrating the effect of electrical stimulation of CNA on PGO wave activity in W, NREM and REM. Each panel contains an excerpt from an episode without stimulation (A) that occurred at approximately thesame time of day during baseline recordings and an excerpt with periodic electrical stimulation indicated by *. No significant effects were found in W. In NREM, PGO wave frequency was reduced by periodic electrical stimulation of CNA. For comparison, examples of PGO waves during NREM without stimulation (A) are indicated by ▲. Two excerpts during REM illustrate the increase in PGO-wave amplitude during periodic stimulation of CNA (B) compared to without stimulation (A). Reprinted with permission of Elsvier Scientific Publisher from De Boer et al., 1998.

link between amygdala activity and PGO waves. In contrast, stimulation of CNA activated only 6% of neurons with positive or negative correlations with the power of delta wave activity in the EEG.

Success in affecting PGO wave production by manipulations in the amygdala, and with a 5-HT agent in particular, stands in striking contrast to a notable lack of success using such substances directly in the pons. For example, 5-HT injected into the cholinergic lateral dorsal tegmental nucleus (LDT) suppressed REM but did not affect REM phasic events independent of the state effects (Horner, Sanford, Annis, Pack and Morrison, 1997). Following the first injection, the amount of REM as a per-cent of the total sleep time was consistently significantly reduced with 5-HT (x %REM = 19.9 ± 2.5 for 5-HT and 26.8 ± 2.4 for saline), and the amount of NREM was significantly increased. Median REM duration was reduced by 30% with 5-HT, but the number of REM episodes per hour was unchanged, suggesting that 5-HT disrupted REM maintenance. With subsequent injections, at approximately one-hour intervals, this same pattern of response was observed but the magnitude of the effects on sleep was reduced. In contrast to expectations from *in vitro* studies (Luebke, Greene, Semba, Kamondi, McCarley and Reiner, 1992), 5-HT had no effect on phasic REM events independent of behavioral state: the mean amplitude and variability of PGO wave activity were unchanged as were the mean and coefficient of variation of REM respiratory rate.

Inhibitory 5-HT modulation of elicited PGO waves (PGO_E) is probably not mediated through a $5\text{-}HT_{1A}$ receptor mechanism in PPT. We (Kaufman and Morrison, 1981; Sanford, Tejani-Butt, Ross and Morrison, 1996) have demonstrated that auditory stimuli elicit waves from the pontine site that yields spontaneous PGO waves during REM sleep in rats. Putative PGO bursting neurons of rat LDT are inhibited *in vitro* by a $5\text{-}HT_1$ agonist (Luebke, Greene, Semba, Kamondi, McCarley and Reiner, 1992). However, lesions of PPT, not LDT, eliminate PGO waves in cats (Shouse and Siegel, 1992). We investigated the inhibition *in vivo* of PGO_E by a $5\text{-}HT_{1A}$ agonist, administered either systemically or locally into PPT (Sanford, Ross, Seggos, Morrison, Ball and Mann, 1994). Rats (N = 9) were implanted with electrodes with the tip in the vicinity of the locus coeruleus for recording PGO_E. Microinjections of 8-OH-DPAT (0.01, 0.1, 1.0 mg/0.5 mg) into PPT (N = 6) did not alter PGO_E response frequency or amplitude across 6 blocks of 40 tones (90 ms, 110 dB, 4 kHz, 2 s ISI). Quantitative autoradiographic ^{3}H-cyanoimipramine analysis of presynaptic uptake sites for 5-HT performed in 19 rats indicated moderate 5-HT innervation in PPT, but there was little ^{3}H-8-OH-DPAT binding (1 nM) to $5\text{-}HT_{1A}$ receptor sites in this region. Inhibitory modulation of PGO_E in rats, as in cats, does not appear to involve $5\text{-}HT_{1A}$ receptors in PPT.

The $5\text{-}HT_2$ receptors in PPT may not be involved in regulating behavioral state. As we have noted, the prevailing explanation for the manner in which 5-HT exerts its influence on behavioral state and PGO waves is that 5-HT DRN cells act on ACh neurons in LDT and PPT (Rye, 1997). We previously showed that the 5-HT$_{1a}$ agonist, 8-OH-DPAT, and a non-specific 5-HT agonist did not affect PGO waves independent of state in cats but that the former did increase the number of abortive REM episodes (Sanford, Ross, Seggos, Morrison, Ball and Mann, 1994). Work in rats has demonstrated 5-HT$_2$ receptors on most ACh cells in PPT/LDT (Morilak and Ciaranello, 1993). We microinfused the relatively specific 5-HT$_2$ agonist, DOI (1-(2,5-dimethoxy-4-iodophenyl)-2-aminopropane), the relatively specific 5-HT$_2$ antagonist, ketanserin, and the non-specifc 5-HT antagonist, methysergide, locally into PPT in cats and monitored behavioral state and PGO waves (Sanford, Hunt, Ross, Pack and Morrison, 1998). Neither drug significantly affected behavioral state or PGO wave activity. These results suggest that 5-HT$_2$ receptors associated with ACh cells are minimally involved in the control of behavioral state. Together with our other findings in rats (Horner, Sanford, Annis, Pack and Morrison, 1997), they suggest that 5-HT may not modulate PGO wave generation via direct action on cholinergic neurons in PPT/LDT, a departure from a long-held (Rye, 1997) but minimally tested view.

REM generation probably requires amygdalar-pontine interactions

Reciprocal connections indicate that the rostral pontine structures can also influence the amygdala (Takeuchi, Megirian and Hopkins, 1982). The major 5-HT innervation of the amygdala arises from the DRN, with additional input from the median raphe nucleus (Fallon and Ciofi, 1992). 5-HT infused into CNA during REM induced arousal more rapidly and more frequently when injected in REM compared with NREM (Sanford, Tejani-Butt, Ross and Morrison, 1995) (Fig. 5). Antagonizing 5-HT with methysergide increased total sleep and released PGO waves in NREM (Sanford, Tejani-Butt, Ross and Morrison, 1995) (Fig. 6).

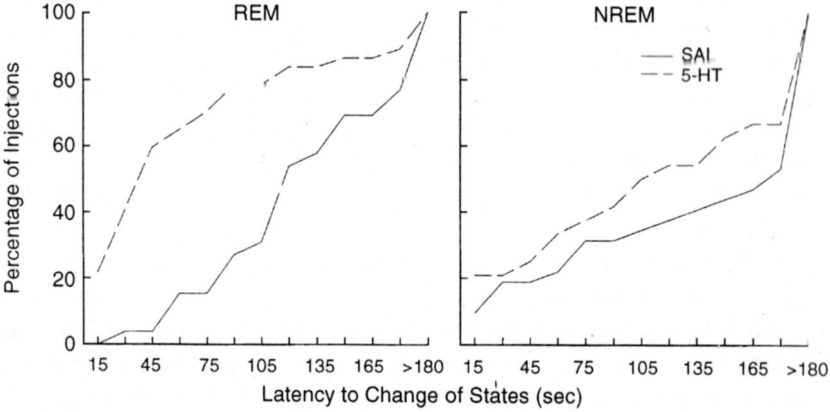

Fig. 5. Percentage of injections in REM and NREM producing a change of state graphed cumulatively across time demonstrating the dramatic effect of 5-HT microinjections in REM compared to 5-HT injections in NREM or saline injections in either REM or NREM. Reprinted with permission of Archives Italienne de Biologie from Sanford et al., 1995.

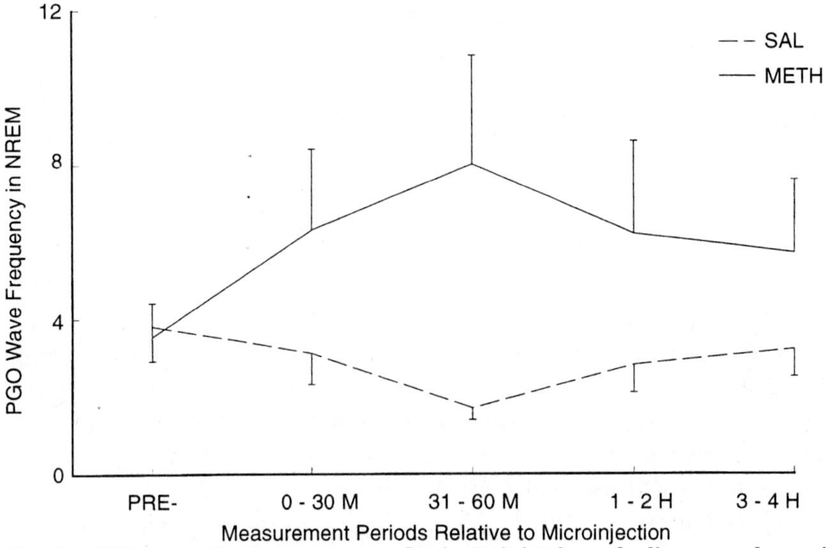

Fig. 6. PGO wave frequency measured prior to injections of saline or methysergide and at 4 post-injection in NREM. Reprinted with permission of Archives Italienne de Biologie from Sanford et al., 1995.

Virtually every region of the amygdala has at least a modest concentration of NA (Fallon and Ciofi, 1992); however, the densest concentrations are found in the medial part of CNA (Fallon and Ciofi, 1992), which projects to the brainstem. Methoxamine, an alpha$_1$ adrenergic-receptor agonist suppressed sleep when injected into CNA (Fuchino, Sanford, Ross and Morrison, 1996). These receptor sites in CNA have been implicated in the cataplexy associated with narcolepsy. Cataplexy has been viewed as an equivalent of the atonia that characterizes normal REM; it is typically triggered by emotional stimuli such as laughter, anger or surprise (Guilleminault, 1976). Prazosin, a selective α_1-adrenergic receptor blocker, exacerbated food-elicited cataplexy in narcoleptic dogs, whereas the drug did not affect normal dogs (Mignot, Guilleminault, Bowersox, Rappaport and Dement, 1988). Increased ^{3}H-prazosin binding was limited to the amygdala in narcoleptic dogs and was not observed in five other brain regions with roles in REM, or in normal dogs (Mignot, Guilleminault, Bowersox et al., 1989). Thus, noradrenergic effects exerted within the amygdala may also regulate the expression of REM, or at least some of its signs. It should be noted, though, that the situation is not simple. Although inhibition of dopaminergic neurons in the ventral tegmental area of narcoleptic dogs enhanced cataplexy, interference with noradrenergic transmission in CNA had no effect (Reid, Tafti, Nishino, Sampathkumaran, Siegel and Mignot, 1996). We suggest that different results may have been obtained had the infusion been paired with an additional significant stimulus.

As we have shown, ACh microinfused into CNA reduces REM in rats, while it has been reported to increase REM and PGO waves in cats (Calvo,

Simón-Arceo and Fernández-Mas, 1996). However, ACh effects on the amygdala are less easily interpreted than those obtained with 5-HT and NA substances because the amygdala has intrinsic ACh neurons (Nitecka and Frotscher, 1989) and receives its major ACh input from the basal forebrain (Woolf and Butcher, 1982) with some from the pontine ACh nuclei, PPT/ LDT (Woolf and Butcher, 1982). The majority of the contributions from LDT/ PPT are of non-ACh origin (Hallanger and Wainer, 1988).

Conclusions

Given all of the above, we believe it would defy logic to maintain that intact, integrated animals would enter the almost defenseless state of REM without prior reorganization of forebrain structures. This process may actually begin in the caudal brainstem, but there is no proof that this is the case. To provide answers will require much more research. The rewards will be great, however, for these studies will also lead to a better understanding of disorders in which REM is altered. They will provide insights in to the pathophysiology of narcolepsy. It is emotion-laden stimuli or situations that induce cataplectic attacks extending in to REM in narcoleptic individuals (Guilleminault, 1976), and it is in the amygdala and not the pons that neuronal degeneration is found in young narcoleptic dogs around the time of development of the disorder (Siegel, Nienhuis, Mignot, Fahringer and Jamgochian, 1996). Also, in related clinical studies we have found that REM is disturbed in posttraumatic stress disorder, which is characterized by hypervigilance to unfamiliar stimuli and stereotypical anxiety dreams (Ross, Ball, Dinges, Kribbs and Morrison, 1994). Long-lasting alterations of the amygdala would seem to be likely contributors to the symptomatology of PTSD.

Acknowledgements

Research reported from this laboratory was supported by grants from the National Institutes of Health (MH 42903, NS 35281 and HL 42236) and the Department of Veterans Affairs. We wish to acknowledge the generous help of our various collaborators listed in the references and to thank Graziella Mann as well as Mark Mallon and Antoinette Mann for their invaluable technical assistance and Eileen Conner for her administrative contributions to our work.

References

Amaral, D.G., Price, J.L., Pitkänen, A. and Carmichael, S.T. (1992). Anatomical organization of the primate amydaloid complex. In: J.P. Aggleton (Ed.), *The Amygdala: Neurobiological Aspects of Emotion, Memory, and Mental Dysfunction*, Wiley-Liss, Inc., New York, pp. 1–66.

Aserinsky, E. and Kleitman, N. Regularly occurring periods of eye motility, and concomitant phenomena during sleep, Science, 118 (1953) 273–274.

Baghdoyan, H.A., Spotts, J.L. and Snyder, S.G. Simultaneous pontine and basal forebrain microinjections of carbachol suppress REM sleep, Journal of Neuroscience, 13 (1993) 229–242.

Bordi, F., LeDoux, J.E., Clugnet, M.C. and Pavlides, C. (1993). Single-unit activity in the lateral nucleus of the amygdala and overlying areas of the striatum in freely behaving rats: Rates, discharge patterns, and responses to acoustic stimuli, Behavioral Neuroscience, 107 (1993) 757–769.

Calvo, J.M., Badillo, S., Morales-Ramirez, M. and Palacios-Salas, P. The role of the temporal lobe amygdala in ponto-geniculo-occipital activity and sleep organization in cats, Brain Research, 403 (1987) 22–30.

Calvo, J.M., Simón-Arceo, K. and Fermández-Mas, R. Prolonged enhancement of REM sleep produced by carbachol microinjection into the amygdala, Neuro Report, 7 (1996) 577–580.

Chase, M.H. and Morales, F.R. (1985). Postsynaptic modulation of spinal cord motoneuron membrane potential during sleep.In: D.J. McGinty, R. Drucker-Colin, A.R. Morrison, and P.L. Parmeggiani (Eds.), *Brain Mechanisms of Sleep*, Raven Press, New York, 1985, pp. 45–61.

DeBoer, T., Sanford, L.D., Ross, R.J. and Morrison, A.R. Effects of electrical stimulation in the amygdala on ponto-geniculo-occipital waves in rats, Brain Research, 793 (1998) 305–310.

DeBoer, T., Sanford, L.D., Ross, R.J. and Morrison, A.R. Electrical stimulation in the amygdala increases the amplitude of elicited PGO waves, Physiology and Behavior, In press (1998).

Dement, W.C. The occurrence of low voltage, fast electroencephalogram patterns during behavioral sleep in the cat, Electroencephalography and Clinical Neurophysiology, 10 (1958) 291–296.

Fallon, J.H. and Ciofi, P. Distribution of monoamines within the amygdala. In: J.P. Aggleton (Ed.), The Amygdala: Neurobiological Aspects of Emotion, Memory and Mental Dysfunction, Wiley-Liss, Inc., New York, 1992, pp. 97–114.

Fuchino, S.H., Sanford, L.D., Ross, R.J. and Morrison, A.R. Effects of microinjections of the alpha 1 adrenergic agonist, methoxamine, into the central nucleus of the amygdala on sleep-wake states, Sleep Research, 25 (1996) 6(abstract).

Guilleminault, C. Cataplexy. In: W.C. Dement, P. Passouant, and C. Guilleminault (Eds.), *Advances in Sleep Research, vol. 3: Narcolepsy*, Spectrum Publications, New York, 1976, pp. 125–143.

Hallanger, A.E. and Wainer, B.H. Ascending projections from the pedunculopontine tegmental nucleus and the adjacent mesopontine tegmentum in the rat, Journal of Comparative Neurology, 274 (1988) 483–515.

Horner, R.L., Sanford, L.D., Annis, D., Pack, A.I. and Morrison, A.R.(1997). Serotonin at the laterodorsal tegmental nucleus suppresses rapid-eye-movement sleep in freely behaving rats, Journal of Neuroscience, 17 (1997) 7541–7552.

Jouvet, M., Cataplexie et sommeil paradoxal réflexes chez le chat pontine, Comptes rendus des séances de la societé de biologie et de ses filiales, 159 (1964) 383–387.

Kapp, B.S., Suppl, W.F. and Whalen, P.J. Effect of electrical stimulation of the amygdaloid central nucleus on neocortical arousal in the rabbit, Behavioral Neuroscience, 108 (1994) 81–93.

Kaufman, L.S. and Morrison, A.R. Spontaneous and elicited PGO spikes in rats, Brain Research, 214 (1981) 61–72.

LeDoux, J.E. Emotion and the amygdala. In: J.P. Aggleton (Ed.), *The Amygdala: Neurobiological Aspects of Emotion, Memory and Mental Dysfunction*, Wiley-Liss, Inc., New York, 1992, pp. 339–351.

Luebke, J.I., Greene, R.W., Semba, K., Kamondi, A., McCarley, R.W. and Reiner, P.B. Serotonin hyperpolarizes cholinergic low-threshold burst neurons in the rat laterodorsal tegmental nucleus in vitro, Proceedings of the National Academy of Science, 89 (1992) 743–747.

Maquet, P., Peters, J., Aerts, J., DelFiore, G., Degueldre, C., Luxen, A. and Franck, G. Functional neuroanatomy of human rapid-eye movement sleep and dreaming, Nature, 383 (1966) 163–166.

Marini, G., Gritti, I. and Mancia, M. The role of some thalamic nuclei in sleep mechanisms: evidence from chemical lesions in the cat. In: M. Mancia and G. Marini (Eds.), *The Diencephalon and Sleep*, Raven Press, New York, 1990, pp. 279–292.

Mignot, E., Guilleminault, C., Bowersox, S., Frusthofer, B., Nishino, S., Maddaluno, J., Ciaranello, R, Dement, W.C. Central alpha1 adrenoceptor subtypes in narcolepsy-cataplexy: a disorder of REM sleep, Brain Research, 490 (1989) 186–191.

Mignot, E., Guilleminault, C., Bowersox, S., Rappaport, A. and Dement, W.C. Role of central alpha-1 adrenoceptors in canine narcolepsy, Journal of Clinical Investigation, 828 (1988) 885–894.

Mirmiran, M., 'Oneiric' behavior during active sleep induced by bilateral lesions of the pontine tegmentum in juvenile rats. In: W.P. Koella (Ed.), *Sleep: 6th Eur. Congr. Sleep Res,* Karger, Basel, Zurich, 1983, pp. 263–239.

Morilak, D.A. and Ciaranello, R.D.5-HT_2 receptor immunoreactivity on cholinergic neurons of the pontomesencephalic tegmentum shown by double immunofluorescence, Brain Research, 627 (1993) 49–54.

Morrison, A.R. Brainstem regulation of behavior during sleep and wakefulness. In: J.M. Sprague and A.W. Epstein (Eds.), *Progress in Psychobiology and Physiological Psychology,* Academic Press, New York, 1979, P. 91–131.

Morrison, A.R. Paradoxical sleep without atonia, Archives Italiennes de Biologie, 126 (1988) 275–289.

Morrison, A.R. and Reiner, P.B. A dissection of paradoxical sleep, In: D.J. McGinty, R. Drucker-Colin, A.R. Morrison and P.L. Parmeggiani (Eds.), *Brain Mechanisms of Sleep*, Raven Press, New York, 1985, pp.97–110.

Nitecka, L. and Frotscher, M. Organization and synaptic interconnections of GABAergic and cholinergic elements in the rat amygdaloid nuclei: single- and double-immunolabeling studies, Journal of Comparative Neurology, 279 (1989) 470–488.

Paré, D. and Llinás, R. Conscious and pre-conscious processes as seen from the standpoint of sleep-waking cycle neurophysiology, Neuropsychologia, 33(1995) 1115–1168.

Parmeggiani, P.L. and Morrison, A.R. Alterations in autonomic functions during sleep. In: A.D. Loewy and K.M. Spyer (Eds.), *Central Regulation of Autonomic Functions*, Oxford University Press, New York, 1991, pp. 367–386.

Portas, C.M. Thakker, M., Rainnie, D. and McCarley, R.W. Microdialysis perfusion of 8-hydroxy-2-(di-n-Propylamino) tetralin (8-OH-DPAT) in the dorsal raphe nucleus decreases serotonin release and increases rapid eye movement sleep in the freely moving cat, Journal of Neuroscience, 16 (1996) 2820–2828.

Portas, C.M., Thakker, M., Rainnie, D.G., Greene, R.W. and McCarley, R.W. Role of adenosine in behavioral state modulation: a microdialysis study in the freely moving cat, Neuroscience, 79 (1997) 225–235.

Price, J.L., Russchen, F.T. and Amaral, D.G. The limbic region. II: The amygdaloid complex. In: A. Bjorklund, T. Hokfelt and L.W. Swanson (Eds.), *Handbook of chemical neuroanatomy. Integrated systems of the CNA, Part I*, Elsevier, New York, 1987, pp. 279–375.

Reid, M.S., Tafti, M., Nishino, S., Sampathkumaran, R., Siegel, J.M. and Mignot, E. Local administration of dopaminergic drugs into the ventral tegmental area modulates cataplexy in the narcoleptic canine, Brain Research, 733 (1996) 83–100.

Roky, R., Valatx, J.L., Paut-Pagano, L. and Jouvet, M. Hypothalamic injection of prolactin or its antibody alters the rat sleep-wake cycle, Physiology and Behavior, 55 (1994) 1015–1019.

Ross, R.J., Ball, W.A., Dinges, D.F., Kribbs, N.B. and Morrison, A.R. Rapid-eye movement sleep disturbance in post-traumatic stress disorder, Biological Psychiatry, 35 (1994) 195–200.

Rye, D.B. Contributions of the pedunculopontine region to normal and altered REM sleep, Sleep, 20 (1997) 757–788.

Sakai, K., Mansari, M.E., Lin, J.S., Zhang, G. and Vanni-Mercier, G. The posterior hypothalamus in the regulation of wakefulness and paradoxical sleep. In: M. Mancia and G. Marini (Eds.), *The Diencephalon and Sleep*, Raven Press, New York, 1990, pp. 171–198.

Sanford, L.D., Hunt, W.K., Ross, R.J., Pack, A.I. and Morrison, A.R. Central administration of a 5-HT_2 receptor agonist and antagonist: lack of effect on rapid eye movement sleep and PGO waves, Sleep Research Online 1 (1998) 80–86.

Sanford, L.D., Morrison, A.R., Ball, W.A., Ross, R.J. and Mann, G.L. The amplitude of elicited PGO waves: a correlate of orienting, *Electroencephalography and clinical Neurophysiology*, 86 (1993) 438–445.

Sanford, L.D., Morrison, A.R. and Ross, R.J. Effects of auditory stimulation on phenomena of rapid eye movement sleep. In: B.N. Mallick and R. Singh (Eds.), *Environment and Physiology*, Narosa Publishing House, New Delhi, 1994, pp. 186–195.

Sanford, L.D., Ross, R.J., Seggos, A.E., Morrison, A.R., Ball, W.A. and Mann, G.L. Central administration of two 5HT receptor agonists: effect on REM sleep initiation and PGO waves, *Pharmacology, Biochemistry and Behavior*, 49 (1994) 93–100.

Sanford, L.D., Tejani-Butt, S.M., Ross, R.J. and Morrison, A.R. Amygdaloid control of alerting and behavioral arousal in rats: Involvement of serotonerg in mechanisms, Archives Italiennes de Biologie, 134 (1995) 81–99.

Sanford, L.D., Tejani-Butt, S.M., Ross, R.J. and Morrison, A.R. Elicited PGO waves in rats: Lack of 5-HT1A inhibition in putative pontine generator region, Pharmacology, Biochemistry and Behavior, 53 (1996) 323–327.

Savander, V., Go, G. G., LeDoux, J.E. and Pitkänen, A. Intrinsic connections of the rat amygdaloid complex: projections originating in the basal nucleus, Journal of Comparative Neurology, 361 (1995) 345–368.

Shouse, M.N. and Siegel, J.M. Differential regulation of REM sleep components by the pedunculo-pontine tegmentum (PPT) and medial-lateral pontine tegmentum (MLP) in cats, Brain Research, 571 (1992) 50–63.

Siaud, P., Manzoni, O., Balmefrezol, M., Barbanel, G., Assenmacher, I. and Alonso, G. The organization of prolactin-like-immunoreactive neurons in the rat central nervous system. Light- and electron-microscopic immunocytochemical studies, Cell Tissue Res., 255 (1989) 107–115.

Siegel, J.M., Nienhuis, R., Mignot, E., Fahringer, H.M. and Jamgochian, G.M. Neuronal degeneration in the amygdala of the narcoleptic dog, Sleep Research, 25 (1996) 370 (abstract).

Silvestri, A.J. and Kapp, B.S. Amygdaloid modulation of mesopontine peribrachial neuronal activity: Implications for arousal, Behavioral Neuroscience, 112 (1998) 571–588.

Steriade, M. and McCarley, R.W. Brainstem Control of Wakefulness and Sleep. New York: Plenum Press, New York, 1990.

Steriade, M., Pare, D., Datta, S., Oakson, G. and Curro Dossi, R. Different cellular types in mesopontine cholinergic nuclei related to ponto-geniculo-occipital waves, Journal of Neuroscience, 10 (1990) 2560–2579.

Takeuchi, Y., Megirian, D. and Hopkins, D.A. Reciprocal connections between the amygdala

and parabrachial nuclei: ultrastructural demonstration by degeneration and axonal transport of horseradish peroxidase in the cat, Brain Research, 239 (1982) 583–588.
Tononi, G., Pompeiano, M., Gianni, S. and Pompeiano, O. Enhancement of desynchronized sleep signs after microinjection of the beta-adrenergic antagonist propranolol in the dorsal pontine tegmentum, Archives Italiennes de Biologie, 126 (1988) 119–123.
Trulson, M.E., Jacobs, B.L. and Morrison, A.R. Raphe unit activity during REM sleep in normal cats and pontine-lesioned cats displaying REM sleep without atonia, Brain Research, 226 (1981) 75–91.
Webster, H.H. and Jones, B.E. Neurotoxic lesions of the dorsolateral pontomesencephalic tegmentum-cholingeric cell area in the cat. II. Effects upon sleep-waking states, Brain Research, 458 (1988) 285–302.
Woolf, N.J. and Butcher, L.L. Cholinergic projections to the basolateral amygdala: a combined Evans Blue and acetylcholinesterase analysis, Brain Research Bulletin, 8 (1982) 751–763.

Rapid Eye Movement Sleep
B.N. Mallick and S. Inoué (Eds)

5. Muscle Atonia in REM Sleep

Yuan-Yang Lai and Jerome M. Siegel

Dept. Psychiat., Sch. Med., UCLA., VAMC, Sepulveda, CA, 91343, USA

SUMMARY

Rapid eye movement (REM) sleep is characterized by skeletal muscle atonia. The neural substrates participating in REM atonia have been identified in the pons and medulla. Electrical and chemical stimulations in these areas induces muscle atonia in both the intact and the decerebrate animals. Cholinergic, glutamatergic and corticotropin-releasing factor (CRF) mechanisms have been found to be involved in muscle atonia. In the medial pons, cholinergic, glutamatergic and CRF agonist injection induces muscle atonia. In the medial medulla, 2 distinct atonia inducing areas, a rostroventral medulla glutamatergic/CRF and a caudal medulla cholinoceptive system mediating muscle atonia have been identified. Microdialysis studies find that acetylcholine release in the pons and caudal medulla increases during natural REM sleep in the intact cat. Active inhibition of motoneurons from the activation of neurons in the pontine inhibitory area and medial medullary reticular formation has a role in muscle atonia. The other systems, including locus coeruleus complex, ventral paralemniscal tegmental field and retrorubral nucleus, may also contribute to muscle tone suppression via inhibitory or disfacilitatory effects on motoneurons.

Key Words: locus coeruleus, medulla, pons, midbrain, glutamate, acetylcholine, microdialysis, motor control.

Introduction

This review concerns experimental studies of REM sleep atonia and the mechanisms underlying atonia. The majority of studies have been done on the cat, a few experiments on the rat are also included. Muscle atonia in REM sleep was first discovered by Jouvet and Michel (1959). REM sleep can be subdivided into two phases—tonic and phasic. Tonic REM sleep includes sustained muscle atonia, whereas phasic REM refers to the occurrence of muscle twitches superimposed on tonic atonia. An understanding of supraspinal regulation of muscle activity in sleep may be relevant to sleep related disorders, such as periodic leg movement, REM behavior disorder and obstructive sleep apnea.

Supraspinal Structures Involved in the Inhibition of Motor Activity

In natural sleep, a complete loss of muscle tone (Jouvet and Michel, 1959) and a suppression of spinal reflexes (Gassel et al., 1964, 1965) in REM sleep have been reported. These muscle atonia and reflexes suppression could be attributed to supraspinal structures. Two areas of the brainstem, the medial pons and medulla, have been extensively studied. Electrophysiological studies on either intact unanesthetized or decerebrate animals have shown that the pontine inhibitory area (PIA) and medial medullary reticular formation play a critical role in postural suppression. A physiological role of muscle tone regulation for 2 other areas, ventro-caudal midbrain and locus coeruleus (LC) complex, have been also discussed.

Pons-midbrain

Pontine inhibitory area

The PIA (Fig. 1) extends from the nucleus reticularis pontis oralis (PoO) and the rostral portion of the nucleus reticularis pontis caudalis (PoC) according to Taber (1961). Sprague et al. (1948) first reported that electrical stimulation of the PoC dorsal to the trapezoid body, produced reflex inhibition bilaterally in the decerebrate cat. In the more rostral pons, electrical stimulation in the PoO produces muscle tone suppression in neck (Lai and Siegel, 1988) and hindlimb muscles (Iwakiri et al. 1994). Fig. 1 shows that inhibitory sites are widely diffused from the dorsal to the ventral portion at the junction of PoO and PoC while a restricted localized area in the medial portion could be found in PoO in the cat. Similar to the cat, electrical stimulation in the PoO and PoC also induced reduction of muscle tone in the decerebrate rat (Hajnik et al., 1995).

Lesion studies in intact animals have localized the neural structures participating in REM atonia. Jouvet (1962) and Carli and Zanchetti (1963) showed that lesions in the pontine reticular formation suppressed REM sleep. Thereafter, Jouvet and Delorme (1965) found that small lesions in this region produced REM sleep without atonia instead of REM sleep suppression. Extensive studies in lesions in the dorsolateral pons were performed by Morrison and colleagues and Sastre et al. (1978). They found that electrolytic lesion in the area ventromedial to the nucleus locus coeruleus (LC) produced REM sleep without atonia (Henley and Morrison, 1974; Sastre et al., 1978). Further experiments have shown that caudal medial pontine lesion produced REM sleep without atonia, while lesions extending more rostroventrally not only produced REM sleep without atonia but also expressed different behaviors (Hendricks et al., 1982; Morrison, 1988). This pontine lesion-induced REM sleep without atonia has been confirmed by other studies using radiofrequency thermal lesion (Friedman and Jones, 1984; Jones, 1979; Shouse and Siegel, 1992). In contrast, neurotoxic kainate lesion in the caudomedial pons fails to produce REM sleep without atonia (Sastre et al., 1981).

The proto-oncogene c-fos is an immediate-early gene expressed in cells in response to stimulation. Fos expressing neurons could be identified by immunohistochemistry. This technique provides another evidence of brainstem participation in muscle atonia. In carbachol-induced REM sleep-like activity, fos-like immunoreactive neurons could be found in the PIA (Shiromani et al., 1992; Yamuy et al., 1993). This result indicates that neurons in the PIA may play an important role in the regulation of muscle activity or in other aspects of REM sleep.

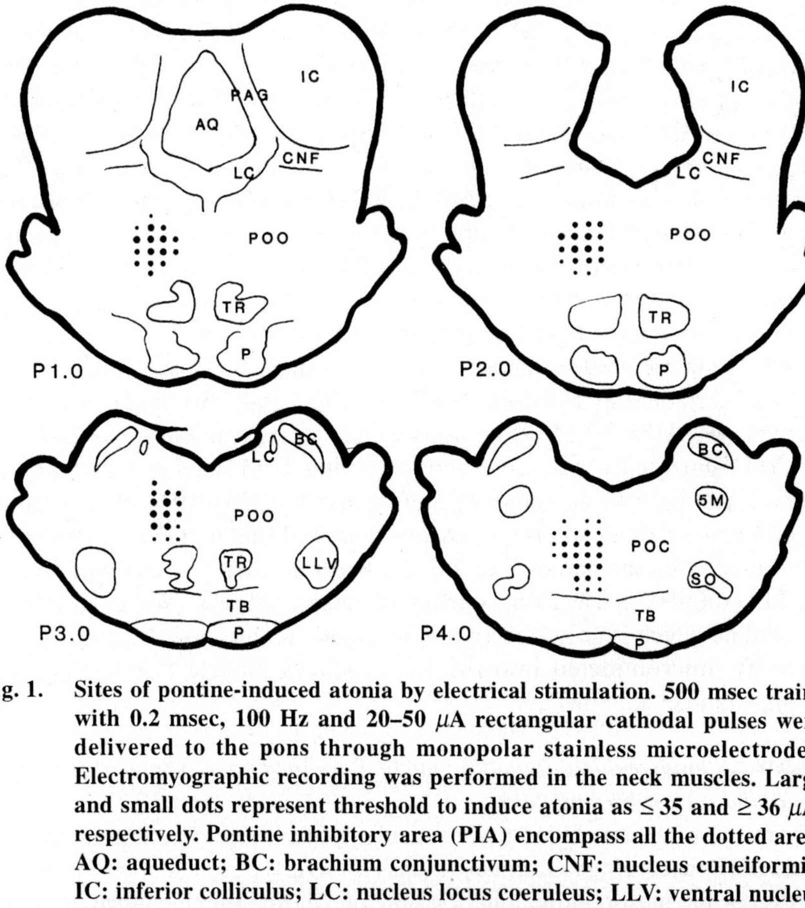

Fig. 1. Sites of pontine-induced atonia by electrical stimulation. 500 msec trains with 0.2 msec, 100 Hz and 20–50 μA rectangular cathodal pulses were delivered to the pons through monopolar stainless microelectrodes. Electromyographic recording was performed in the neck muscles. Large and small dots represent threshold to induce atonia as ≤ 35 and ≥ 36 μA, respectively. Pontine inhibitory area (PIA) encompass all the dotted area. AQ: aqueduct; BC: brachium conjunctivum; CNF: nucleus cuneiformis; IC: inferior colliculus; LC: nucleus locus coeruleus; LLV: ventral nucleus of the lateral lemniscus; P: pyramid tract; PAG: periaqueductal gray; POC: nucleus reticularis pontis caudalis; POO: nucleus reticularis pontis oralis; SO: superior olivary nucleus; TB: trapezoid body; TR: tegmental reticular nucleus: 5M: trigeminal motor nucleus.

Extracellular recording studies have shown that firing rate increases in the PIA neuron in parallel with postural atonia. Pompeiano and Hoshino (1976a, 1976b) demonstrated that certain neurons in the PoO increased firing during systemic anticholinesterase injection-induced postural atonia in the decerebrate animals. REM-on cells which increase firing rate during REM sleep, have been found in the LCα and peri-LCα in natural sleep (Sakai, 1988; Sakai and Koyama, 1996). Some REM-on cells have a descending projection to the NMC suggesting these REM-on cells in the pons are involved in muscle atonia. Injection of cholinergic agonists into the area medioventral to the LC, induce postural atonia in unanesthetized (George et al., 1964; van Dogen et al., 1978) and decerebrate animals (Lai and Siegel, 1988; Hajnik et al., 1996).

Locus coeruleus complex

Physiologically, LC has been known to modulate the spinal motor system (Fung and Barnes, 1981, 1987), nociception (Jones, 1991; Margalit and Segal, 1979; Mokha et al., 1985; West et al., 1993), sensory (Foote et al., 1980; Swick et al., 1994), EEG activation (Foote et al., 1980), cardiovascular regulation (Ward and Gunn, 1976; Chalmers et al., 1981; Gurtu et al., 1984), fear and anxiety (Aston-Jones et al., 1994) and memory (Sara et al., 1988). However, the functional aspects of the LC effect on muscle atonia have been neglected. Extracellular recording from LC neurons has shown these neurons tend to be silent during cholinergic induced muscle atonia/REM sleep in the decerebrate cat (Pompeiano and Hoshino, 1976a, 1976b). The decrease of neural activity of the LC neurons starts prior to the onset of slow wave sleep, and there is a complete absence of activity in REM sleep (Hobson et al., 1975; Steriade and Hobson, 1976). Further study by Sakai (1980) has demonstrated that REM-off neurons which ceases firing during REM sleep in the LC, markedly increase in activity during REM sleep-waking transition period, just prior to the return of muscle activity. This REM-off neuron has been identified as noradrenergic (Aston-Jones and Bloom, 1981). Extracellular recording from the narcoleptic dog demonstrates that all REM-off neurons in LC complex cease firing during cataplexic attacks (Wu et al., 1996). Clonidine which is an α_2 receptor agonist and inhibits noradrenergic neuron activity, microinjected into the LC produces muscle tone suppression (Pompeiano et al., 1987).

Other systems (pedunculopontine nucleus, ventral paralemniscal tegmental field and retrorubral nucleus)

The most rostral part of the pontine reticular formation, paralemniscal tegmental field (FTP) according to Berman's Atlas (1968) and pedunculopontine nucleus (PPN) of the mesopontine junction may play a role in the control of muscle tone. Electrical stimulation of the ventral portion of the FTP (VFTP) and PPN generates muscle tone suppression bilaterally in the decerebrate animal (Lai and Siegel, 1990; Hajnik et al., 1995). The latency (Table 1) of VFTP— (31 msec) and PPN-induced (27.1 msec) muscle tone suppression is short and comparable to those of the pons (18.7 msec) and medulla (16.5 msec). Injection of cholinergic agonists into the VFTP induces increases in REM sleep in the intact animals (Reinoso-Suarez et al., 1990). Non-NMDA agonists injection into the VFTP produces reduction of muscle tone in the decerebrate rat (Hajnik, et al., 1996). At the midbrain level, the so called A8 area of the retrorubral nucleus (RRN), located in between the substantia nigra and ventral tegmental area, has also been found to be involved in muscle tone regulation. The caudal portion of the RRN forms a continuation of the VFTP. Electrical stimulation in the RRN elicits muscle tone suppression with a latency of 36.5 msec (Lai and Siegel, 1990).

Table 1. Latency of muscle tone suppression by electrical stimulation in midbrain, pons and medulla

Nucleus	RRN	VFTP	PPN	PIA	NMC	NPM
Latency (ms)	36.5	31.0	27.1	18.7	16.5	15.4

NMC: nucleus magnocellularis;NPM: nucleus paramedianus; PIA: pontine inhibitory area; PPN: pedunculopontine nucleus; RRN: retrorubral nucleus; VFTP: ventral paralemniscal tegmental field.

We have found that repetitive electrical stimulation in the VFTP and PPN elicits reduction of muscle tone during stimulation, while elicited locomotion could be seen during the inter-stimulation interval (Lai and Siegel, 1990). Neurotoxic NMDA lesions of the ventral mesopontine junction including the VFTP and the caudal RRN, induces muscle twitching in decerebrate animals (Lai and Siegel, 1997a). We have hypothesized that this ventral mesopontine junction area may participate in the regulation of phasic events of REM sleep mediated through the nucleus magnocellularis (NMC) in the medulla (detail discussed below).

Medulla

In 1946, Magoun and Rhines described a global inhibitory effect from the medullary reticular formation on spinal motor activity in the decerebrate cat. Unilateral electrical stimulation in the medial medulla suppressed spinal reflexes of both hindlimbs. The bulbar sites which produce reflex inhibition are located in the NMC, nuclei gigantocellularis (NGC) and paramedianus (NPM). Further studies have found that stimulation in this medial medulla region inhibits visceral reflexes (Alderson and Downman, 1966), sensory-induced reciprocal inhibition (Bach, 1950), basal muscle tone (Lai et al., 1987), muscle twitches (Lai and Siegel, 1997b), and movement in the intact cat (Schenkel et al., 1998), movement in the anesthetized animal evoked by cortical stimulation (Niemer and Magoun, 1947) and in the decerebrate cat (Noga et al., 1988; Kinjo et al., 1990). In the rodent, electrical stimulation in the medial medulla of nuclei dorsal paragigantocelluaris, gigantocellularies and gigantocellularis ventralis also elicits muscle tone suppression (Hajnik et al., 1995).

The most potent inhibitory area in the medial medulla in the cat has been reported to be located in the rostral medial medulla. However, whether the dorsal medulla of the NGC or the ventral NMC generates the maximal effect on muscle activity remains unclear. Using orthodromic stimulation and extracellular recording techniques, Mori and colleagues find that NGC neurons respond to stimulation of the pontine-induced postural suppression area, while NMC neurons respond to stimulation applied to the mesencephalic locomotor region (Iwakiri et al., 1995). In contrast, Magoun (1950) and Engberg et al. (1968a) have reported that the maximal suppression effect on spinal reflexes was originated from NMC stimulation. Our work in the decerebrate cat has been consistent that of Magoun (1950) and Engberg et al., (1968a). As shown on Table 2, the lowest threshold area to induce muscle atonia has been found to be located in the NMC by electrical stimulation.

Table 2. Threshold for eliciting muscle tone suppression in the medial medulla

	Threshold (μA)
NGC	32 ± 14
NMC	21 ± 10
NPM	27 ± 12

Electrical stimulation studies have shown that muscle tone suppression and reflex inhibition were "state-dependent" in intact animals. Stimulation in the NGC produces facilitation of masseteric reflex in waking and quiet sleep, while the same stimulation in the same site produces reflex inhibition in REM sleep (Chase and Wills, 1979). A recent study in our laboratory has found that medial medullary stimulation induces either no effect or a short inhibition followed by facilitation of muscle tone in waking, while the same stimulation produces muscle tone suppression in non-REM sleep (Schenkel et al., 1990). In decerebrate animals, stimulation in the NGC produced no effect on monosynaptic reflex and membrane potential of the motoneuron before intrapontine carbachol injection. However, suppression of monosynaptic reflex and hyperpolarization of spinal motoneuron could be seen after carbachol injection into the pons (Pereda et al., 1990).

Neuronal activity in the NGC and NMC increases during pontine carbachol-induced atonia and natural REM sleep. Extracellular recording study demonstrates that increases in firing rate in NGC neurons parallel the decreases in soleus muscle activity during intrapontine carbachol injection (Takakusaki et al., 1994). REM-on neurons have been identified in NMC in the cat (Sakai et al., 1981). Cataplexy-REM-on neurons have been found in the NMC in the narcoleptic dog (Siegel et al., 1991). Proto-oncogene c-fos immunoreactive neurons could be found in the medial medulla, especially in the NMC, in pontine carbochol-induced atonia (Yamuy et al., 1993). On the other hand, lesions in the medial medulla produces REM sleep without atonia (Sastre et al., 1978; Schenkel and Siegel, 1989), indicating that the medial medulla is a final common pathway to regulate muscle atonia in REM sleep.

Neural pathway

Figure 2 summarizes the hypothesized neural pathway in control of muscle atonia. Neurons from the limbic system including bed nucleus of the stria terminalis, amygdala, hypothalamus-preoptic region and the anterior and mediodorsal thalamic nuclei (Holstege, 1991) project to the RRN, VFTP, periaqueductal gray (PAG), PPN, LC complex and NMC (Hopkins and Holstege, 1978; Saper et al., 1979; Price and Amaral, 1981; Chiba and Murata, 1985; Holstege et al., 1985; Hosoya, 1985; Holstege, 1987, 1991; Deutch et al., 1988; Wallace et al., 1989; Shammah-Lagnado et al., 1993; Lai et al., 1994; Zardetto-Smith and Johnson, 1995; Rizvi et al., 1996). A direct projection from the limbic system to the spinal cord have also been found (Kuypers and Maisky, 1975; Holstege, 1987). Neurons from the RRN, VFTP, PPN and PAG have been found projecting to the PIA and NMC by retrograde transport tracing studies (Lai et al., 1993a, 1993b).

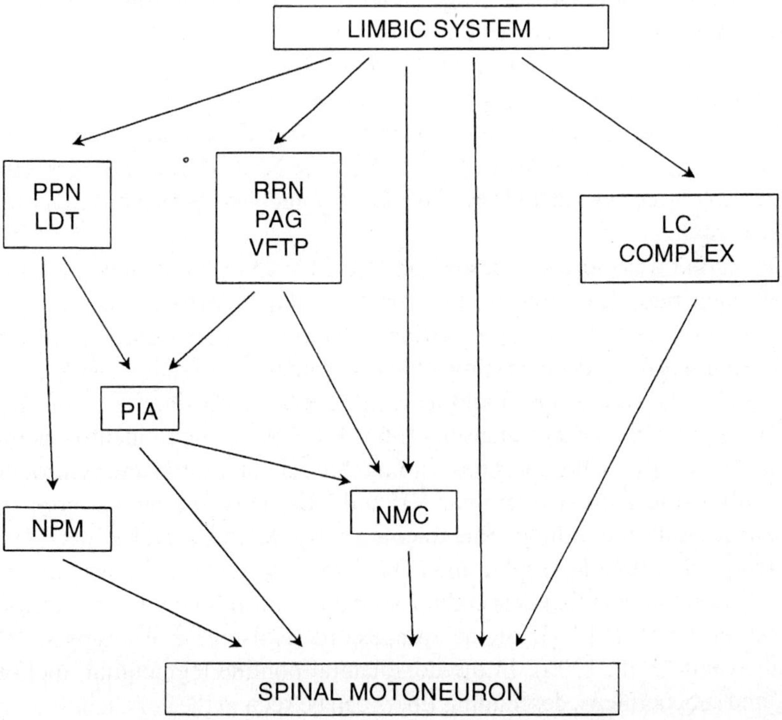

Fig. 2. Hypothesized neural structures and pathways involved in muscle atonia. Neurons in the limbic system project to the bainstem and exert either facilitatory or inhibitory effects on the target neurons, which in turn project to the medullary relay nuclei and spinal cord. Spinal motoneurons receive multiple innervations from pontomedullary neural structures. Muscle atonia resulted from the summation of input fiber activity. LDT: dorsolateral tegmental nucleus; PAG: Periagueductal gray.

Neurons in the pons project to the NGC (Matsuyama et al., 1993), NMC

(Sakai et al., 1979; Lai and Siegel, 1988; Lai et al., 1993) and NPM (Shiromani et al., 1988) which in turn project to the spinal cord (Tohyama et al., 1979; Newman, 1985). However, a direct projection from the PIA to the spinal cord has also been reported (Nyberg-Hansen, 1966; Kuypers and Maisky, 1975, 1977; Matsuyama et al., 1997). Anatomical and electrophysiological studies find the neurons in the pons send their fibers through the dorsal tegmental field and terminate in the NGC (Ohta et al., 1988). Nucleus magnocellularis neurons also receive fiber inputs from the pons. Orthodromic stimulation in the pons excitates NMC neurons (Kanamori et al., 1980), and pontine neurons can be antidromically activated from the NMC (Sakai et al., 1981) indicating neurons from pons project to the NMC. Labelled neurons have been found in the pons (Sakai et al., 1979; Luppi et al., 1988; Lai et al., 1993b) when retrograde tracers were injected into the NMC in the cat, as well as in the nucleus gigantocellularis alpha and ventralis in the rat (Gallager and Pert, 1978; Bayev et al., 1988). Anterograde PHA-L and autoradiographic studies have demonstrated that the nucleus gigantocellularis alpha and ventralis in the rat receive afferents from the pons (Jones and Yang, 1985). In our physiological study in the decerebrate cat, we found that pontine carbachol-induced muscle atonia could be reversed by unilateral injection of the glutamate antagonist, gamma-D-glutamyglycine, into the NMC (Lai and Siegel, 1988). This result indicates that the NMC is the relay nucleus for the pons in inducing muscle atonia.

Funicular trajectories of descending fibers from the brainstem nuclei have been identified. The pioneer study on funicular trajectory was done using degeneration and silver impregnation techniques. After lesions in different brainstem regions, degenerating fibers, terminal arborizations and terminal boutons in the spinal cord could be identified by silver impregnation. Using these techinques, Nyberg-Hansen (1965, 1966) was able to identify neurons from the PoC that send fibers descending through the ventral and ventrolateral funiculi and terminate in laminae VII and VIII. This finding was confirmed by anterograde autoradiographic tracing study (Holstege and Kuypers, 1982). Anterograde autoradiographic and PHA-L tracing studies have also identified fibers from the PoO that descend through the ventral funiculus and terminate in laminae VII and VIII of the spinal cord (Holstege and kuypers, 1982; Matsuyama et al., 1993). In the dorsolateral pontine tegmentum, including LC and subcoeruleus, descending fibers can be seen in the dorsal, dorsolateral ventrolateral and ventral funiculi (Holstege et al., 1979; Bjorklund and Skagerberg, 1982; Kuypers and Huisman, 1982; Carlton et al., 1985). These dorsolateral pontine projection fibers terminate in laminae V–VIII and X, throughout all levels of the spinal cord (Holstege et al., 1979).

In the medial medulla, neurons from both the NGC and NMC send descending fibers through the ventral and ventrolateral funiculi through all levels of the spinal cord, and terminate in laminae VI–IX in the cat (Nyberg-Hansen, 1965, 1966; Tohyama et al., 1979; Johannessen et al., 1984; Matsuyama et al., 1988, 1995; Mitani et al., 1988). However, a different spinal pathway from the NMC has been reported. Descending fibers from the NMC travel through ventral and ventrolateral funiculi at the cervical and thoracic cord and then over the dorsolateral funiculus at the lumbosacral level, terminate in all levels of spinal cord in laminae V–VIII and X (Holstege et al., 1979; Holstege and Kuypers, 1982, 1987). Finally, descending projections from the NPM pass via dorsal and lateral funiculi in the spinal cord (Tohyama et al., 1979).

Mechanism of supraspinal inhibition of motor activity

Stimulation in the medial medulla produces IPSP in both of the extensor and flexor motoneurons (Llinas and Terzuolo, 1964, 1965; Jankowska et al., 1968). Since a long slow positive potential in the motoneuronal pool accompanying with the suppression has been recorded, Lettvin and Dell (1950) suggest that this medullary inhibitory effect on motoneurons is mediated via the interneurons in the spinal cord. Subsequent studies by either intracellular or extracellular recording from spinal interneurons have confirmed this hypothesis (Lungberg and Vyklicky 1966; Engberg et al., 1968a, 1968b; Jankowska et al., 1968). Using the spike-trigger averaging technique, Takakusaki et al. (1989) found that stimulation in the medial medulla generated IPSPs in spinal motoneurons with a segmental delay of 1.3 ms, indicating at least one interneurons is involved.

Electrical stimulation in the medial pons (Fung et al., 1982) and NGC (Chase et al., 1986) induces hyperpolarization of spinal motoneuron in intact animals. This brainstem-induced hyperpolarization of spinal motoneuron has been hypothesized to be "state-dependent". In waking and quiet sleep, stimulation in the pons produced a short latency of hyperpolarizing potential of hindlimb motoneuron. However, this same stimulation elicits not only a short latency IPSP but also a following longer latency IPSP in the same motoneuron during REM sleep (Fung et al., 1982). Similar findings have been reported from stimulation in the NGC. Chase et al., (1986) report that 59% of spinal motoneurons produced a small IPSP and another 41% of the cells produced no apparent change in membrane potential with NGC stimulation in waking and quiet sleep. On the other hand, the same NGC stimulation in REM sleep produced a potent IPSP in all recorded motoneurons.

For decades, muscle tone inhibition has been hypothesized due to active inhibition instead of disfacilitation (Giaquinto et al., 1964; Gottesmann, 1997). Glenn and Dement (1981) found that an increases in IPSPs but no decreases in EPSP in the spinal motoneurons in REM sleep. This finding led them to suggest that muscle atonia during REM sleep is mediated by the activation of the medial medullary inhibitory system. However, monoaminergic neurons in the pons may play a role in muscle tone suppression through a disfacilitatory effect on spinal motoneurons. It has been shown that blockage of LC neuron activity by local injection of clonidine produced muscle tone suppression, ipsilateral to the injection side (Pompeiano et al., 1987). Unilateral electrolytic lesion of the LC also produces a reduction of muscle tone ipsilateral to the lesion side (D'Ascanio et al., 1989). This LC effect on muscle activity led the authors to suggest that decreases in LC neuron activity lead to increases in pontine cholinoceptive neuron activity, which in turn increases the firing of medullary inhibitory neurons thereby producing muscle atonia (Horn et al., 1987). However, LC has a strong projection to the spinal cord (Hancock and Fougerousse, 1976; Nygren and Olson, 1977; Westlund and Coulter, 1980; Bjouklund and Skagerberg, 1982; Carlton et al., 1985; Lai and Barnes, 1985; Zhuo et al., 1992; Fung et al., 1994; Liu et al., 1995), and exerts facilitatory effect on Motoneuron (Fung and Barnes, 1981, 1987; Lai et al., 1989). Therefore, reduction of muscle tone could result directly from withdrawal of the LC facilitatory effect on α motoneurons.

Neurotransmitters Involved in Inhibition of Motor Activity

Table 3 and Fig. 3 summarize neurotransmitters involved in muscle tone suppression in different areas of the brainstem.

Table 3. Effect of chemical injection in the brainstem on muscle activity

	Pons	NMC	NPM
Ach	—	0	—
CRF	—	—	0
nonNMDA	—	—	0
NMDA	+	+	0

+: facilitation; —: inhibition; 0: no effect.
Ach: acetylcholine; CRF: corticotropin-releasing factor.

Cholinergic mechanism

It has been well established that cholinergic mechanisms participate in muscle tone suppression. This cholinergic induced muscle atonia can be achieved by either systemic administration or local microinjection of cholinergic agonists. Systemic injection of cholinergic agonists (Matsuzaki et al., 1968; Matsuzaki, 1969) or of an anticholinesterase (Pompeiano and Hoshino, 1976a, 1976b) in the decerebrate cat induced muscle atonia and eye movement. This systemic cholinergic effect on muscle tone suppression has been suggested to be mediated by activation of PIA neurons.

In 1964, George et al. localized the atonia area by local injection of cholinergic agonists into the brainstem in the intact cat. Oxotremorine or carbachol injected into the dorsal pons produced muscle atonia, and this cholinergic effect on muscle activity could be blocked by prior injection of atropine. Behaviorally, animals were awake despite their muscle atonia. Subsequent studies by van Dongen et al. (1978) confirmed that carbachol

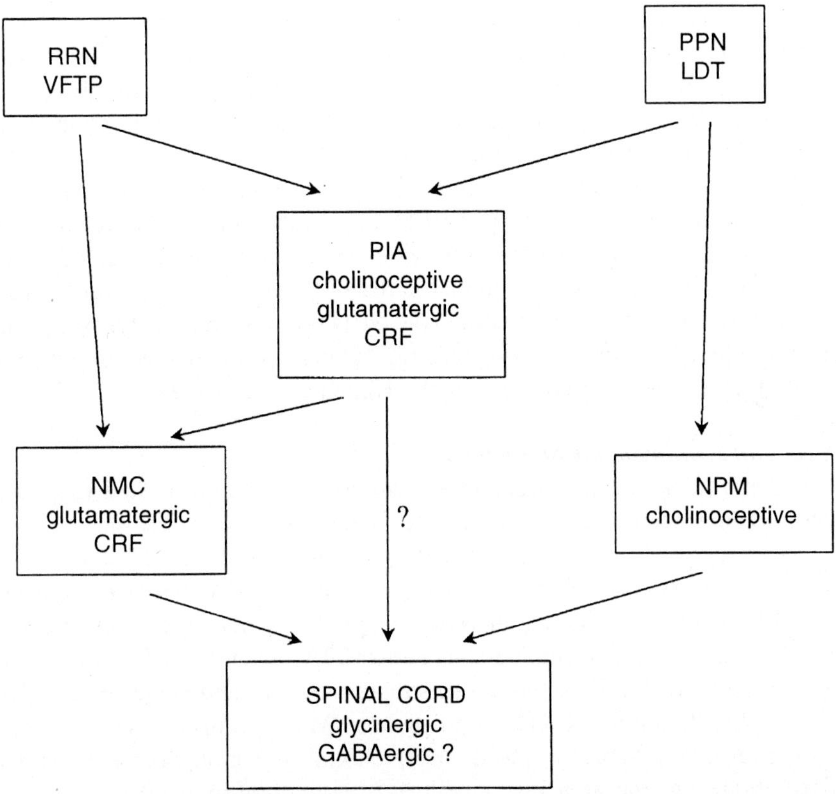

Fig. 3. Putative neurotransmitters involved in muscle atonia. The PIA received glutamatergic fibers from the RRN, VFTP, PPN and LDT, while cholinergic and CRF input fibers may originate from PPN and LDT. The PPN and LDT also send fibers to NPM. In NMC, glutamatergic innervation comes from RRN, VFTP and PIA. In spinal cord, both glycinergic and GABAergic inputs originate from NMC, and also possible from PIA. The question mark from PIA to spinal cord represents an unknown inhibitory pathway.

injection into the area ventromedial to the LC produced waking with atonia. Similar findings have been also reported in the cat and rat (Baghdoyan et al., 1984; Katayama et al., 1984; Gnadt and Pegram, 1986; Taguchi et al., 1992; Lopez-Rodriguez et al., 1995). These results indicate that muscarinic cholinergic receptors are involved in pontine-induced atonia. Further studies using specific agonists and antagonists microinjected into the pons, have determined that the M_2 muscarinic system plays a critical role in pontine-induced atonia (Velazquez-Moctezuma et al., 1989, 1991; Imeri et al., 1992). In the more rostroventral pons, carbachol injection not only produces long lasting muscle atonia but also increases in REM sleep (Reinoso-Suarez et al., 1990). Atonia with waking will be found when cholinergic agonists injected into the dorsal pons, while ventral pontine injection produces REM sleep with atonia (Rodrigo-Angulo et al., 1994).

The NPM of the caudomedial medulla also generates muscle tone suppression when cholinergic agonists are injected into it (Lai and Siegel, 1988). This cholinergic induced muscle atonia could be blocked by prior local injection of atropine, suggesting that a muscarinic receptor is involved (Lai and Siegel, 1988). The NPM area has been demonstrated to receive cholinergic innervation from the PPN (Shiromani et al., 1988).

The microdialysis technique has provided another line of evidence regarding the neurotransmitters involving in REM sleep and atonia. In the free moving animals, acetylcholine release in the pons is increased during natural REM sleep (Kodama et al., 1990) or pontine carbachol-induced muscle atonia (Lydic et al., 1991). Acetylcholine release in the NPM is also increased during muscle atonia induced by either pontine electrical stimulation or carbachol injection in both intact and decerebrate animals (Kodama et al., 1992).

Excitatory amino acids and peptide

Our experiments on the decerebrate animals have demonstrated that non-cholinergic mechanism also participates in muscle atonia in the medial pons. Microinjection of corticotropin-releasing factor (CRF; Lai and Siegel, 1992) and glutamate (Lai and Siegel, 1988) agonists into the PIA, in which carbachol and acetylcholine injection induced atonia, produces muscle tone suppression. Further study has determined that nonNMDA but not NMDA glutamate agonists are involved in muscle tone suppression (Lai and Siegel, 1991). The latency and duration of CRF, glutamate nonNMDA agonists-induced muscle atonia are comparable to that of acetylcholine injection (Table 4). In the decerebrate rat, the area responsible for cholinergic-induced muscle tone suppression in the pons is also responsible for nonNMDA agonists inducing atonia (Hajnik et al., 1996). This pontine nonNMDA effect on muscle activity has been confirmed by Onoe and Sakai (1995). They find that the nonNMDA agonists, kainate and AMPA, but not NMDA agonists microinjected into the pons induce increases in REM sleep and atonia in intact animals.

Table 4. Latency and duration of muscle tone suppression induced by agonist injection in the pons and medulla

	Latency (sec)			Duration (min)		
	Pons	NMC	NPM	Pons	NMC	NPM
Ach	25	—	34	7.7	—	4.1
CRF	27	19	—	10.6	4.1	—
Glut	24	18	—	12.9	4.2	—
KA	37	29	—	9.7	4.5	—
QA	26	21	—	14.3	13.3	—

Glut: glutamate; KA: kainic acid; QA: quesqualic acid.

In the medial medulla, we have identified 2 distinct areas, a rostral NMC glutamatergic/CRF and a caudal NPM cholinoceptive (see above), participating in muscle tone suppression (Lai and Siegel, 1988; 1992). As in the pons, the NMC area also responses to glutamate and CRF injection. Glutamate agonists injected into the NMC but not the NPM produce muscle tone suppression bilaterally, and this effect can be blocked by glutamate antagonists, L-glutamic acid diethylester and gamma-D-glutamyglycine. As in the pons, nonNMDA agonists kainate, quisqualate and willardine but not NMDA agonists microinjected into the NMC elicit muscle atonia (Lai and Siegel, 1991). This nonNMDA agonist-induced atonia can be blocked by specific nonNMDA antagonists, 6-cyano-7-nitroquinoxaline-2, 3-dione and 6,7-dinitroquinoxaline-2, 3-dione. As with the nonNMDA agonist, CRF injected into the NMC but not the NPM elicits muscle tone suppression (Lai and Siegel, 1992). On the other hand, cholinergic agonists microinjected into the NMC fails to induce muscle tone suppression (Lai and Siegel, 1988).

In contrast to the nonNMDA agonist-induced muscle atonia, NMDA agonists microinjected into the PIA and NMC produced either increases in muscle tone or locomotion (Lai and Siegel, 1991). We have hypothesized that imbalance of NMDA and nonNMDA receptor activity in the NMC could contribute to increases in the generation of phasic events superimposed on the muscle atonia of REM sleep. Further support for this hypothesis has come from our recent study showing that ventral mesopontine junction lesion-induced muscle twitches (Lai and Siegel, 1997a) can be attenuted or blocked by nonNMDA agonists injection into the NMC (Lai and Siegel, 1997b). Anatomically, the NMC received glutamatergic afferent projections from the ventral mesopontine junction (Lai et al., 1993b).

Inhibitory amino acid

At the spinal cord level, extensive studies on neurotransmitters participating in motoneuron inhibition has been performed by Chase and colleagues. With intracellular recording from the lumbar motoneurons and iontophoretic application of chemicals, they found that the medulla-induced IPSPs in the motoneurons during active sleep were abolished or partially suppressed by strychnine, a glycine antagonist (Soja et al., 1987, 1991). This glycinergic effect in spinal motoneurons could be also seen in decerebrate animals during carbachol-induced REM-like activity (Morales et al., 1987). On the other hand, GABA antagonists failed to block these motoneuron IPSPs.

Spinal interneurons including Renshaw cells and I_a inhibitory neurons exert inhibitory effect on motoneurons mediated by GABA and glycine (Cullheim and Kellerth, 1981). Glycinergic spinal Renshaw cells and I_a inhibitory interneurons have been identified (Fyffe, 1991). However, the glycinergic mechanism involved in motoneuron IPSP generation in REM sleep may primarily originate from supraspinal structures since activity of the majority of Renshaw cell (85%) was suppressed (Morales et al., 1988), and only a few glycinergic spinal interneurons were double labelled with c-fos (Rampon et al., 1997) by carbachol administered into the pons which induced muscle atonia. Glycinergic neurons have been identified in the ventromedial medulla (Fort et al., 1993). Fifteen percent of the descending terminals originating from the NMC contained glycine (Holstege and Bongers, 1991). Although electrophysiological study failed to demonstrate that GABAergic system involved in motoneuron suppression (Soja et al., 1987), immunohistochemical study has identified that 40% of the NMC descending projection terminals was GABAergic (Holstege, 1991). Meanwhile, co-localization of glycine and GABA could be seen in terminals on cat lumbar motoneurons (Taal and Holstege, 1994).

Conclusions

Two mechanisms, active inhibition and disfacilitation, mediate muscle atonia during REM sleep. Active inhibition originates from the medial reticular formation including PIA, NGC, NMC and NPM, while disfacilitatory mechanism originates from the dorsolateral pons in the LC complex. REM-on cells have been found in the medial reticular formation and REM-off cells have been recorded from the LC complex. At the spinal cord level, hyperpolarization of the spinal motoneuron in REM sleep results partially from glycinergic inhibition triggered by neurons in the NMC and NPM. Simultaneous activates of NMDA and nonNMDA receptors may be responsible for phasic activity superimposed on atonia during REM sleep.

References

Alderson, A.M. and Downman, C.B.B. Supraspinal inhibition of thoracic reflexes of somatic and visceral origin, Arch. ital. Biol., 104 (1966) 309–327.

Aston-Jones, G. and Bloom, F.E. Activity of norepinephrine-containing locus coeruleus neurons in behaving rats anticipates fluctuations in the sleep-waking cycle, J. Neurosci., 1 (1981) 876–886.

Aston-Jones, G., Valentino, R.J., Van Bockstaele, E. and Meyerson, A. Locus coeruleus, stress, and PTSD: Neurobiological and clinical parallels. In: M. Murburg (Ed.), *Catecholamine function in post-traumatic stress disorder,* American Psychiatric Press, Washington DC, 1994, pp. 17–62.

Bach, L.M.N. Effect of bulbar facilitation and inhibition on peripheral reflex inhibition, J. Neurophysiol., 13 (1950) 259–263.

Baghdoyan, H.A., Rodrigo-Angulo, M.L., McCarley, R.W. and Hobson, J.A. Site-specific enhancement and suppression of desynchronized sleep signs following cholinergic stimulation of three brainstem regions, Brain Res., 306 (1984) 39–52.

Bayev, K.V., Beresovskii, V.K., Kebkalo, T.G. and Savoskina, L.A. Afferent and efferent connections of brainstem locomotor regions: Study by means of horseradish peroxidase transport technique, Neurosci., 26 (1988) 871–891.

Berman, A.L., The brain stem of the cat. Univ. Wisconsin Press, Madison, Wisconsin, 1968.

Bjorklund, A. and Skagerberg, G. Descending monoaminergic projections to the spinal cord. In: B. Sjolund and A. Bjorklund (Eds.) *Brain stem control of spinal mechanisms,* Elsevier, Amsterdam, 1982, pp. 55–88.

Carli, G. and Zanchetti, A. A study of pontine lesions suppressing deep sleep in the cat, Arch. ital. Biol., 103 (1965) 751–788.

Carlton, S.M., Chung, J.M., Leonard, R.B. and Willis, W.D. Funicular trajectories of brainstem neurons projecting to the lumbar spinal cord in the monkey (*Macaca fascicularis*): A retrograde labeling study, J. Comp. Neurol., 241 (1985) 382–404.

Chalmers, J.P., Blessing, W.W., West, M.J., Howe, P.R.C., Costa, M. and Furnes, J.B. Importance of new catecholamine pathway in control of blood pressure, Clin. Exp. Hypertension, 3 (1981) 393–416.

Chase, M.H., Morales, F.R., Boxer, P.A., Fung, S.J. and Soja, P.J. Effect of stimulation of the nucleus reticularis gigantocellularis on the membrane potential of cat lumbar motoneurons during sleep and wakefulness, Brain Res., 386 (1986) 237–244.

Chase, M.H. and Wills, N. Brain stem control of masseteric reflex activity during sleep and wakefulness: Medulla, Exp. Neurol., 64 (1979) 118–131.

Chiba, T. and Murata, Y. Afferent and efferent connections of the medial preoptic area in the rat: A WGA-HRP study, Brain Res. Bull., 14 (1985) 261–272.

Cullheim, S. and Kellerth, J.O. Two kinds of recurrent inhibition of cat spinal alpha-motoneurones as differentiated pharmacologically, J. Physiol., 312 (1981) 209–224.

D'Ascanio, P., Pompeiano, M. and Tononi, G. Inhibition of vestibulospinal reflexes during the episodes of postural atonia induced by unilateral lesion of the locus coeruleus in the decerebrate cat, Arch. ital. Biol., 127 (1989) 81–97.

Deutch, A.Y., Goldstein, M., Baldino, F., Jr. and Roth, R.H. Telencephalic projections of the A8 dopamine cell group. In: P.W. Kalivas and C.B. Nemeroff (Eds.), *The mesocorticolimbic dopamine system, Ann. New York Acad. Sci. Vol. 537,* New York, 1988, pp. 27–50.

Engberg, I., Lundberg, A. and Ryall R.W. Reticulospinal inhibition of transmission in reflex pathways, J. Physiol., 194 (1968a) 201–233.

Engberg, I., Lundberg, A. and Ryall, R.W. Reticulospinal inhibition of interneurones, J. Physiol., 194 (1968b) 225–236.

Foote, S.L., Aston-Jones, G. and Bloom, F.E. Impulse activity of locus coeruleus neurons in awake rats and monkeys is a function of sensory stimulation and arousal, Proc. Natl. Acad. Sci. USA, 77 (1980) 3033–3037.

Fort, P., Luppi, P.H. and Jouvet, M. Glycine-immunoreactive neurones in the cat brain stem reticular formation, NeuroReport, 4 (1993) 1123–1126.

Friedman, L. and Jones, B. E. Computer graphics analysis of sleep-wakefulness state changes after pontine lesions, Brain Res. Bull., 13 (1984) 53–68.

Fung, S.J. and Barnes, C.D. Evidence of facilitatory coerulospinal action in lumbar motoneurons of cats, Brain Res., 216 (1981) 299–311.

Fung, S.J. and Barnes, C.D. Membrane excitability changes in hindlimb motoneurons induced by stimulation of the locus coeruleus in cats, Brain Res., 402 (1987) 230–242.

Fung, S.J., Boxer, P.A., Morales, F.R. and Chase, M.H. Hyperpolarizing membrane responses induced in lumber motoneurons by stimulation of the nucleus reticularis pontis oralis during active sleep, Brain Res., 24: (1982) 267–273.

Fung, S.J., Reddy, V.K., Zhuo, H., Liu, R.–H. and Barnes, C.D. Anatomical evidence for the presence of glutamate or enkephalin in noradrenergic projection neurons of the locus coeruleus, Microscopy Res. Tech., 29 (1994) 219–225.

Gallager, D.W. and Pert, A., Afferents to brain stem nuclei (brain stem raphe, nucleus reticularis pontis caudalis and nucleus gigantocellularis) in the rat as demonstrated by microiontophoretically applied horseradish peroxidase, Brain Res., 144 (1978) 257–275.

Gassel, M.M., Marchiafava, P.L. and Pompeiano, O. Tonic and phasic inhibition of spinal reflexes during deep, desynchronized sleep in unrestrained cats, Arch. ital. Biol., 102 (1964) 471–499.

Gassel, M.M., Marchiafava, P.L. and Pomeiano, O. An analysis of the supraspinal influences acting on motoneurons during sleep in the unrestrained cat. Modification of the recurrent discharge of the alpha motoneurons during sleep, Arch. ital. Biol., 103 (1965) 25–44.

George, R., Haslett, W.L. and Jenden, D.J. A cholinergic mechanism in the brainstem reticular formation: Induction of paradoxical sleep, Int. J. Neuropharmacol., 3 (1964) 541–552.

Giaquinto, S., Pompeiano, O. and Somogyi, I. Descending inhibitory influences on spinal reflexes during natural sleep, Arch. ital. Biol., 102 (1964) 282–307.

Glenn, L.L. and Dement, W.C. Membrane resistance and rheobase of hindlimb motoneurons during wakefulness and sleep, J. Neurophysiol., 46 (1981) 1076–1088.

Gnadt, J.W. and Pegram, G.V. Cholinergic brainstem mechanisms of REM sleep in the rat, Brain Res., 384 (1986) 29–41.

Gottesmann, C. Introduction to the neurophysiological study of sleep: Central regulation of skeletal and ocular activities, Arch. ital. Biol., 135. (1997) 279–314.

Gurtu, S., Pant, K.K., Sinha, J.N. and Bhargava, K.P. An investigation into the mechanism of cardiovascular responses elicited by electrical stimulation of locus coeruleus and subcoeruleus in the cat, Brain Res., 301 (1984) 59–64.

Hajnik, T., Lai, Y.Y. and Siegel, J.M. Atonia related regions in the rodent pons and medulla, Sleep Res., 24A (1995) 26.

Hajnik, T., Lai, Y.Y. and Siegel, J.M. The atonia related regions in the rodent pons are glutamate sensitive, APSS Abstr., (1996) 133.

Hancock, M.B. and Fougerousse, C.L. Spinal projections from the nucleus locus coeruleus and nucleus subcoeruleus in the cat and monkey as demonstrated by the retrograde transport of horseradish peroxidase, Brain Res. Bull., 1 (1976) 229–234.

Hendricks, J.C., Morrison, A.R. and Mann, G.L. Different behaviors during paradoxical sleep without atonia depend on pontine lesion site, Brain Res., 239 (1982) 81–105.

Henley, K. and Morrison, A.R. A re-evaluation of the effects of lesions of the pontine tegmentum and locus coeruleus on phenomena of paradoxical sleep in the cat, Acta Neurobiol. Exp., 34 (1974) 215–232.

Hobson, J.A., McCarley, R.W. and Wyzinski, P.W. Sleep cycle oscillation: Reciprocal discharge by two brainstem neuronal groups, Science, 189 (1975) 55–58.

Holstege, G. Some anatomical observations on the projections from the hypothalamus to brainstem and spinal cord: An HRP and autoradiographic tracing study in the cat, J. Comp. Neurol., 260 (1987) 98–126.

Holstege, G., Descending motor pathways and the spinal motor system: Limbic and non-limbic components. In: G. Holstege (Ed.), *Role of the forebrain in sensation and behavior, Prog. Brain Res., 87,* Elsevier, Amsterdam, 1991, pp.307–421.

Holstege, G. and Kuypers, H.G.J.M. The anatomy of brain stem pathways to the spinal cord in cat. A labeled amino acid tracing study. In: H.G.J.M. Kuypers and G.F. Martin (Eds.), *Descending pathways to the spinal cord, Prog. Brain Res., 57,* Elsevier, Amsterdam, 1982, pp. 145–175.

Holstege, G., Kuypers, H.G.J.M. and Boer, R.C. Anatomical evidence for direct brain stem projections to the somatic motoneuronal cell groups and autonomic preganglionic cell groups in cat spinal cord, Brain Res., 171 (1979) 329–333.

Holstege, G., Meiners, L. and Tan, K. Projections of the bed nucleus of the stria terminalis to the mesencephalon, pons, and medulla oblongata in the cat, Exp. Brain Res., 58 (1985) 379–391.

Holstege, J.C. Ultrastructural evidence for GABAergic brain stem projections to spinal motoneurons in the rat. J. Neurosci., 11 (1991) 159–167.

Holstege, J.C. and Bongers, C.M.H.A glycinergic projection from the ventromedial lower brainstem to spinal motoneurons. An ultrastructural double labeling study in rat, Brain Res., 566 (1991) 308–315.

Holstege, J.C. and Kuypers, H.G.J.M. Brainstem projections to spinal motoneurons: An update, Neurosci., 23 (1987) 809–821.

Hopkins, D.A. and Holstege, G. Amygdaloid projections to the mesencephalon, pons and medulla oblongata in the cat, Exp. Brain Res., 32 (1978) 529–547.

Horn, E., D'Ascanio, P., Pompeiano, O. and Stampacchia, G. Pontine reticular origin of cholinergic excitatory afferents to the locus coeruleus controlling the gain of vestibulospinal and cervicospinal reflexes in decerebrate cats, Arch. ital. Biol., 125 (1987) 273–304.

Hosoya, Y. Hypothalamic projections to the ventral medulla oblongata in the rat, with special reference to the nucleus raphe pallidus: A study using autoradiographic and HRP techniques, Brain Res., 344 (1985) 338–350.

Imeri, L., Bianchi, S., Angeli, P. and Mancia, M. Differential effects of M_2 and M_3 muscarinic antagonists on the sleep-wake cycle, NeuroReport, 2 (1992) 383–385.

Iwakiri, H., Oka, T. and Mori, S. Multi-segmental and generalized suppression of postural muscle tone evoked by stimulating the dorsal tegmental field and the medial pontine reticular formation in acute decerebrate cats, Neurosci. Lett., 182 (1994) 83–86.

Iwakiri, H., Oka, T., Takakusaki, K. and Mori, S. Stimulus effects of the medial pontine reticular formation and the mesencephalic locomotor region upon medullary reticulospinal neurons in acute decerebrate cats, Neurosci. Res., 23 (1995) 47–53.

Jankowska, E., Lund, S., Lundberg, A. and Pompeiano, O. Inhibitory effects evoked through ventral reticulospinal pathways, Arch. ital. Biol., 106 (1968) 124–140.

Johannessen, J.N., Watkins, L.R. and Mayer, D.J. Non-serotonergic origins of the dorsolateral funiculus in the rat ventral medulla, J. Neurosci., 4 (1984) 757–766.

Jones, B.E., Elimination of paradoxical sleep by lesions of the pontine gigantocellular tegmental field in the cat, Neurosci. Lett., 13 (1979) 285–293.

Jones, B.E. and Yang, T.-Z. The efferent projections from the reticular formation and the locus coeruleus studied by anterograde and retrograde axonal transport in the rat, J. Comp. Neurol., 242 (1985) 56–92.

Jones, S.L., Descending noradrenergic influences on pain. In: C.D. Barnes and O. Pomeiano (Eds.), *Neurobiology of the locus coeruleus, Prog. Brain Res. Vol 88*, Elsevier, Amsterdam, 1991, pp. 381–394.

Jouvet, M. Recherches sur les structures nerveuses et les mecanismes responsables des differentes phases du sommeil physiologique, Arch. ital. Biol., 100 (1962) 125–206.

Jouvet, M. and Delorme, F. Locus coeruleus et sommeil paradoxal, C.R. Soc. Biol., 159 (1965) 895–899.

Jouvet, M. and Michel, F. Correlations electromyographiques du sommeil chez le chat, C.R. Soc. Biol., Paris, 153 (1959) 422–425.

Kanamori, N., Sakai, K. and Jouvet, M. Neuronal activity specific to paradoxical sleep in the ventromedial medullary reticular formation of unrestrained cats, Brain Res., 189 (1980) 251–255.

Katayama, Y., DeWitt, D.S., Becker, D.P. and Hayes, R.L. Behavioral evidence for a cholinoceptive pontine inhibitory area: Descending control of spinal motor output and sensory input, Brain Res., 296 (1984) 241–262.

Kinjo, N., Atsuta, Y., Webber, M., Kyle, R., Skinner, R.D. and Garcia-Rill, E. Medioventral medulla-induced locomotion, Brain Res. Bull., 24 (1990) 509–516.

Kodama, T., Lai, Y.Y. and Siegel, J.M. Enhancement of acetylcholine release during REM sleep in the caudomedial medulla as measured by in vivo microdialysis, Brain Res., 580 (1992) 348–350.

Kodama, T., Takahashi, Y. and Honda, Y. Enhancement of acetylcholine release during paradoxical sleep in the dorsal tegmental field of the cat brain stem, Neuronsci. Lett., 144 (1990) 277–282.

Kuypers, H.G.J.M. and Huisman, A.M. The new anatomy of the descending brain pathways. In: B. Sjolund and A. Bjorklund (Eds.), *Brain stem control of spinal mechanisms,* Elsevier, Amsterdam, 1982, pp. 29–54.

Kuypers, H.G.J.M. and Maisky, V.A. Retrograde axonal trasport of horseradish peroxidase from spinal cord to brain stem cell groups in the cat, Neurosci. Lett., 1 (1975) 9–14.

Kuypers, H.G.J.M. and Maisky, V.A. Funicular trajectories of descending brain stem pathways in cat, Brain Res., 136 (1977) 159–165.

Lai, Y.Y. and Barnes, C.D. A spinal projection of serotonergic neurons of the locus coeruleus in the cat, Neurosci. Lett., 58 (1985) 159–164.

Lai, Y.Y. and Siegel,, J.M. Medullary regions mediating atonia, J. Neurosci., 8 (1988) 4790–4796.

Lai, Y.Y. and Siegel, J.M. Muscle tone suppression and stepping produced by stimulation of midbrain and rostral pontine reticular formation, J. Neurosci., 10 (1990) 2727–2734.

Lai, Y.Y. and Siegel, J.M. Pontomedullary glutamate receptors mediating locomotion and muscle tone suppression, J. Neurosci., 11 (1991) 2931–2937.

Lai, Y.Y. and Siegel, J.M. Corticotropin-releasing factor mediated muscle atonia in pons and medulla, Brain Res., 575 (1992) 63–68.

Lai, Y.Y. and Siegel, J.M. Brainstem-mediated locomotion and myoclonic jerks. I. Neural substrates, Brain Res., 745 (1997a) 257–264.

Lai, Y.Y. and Siegel, J.M. Brainstem-mediated locomotion and myoclonic jerks. II. Pharmacological effects, Brain Res., 745 (1997b) 265–270.

Lai, Y.Y., Clemente, J.R. and Siegel, J.M. Glutamatergic and cholinergic projections to the pontine inhibitory area identified with horseradish peroxidase retrograde transport and immunohistochemistry, J. Comp. Neurol., 336 (1993a) 321–330.

Lai, Y.Y., Clemente, J.R. and Siegel, J.M. Afferents to the nucleus magnocellularis of the medulla, Soc. Neurosci. Abstr., 19 (1993b) 1437.

Lai, Y.Y., Siegel, J.M. and Wilson, W.J. Effect of blood pressure on changes in muscle tone poduced by stimulation of the medial medulla, Am. J. Physiol., 252 (1987) H1249–H1257.

Lai, Y.Y., Strahlendorf, H.K., Fung, S.J. and Barnes, C.D. The actions of two monoamines on spinal motoneurons from stimulation of the locus coeruleus in the cat, Brain Res., 484 (1989) 268–272.

Lai, Y.Y., Wu, J.-P., Horng, H.T., Wang, S.D., Chai, C.Y., Kuo, J.S. and Siegel, J.M. Projections to the pontine reticular formation from the diencephalon and medulla: WGA-HRP and catecholamine immunohistochemistry studies, Sleep Res., 23 (1994) 50.

Lettvin, J.Y. and Dell, P.C. Positivity in ventral horn during bulbar reticular inhibition of motoneurons, Fed. Proc., 9 (1950) 77.

Liu, R.-H., Fung, S.J., Reddy, V.K. and Barnes, C.D. Localization of glutamatergic neurons in the dorsolateral pontine tegmentum projecting to the spinal cord of the cat with a proposed role of glutamate on lumbar motoneuron activity, Neurosci., 64 (1995) 193–208.

Llinas, R. and Terzuolo, C.A. Mechanisms of supraspinal actions upon spinal cord activities. Reticular inhibitory mechanisms on alpha-extensor motoneurons, J. Neurophysiol., 27 (1964) 579–591.

Llinas, R. and Terzuolo, C.A. Mechanisms of supraspinal actions upon spinal cord activities. Reticular inhibitory mechanisms upon flexor motoneurons, J. Neurophysiol., 28 (1965) 413–422.

Lopez-Rodroguez, F., Kohlmeier, K.A., Yamuy, J., Morales, F.R. and Chase, M.H. Muscle atonia can be induced by carbachol injections into the nucleus pontis oralis in cats anesthetized with α-chloralose, Brain Res., 699 (1995) 201–207.

Lungberg, A. and Vyklicky, L. Inhibition of transmission to primary afferents by electrical stimulation of the brain stem, Arch. ital. Biol., 104 (1966) 86–97.

Lydic, R., Baghdoyan, H.A. and Lorinc, Z. Microdialysis of cat pons reveals enhanced acetylcholine release during state-dependent respiratory depression, Am. J. Physiol., 261 (1991) R766–R770.

Luppi, P.-H., Sakai, K., Fort, P., Salvert, D. and Jouvet, M. The nuclei of origin of monoaminergic, peptidergic, and cholinergic afferents to the cat nucleus reticularis magnocellularis: A double-labeling study with cholera toxin as a retrograde tracer, J. Comp. Neurol., 277 (1988) 1–20.

Magoun, H.W. Caudal and cephalic influences of the brain stem reticular formation, Physiol. Rev., 30 (1950) 459–474.

Magoun, H.W. and Rhines, R. An inhibitory mechanism in the bulbar reticular formation, J. Neurophysiol., 9 (1946) 165–171.

Margalit, D. and Segal, M. A pharmacologic study of analgesia produced by stimulation of the nucleus locus coeruleus, Psychopharmacol., 62 (1979) 169–183.

Matsuyama, K., Kobayashi, Y., Takakusaki, K., Mori, S. and Kimura, H. Termination mode and branching patterns of reticuloreticular and reticulospinal fibers of the nucleus reticularis pontis oralis in the cat: an anterograde PHA-L tracing study, Neurosci. Res., 17 (1993) 9–21.

Matsuyama, K., Nakajima, K., Miyashita, E. and Mori, S. Spinal innervation patterns of the nucleus reticularis magnocellularis in the cat, Fourth IBRO World Congr. Neurosci., (1995) 331.

Matsuyama, K., Ohta, Y. and Mori, S. Ascending and descending projections of the nucleus reticularis gigantocellularis in the cat demonstrated by the anterograde neural tracer, Phaseolus vulgaris leucoagglutinin (PHA-L), Brain Res., 460 (1988) 124–141.

Matsuyama, K., Takakusaki, K., Nakajima, K. and Mori, S. Multi-segmental innervation of single pontine reticulospinal axons in the cervico-thoracic region of the cat: Anterograde PHA-L tracing study, J. Comp. Neurol., 377 (1997) 234–250.

Matsuzaki, M. Differential effects of sodium butyrate and physostigmine upon the activities of para-sleep in acute brain stem preparations, Brain Res., 13 (1969) 247–265.

Matsuzaki, M., Okada, Y. and Shuto, S. Cholinergic agents related to para-sleep state in acute brain stem preparations, Brain Res., 9 (1968) 253–267.

Mitani, A., Ito, K., Mitani, Y. and McCarly, R.W. Descending projections from the gigantocellular tegmental field in the cat: Cells of origin and their brainstem and spinal cord trajectories, J. Comp. Neurol., 268 (1988) 546–566.

Mokha, S.S., McMillan, J.A. and Iggo, A. Descending control of spinal nociceptive transmission: actions produced on spinal multireceptive neurons from the nuclei locus coeruleus (LC) and raphe magnus (NRM), Exp. Brain Res., 58 (1985) 213–225.

Morales, F.R., Engelhardt, J.K., Soja, P.J., Pereda, A.E. and Chase, M.H. Motoneuron properties during motor inhibition produced by microinjection of carbachol into the pontine reticular formation of the decerebrate cat, J. Neurophysiol., 57 (1987) 1118–1129.

Morales, F.R., Engelhardt, J.K., Pereda, E.P., Yamuy, J. and Chase, M.H. Renshaw cells are inactive during motor inhibition elicited by pontine micorinjection of carbachol, Neurosci, Lett., 86 (1988) 289–295.

Morrison, A.R. Paradoxical sleep without atonia, Arch. ital. Biol., 126 (1988) 275–289.

Newman, D.B., Distinguishing rat brainstem reticulospinal nuclei by their neuronal morphology. I. Medullary nuclei, J. Hirnforsch., 26 (1985) 187–226.

Niemer, W.T. and Magoun, H.W. Reticulo-spinal tracts influencing motor activity, J. Comp. Neurol., 87 (1947) 367–379.

Noga, B.R., Kettler, J. and Jordan, L.M. Locomotion produced in mesencephalic cats by injections of putative transmitter substances and antagonists into the medial reticular formation and the pontomedullary locomotor strip, J. Neurosci., 8 (1988) 2074–2086.

Nyberg-Hansen, R. Sites and mode of termination of reticulo-spinal fibers in the cat. An experimental study with silver impregnation methods, J. Comp. Neurol., 124 (1965) 71–100.

Nyberg-Hansen, R. Functional organization of descending supraspinal fibre systems to the spinal cord. Anatomical observations and physiological correlations, Ergebn, Anat. Entwickl., 39 (1966) 6–47.

Nygren, L.-G. and Olson, L. A new major projection from locus coeruleus: The main source of noradrenergic nerve terminals in the ventral and dorsal coulmns of the spinal cord, Brain Res., 132 (1977) 85–93.

Ohta, Y., Mori, S. and Kimura, H. Neuronal structures of the brainstem participating in postural suppression, Neurosci. Res., 5 (1988) 181–202.

Onoe, H. and Sakai, K. Kainate receptors: A novel mechanism in paradoxical (REM) sleep generation, NeuroReport, 6 (1995) 353–356.

Pereda, A.E., Morales, F.R. and Chase, M.H. Medullary control of lumbar motoneurons during carbachol-induced motor inhibition, Brain Res., 514 (1990) 175–179.

Pompeiano, O. and Hoshino, K. Central control of posture: reciprocal discharge by two pontine neuronal groups leading to suppression of decerebrate rigidity, Brain Res., 116 (1976a) 131–138.

Pompeiano, O. and Hoshino, K. Tonic inhibition of dorsal pontine neurons during the postural atonia produced by an anticholinesterase in the decerebrate cat, Arch. ital. Biol., 114 (1976b) 310–340.

Pompeiano, O., D' Ascanio, P., Horn, E. and Stampacchia, G. Effects of local injection of the a_2-adrenergic agonist clonidine into the locus coeruleus complex on the gain of vestibulospinal and cervicospinal reflexes in decerebrate cats, Arch. ital. Biol., 125 (1987) 225–289.

Price, J.L. and Amaral, D.G. An autoradiographic study of the projections of the central nucleus of the monkey amygdala, J. Neurosci., 1 (1981) 1242–1259.

Rampon, C., Morales, F.R., Fort, P., Luppi, P.H., Sampogna, S. and Chase, M. H. Glysine immunohistochemistry and c-fos expression in the cervical spinal cord of cats during carbachol-induced active sleep, Soc. Neurosci. Abstr., 23 (1997) 795.

Reinoso-Suraez, F., De Andres, I., Rodrigo-Angulo, M.L. and Rodriguez-Viega, E. Location and anatomical connections of a paradoxical sleep induction site in the cat ventral pontine tegmentum, Eur. J. Neurosci., 6 (1994) 1829–1836.

Reinoso-Suarez, F., Rodrigo-Angulo, M.L., Rodriguez-Veiga, E. and De Andres, I. Thalamic connections of the oral pontine tegmentum sites whose cholinergic stimulation produces enhancement of paradoxical sleep signs. In: M. Mancia and G. Marini (Eds.), *The diencephalon and sleep*, Raven Press, New York, 1990, pp. 49–63.

Rizvi, T.A., Murphy, A.Z., Ennis, M., Behbehani, M.M. and Shipley, M.T. Medial preoptic area afferents to periaqueductal gray medullo-output neurons: A combined fos and tract tracing study, J. Neurosci., 16 (1996) 333–344.

Sakai, K., Some anatomical and physiological properties of ponto-mesencephalic tegmental neurons with special references to the PGO waves and postural atonia during paradoxical sleep in the cat. In: J.A. Hobson and M.A.B. Brazier (Eds.), *The reticular formation revisited,* Raven Press, New York, 1980, pp. 427–447.

Sakai, K., Executive mechanisms of paradoxical sleep, Arch. ital. Biol., 126 (1988) 239–257.

Sakai, K. and Koyama, Y. Are there cholinergic and non-cholinergic paradoxical sleep-on neurones in the pons?, NeuroReport, 7 (1996) 2449–2453.

Sakai, K., Sastre, J.P., Salvert, D., Touret, M., Tohyama, M. and Jouvet, M. Tegmentoreticular projections with special reference to the muscular atonia during paradoxical sleep in the cat: An HRP study, Brain Res., 176 (1979) 233–254.

Sakai, K., Sastre, J.-P., Kanamore, N. and Jouvet, M. State-specific neurons in the ponto-medullary reticular formation with special reference to the postural atonia during paradoxical sleep in the cat. In: O. Pompeiano and C.A. Marsan (Eds.), *Brain mechanisms of perceptual awareness and purposeful behavior*, Raven Press, New York, 1981, pp. 405–429.

Saper, C.B., Swanson, L.W. and Cowan, W.M. An autoradiographic study of the efferent connections of the lateral hypothalamic area in the rat, J. Comp. Neurol., 183 (1979) 689–706.

Sara, S.J., Devauges, V. and Segal, M. Locus coeruleus engagement in memory retrieval and attention. In: M. Sandler, A. Dahlstrom and R.H. Belmaker (Eds.), *Prog. in catecholamine research, part B: Central aspects*, Alan R Liss, New York, 1988, pp. 155–161.

Sastre, J.-P., Sakai, K. and Jouvet, M. Bilateral lesions of the dorsolateral pontine tegmentum. II. Effects upon the muscle atonia, Sleep Res., 7 (1978) 44.

Schenkel, E. and Siegel, J.M. REM sleep without atonia after lesions of the medial medulla, Neurosci. Lett., 98 (1989) 159–165.

Schenkel, E., Lai, Y.Y. and Siegel, J.M. Cataplexy-like behavior after electrical or chemical stimulation of the medial medulla in the intact cat, APSS Abstr., (1990) 8.

Shammah-Lagnado, S.J., Costa, M., Goto, M., Peracoli, A. and Canteras, N.S. Efferent connections of the retrorubral field (A8): A PHA-L study in the rat, Soc. Neurosci. Abstr., 19 (1993) 1434.

Shiromani, P.J., Kilduff, T.S., Bloom, F.E. and McCarley, R.W. Cholinergically induced REM sleep triggers Fos-like immunoreactivity in dorsolateral pontine regions associated with REM sleep, Brain Res., 580 (1992) 351–357.

Shiromani, P.J., Lai, Y.Y. and Siegel, J.M. Descending projections from the dorsolateral pontine tegmentum to the paramedian reticular nucleus of the caudal medulla in the cat, Brain Res., 517 (1988) 224–228.

Shouse, M.N. and Siegel, J.M. Pontine regulation of REM sleep components in cats: Integrity of the pedunculopontine tegmentum (PPT) is important for phasic events but unnecessary for atonia during REM sleep, Brain Res., 571 (1992) 50–63.

Siegel, J.M., Nienhuis, R., Fahringer, H.M., Paul, R., Shiromani, P., Dement, W.C., Mignot, E. and Chiu, C. Neuronal activity in narcolepsy: identification of cataplexy-related cells in the medial medulla, Science, 252 (1991) 1315–1318.

Sprague, J.M., Schreiner, L.H., Lindsley, D.B., Magoun, H.W. Reticulo-spinal influences on stretch reflexes, J. Neurophysiol. 11: (1948) 501–507.

Soja, P.J. Morales, F.R., Baranyi, A. and Chase, M.H. Effect of inhibitory amino acid antagonists on IPSPs induced in lumbar motoneurons upon stimulation of the nucleus reticularis gigantocellularis during active sleep, Brain Res., 423 (1987) 353–358.

Soja, P.J., Lopez-Rodriquez, F., Morales, F.R. and Chase, M.H. The postsynaptic inhibitory control of lumbar motoneurons during the atonia of active sleep: Effect of strychnine on motoneuron properties, J. Neurosci., 11 (1991) 2804–2811.

Steriade, M. and Hobson, J.A. Neuronal activity during the sleep-waking cycle, Prog. Neurobiol., 6 (1976) 155–376.

Swick, D., Pineda, J.A., Schacher, S. and Foote, S.L. Locus coeruleus neuronal activity in awake monkeys: Relationship to auditory P300-like potentials and spontaneous EEG, Exp. Brain Res., 101 (1994) 86–92.

Taal, W. and Holstege, J.C. GABA and glycine frequently colocalize in terminals on cat spinal motoneurons, NeuroReport, 5 (1994) 2225–2228.

Taber, E. The cytoarchitecture of the brain stem of the cat. I. Brain stem nuclei of cat, J. Comp. Neurol., 116 (1961) 27–70.

Taguchi, O., Kubin, L. and Pack, A.I. Evocation of postural atonia and respiratory depression by pontine carbachol in the decerebrate rat, Brain Res., 595 (1992) 107–115.

Takakusaki, K., Ohta, Y. and Mori, S. Single medullary reticulospinal neurons exert postsynaptic inhibitory effects via inhibitory interneurons upon alpha-motoneurons innervating cat hindlimb muscles, Exp. Brain Res., 74 (1989) 11–23.

Takakusaki, K., Shimoda, N., Matsuyama, K. and Mori, S. Discharge properties of medullary reticulospinal neurons during postural changes induced by intrapontine injections of carbachol, atropine and serotonin, and their functional linkages to hindlimb motoneurons in cats, Exp. Brain Res., 99 (1994) 361–374.

Tohyama, M., Sakai, K., Touret, M., Salvert, D. and Jouvet, M. Spinal projections from the lower brain stem in the cat as demonstrated by the horseradish peroxidase technique. I. Origins of the reticulospinal tracts and their funicular trajectories, Brain Res., 173 (1979) 383–404.

Van Dongen, P.A.M., Broekkamp, C.L.E. and Cools, A.R. Atonia after carbachol microinjections near the locus coeruleus in cats, Pharmacol. Biochem. Behav., 8 (1978) 527–532.

Velazquez-Moctezuma, H., Gillin, J.C. and Shiromani, P.J. Effect of specific M1, M2 muscarinic receptor agonists on REM sleep generation, Brain Res., 503 (1989) 128–131.

Velazquez-Moctezuma, J., Shalauta, M. Gillin, J.C. and Shiromani, P.J. Cholinergic antagonists and REM sleep generation, Brain Res., 543 (1991) 175–179.

Wallace, D.M., Magnuson, D.J. and Gray, T.S. The amygdalo-brainstem pathway: Selective innervation of dopaminergic, noradrenergic and adrenergic cells in the rat, Neurosci. Lett., 97 (1989) 252–258.

Ward. D.G. and Gunn, C.G. Locus coeruleus complex: Elicitation of a pressor response and a brain stem region necessary for its occurrence, Brain Res., (1976) 401–406.

West, W.L., Yeomans, D.C. and Proudfit, H.K. The function of noradrenergic neurons in mediating antinociception induced by electrical stimulation of the locus coeruleus in two different sources of Sprague-Dawley rats, Brain Res., 626 (1993) 127–135.

Westlund, K.N. and Coulter, J.D. Descending projections of the locus coeruleus and subcoeruleus/medial parabrachial nuclei in monkey: Axonal transport studies and dopamine-β-hydroxylase immunocytochemistry, Brain Res. Rev., 2 (1980) 235–264.

Wu, W.-F., Gulyani, S., Mignot, E. and Siegel, J.M. Activity of REM-off cells during cataplexy in narcoleptic dog, APSS Abstr., (1996) 3.

Yamuy, J., Mancillas, J.R., Morales, F.R. and Chase, M.H. C-fos expression in the pons and medulla of the cat during carbachol-induced active sleep, J. Neurosci., 13 (1993) 2703–2718.

Zardetto-Smith, A.M. and Johnson, A.K. Chemical topography of efferent projections from the median preoptic nucleus to pontine monoaminergic cells groups in the rat, Neurosci. Lett., 199 (1995) 215–219.

Zhuo, H., Fung, S.J., Reddy, V.K. and Barnes, C.D. Immunohistochemical evidence for coexistence of methionine-enkephalin and tyrosine hydroxylase in neurons of the locus coeruleus complex projecting to the spinal cord of the cat, J. Chem. Neuroanat., 5 (1992) 1–10.

Rapid Eye Movement Sleep
B.N. Mallick and S. Inoué (Eds)

6. PGO Wave Generation: Mechanism and functional significance

Subimal Datta

Sleep Research Laboratory, Center for Behavioral Development,
Department of Psychiatry, Boston University School of Medicine,
715 Albany St., Building M923, Boston, Massachusetts 02118, USA

SUMMARY

Ponto-geniculo-occipital (PGO) waves are recorded during rapid eye movement (REM) sleep from the pons. PGO wave field potentials can also be recorded in other parts of the brain in addition to the pontine sites, and their distribution is different in different species. Species differences are due to variation in species-specific post-synaptic target sites of the pontine PGO generator. The triggering neurons of the pontine PGO wave generator are located within the caudo-lateral peribrachial (C-PBL) and the nucleus subcoeruleus areas. The transferring neurons of the pontine PGO generator are located within the cholinergic neurons of the laterodorsal tegmentum and the pedunculopontine tegmentum. The triggering and transferring neurons of the pontine PGO wave generator are modulated by aminergic, cholinergic, nitroxergic, GABA-ergic, and glycinergic cells of the brainstem. The PGO system is also modulated by suprachiasmatic, amygdaloid, vestibular, and brainstem auditory cell groups. The activation of pontine PGO wave generating cells is critical for REM sleep dependent memory consolidation.

Key Words Ponto-geniculo-occipital waves, rapid eye movement sleep, learning and memory, peribrachial area, subcoeruleus nucleus, pedunculopontine tegmentum; locus coeruleus, dorsal raphe.

During the last decade, many studies have attempted to elucidate the mechanism and functions of REM sleep. This chapter focuses on one of the distinguishing signs of REM sleep, specifically, the description, cellular and molecular mechanisms, and function of pontine PGO waves. This chapter will consider the hypothesis that the activation of PGO generating cells in the brain stem are a critical input for REM sleep dependent learning and memory processes.

Polygraphic Signs of Rapid Eye Movement Sleep

Rapid eye movement sleep (REM) of mammalian species is characterized by a constellation of events including: (1) a desynchronized pattern of cortical

EEG activity; (2) marked atonia of the postural muscles; (3) rapid eye movements (REM); (4) a theta rhythm within the hippocampus; (5) ponto-geniculo-occipital (PGO) spikes; (6) myoclonic twitches, most apparent in the facial and distal limb musculature; and (7) pronounced cardiorespiratory fluctuations (Fig. 1). During the last decade, evidence has accumulated suggesting that each of the events of REM sleep is executed by distinct cell groups in the brainstem (Datta, 1995 for reviews). They are discrete components of a widely distributed network rather than being a single REM sleep "center". The cortical EEG desynchronization of REM sleep is executed by the mesencephalic reticular formation; muscle atonia by peri-Locus Coeruleus; rapid eye movements by periabducens reticular formation; PGO waves by peribrachial area and nucleus subcoeruleus; hippocampal theta rhythm by pontis oralis; muscle twitches by nucleus gigantocellularis especially the caudal part; increased brain temperature and cardio-respiratory fluctuations by the parabrachial nucleus.

Description of Phasic PGO Waves

Field potentials in the pontine tegmentum which begin just prior to the onset of REM sleep are prominent phasic events of REM sleep and continue throughout that state (Fig. 1). These field potentials have been recorded from the lateral geniculate body (LGB) and the occipital cortex of the cat as well.

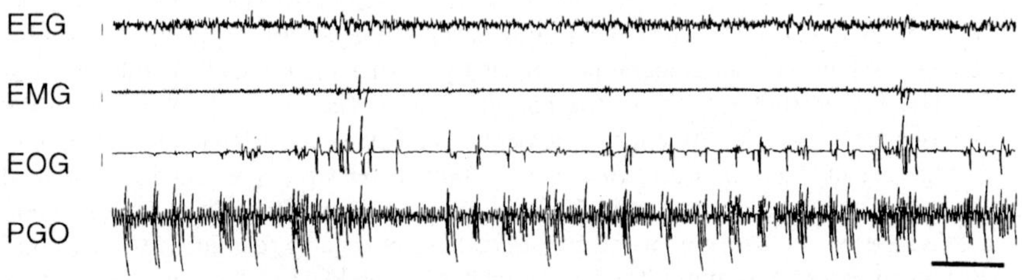

Fig. 1. Polygraphic records showing characteristic electrographic signs of REM sleep in the cat: low-voltage, high-frequency or desynchronized waves recorded from frontal cortex (EEG); muscular atonia (EMG); rapid eye movements (EOG); and spiky PGO waves in lateral geniculate body EEG (LGB). Calibration bar: 5 Sec.

Since, in the cat, these field potentials originate in the pons (P) and then propagate to the geniculate (G) and occipital cortex (O)), they were called PGO waves (Datta, 1997 for reviews). Subsequent studies found that PGO waves in the cat could also be recorded from points throughout the extent of the thalamus and cortex. However, such PGO waves reach their highest amplitude in the LGB, primary visual cortex, and association visual cortex. In addition to the pons, thalamus, and cortex, phasic potentials have been recorded in the oculomotor nuclei and the cerebellum of the cat. Phasic potentials of pontine origin have also been recorded from the amygdala, cingulate gyrus, and hippocampus, suggesting that PGO waves also occur in the limbic system. More importantly, all of these PGO wave mapping studies in the cat have demonstrated that the pons is the primary site in which PGO wave activity is generated (Datta, 1997).

PGO waves have been studied most extensively in the cat. However, they have also been documented and studied in other mammalian species including non-human primates, humans, and rodents. In non-human primates, PGO wave-like phasic field potentials have been recorded from the LGB and pons of macaques (Cohen and Feldman, 1968) and in the LGB of baboons (Vuillon-Cacciuttolo and Seri, 1978). Phasic potentials also have been recorded in the striate cortex of humans during REM sleep (Salzarulo et al., 1975). Such field potentials recorded from the striate cortex were probably cortical components of state-specific phasic potentials of pontine origin. Also in humans, the observation of phasic scalp potentials associated with eye movements during REM sleep suggests that PGO wave-like activity may be present (McCarley et al., 1983; Miyauchi et al., 1987). PGO-like waves are also recorded in the pons and creebellum of rats (Farber et al., 1980; Marks et al., 1980; Sanford et al., 1995).

The waveform, amplitude, and frequency characteristics of PGO waves recorded from the pons, geniculate body, and cortex have been most intensively examined in the cat. These studies noted that PGO waves are biphasic in shape with a duration of 60–120 mSec and an amplitude of 200–300 μV. These waves occur as singlets and as clusters containing a variable number of PGO waves (3–5 waves/burst) at a density range of 30–60 spikes/min during REM sleep (see Datta, 1997 for references). Pontine PGO-like waves recorded in the rat have amplitudes (150 μV) and durations (100 mSec) similar to those of PGO waves recorded in the cat (Farber et al., 1980; Marks et al., 1980). Singlet PGO waves occur commonly in non-REM sleep, whereas clusters of PGO waves are typically recorded during REM sleep. Singlet PGO waves have been shown to be independent of eye movement and are called type-I waves, while clusters of PGO waves are associated with the eye movement bursts of REM sleep and are called type-II waves (Morrison and Pompeiano, 1966). Type-II PGO wave activity accounts for 55-65% of the total number of PGO waves recorded during REM sleep (Datta et al., 1992).

Mechanism of PGO Wave Generation

Anatomical and physiological studies strongly suggest that the critical structures for triggering PGO waves are localized in the pons (Datta, 1995; Datta and Hobson, 1994, 1995). These structures are discrete components of a widely distributed network that includes two basic elements: (1) executive neurons and (2) modulatory neurons.

1. Executive Neurons

The executive elements of PGO wave generation are a collection of neurons which initiate PGO waves. This group of neurons is comprised of two functionally different subgroups: (a) triggering neurons and (b) transferring neurons.

(a) Triggering Neurons

Triggering neurons in the PGO generator network of the cat are located in the caudolateral peribrachial area (C-PBL) (Datta et al., 1992; Data and Hobson, 1994, 1995). In this species, PGO triggering neurons are lightly stained ChAT-positive cholinergic neurons (Vincent and Reiner, 1987). A majority of the C-PBL neurons begin to fire or increase their firing rates several seconds before the first PGO wave, and these rates remain high until the end of the PGO-related state (Datta and Hobson, 1994). During slow-wave sleep with PGO and REM sleep, these PGO state-on burst cells fire high-frequency bursts on a background of tonic activity in association with each ipsilateral primary PGO wave. The first spike of a burst precedes the beginning of the negative component of the ipsilateral LGB PGO wave by 25 ± 7.5 mSec (Fig. 2). Depending on the type of species (prey or predator animal), triggering neurons may be in slightly different locations. For example, in the rat (prey animal), these PGO triggering neurons are located in the nucleus subcoeruleus (Datta et al., 1998) which is proximate to the locus coeruleus, an area known to be important for alerting behavior. The chemical identity of PGO triggering neurons in the rat is unknown and suspected to be glutamatergic (Datta et al., 1998).

(b) Transferring Neurons

The transferring neurons of the PGO circuit are located in the rostral part of the peribrachial area (R-PBL; Datta, 1997). These PGO transferring cells in the pedunculopontine tegmentum (PPT) and laterodorsal tegmentum (LDT) are cholinergic and are also capable of synthesizing nitric oxide (Datta, 1995; Vincent and Reiner, 1987).

2. *Modulatory Neurons*

The executive elements of the PGO wave generator are modulated by many excitatory and inhibitory inputs from the brainstem as well as by forebrain structures. These modulatory structures are not directly responsible or PGO wave generation but they are instrumental in the final occurrence of the PGO wave by disinhibiting, inhibiting, or exciting the executive elements of PGO waves. Available pharmacological and physiological studies indicate that the PGO generating network could be modulated by the aminergic, cholinergic, nitroxergic, and GABA-ergic cell groups of the brainstem. In addition to these above mentioned cell groups, the PGO system is also positively modulated by amygdaloid, vestibular, and brainstem auditory cell groups, (see Datta, 1997 for references).

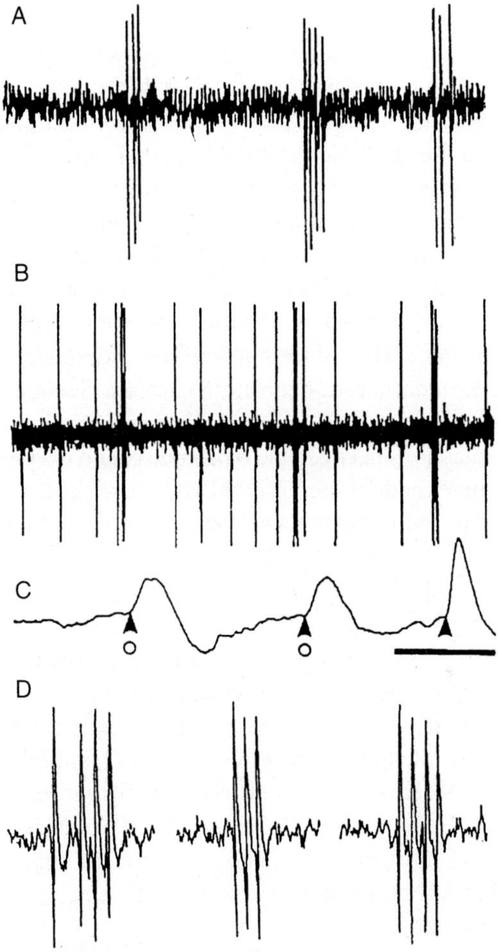

Fig. 2. Temporal relationships of the burst type of discharge in a triggering cell and a transferring cell in the brainstem PGO wave generator of the cat with illustration of ipsilateral LGB PGO waves. (A) A trace of action potentials for a PGO transferring neuron in rostral peribrachial area (recorded from the pedunculopontine tegmentum) showing low frequency bursting in relation to each ipsilateral LGB PGO wave. (B) A train of action potentials from a PGO triggering neuron in the caudolateral peribrachial area showing recurrent high-frequency bursts in relation to ipsilateral PGO waves on a background of tonic action potentials. (C) A cluster of ipsilateral LGB PGO waves recorded simultaneously with cells A and B. (D) The 3 high-frequency bursts seen in B, the triggering neuron, are displayed on an expanded time scale. Note that the latency of the burst and the onset of the LGB PGO waves is significantly longer in the triggering neuron (B) compared to transferring neuron (A). Also note the higher burst discharge frequency in the triggering neuron compared to transferring neuron. The time scale in A, B, and C is 100 mSec; in D. 10 mSec. (Datta, 1997).

Aminergic Neurons

The executive neuronal population of the PGO generating network described above appears to be held in inhibitory restraint during waking and slow wave sleep by two modulatory neuronal populations: the noradrenergic (NE) locus coeruleus (LC) and the serotonergic (5-HT) dorsal raphe (DRN). Three lines of experimental evidence support this statement. First, pharmacological manipulations reducing the availability of 5-HT and NE in the brain have been shown to promote the appearance of PGO waves, while administration of pharmacological agents which increase brain 5-HT or NE have been shown to suppress PGO waves in the cat (see Datta, 1997 for references). The activity of the PGO wave generator is thus negatively correlated with increased serotonergic and noradrenergic neurotransmission. Second, both a midline lesion which destroyed the raphe nuclei and parasaggital transections which interrupted the connections between the raphe nuclei and the pontine tegmentum produced a state-independent increase in PGO wave discharge (Simon et al., 1973). Analogous to the results of knife cut lesions mentioned above, some success in increasing PGO wave frequency with localized cooling of the DRN has also been reported (Cespuglio et al., 1979). Similarly, REM sleep signs in the cat, including PGO waves, have been evoked by selective cooling of the LC (Cespuglio et al., 1982). These findings support the hypothesis that the DRN and LC provide inhibitory control or gating of the PGO wave generator. The third line of evidence derives from single cell recording studies which indicated that most of the neurons of the LC and DR nuclei decrease their firing rates at sleep onset and nearly stop firing during SP and REM sleep (Datta, 1997; Hobson et al., 1975; Jacobs 1986; McGinty and Harper, 1976). Since a cause must precede an effect, these findings strongly support the hypothesis that LC and DRN discharge levels are causally involved in regulating PGO waves.

Cholinergic Neurons

The role of cholinergic mechanisms in the modulation of PGO wave activity generation has been explored by a number of investigators using the pharmacological techniques (Datta et al., 1992, 1993; Henriksen et al., 1972; Jacobs et al., 1972). Henriksen et al., (1972) demonstrated that, when cats were deprived of REM sleep for five days, atropine did block the bursts of PGO waves which normally accompany rapid eye movements during REM sleep recovery. Although isolated PGO waves still occurred during the recovery of slow wave sleep with PGO and REM sleep, the frequency of these isolated waves was reduced by atropine. In a related study, Jacobs et al. (1972) showed that atropine blocked the continous PGO waves induced by pretreatment with para-chloro-phenyl-alanine (PCPA), and that eserine partially reversed this atropine induced blockade. The cholinoceptive PGO trigger zone of the brain stem has been mapped with the use of the intracerebral microinjection method (Datta et al., 1992, 1993). This mapping study demonstrated that cholinergic stimulation of the caudolateral peribrachial area induced short-latency state-independent ipsilateral LGB PGO waves.

Single unit discharge patterns of putatively cholinergic cells in the PPT and LDT constitute the most physiologically significant correlation between the activity of the cholinergic systems and PGO waves. These cholinergic neurons' discharge properties show a marked heterogeneity in behaving animals (Datta, 1995). Recordings from the PPT and LDT of behaving cats have led to the identification of several different classes of cells whose firing rates correlate with both phasic and tonic components of REM sleep (Datta, 1995). One group of tonicaly discharging cholinergic neurons progressively increases its firing rate as the animal moves from waking to slow wave sleep and then to REM sleep. It is thus possible that the tonic increase of this neuronal activity slowly increases the level of acetylchline in the C-PBL during slow wave with PGO and REM sleep. Therefore, it is tentatively speculated that during slow wave sleep with PGO and REM sleep which normally accompany PGO waves, increased release of acetylcholine from the PPT and LDT acts as an excitatory neuromodulator at the level of C-PBL PGO triggering neurons.

Nitroxergic Neurons

All of the PPT and LDT cholinergic cells of the brainstem also displayed intense NADPH-diaphorase activity (Vincent and Reiner, 1987). Recent studies have shown that NADPH-diaphorase stained PPT and LDT cells contain nitric oxide synthase (NOS) mRNA (Bredt and Snyder, 1992). The presence of NADPH-diaphorase together with NOS indicates that the PPT and LDT cholinergic cells are capable of synthesizing NO. Excitation and release of acetylcholine by PPT and LDT cells is important for the modulation of PGO wave generation as discussed above. It is possible that the release of NO into the PPT and LDT leads to the release of more acetylcholine from the PPT and LDT (Leonard and Lydic, 1995) to PGO triggering neurons in the C-PBL. It is also possible that the release of NO from the PPT and LDT directly modulates C-PBL triggering neurons. Tentatively, NO acts as an excitatory neuromodulator in the generation of pontine PGO waves.

GABA-ergic Neurons

GABA-ergic neurons of the brainstem and substantia nigra reticulata modulate PGO generation by actively inhibiting inhibitory modulating aminergic neurons of the brainstem (see Datta, 1997 or references). The first line of evidence indicating that the GABA-ergic neurotransmission is involved in the modulation of PGO wave generation comes from the intravenous and intracerebroventricular administration of γ-aminobutyric acid (GABA) and GABA mimetic agents. For example, intraventricular administration of GABA and intravenous administration of GABA mimetic agents (benzodiazepines) increased the frequency of PCPA-induced PGO waves. These results were further confirmed

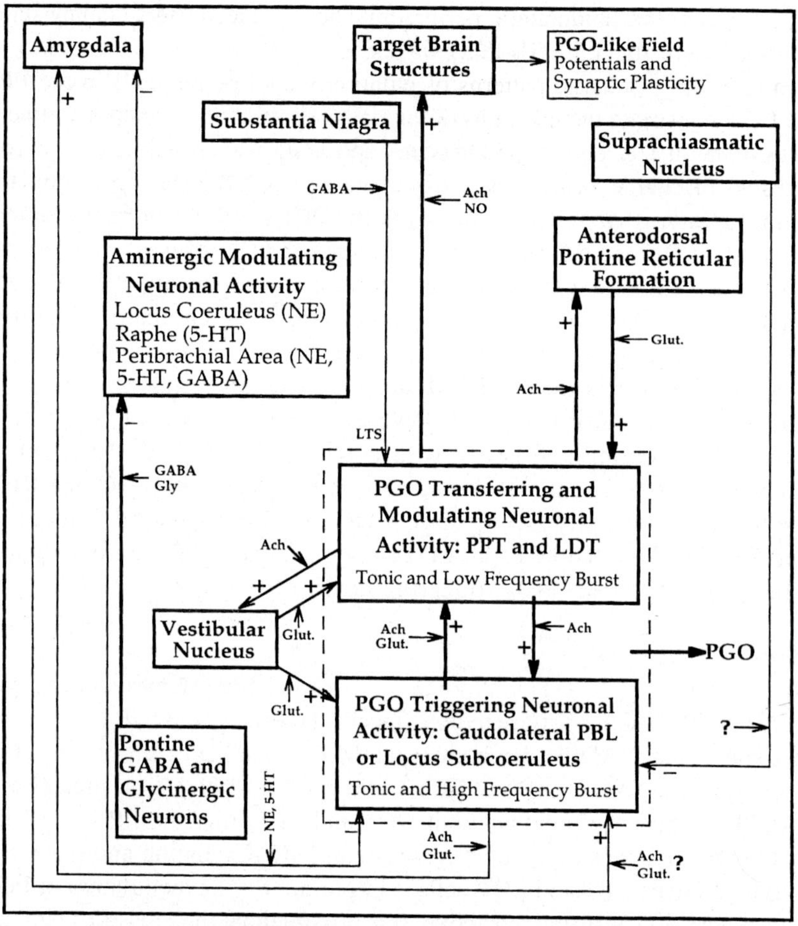

Fig. 3. Cellular, molecular, and network (CMN) model of brainstem PGO wave generation. Increased neuronal activity of brainstem GABA-ergic and glycinergic neurons actively inhibits aminergic cells of the locus coeruleus, dorsal raphe, and peribrachial area. Decreased activity of these aminergic neurons then leads to decreased release of norepinephrine (NE) and serotonin (5-HT) in the caudolateral peribrachial area. Reduced availability of aminergic neurotransmitters at the brainstem PGO triggering cells in the caudolateral peribrachial area increases the excitability of these cells as the tonic inhibition by aminergic cells is withdrawn. Subsequently, increased activity of the PGO triggering neurons in the caudolateral peribrachial area initiates pontine PGO waves. These PGO triggering neurons activate rostral peribrachial area (PPT and LDT) PGO transferring neurons which, in turn, transfer brainstem PGO information to the other target structures of the brain. Input and output connections of the PGO generator and modulators are indicated connecting two structures with arrow signs while involved neurotransmitters are labeled with short arrow and smaller font. Abbreviations: +, excitatory input; –, inhibitory input; Ach, acetylcholine; GABA, gama-amino butyric acid; Glut., glutamate; Gly, glycine; LDT, laterodorsal tegmentum; LTS, indicates activation of low-thredshold calcium spike in bursting cell; NO, nitric oxide; PPT, pedunculopontine tegmentum; ?, unknown/suspected. Further details in the text.

with other direct GABA agonists and GABA transaminase inhibitors in PCPA, Ro 4-1284, and reserpine-treated cats (Neal and Bond, 1983). The second line of evidence comes from the direct measurement of GABA release into the PGO generating region of the pons (Nitz and Siegel, 1993). This study indicates that the release of GABA is greater during the PGO related state of REM sleep. Lastly, we have demonstrated that the GABA-ergic cells projecting to the pontine PGO generator increased their firing activity before and during PGO waves (Datta et al., 1991).

Cellular, Molecular and Network (CMN) Model of Pontine PGO Wave Generation

The PGO-state-on cells in the C-PBL of the cat and the subcoeruleus cells of the rat are the triggering elements of pontine PGO wave generation. During waking and slow-wave sleep all of these cells remain silent due to presynaptic inhibitory inputs from the aminergic cells of the LC, DRN, and the PBL PGO-state-off cells. Local GABA-ergic cells located within the PGO trigger zone are also involved in the inhibition of these triggering neurons. During slow wave sleep with PGO and REM sleep, however, these aminergic and GABA-ergic cells are actively inhibited by GABA-ergic and glycinergic neurons of the LC, DRN, and substantia nigra reticulata. When such aminergic and local GABA-ergic inhibitions are removed from the PGO-state-on neurons of the C-PBL of the cat and nucleus subcoeruleus of the rat, these cells start discharging. Active PGO-stage-on cells (triggering elements) activate R-PBL PGO-on cholinergic/nitroxergic cells (transferring elements) synaptically. Activated R-PBL PGO-on cells ultimately act as pontine output and transferring elements for pontine PGO waves. The interaction of these triggering and transferring elements generates the pontine PGO wave.

Current Hypotheses on the Functions of REM Sleep

It is clear that REM sleep is a crucial aspect of mammalian integrative regulatory physiology, however, it is still not clear exactly what the function of REM sleep is. Rats which are consistently deprive of REM sleep will eventually perish. Their metabolism soars out of control and no matter how much they eat, they cannot sustain life.

There are many speculations on the functions of REM sleep, the most common of which are enumerated below. One hypothesis is that dreams during REM sleep are part of a reverse learning process to rid the brain of unnecessary information (Crick and Mitchison, 1983). Unfortunately, the reverse learning hypothesis is not testable. A contrasting hypothesis, the sleep-learning hypothesis, proposes that certain types of memory are consolidated during REM sleep (Dement, 1965; Fishbein and Gutwein, 1977).

Another hypothesis on the functions of REM sleep comes from an evolutionary perspective. This hypothesis suggests that sleep, and especially REM sleep, is part of a mechanism to maintain and enhance synaptic eicacy by a process of cortical activation called dynamic stabilization (Kavanau, 1994). Dynamic stabilization is the repetitive activation of synapses resulting from use or from spontaneous neural oscillatory activity. Repetitive activation of a synapse leads to its strengthening, and it is hpothesized that during REM sleep, frequently and infrequently used neural circuits are activated to maintain their connections. A fourth hypothesis suggests that the neuronal activities during sleep are an adaptation to preserve essential synapses that are insufficiently utilized during wakefulness (Krueger et al., 1995). This hypothesis suggests that sleep is a use-driven rather than a wakefulness-dependent phenomenon. The central idea is that locally produced use-dependent substances in the brain regulate the activation of vital synapses that are insufficiently used during wakefulness. When certain synapses are used disproportionately more than others, a disjunctive state is reached. When several neuronal groups are in this disjunctive state, sleepiness is perceived. By this mechanism, synapses that are insufficiently stimulated during waking are stimulated during sleep in order to maintain their efficacy. Another hypothesis suggests that the primary function of REM sleep is to ensure that a minimum level of activation of the brain is maintained to enable recovery from deep rest (Vertes, 1986). It is hypothesized that the brain may not be equipped to survive extended periods of low levels of activity, such as those during slow wave sleep. Thus, REM sleep may be an adaptive mechanism by which inactivation is interrupted.

REM Sleep Dependent Learning and Memory

In this section we will focus primarily on the hypothesis that REM sleep and learning are mechanistically connected. The sleep-learning hypothesis proposes that REM sleep is necessary for memory consolidation. It derives most of its support from human and animal studies as discussed below.

It has been shown that the length and number of REM sleep episodes increases following the acquisition of complex skills in humans and animals (reviewed in Fishbein and Gutwein, 1977; Smith, 1995). For example, college students were monitored before and after examination periods, and it was found that after exams the number of REM sleep episodes increased. Another study demonstrated higher percentages of REM sleep in aphasic patients who showed measurable improvements during a course of speech pathology treatments compared to aphasic patients who did not show improvement (Greenberg and Dewan, 1969). Similar REM sleep increases have been documented following intensive language learning (De Koninck et al., 1975). Furthermore, it has been shown that the percentage of REM sleep on nights following four hours of visual or auditory learning tasks increased compared to control nights (Verschoor and Holdstock, 1984). Based on these and other studies, it is widely hypothesized that this increased REM sleep is neccessary for learning and memory processes.

Studies that deprive animals and humans of REM sleep following intensive learning periods have shown that subsequent performance on the learned task is impaired under various circumstances (Smith, 1995). In humans, some types of memory are affected by REM sleep deprivation and some are not. Specifically, declarative or explicit memory is not impaired by REM sleep deprivation, but cognitive procedural or implicit memory is negatively impacted (Smith, 1995). One study showed that human performance on a recently learned visual discrimination task improved after an undisturbed night of sleep (Karni et al., 1994). Deprivation of REM sleep disrupted the post-sleep performance improvement, whereas disruption of nonREM sleep did not. REM sleep deprivation did not negatively impact performance on previously learned tasks (i.e. presumably already consolidated material). Thus, it appears that REM sleep is specifically involved in memory consolidation of newly learned material.

Wilson and McNaughton (1994) have shown a correlation between single cell activity during waking and subsequent sleep in the rat. Place cells are selectively activated when a waking animal is in a specific location in a given environment. During subsequent sleep these same place cells were again selectively activated. The major impact of this finding is that it provides direct physiological evidence that the sleeping brain replays neuronal activity experienced during wakefulness. The mechanistic significance of this replay still needs to the investigated, as does the role of REM sleep.

Role of PGO Waves in REM Sleep Dependent Learning

It is well established that there is a significant relationship between REM sleep and learnign, however, the mechanism is still unclear. Based on our ongoing research and other evidence discussed below, we believe that the phasic excitation of PGO wave generating cells in the pons is critical for REM sleep dependent memory consolidation (Datta, 1997).

Several studies indicate that rapid eye movements may represent the element of REM sleep that is crucial for memory consolidation (Smith and Weeden, 1990; Verschor and Holdstock, 1984; Mandai et al., 1989; Herman and Roffwarg, 1983). For example, when a background clicking noise was presented during acquisition of a learned skill, presentation of the same auditory stimulus during subsequent eye movements during REM sleep (cueing), was correlated with a 23% improvement on retest performance one week later. The same cueing applied during non-eye movement REM sleep episodes correlated with only an 8.8% retest improvement. It is hypothesized, thereore, that the eye movements (or at least that segment of REM sleep in which they occur) are selectively important in REM sleep dependent memory consolidation (Smith and Weeden, 1990). Visual learning tests in human volunteers showed that in addition to increases in percentage

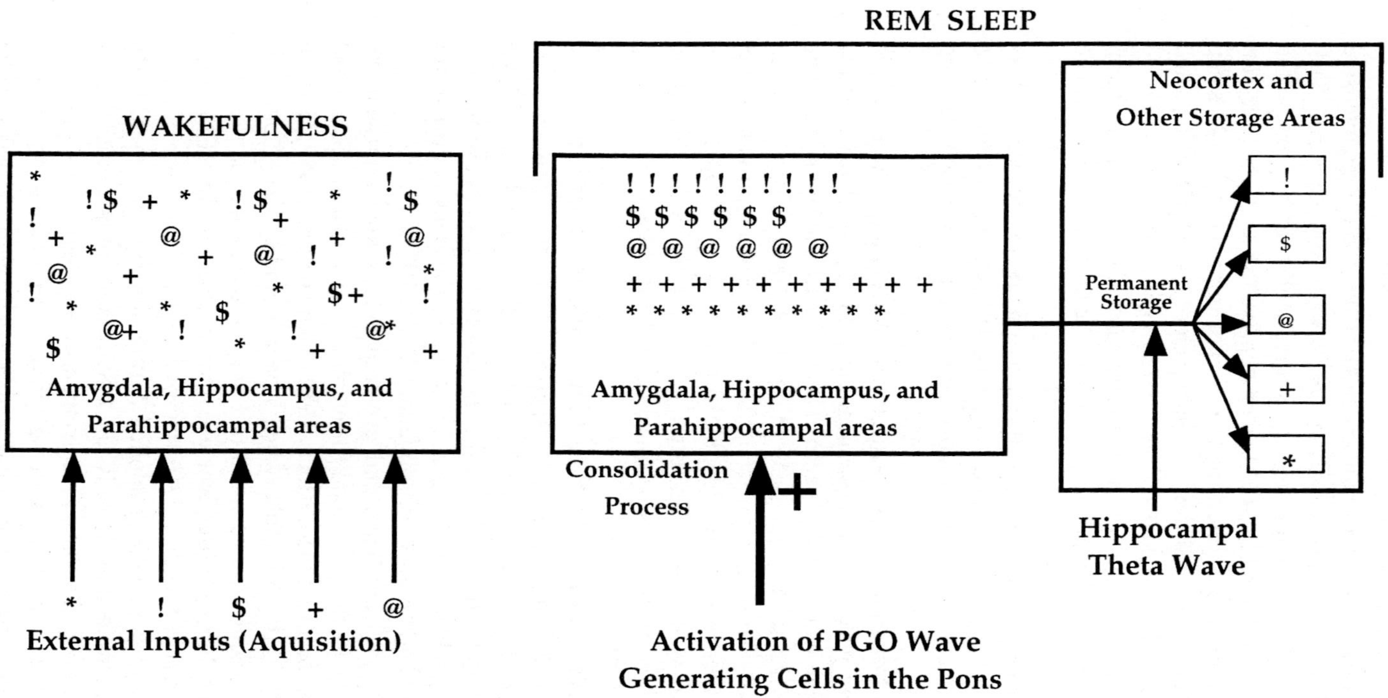

Fig. 4. **Model of REM sleep dependent memory consolidation and long term storage. During wakefulness, external information is randomly input into the brain and stored in the amygdala, hippocampus, and parahippocampal areas. This process is called aquisition of information. The activation of PGO wave generating cells in the pons during REM sleep stimulates the amygdala, hippocampus, and parahippocampal areas to organize and consolidate the random information aquired during wakefulness. Almost at the same time theta-frequency waves help to bind that consolidated information into the long term memory storage in the neocortex and other storage areas. Symbols: *, !, @, $, +, different types of external information; larger addition sign outside the box, excitatory input.**

of REM sleep, the percentage of eye bursts during post-training REM sleep increased (Verschoor and Holdstock, 1984). These researchers hypothesize that these augmented eye bursts represent the scanning of visual stimuli encountered during the learning task, as part of the process of sorting, organizing and consolidating daily input (Verschoor and Holdstock, 1984). A study of Morse language learning in humans provides further evidence for an eye movement role in learning and memory processing during REM sleep. After a 90 min Morse language learning session immediately prior to bedtime, subjects who had the greatest success had the most dense rapid eye movements (Mandai et al., 1989). It is well established that the occurence and direction of rapid eye movements during REM sleep depends exclusively on the excitation of the pontine PGO wave generating cells (Datta and Hobson, 1994). Therefore, the studies described above indirectly substantiate our hypothesis that the excitation of pontine PGO wave generating cells are critical for REM sleep dependent memory consolidation.

To account for REM sleep dependent memory processing and learning, the following hypothesis has been suggested. During wakefulness, acquired information is stored temporarily in the hippocampus. During subsequent REM sleep the hippocampus is reactivated to consolidate acquired information into more permanent storage in the neocortex (Pavlides and Winson, 1989; Karni et al., 1994; Wilson and McNaughton, 1994; Datta, 1997). For this type of activity dependent plasticity, this proposed hippocampal mechanism would require triggering inputs during REM sleep (Datta, 1997). We believe these triggering inputs to the hippocampus and other brain structures involved in the consolidation of learning and memory processes stem from the pontine PGO wave generating cells (Fig. 4). If PGO waves are involved in learning and memory processes, the primary brain structures involved in this function would receive anatomical projections from PGO generating cells in the pons. Indeed, many amygdaloid, parahippocampal, and hippocampal structures receive direct input from the cells in the pontine PGO generation site (Datta et al., 1998). Thus, there is anatomical, physiological, and behavioral evidence substantiating the hypothesis that the PGO trigger zone in the pons is part of the physiological substrate of REM sleep dependent memory consolidation.

Acknowledgements

This work was supported by USPHS grant NS 34004 to S. Datta from National Institute of Health.

References

Bredt, D.S. and Snyder, S.H. Nitric oxide, a novel neuronal messenger, Neuron, **8** (1992) 3–11.

Cespuglio, R., Gomez, M.E., Faradji, H. and Jouvet, M. Alterations in the sleep-waking cycle induced by cooling of the locus coeruleus area, *Electroenceph. Clin. Neurophysiol,* **54** (1982) 570–578.

Cespuglio, R., Gomez, M.E., Walker, E. and Jouvet, M. Effets du refroidessement et de la stimulation des moyaux du systeme du raphe sur les etats de vigilance chez le chat, Electroenceph. Clin. Neurophysiol., **47** (1979) 289–308.

Cohen, B. and Feldman, M Relationship to electrical activity in the pontine reticular formation and lateral geniculate body to rapid eye movements, J. Neurophysiol. **31** (1968) 807–817.

Crick, E. and Mitchison, G. The function of dream sleep, Nature, **304** (1983) 111–114.

Datta, S. Neuronal activity in the peribrachial area: relationship to behavioral state control, Neurosci. Biobehav. Rev., **19** (1995) 67–84.

Datta, S., Cellular basis of pontine ponto-geniculo-occipital wave generation and modulation, Cell. Mol. Neurobiol., **17** (1997) 341–365.

Datta, S., Calvo, J.M., Quatrochi, J.J. and Hobson, J.A. Cholinergic microstimulation of the peribrachial nucleus in the cat. I. immediate and prolonged increases inponto-geniculo-occipital waves, Arch. Ital. Biol., **130** (1992) 263–284.

Datta, S., Currodossi, R., Pare, D., Oakson, G. and Steriade, M. Substantia nigra reticulata neurons during sleep-waking states: relation with Ponto-geniculo-occipital waves, Brain Res., **566** (1991) 344–347.

Datta, S. and Hobson, J.A., Neuronal activity in the caudo-lateral peribrachial pons: relationship to PGO waves and rapid eye movements, J. Neurophysiol., **71** (1994) 95–109.

Datta, S. and Hobson, J.A. Suppression of ponto-geniculo-ocipital waves by neurotoxic lesions of pontine caudo-lateral peribrachial cells, Neurosci., **67** (1995) 703–712.

Datta, S., Quattrochi, J.J. and Hobson, J.A., Effect of specific muscarinic M2 receptor antagonist on carbachol induced long-term REM sleep, Sleep, **16** (1993) 8–14.

Datta, S., Siwek, D.F., Patterson, E.H. and Cipolloni, P.B., Localization of pontine PGO wave generation sites and their anatomical projections in the rat, Synapse (1998), In Press.

DeKoninck, J., Proulx, G., Healey, T., Arsenault, R. and Prevost, F. Intensive language learning and REM sleep, Sleep Research, **4** (1975) 150.

Dement, W.C., Recent studies on the biological role of REM sleep, Amer. Journ. Psychia., **122** (1965) 404–408.

Farber, J., Marks, G.A. and Roffwarg, H.P. Rapid eye movement sleep PGO-type waves are present in the dorsal pons of the albino rat, Science, **209** (1980) 615–517.

Fishbein, W. and Gutwein, B.M., Paradoxical sleep and memory storage processes, Behav. Biol., **19** (1977) 425–464.

Greenberg, R. and Dewan, E.M., Aphasia and rapid eye movement sleep, Nature, **223** (1969) 183–184.

Henriksen, S.J., Jacobs, B.L. and Dement, W.C., Dependence of REM sleep PGO waves on cholineargic mechanisms, Brain Res., **48** (1972) 412–416.

Herman, J.H. and Roffwarg, H.P., Modifying oculomotor activity in awake subjects increases the amplitude of eye movements during REM sleep, Science, **220** (1983) 1074–1076.

Hobson, J.A., McCarley, R.W. and Wyzinski, P.W. Sleep cycle oscillations: reciprocal discharge by two brainstem neuronal groups, Science, **189** (1975) 55–58.

Jacobs, B.L., Single unit activity of locus coeruleus neurons in behaving animals, Prog. Neurobiol., **27** (1986) 183–194.

Jacobs, B.L., Henriksen, S.J. and Dement, W.C., Neurochemical basis of the PGO wave, Brain Res., **48** (1972) 406–411.

Karni, A., Tanne, D., Rubenstein, B.S., Askenasy, J.J.M. and Sagi, D. dependence on REM sleep of overnight improvement of a perceptual task, Science, **265** (1994) 679–682.

Kavanau, J.L., Sleep and dynamic stabilization of neural circuitry: a review and synthesis, Behav. Brain Res., **63** (1994) 111–126.

Krueger, J.M., Obal, F., Jr., Kapas, L. and Fang, J., Brain organization and sleep function, Behav. Brain Res., **69** (1995) 177–185.

Leonard, T.O. and Lydic, R. Nitric oxide synthase inhibition decreases pontine acetylcholine release, NeuroReport, **6** (1995) 1525–1529.

Mandai, O., Guerrien, A., Sockeel, P., Dujardin, K. and Leconte, P. REM sleep modifications after a Morse language learning session, Physiology and Behavior, **46** (1989) 639–642.

Marks, G.A., Farber, J., Rubinstein, M. and Roffwarg, H.P. Demonstration of ponto-geniculo-occipital waves in the albino rat, Exp. Neurol., **69** (1980) 648–655.

McCarley, R.W., Winkelman, J.W. and Duffy, F.H. Human cerebral potentials associated with REM sleep rapid eye movements: links to PGO waves and waking potentials, Brain Res., **274** (1983) 359–364.

McGinty, D.J. and Harper, R.M., Dorsal raphe neurons: depression of firing during sleep in cats, Brain Res., **101** (1976) 569–575.

Miyauchi, S., Takino, R., Fokuda, H. and Torii, S., Electrophysiological evidence for dreaming: human cerebral potentials associated with rapid eye movement during REM sleep, Electroenceph. Clin. Neurophysiol., **66** (1987) 383–390.

Morrison, A.R. and Pompeiano, Q. Vestibular influences during sleep. IV. functional relations between the vestibular nuclei and lateral geniculate nucleus during desynchronized sleep, Arch. Ital. Biol., **104** (1966) 425–458.

Neal, H. and Bond, A. The influence of GABAergic drugs on PGO activity in the cat, Neuropharmacol., **22** (1983) 881–886.

Nitz, D.A. and Siegel, J.M. GABA release in the mesopontine central gray as a function of sleep state, Sleep Res., **22** (1993) 447.

Pavlides, C. and Winson, J. Influences of hippocampal cell firing in the awake state on the activity of these cells during subsequent sleep episodes, J. Neurosci. **9** (1989) 2907–2918.

Salzarulo, P., Lairy, G.C., Bancaud, J. and Munari, C., Direct depth recording of the striate cortex during REM sleep in man: Are there PGO potentials?, Electroenceph. Clin. Neurophysiol., **38** (1975) 199–202.

Sanford, L.D., Tejani-Butt, S.M. Ross, R.J. and Morrison, A.R. Amygdaloid control of alerting and behavioral arousal in rats: involvement of serotonergic mechanisms. Arch. Ital. Biol., **134** (1995) 81–99.

Simon, R., Gershon, M.D. and Brooks, D.C., The role of the raphe nuclei in the regulation of PGO wave activity, Brain Res., **58** (1973) 313–330.

Smith, C. Sleep and memory, Sleep Res. Soc. Bull., **1** (1995) 50–53.

Smith, C. and Weeden, K., Post training REMs concident auditory stimulation enhances memory in humans, Psychiatric Journal of the University of Ottowa, **15** (1990) 85–90.

Verschoor, G.J. and Holdstock, T.L., REM bursts and REM sleep following visual and auditory learning, South African Journal of Psychology, **14** (1984) 69–74.

Vertes, R.P., A life-sustaining function for REM sleep: a theory, Neurosci. Biobeh. Rev., **10** (1986) 371–376.

Vincent, S.R. and Reiner, P.B., The immunohistochemical localization of choline acetyltransferase in the cat brain, Brain Res. Bull., **18** (1987) 371–415.
Vuillon-Cacciuttolo, G. and Seri, B. Effets de la section des nerfs optiques chez le babouin sur l'activite A Type de pointes genouilles et corticales au cours des divers etats de vigilance, Electroencephalogr. Clin. Neurophysiol., **44** (1978) 754–768.
Wilson, M.A. and McNaughton, B.L. Reactivation of hippocampal ensemble memories during sleep, Science, **265** (1994) 676–679.

Rapid Eye Movement Sleep
B.N. Mallick and S. Inoué (Eds)

7. Norepinephrine and REM Sleep

Pierre-Hervé Luppi, Damien Gervasoni, Christelle Peyron, Claire Rampon, Bruno Barbagli, Romuald Boissard and Patrice Fort

INSERM U480, Faculté de Médecine, Université C. Bernard, LYON I, 8 avenue Rockefeller, 69373 LYON cedex 08, FRANCE

Introduction

Using transection and lesions, Jouvet and coworkers (Review in Jouvet, 1972) demonstrated that the structures responsible for the genesis of paradoxical sleep (PS) are localized in the pontine reticular formation. The subsequent identification in this region of the noradrenergic (NA) neurons in the nucleus locus coeruleus (LC) (Shimizu et al., 1959; Dahlström and Fuxe, 1964) and pharmacological results showing that manipulation of Noradrenaline modify the occurrence of PS led Jouvet to hypothesize that the LC neurons might be responsible for the genesis of PS. Supporting this idea, Jouvet and Delorme (1965) found that electrolytic lesions involving the LC localized in the cat dorsolateral tegmentum are followed by a loss of PS and/or the musle atonia that accompanies it.

However, subsequent lesion, pharmacological and electrophysiological studies showed that the region ventral to the LC rather than the LC itself contains the neurons responsible for the genesis of PS. Indeed, lesions more limited to the NA neurons did not have a profound effect on PS (Henley and Morrison, 1974; Jones et al., 1977). In addition, pharmacological depletion of the catecholamines with α-methyl-para-tyrosine (AMPT) increases PS and reduces waking (King and Jewet, 1971). Moreover, 6-hydroxydopamine-induced lesions of the ascending noradrenergic bundle induced a significant reduction in waking but not of PS (Lidbrink, 1974). In addition, electrophysiological recordings have shown that the NA neurons of the LC are active during waking, decrease their activity during slow wave sleep and cease firing during PS (Hobson et al., 1975; Aston-Jones and Bloom, 1981). Altogether these data indicate that the LC neurons by means of their widespread projections throughout the entire brain constitute one of the structures involved in the control of wakefulness (Jouvet, 1972; Hobson et al., 1975). It has further been shown in rats and cats that LC cells display a burst of activity when the animals are subjected to new stimuli. These neurons might therefore be more specifically implicated in arousal and vigilance that in the tonic maintenance of waking. This view is supported by results showing that lesions of LC or the noradrenergic bundle do not induce a long lasting decrease of wakefulness (Lidbrink, 1974).

Hobson et al. (1975) in the first report on the activity of LC neurons across the sleep-waking proposed that they might gate the appearance of PS. Indeed, they found in the gigantocellular tegmental field (FTG) neurons with a dramatic and state-specific discharge rate increase during PS (McCarley and Hobson, 1971). These findings led them to postulate that these FTG neurons may be a critical part of an executive system for PS. They further made the hypothesis that their periodic activation might be due to decreased activity of an inhibitory population. Supporting this proposal, they made the first demonstration that LC neurons are nearly silent during PS and have therefore an inverse discharge rate selectivity of FTG cells (Hobson et al., 1975). The LC neurons were named "REM-off" or "PS-off" cells. Although this denomination is not inaccurate, it does not reflect that the main role of noradrenergic LC neurons seems to be related to wakefulness. Indeed, the inhibition of PS by stimulation of the noradrenergic system could be secondary to the induction of wakefulness rather that due to a direct inhibition of the PS generating systems. However, supporting the hypothesis that such inhibition occurs, Sakai and Koyama (1996) recently reported that cholinoceptive PS-on cells localized in the peri-LCα are inhibited by noradrenaline. Furthermore, Sakai and Onoe (unpublished observations) and Tononi et al. (1989) showed that application of $\alpha 2$ adrenergic agonists in the peri-LCα region suppressed PS. In conclusion, they are some arguments that PS-off noradrenergic cells of the LC could inhibit the PS-on cells. The classical "reciprocal interactions" models also implied that the cessation of firing of LC neurons during PS is the result of active PS-specific inhibitory processes, originating from the pontine neurons responsible for the PS onset and maintenance ("PS-on" cells) (Hobson et al., 1975; Sakai, 1985). These neurons were first thought to be the PS-on neurons presumed cholinergic localized in the dorsal pons (laterodorsal tegmental, pedunculopontine and peri-LCα nuclei). It has later been suggested that they might use GABA or glycine rather than acetylcholine as an inhibitory neurotransmitter (Luppi et al., 1991; Jones, 1991). Indeed, acetylcholine excites LC neurons (Guyenet and Aghajanian, 1979; Koyama and Kayama, 1993). In contrast, in anesthetized rats, iontophoretic application of GABA or glycine (GLY) strongly inhibits LC neurons and co-iontophoresis of bicuculline or strychnine, respectively $GABA_A$ and GLY antagonists, antagonize this effect (Gallager and Aghajanian, 1976; Gallager, 1978; Ennis and Aston-Jones, 1989; Luppi et al., 1991). Furthermore, *in vitro* studies on slices using focal stimulation and bath-applied bicuculline or strychnine revealed GABA- and GLY- mediated IPSP's in LC neurons (Cherubini et al., 1988; Williams et al., 1991; Osmanovic and Shefner, 1990; Pan and Williams, 1989). In agreement with these results, GABA- and GLY-immunoreactive varicose fibers and $GABA_A$ and GLY receptors have been found in the rat LC (Zarbin et al., 1981; Jones, 1991; Luppi et al., 1991; Wang et al., 1992; Luque et al., 1994). Supporting the hypothesis that glycinergic neurons could be responsible for the inhibition of monoaminergic neurons during PS, it has been shown that glycine was responsible for the inhibition of the motoneurons during PS (Chase et al., 1989). In contrast, rather supporting a role for GABA, microinjections of picrotoxin (a $GABA_A$ antagonist) in rat LC significantly reduced the duration of PS episods (Kaur et al., 1997).

In addition to a near cessation of activity during PS, the LC neurons greatly reduce their activity during SWS. This reduction implies either that a deactivation or an active inhibition takes place at the transition between W and SWS and during SWS. The classical view is that the GABA transmission is increased during SWS as supported by the strong hypnotic properties of the benzodiazepines acting on the $GABA_A$ receptors (reviews in Mendelson, 1992; Gaillord, 1994). Moreover, Nitz and Siegel (1997) recently found an increase of the amount in GABA in the cat's LC during SWS compared to W. A GABA inhibition might therefore likely be responsible for the decrease of activity of LC neurons during SWS. Another possibility is that the LC neurons are deactivated during SWS. Against this hypothesis, it has been proposed that their spontaneous tonic firing during W, which is close to that observed in anesthetized rats is mainly due to their intrinsic pacemaker properties, as revealed by intracellular recordings in slices (Aghajanian and Vandermaelen, 1982; Williams et al., 1984) and in cultures of LC cells (Masuko et al., 1986). This implies that at least during quiet W the LC cells are not tonically activated by excitatory influences.

In conclusion, a number of arguments suggest that GABA and/or GLY might be responsible for the inhibition of LC neurons during SWS and PS. To test this hypothesis: (i) we applied bicuculline or strychnine on LC cells during SWS, PS and W using a new method which allows extracellular single-unit recordings combined with iontophoresis, in the head restrained awake rat (Darracq et al., 1996) (Fig. 1) and (ii) we localized in rats the glycinergic and GABAergic neurons potentially responsible for the inhibition of the NA neurons, combining injections of the retrograde tracer cholera-toxin B subunit (CTb) in LC and immunohistochemistry of GLY or GAD (GABA enzyme of synthesis).

Effect of the Application of GABA and Glycine Antagonists on the Activity of the Rat LC Neurons During Sleep

Like in the study of Aston-Jones and Bloom (1981), noradrenergic LC neurons were identified by their typical broad action potentials (1.5–2 msec duration), and a phasic excitatory response to a sensory stimulus (auditive click) immediately followed by an inhibition (Chouvet et al., 1988; Akaoka et al.,

1992). Their mean discharge rate was of 1.39 Hz during quiet W. During SWS, LC cells showed a decrease in their firing rate (0.56 Hz) while during PS episods they were nearly quiescent (0.01 Hz) showing only occasional single spikes (Gervasoni et al., 1998).

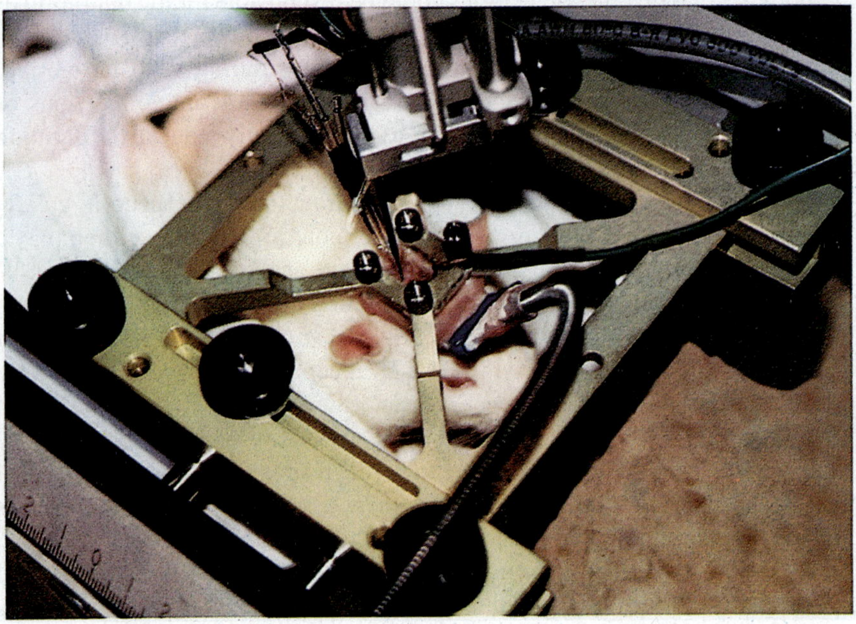

Fig. 1. Photograph illustrating rat in the restraining device. Rat is fixed to the specially designed restraining frame with a U-shaped piece of metal bearing four vertical screws glued together with the polygraphic electrodes and a connector for EEG/EMG recording in dental cement. The multibarrel electrode is inserted into the brain through a trephine hole made above the LC.

Iontophoretic ejections of strychnine induced a sustained increase in the discharge rate of LC neurons regardless of the vigilance state: from 1.70 to 11.41 Hz during W, 0.39 to 9.34 Hz during SWS (Fig. 2A) and 0.06 to 14.13 Hz during PS (Fig. 2B) (Darracq et al., 1996). For some neurons, the rat displayed successive short periods of SWS and W during the effect of strychnine (Fig. 2A). In these neurons, we compared their increase in discharge rate during successive SWS and W periods. We found that their increase of activity was 1.93 time superior during W that during SWS periods. This difference was statistically significant.

Like in anesthetized animals (Ennis and Aston-Jones, 1989; Luppi et al., 1991), iontophoretic applications of glycine or GABA suppressed the spontaneous discharge of LC neurons during W or SWS, and glycine- but not GABA-induced inhibitions were antagonized by co-iontophoresis of strychnine.

Iontophoretic ejections of bicuculline during W, SWS and PS induced a progressive and sustained increase of the discharge rate of LC neurons without

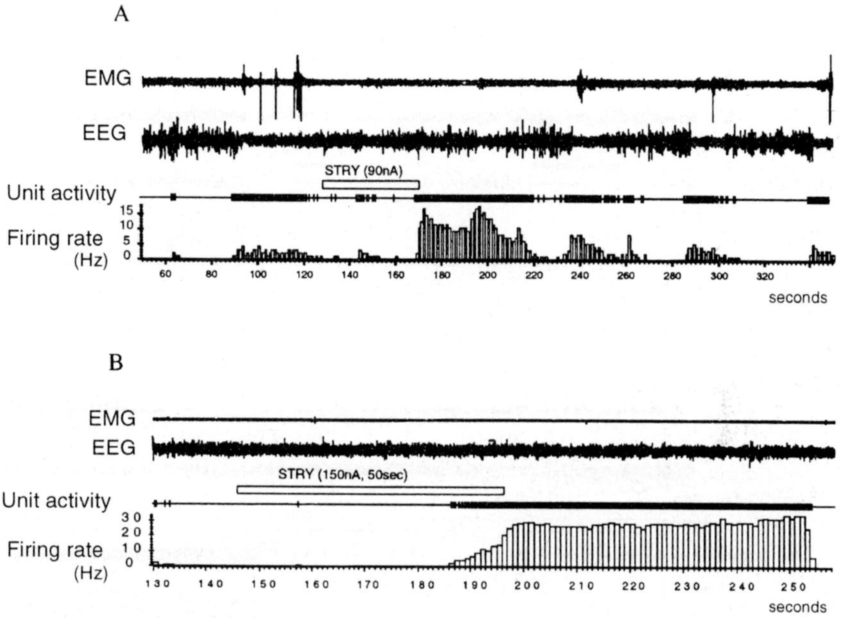

Fig. 2. **(A) Effect of an iontophoretic application of ṣtrychnine (90 nA). This application induced an activation of the neuron during SWS (EEG with high voltage slow activity and spindles), from 0.16 to 8.9 Hz starting 42 sec after the onset of the drug application. During the following short period of W (193–200 sec), the discharge rate of the neuron reached 14 Hz. Then, with the return of slow waves, it strongly decreased up to the next W phase (235–250 sec) during which the discharge rate of the neuron re-increased largely above the normal W values. The persistence of a decrease of activity during SWS of this LC cells during the strychnine effect suggest that another inhibitory neurotransmitter that GLY specifically inhibits LC cells during SWS. (B): Effect of an iontophoretic application of strychnine during PS (150 nA, 50 sec). Note the sustained activation of the LC unit starting about 40 sec after the onset of the strychrine application.**

inducing a change in the vigilance state (Gervasoni et al., 1998). The firing frequency increased from 1.09 to 8.72 Hz during W (Fig. 3B) and from 0.34 to 6.91 Hz during SWS (Fig. 3A). During PS, while LC neurons were practically silents (0.01 Hz), they showed following bicuculline application a remarkable firing rate of 6.59 Hz (Fig. 3A, B). For some neurons, the rat displayed successive short periods of SWS and W during the effect of bicuculline. In these neurons, we compared their increase in discharge rate during successive SWS and W periods. We found that their increase of activity was 1.23 time superior during W that during SWS periods. In contrast to strychnine, this difference was not statistically significant.

During W or SWS, GABA-induced inhibitions were fully antagonized by bicuculline co-iontophoresis, while GLY-induced inhibitions were unaffected (Fig. 3).

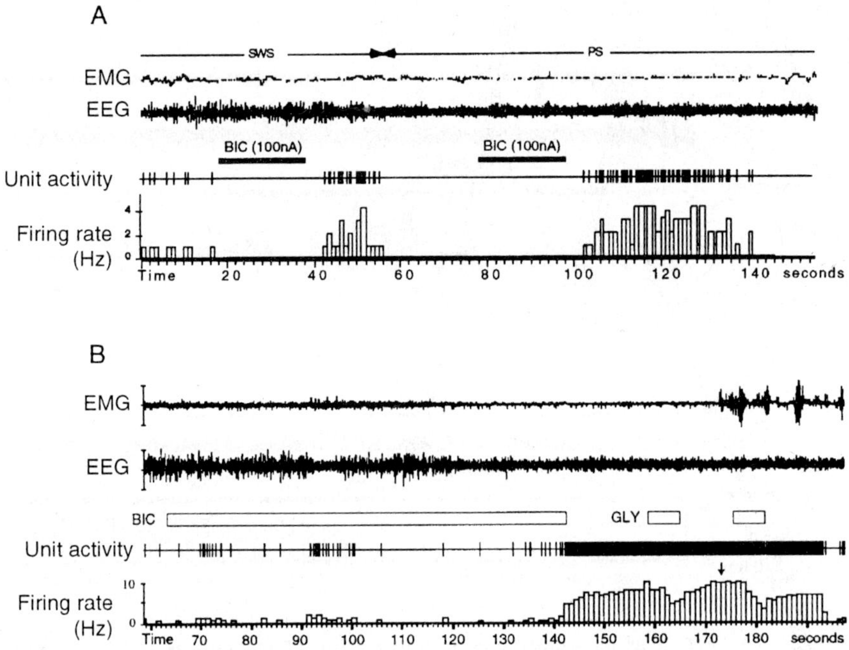

Fig. 3 (A) Effect of two successive iontophoretic applications of bicuculline (100 nA, 20 sec) during SWS and PS (characterized by flat EMG and desynchronized EEG). Note the small increase in the discharge rate of the LC neuron, from 0.6 to 1.8 Hz during SWS, and 0 to 3.6 Hz during PS. The effect appeared around 25 sec after the onset of the ejections. The increase in firing rate did not alter the vigilance state of the animal. (B): Polygraphic recordings displaying the EMG, EEG, the unit activity of a LC neuron, its firing rate, and the effect of the iontophoresis of bicuculline (50 nA, 52 sec) during PS. GLY still inhibits this LC neuron during the bicuculline-induced activation. Note that during the effect of bicuculline, the firing rate of the LC cell did not further increase when the rat woke up (arrow). This result suggest that GABA is responsible for the inhibition of LC cells during PS.

Glycinergic and GABAergic Afferent Projections to the Rat LC

Afferent projections to the LC

The procedure used for retrograde tracing with cholera-toxin B subunit (CTb) and anterograde tracing with PHAL has been described previously in details (Luppi et al., 1990, 1995; Peyron et al., 1996). The high sensitivity of CTb allowed us to demonstrate pathways not seen before using HRP, WGA-HRP or fluorescent tracers.

Following iontophoretic injections of CTb into the core of the LC, we observed a substantial number of retrogradely-labeled cells (CTb+) in the lateral and dorsal paragigantocellular nuclei as previously described (Aston-Jones et al., 1986).

We also saw a substantial number of retrogradely-labeled neurons in: (1) the preoptic area dorsal to the supraoptic nucleus; (2) areas of the posterior hypothalamus; (3) the Kölliker-Fuse nucleus; and (4) and the mesencephalic reticular formation. Fewer labeled cells were also observed in other regions including the hypothalamic paraventricular nucleus, nucleus raphe dorsalis, median raphe nucleus, dorsal and ventral parts of the periaqueductal gray, the area of the noradrenergic A5 group, the lateral parabrachial nucleus and the caudoventrolateral reticular nucleus.

These results were confirmed and extended with anterograde transport of CTb or PHAL. Injections of these tracers in the lateral paragigantocellular nucleus, preoptic area dorsal to the supraoptic nucleus, the posterior hypothalamic areas, the ventrolateral part of the periaqueductal gray, the Kölliker-Fuse nucleus yielded a substantial to large number of labeled fibers in the nuclear core of the LC.

In conclusion, our results indicate that the LC receives afferents from a very large number of structures from the forebrain to the medulla. Failure to demonstrate these afferents in earlier works (Cedarbaum and Aghajanian,1978; Aston-Jones et al., 1986) was certainly due to the poor sensibility of the retrograde tracers used (HRP, WGA-HRP and fluorogold).

Origin of glycinergic inputs to the LC

Following CTb injections centered on the LC, a large number of double-labeled cells (CTb+/GLY+) were observed in the ventrolateral and lateral parts of the periaqueductal gray where they represented a small proportion of the CTb+ cells (Fig. 4A). At the same level, a substantial number of CTb+/GLY+ neurons were also detected in the mesencephalic reticular formation. A moderate number of double-labeled neurons were also seen in the raphe magnus, gigantocellular alpha, lateral and dorsal paragigantocellular nuclei (Rampon et al.,1996). These results are summarized in Fig. 4.

Origin of GABAergic inputs to the LC

Following CTb injections in the LC, many brain regions contained GAD+/CTb+ neurons. A large number of double-labeled cells were localized in the lateral preoptic area, the lateral hypothalamic area, the lateral and ventrolateral parts of the periaqueductal gray and the dorsal paragigantocellular nucleus. A small number of double-labeled cells were also seen in the dorsal hypothalamic area, the tuberomamillary nucleus, the mesencephalic reticular formation, the lateral parabrachial nucleus and the nucleus raphe magnus (Peyron et al., 1995). These results are summarized in Fig. 5.

Physiological Role of Glycinergic Inputs to the LC

The effect of strychnine indicates the existence of a tonic glycinergic input

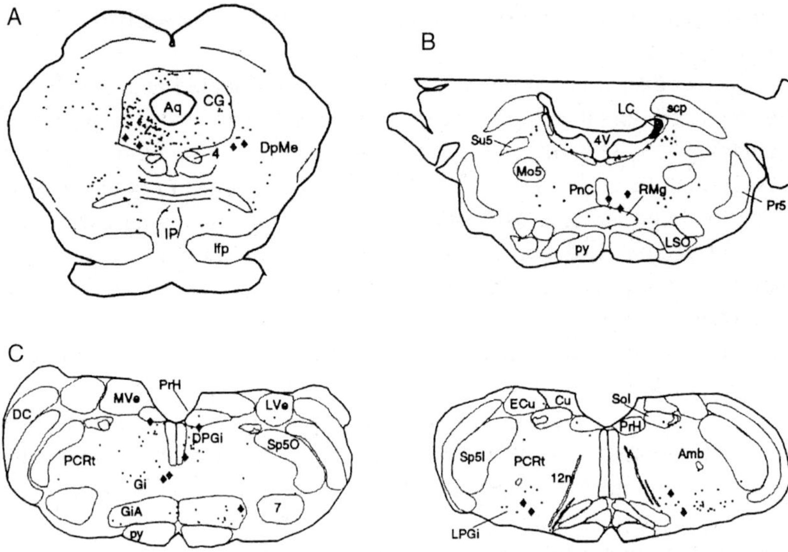

Fig. 4. Rostro-caudal schematic drawings of 20 μm frontal sections from a rat injected in the LC. Each losange represents a double-labeled neuron (GLY+/CTb) and each dot a singly CTb labeled neuron.

on LC cells during W, SWS and PS. We further found that when the same neuron was recorded during short successive periods of SWS and W during the pharmacological effect, its increases in discharge rate was strongly superior during W than SWS. These results demonstrate that when the tonic inhibition by GLY is completely removed by strychnine, the LC cells are still under a specific inhibition by another neurotransmitter during SWS. Furthermore, our results with bicuculline indicate that GABA appears to be this neurotransmitter (see next chapter). Unfortunately, due to the smaller number of LC cells successively recorded during SWS and PS, we were not able to make the same comparison between these two sleep states. However, with the microdialysis technique, Nitz and Siegel (1997) recently found no detectable change in the GLY concentration in the cat LC during SWS and PS as compared to waking values. The glycinergic tonic inhibition might therefore be kept constant across the sleep-waking cycle and control the general excitability of LC neurons. This control could either arise from the major glycinergic input to the LC from the ventrolateral and lateral parts of the periaqueductal gray (CG) and the adjacent mesencephalic reticular formation or the smaller inputs from the lateral paragigantocellular, raphe magnus, dorsal paragigantocellular and gigantocellular alpha nuclei.

Of interest regarding the periaqueductal gray input, Sastre et al. (1996) in cats recently showed that the inactivation of its ventrolateral part by muscimol (a $GABA_A$ agonist) induced a dramatic increase of PS. This result might be explained by the fact that the glycinergic neurons constitute only a minor

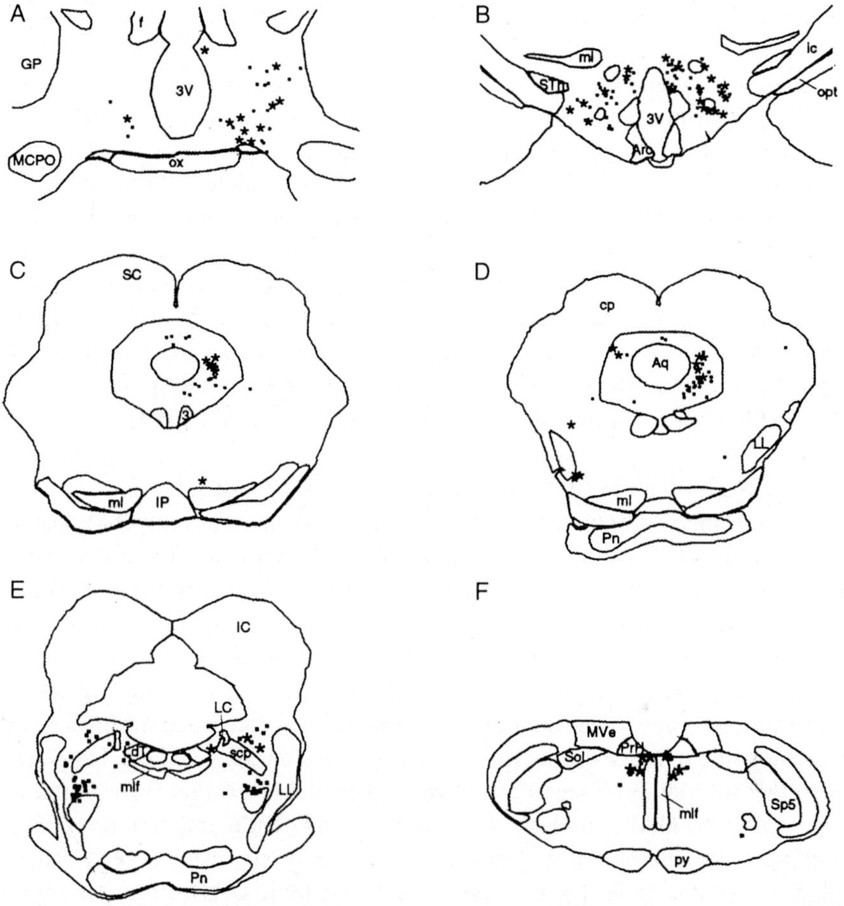

Fig. 5. Rostro-caudal schematic drawings of 20 μm frontal sections from a rat injected in the LC. Each star represents a double-labeled neuron (GAD+/CTb) and each dot a small square CTb labeled neuron.

proportion of the neurons from the ventrolateral periaqueductal gray. The effect seen by Sastre et al. (1996) could therefore be due to the inhibition of other types of neurons (in particular glutamatergic or GABAergic) (Beitz, 1995).

Besides, we observed a small number of glycinergic neurons in the nucleus gigantocellular alpha and the adjacent raphe magnus projecting to the LC. Based on a number of studies in cats, we made the hypothesis that during PS, the monoaminergic neurons and the cranial and spinal somatic motoneurons might be inhibited by a single population of glycinergic neurons located in the magnocellular reticular nucleus (cat's equivalent of gigantocellular alpha and ventral nuclei, Giα and GiV). However, the present results and those we recently obtained on the glycinergic afferents to the motor trigeminal nucleus (Rampon et al., 1996) indicate at least in rats that the GiV and Giα provide only a limited glycinergic input to the LC and Mo5 and are therefore unlikely to contain a single population of neurons responsible for the inhibition of monoaminergic and all somatic motoneurons during PS. In fact, in rats, the only motoneurons receiving a strong glycinergic projection from the Gia and GiV are those of the spinal cord as demonstrated by Holstege and Bongers (1991). In view of these results in rats, the cranial motoneurons might be rather inhibited during PS by neurons from the parvocellular and parvocellular alpha medullary reticular nuclei and glycine might not be specifically involved in the inhibition of LC neurons during sleep. Additional experiments are nevertheless necessary to confirm such conclusion in particular in cats in which it has been shown that the magnocellular but not the parvocellular reticular nuclei contains c-Fos positive cells following long periods of PS induced by carbachol injections in the pons (Yamuy et al., 1993)

Physiological Role of GABAergic Inputs to the LC

Our electrophysiological data indicates the existence of a tonic GABAergic input on the LC cells during W, SWS and PS. Further, we found that when the same neuron was recorded during short successive periods of SWS and W during the bicuculline effect, its discharge rate was no longer statistically different between W and SWS. These results highly suggest that a GABA release is responsible for the inactivation of LC neurons during SWS. Unfortunately, due to the smaller number of LC cells recorded during PS, we were not able to make the same comparison between SWS and PS. However, with the microdialysis technique, Nitz and Siegel (1997) recently found an increase of the GABA release in the cat LC during SWS and moreover PS as compared to waking values. Based on these and our results, we therefore suggest that during W, the LC cells are under a tonic GABAergic inhibition which increases during SWS and even more during PS and is at least partly responsible for their inactivation during these states.

Our results indicate that the LC receives GABAergic inputs from neurons located in a large number of distant regions from the forebrain to the medulla. We also observed a substantial number of GAD+/CTb+ neurons in the pontine and mesencephalic periaqueductal gray. These results indicate that the GABA innervation of this nucleus arises from multiple distant GABAergic groups in addition to local GABAergic groups. Such results contrast with the classical concept that GABA is mainly contained in interneurons. They suggest that the NA neurons of the LC could be inhibited by multiple populations of GABAergic neurons located in different structures and raises the question of the functional role of such complexity. One possibility is that only some of these GABAergic afferents are destined to the noradrenergic neurons of the LC. This seems unlikely because the LC in rats contains nearly exclusively noradrenergic cells. Another possibility is that some of these afferents are postsynaptic and others presynaptic, but the more likely explanation is that each of these afferents is active only under specific physiological conditions.

Based on physiological and electrophysiological data (see above), we expect that one or several of these GABAergic afferents are "turned on" specifically during SWS and others during PS and are responsible for the progressive decrease of activity of noradrenergic neurons during these sleep states.

The most likely candidate for the inhibition of the noradrenergic neurons during SWS is the lateral preoptic area. Indeed, lesion of this structure in cats and rats induces an insomnia while its stimulation induced SWS (Asala et al., 1990; Lucas and Sterman, 1975; McGinty and Sterman, 1968; Sallanon et al., 1989; John et al., 1994; Sterman and Clemente, 1962). Neurons increasing their activity during SWS have been recorded in this area (Kaitin, 1984; Szymusiak and McGinty, 1986; Koyama and Hayaishi, 1994). Moreover, c-Fos positive cells were observed in the lateral preoptic area after long periods of SWS (Sherin et al., 1996) and it has been further shown that these neurons are in part GABA- and galanin-positive and project to the tuberomamillary nucleus which contains waking active neurons presumably histaminergic (Vanni-Mercier et al., 1984). From these and our results, we can therefore propose that GABAergic neurons in the lateral preoptic area increase their firing just before the onset and during SWS and induce SWS via their inhibitory projections to waking-inducing structures (tuberomamillary nucleus, DRN and LC among others).

Besides the lateral preoptic area, the posterior hypothalamic areas, the periaqueductal gray and the dorsal paragigantocellular nucleus provide substantial GABA inputs to the LC.

The strong GABAergic projection from the lateral hypothalamic area is rather puzzling. Indeed, since the initial demonstration that the lesion of the posterior hypothalamus induces somnolence (Von Economo, 1926), the histaminergic neurons of the posterior hypothalamus have been more specifically implicated in waking (Lin et al., 1996). These neurons are located in the tuberomamillary nucleus ventral and caudal to the GABAergic neurons from the lateral hypothalamic area projecting to the LC. Nevertheless, muscimol injections in cats in the lateral hypothalamic area in addition to those in the tuberomamillary nucleus induced hypersomnia (Lin et al., 1989). Additional experiments are therefore needed to determine whether the lateral hypothalamic area and its GABAergic neurons play a role in vigilance via their projections to the LC.

The GABAergic afferents responsible for the inhibition of noradrenergic neurons during PS should be located in the brainstem. Indeed, it is well known that PS-like episods occur in pontine or decerebrate cats (Jouvet, 1972). Moreover, it has recently been shown that, in decerebrate animals, PS episods induced by carbachol injections in the pons are still associated with a cessation of activity of serotonergic neurons of the raphe obscurus and pallidus nuclei (Woch et al., 1996).

In the brainstem, we observed GABAergic projections to the LC from the periaqueductal gray and the dorsal paragigantocellular nucleus. In agreement with this last result, local application of bicuculline blocked the dorsal paragigantocellular-evoked inhibition of LC neurons (Ennis and Aston-Jones, 1989). The GABAergic afferents from the periaqueductal gray and the dorsal paragigantocellular nucleus could therefore be responsible for the inhibition of noradrenergic neurons of the LC during PS.

The hypothesis that this inhibition is arising from neurons located in the periaqueductal gray is further supported by two recent studies. Indeed, Yamuy et al. showed that after a long period of PS induced by pontine injection of carbachol, a large number of 5HT negative c-Fos positive cells are visible in the DRN and lateral to it (Yamuy et al., 1995). Moreover, Maloney and Jones (1997) observed after a PS rebound induced by deprivation an increase in c-Fos+/GAD+ neurons in the periaqueductal gray and the lateral tegmental nucleus.

Finally, it must be precised that although a number of arguments are in favor of a role of GABA in the inhibition of noradrenergic neurons of the LC during SWS and PS, other inhibitory neuroactive substances might also participate in this inhibition. Indeed, it has for example been shown that LC cells are inhibited by local application of enkephalin (Bird and Kuhar, 1977; Guyenet and Aghajanian, 1979). Moreover, it has been shown that adenosine release increases during wakefulness and decreases during SWS (Porkka-Heiskanen et al., 1997) and adenosine application is inhibitory on LC noradrenergic cells (Pan et al., 1995; Shefner and Chiu, 1986).

The removal during SWS or PS of some tonic excitatory inputs present during W (like acetylcholine or glutamate) might also participate in the decrease of activity of LC neurons during sleep.

Conclusions and New Hypothesis

During SWS, the noradrenergic neurons of the LC would be inhibited by GABAergic neurons located in the lateral preoptic area. At the entrance and during PS, a second population of GABAergic neurons located in the periaqueductal gray or the dorsal paragigantocellular nucleus would be responsible for the complete cessation of the noradrenergic neurons of the LC during this sleep state. The cranial motoneurons would be inhibited by glycinergic neurons of the parvocellular nucleus and those of the spinal cord by glycinergic neurons from the gigantocellular ventral nucleus and glycinergic interneurons of the intermediate zone of the spinal cord.

Acknowledgements

This work was supported by INSERM (U480), CNRS (ERS 5645), Université Claude Bernard Lyon I and the 1996 ESRS-Synthélabo European Research Grant. The authors wish to thank C. Guillemort (GFG Co, Pierre-Bénite) for his help in designing the head-restraining system.

Abbreviations used in the text

CNS	central nervous system
CTb	cholera toxin B subunit
DRN	dorsal raphe nucleus
EEG	electroencephalogram
EMG	electromyogram
GABA	gamma amino-butyric acid
GAD	glutamate decarboxylase
Giα	nucleus gigantocellular alpha
GiV	nucleus gigantocellular ventral;
GLY	glycine
IPSP	inhibitory post-synaptic potential
LC	locus coeruleus
PS	paradoxical sleep
SWS	slow wave sleep
W	wakefulness

References

Aghajanian, G.K. and VanderMaelen, C.P. Intracellular identification of central noradrenergic and serotonergic neurons by a new double labeling procedure, J. Neurosci., **2** (1982) 1786–1792.

Akaoka, H., Charléty, P.J., Saunier, C.F., Buda, M. and Chouvet, G. Combining in vivo volume-controlled pressure microinjection with extracellular unit recording, J. Neurosci. Meth., **42** (1992) 119–128.

Asala, S.A., Okano, Y., Honda, K. and Inoué, S. Effects of medial preoptic area lesions on sleep and wakefulness in unrestrained rats, Neurosci. Lett., **114** (1990) 300–304.

Aston-Jones, G., Ennis, M., Pieribone, V.A., Nickell, W.T. and Shipley, M.T. The brain nucleus locus coeruleus: restricted afferent control of a broad efferent network, Science, **234** (1986) 734–737.

Aston-Jones, G. and Bloom, F.E. Activity of norepinephrine-containing locus coeruleus neurons in behaving rats anticipates fluctuations in the sleep-waking cycle, J. Neurosci., **1** (1981) 876–886.

Beitz, A., *The Rat Nervous System*, Second Edition, G. Paxinos (Ed.), San Diego, 1995, pp. 173–182.

Bird, S.J. and Kuhar, M.J. Iontophoretic applications of opiates to the locus coeruleus, Brain Res., **122** (1977) 523–533.

Cedarbaum, J.M. and Aghajanian, G.K. Afferent projections to the rat locus coeruleus as determined by a retrograde tracing technique, J. Comp. Neurol., **178** (1978) 1–16.

Chase, M.H., Soja, P.J. and Morales, F.R. Evidence that glycine mediates the postsynaptic potentials that inhibit lumbar motoneurons during the atonia of active sleep, J. Neurosci., **9** (1989) 743–751.

Cherubini, E., North, R.A. and Williams, J.T. Synaptic potentials in rat locus coeruleus neurons, J. Physiol. (London), **406** (1988) 431–442.

Chouvet, G., Akaoka, H. and Aston-Jones, G. Serotonin selectively decreases glutamate-induced excitation of locus coeruleus neurons, C.R. Acad. Sci., Paris, **306** (1988) 339–344.

Dahlström, A. and Fuxe, K. Evidence for the existence of monoamine-containing neurons in the central nervous system. I. Demonstration in the cell bodies of brain stem neurons, Acta Physiol, Scand. (Suppl.), **232** (1964) 1–55.

Darracq, L., Gervason, D., Soulière, F., Lin, J.S., Fort, P., Chouvet, G. and Luppi, P-H. Effect of strychnine on rat locus coeruleus neurons during sleep and wakefulness, NeuroReport, **8** (1996) 351–355.

Ennis, M. and Aston-Jones, G. GABA-mediated inhibition of locus coeruleus from the dorsomedial rostral medulla, J. Neurosci., **9** (1989) 2973–2981.

Gaillard, J.M., *Principles and Practice of Sleep Medicine*, M.H. Kryger, T. Roth and W.C. Dement (Eds.), Philadelphia, 1994, pp. 349–354.

Gallager, D.W. and Aghajanian, G.K. Effect of antipsychotic drugs on the firing of dorsal raphe cells. II. Reversal by picrotoxin, Eur. J. Pharmacol., **39** (1976) 357–364.

Gallager, D.W. Benzodiazepines: Potentiation of a GABA inhibitory response in the dorsal raphe nucleus, Eur. J. Pharmacol., **49** (1978) 133–143.

Gervasoni, D., Darracq, L., Fort P., Soulière F., Chouvet, G. and Luppi, P.H. Electrophysiological evidence that noradrenergic neurones of the rat locus coeruleus are tonically inhibited by GABA during sleep, Eur. J. Neurosci., **10** (1998) 964–970.

Guyenet, P.G. and Aghajanian, G.K. Ach, Substance P and Met-Enkephalin in the locus coeruleus: pharmacological evidence for independent sites of action, Eur. J. Pharmacol., **53** (1979) 319–328.

Henley, K. and Morrison, A.R. A re-evaluation of the effects of lesions of the pontine tegmentum and locus coeruleus on phenomena of paradoxical sleep in the cat, Acta Neurobiol. Exp. Warszawa, **34** (1974) 215–232.

Hobson, J. McCarley, R. and Wyzinski, P. Sleep cycle oscillation: reciprocal discharge by two brainstem groups, Science, **189** (1975) 55–58.

Holstege, J.C. and Bongers, C.M.H. A glycinergic projection from the ventromedial lower brainstem to spinal motoneurons. An ultrastructural double labeling study in rat, Brain Res., **566** (1991) 308–315.

John, J., Kumar, V.M., Gopinath, G., Ramesh, V. and Mallick, H. Changes in sleep-wakefulness after kainic acid lesion of the preoptic area in rats, Jpn. J. Physiol., **44** (1994) 231–242.

Jones, B.E., Harper S.T. and Halaris A.E. Effects of locus coeruleus lesions upon cerebral monoamines content, sleep-wakefulness states and the response to amphetamine in the cat, Brain Res., **124** (1977) 473–496.

Jones, B.E. Noradrenergic locus coeruleus neurons: their distant connections and their relationship to neighboring, including cholinergic and GABAergic neurons of the central gray and reticular formation. Prog. Brain Res., **88** (1991) 15–30.

Jouvet, M. The role of monoamines and acetylcholine-containing neurons in the regulation of the sleep, Ergebn. Physiol., **64** (1972) 166–307.

Jouvet, M. and Delorme, F. Locus coeruleus et sommeil paradoxal, C.R. Soc. Biol., Paris, **159** (1965) 895–899.

Kaitin, K.I. Preoptic area unit activity during sleep and wakefulness in the cat, Exp. Neurol., **83** (1984) 347–357.

Kaur, S., Saxena, R.N. and Mallick, B.N. GABA in locus coeruleus regulates spontaneous rapid eye movement sleep by acting on $GABA_A$ receptors in freely moving rats, Neurosci, Lett., **223** (1997) 105–108.

King, C.D. and Jewet, R.E. The effects of α-methyl-tyrosine on sleep and brain norepinephrine in cats, J. Pharmacol. Exp. Ther., **177** (1971) 188–195.

Koyama, Y. and Kayama, Y. Mutual interactions among cholinergic, noradrenergic and serotonergic neurons studied by iontophoresis of these transmitters in rat brainstem nuclei, Neuroscience, **55** (1993) 117–126.

Koyama, Y. and Hayaishi, O. Firing of neurons in the preoptic/anterior hypothalamic areas in rat: its possible involvement in slow wave sleep and paradoxical sleep, Neurosci, Res., **19** (1994) 31–38.

Lidbrink, P. The effect of lesions of ascending noradrenaline pathways on sleep and waking in the rat, Brain Res., **74** (1974) 19–40.

Lin, J.S., Hou, Y., Sakai, K. and Jouvet, M. Histaminergic inputs to the mesopontine tegmentum and their role in the control of cortical activation and wakefulness in the cat, J. Neurosci., **15** (1996) 1523–1537.

Lin, J.S., Sakai, K., Vanni-Mercier, G. and Jouvet, M. A critical role of the posterior hypothalamus in the mechanisms of wakefulness determined by microinjections of muscimol in freely moving cats, Brain Res., **429** (1989) 225–240.

Lucas, E.A. and Sterman, M.B. Effect of a forebrain lesion on the polycyclic sleep wake patterns in the cat, Exp. Neurol., **46** (1975) 368–388.

Luppi, P-H., Charléty, P.J., Fort, P., Akaoka,H., Chouvet, G. and Jouvet, M. Anatomical and electrophysiological evidence for a glycinergic inhibitory innervation of the rat locus coeruleus, Neurosci. Lett., **128** (1991) 33–36.

Luppi, P-H., Aston-Jones, G., Akaoka, H., Chouvet, G. and Jouvet, M. Afferent projections to the rat locus coeruleus demonstrated by retrograde and anterograde tracing with cholera-toxin B submit and Phaseolus vulgaris leucoagglutinin, Neuroscience, **65** (1995) 119–160.

Luppi, P-H., Fort, P. and Jouvet, M. Iontophoretic application of unconjugated cholera toxin B subunit CTb combined with immunohistochemistry of neurochemical substances: a method for transmitter identification of retrogradely labeled neurons, Brain Res., **534** (1990) 209–224.

Luque, J.M., Malherbe, P. and Richards, J.G. Localization of $GABA_A$ receptor subunit mRNAs in the rat locus coeruleus, Mol. Brain Res., **24** (1994) 219–226.

Maloney, K.J. and Jones, B.E. C-Fos expression in cholinergic, GABAergic and monoaminergic cell groups during paradoxical sleep deprivation and recovery, *Soc.* Neurosci., Abstr. **23** (1997) 2131.

McCarley, R. and Hobson, J.A. Single neuron activity in cat gigantocellular tegmental field: Selectivity of discharge in desynchronized sleep, Science, **174** (1971) 1250–1252.

McGinty, D.J. and Stermann, M.B. Sleep suppression after basal forebrain lesions in the cat, Science, **160** (1968) 1253–1255.

McGinty, D.J. and Harper, R.M. Dorsal raphe neurons: depression of firing during sleep in cats, Brain Res., **101** (1976) 569–575.

Masuko, S., Nakajima, Y., Nakajima, S. and Yamaguchi, K. Noradrenergic neurons from the locus coeruleus in dissociated cell culture: culture methods, morphology and electrophysiology, J. Neurosci., **6** (1986) 3229–3241.

Mendelson, W.B. Neuropharmacology of sleep induction by benzodiazepines, Crti. Rev. Neurobiol., **622** (1992) 1–232.

Nitz, D.A. and Siegel, J.M. GABA release in the locus coeruleus as a function of sleep/wake state, Neuroscience, **78** (1997) 795–801.

Osmanovic, S.S. and Shefner, S.A. γ-aminobutyric acid responses in rat locus coeruleus neurons in vitro: a current-clamp and voltage-clamp study, J. Physiol., (London), **421** (1990) 151–170.

Pan, W.J., Osmanovic, S.S. and Shefner, S.A. Characterization of the adenosine A1 receptor-activated potassium current in rat locus coeruleus neurons, J. Pharmacol. Exp. Ther., **273** (1995) 537–544.

Pan, Z.Z. and Williams, J.T. GABA- and glutamate-mediated synaptic potentials in rat dorsal raphe neurons in vitro, J. Neurophysiol., **61** (1989) 719–726.

Peyron, C., Luppi, P-H., Fort, P., Rampon, C. and Jouvet, M. Lower brainstem catecholamine afferents to the rat dorsal raphe nucleus, J. Comp. Neurol., **364** (1995) 402–413.

Peyron, C., Luppi, P-H., Rampon, C. and Jouvet, M. Location of the GABAergic neurons projecting to the dorsal raphe nucleus and the locus coeruleus of the rat, Soc. Neurosci, Abstr., **21** (1995) 373.

Porkka-Heiskanen, T., Strecker, R.E., Thakkar, M., Bjorkum, A.A., Greene, R.W. and

McCarley, R.W. Adenosine: a mediator of the sleep-inducing effects of prolonged wakefulness, Science, **276** (1997) 1265–1268.

Rampon, C., Peyron, C., Petit, J.M., Fort, P., Gervasoni, D. and Luppi, P-H. Origin of the glycinergic innervation of the rat trigeminal motor nucleus, NeuroReport, **7** (1996) 3081–3085.

Rampon, C., Peyron, C., Gervasoni, D., Cespuglio, R., Fort, P. and Luppi, P-H. Localization of the glycinergic neurons projecting to the rat locus coeruleus, dorsal raphe and trigeminal motor nuclei, Soc. Neurosci, Abstr., **22** (1996) 1838.

Sakai, K., *Sleep: Neurotransmitters and Neuromodulators*, A. Wauquier, J.M. Monti, J.M. Gaillard and M. Radulovacki (Eds.), New York, 1985, pp. 29–42.

Sastre, J.P., Buda, C., Kitahama, K. and Jouvet, M. Importance of the ventrolateral region of the periaqueductal gray and adjacent tegmentum in the control of paradoxical sleep as studied by muscimol microinjection in the cat, Neuroscience, **74** (1996) 415–426.

Sallanon, M., Denoyer, M., Kitahama, K., Aubert, C., Gay, N. and Jouvet, M. Long lasting insomnia induced by preoptic neuron lesions and its transient reversal by muscimol injection into the posterior hypothalamus in the cat, Neuroscience, **32** (1989) 669–683.

Shefner, S.A. and Chiu, T.H. Adenosine inhibits locus coeruleus neurons: an intracellular study in a rat brain slica preparation, Brain Res., **366** (1986) 364–368.

Sherin, J.E., Shiromani, P.J., McCarley, R.W. and Saper, C.B. Activation of ventrolateral preoptic neurons during sleep, Science, **271** (1996) 216–219.

Shimizu, N., Morikawa, N. and Hokada, M. Histochemical studies of monoamine oxidase of the brain of rodents, Zeitschrift für Zellforschung, **49** (1959) 389–400.

Sterman, M.B.and Clemente, C.D. Forebrain inhibitory mechanisms, sleep pattern induced by basal forebrain stimulation in behaving cat, Exp. Neurol., **6** (1962) 103–117.

Szymusiak, R. and McGinty, D.J. Sleep-related discharge in the basal forebrain of cats, Brain Res., **370** (1986) 82–92.

Tononi, G., Pompeiano, M. and Pompeiano, O. Modulation of desynchronized sleep through microinjection of beta-adrenergic agonists and antagonists in the dorsal pontine tegmentum of the cat, Pflugers Arch., **415** (1989) 142–149.

Vanni-Mercier, G., Sakai, K., Salvert, D. and Jouvet, M. Waking-state specific neurons in the caudal hypothalamus of the cat, C.R. Acad. Sci., Paris, **298** (1984) 195–200.

Von Economo, C., *Handbuch des Normalen und Patholigischen Physiologie*, A. Von Bethe G.V., Bergman G., Embden and U.A. Ellinger (Eds.), Berlin, 1936, pp. 291–610.

Wang, Q.P., Ochiai, H. and Nakai, Y. GABAergic innervation of serotonergic neurons in the dorsal raphe nucleus of the rat studied by electron microscopy double immunostaining, Brain Res. Bull., **6** (1992) 943–948.

Williams, J.T., Bobker, D.H. and Harris, G.C. Synaptic potentials in locus coeruleus neurons in brain slices, Prog. Brain Res., **88** (1991) 167–172.

Williams, J.T., North, R.A., Shefner, A., Nishi, S. and Egan, T.M. Membrane properties of rat locus coeruleus neurons, Neuroscience, **13** (1984) 137–156.

Woch, G., Davies, R.O., Pack, A.I. and Kubin, L. Behavior of raphe cells projecting to the dorsomedial medulla during carbachol-induced atonia in the cat, J. Physiol. (London) **490** (1996) 745–758.

Yamuy, J., Mancillas, J.R., Morales, F.R. and Chase, M.H. C-Fos expression in the pons and medulla of the cat during carbachol-induced active sleep, J. Neurosci., **13** (1993) 2703–2718.

Yamuy, J., Sampagna, S., Lopez-Rodriguez, F., Luppi, P-H., Morales, F.R. and Chase, M.H. Fos and serotonin immunoreactivity in the raphe nuclei of the cat during carbachol-induced active sleep: a double-labeling study, Neuroscience, **67** (1995) 211–233.

Zarbin, M.A., Wamsley, J.K. and Kuhar, M.J. Glycine receptors: light microscopic autoradiographic localization with [3H] strychnine, J. Neurosci., **5** (1981) 5332–547.

Rapid Eye Movement Sleep
B.N. Mallick and S. Inoué (Eds)

8. New Directions for the Study of Cholinergic REM Sleep Generation: Specifying pre- and postsynaptic mechanisms

M. Luisa Capece, Helen A. Baghdoyan and R. Lydic

Department of Anesthesia, The Pennsylvania State University,
College of Medicine, Hershey, PA 17033, USA

SUMMARY

Cholinergic mechanisms in the pontine brain stem play a major role in rapid eye movement (REM) sleep generation. Cholinergic neurons located within the pontomesencephalic tegmentum provide acetylcholine (ACh) to cholinoceptive pontine reticular formation neurons. Direct administration of cholinergic agonists into the medial pontine reticular formation (mPRF) of cat or the nucleus pontis oralis (PnO) of rat induces a REM sleep-like state. The ability of cholinomimetics to cause REM sleep has permitted the testing of causal hypotheses concerning the pre- and postsynaptic mechanisms regulating REM sleep generation. Muscarinic cholinergic receptors (mAChRs) have been localized to the pontine reticular formation, and *in vivo* pharmacological evidence supports the view that activation of mAChRs is key to REM sleep generation. M2 autoreceptors regulate ACh release within the mPRF, and mPRF microinjections have shown that components of the signal transduction pathway modulated by m2/m4 receptors participate in cholinergic REM sleep generation. *In vitro* localization and quantification of M2 receptors and muscarinically activated *G* proteins within REM sleep-related areas of the brain, in addition to *in vivo* studies linking specific intracellular proteins with the behavioral state of REM sleep, have begun to elucidate the cellular and molecular mechanisms modulating REM sleep.

Key Words: acetylcholine, muscarinic receptor, pons.

Introduction

Landmark studies in the early 1960s suggested that acetylcholine (ACh) plays a key role in controlling sleep and wakefulness. Intracerebral application of ACh in cat produces changes in cortical arousal (Cordeau et al., 1963) and induces sleep (Hernandez-Peon et al., 1963). Subsequent studies revealed that cholinergic agonists microinjected into the pontine reticular formation cause a rapid eye movement (REM) sleep-like state that is polygraphically and behaviorally similar to natural REM sleep (Baxter, 1969; George et al.,

1964; Jouvet, 1962). Even a quarter of a century ago, the view that pontine cholinergic neurotransmission contributes to REM sleep generation was supported by a large body of evidence (Jouvet, 1972). These pioneering studies of the neuronal networks and neurotransmitters regulating sleep encouraged efforts in the 1980s to characterize the effects of sleep on systems physiology (Orem and Barnes, 1980) and to understand the clinical significance of sleep-dependent changes in autonomic control (Kryger et al., 1989; Lydic and Biebuyck, 1988). The importance of pontine cholinergic neurotransmission for REM sleep generation featured for the first time in the American Physiological Society *Handbook of Physiology* (Hobson and Steriade, 1986). One result of these diverse lines of inquiry is that more is presently known about the neurobiology of REM sleep than any other complex behavioral state (Steriade and McCarley, 1990). Interested readers are referred elsewhere for detailed reviews of the cholinergic basis of REM sleep generation (Baghdoyan, 1997a; Cuello, 1993; Lydic and Baghdoyan, 1994; Lydic and Baghdoyan, 1998; McCarley et al., 1995). The goal of this chapter is to selectively highlight recent progress aimed at differentiating presynaptic and postsynaptic cholinergic mechanisms generating REM sleep.

Defining Pre- and Postsynaptic Components Generating REM Sleep

In cat, cholinomimetics elicit a REM sleep-like state most reliably when microinjected into the rostrodorsal portion of the medial pontine reticular formation (mPRF) (reviewed in Table 4 of Baghdoyan et al., 1993), and suppress REM sleep when microinjected into the midbrain or medullary reticular formation (Baghdoyan et al., 1984). It rat, the cholinergically induced REM sleep-like state is produced most reliably when cholinergic agonists are microinjected into the caudal portion of the oral pontine reticular formation (Bourgin et al., 1995). This mPRF region is referred to as the gigantocellular tegmental field (FTG) in cat (Berman, 1968), and the homologous area in rat is the nucleus pontis oralis (PnO) (Paxinos and Watson, 1986). Neurons in the mPRF and PnO are not cholinergic but are cholinoceptive. Neuroanantomical studies demonstrated that projections to the mPRF arise from cholinergic neurons in the laterodorsal (LDT) and pedunculopontine tegmental (PPT) nuclei (Jones, 1990; Mitani et al., 1988; Semba et al., 1990; Shiromani et al., 1988). Furthermore, microdialysis data demonstrated that ACh release in the mPRF is regulated by LDT/PPT neurons (Lydic and Baghdoyan, 1993). *In vivo* extracellular recordings revealed that LDT/PPT neurons fire tonically during wakefulness and increase firing before and throughout REM sleep (El Mansari et al., 1989; Kayama et al., 1992). Injections of kainic acid into the LDT, which selectively lesion cell bodies, inhibit REM sleep and cause a loss of muscle atonia in those animals that are still able to initiate REM sleep (Webster and Jones, 1988).

The LDT/PPT has been shown to modulate key electrographic and autonomic characteristics of REM sleep. LDT/PPT neurons also are referred to as mesopontine cholinergic cells and have been suggested to modulate electroencephalographic (EEG) arousal via projections to the thalamus (Semba et al., 1990; Wainer et al., 1993). The LDT/PPT and surrounding pontomesencephalic tegmentum have been implicated in modulating the phasic and electrographic signs of REM sleep such as ponto-geniculo-occipital (PGO) waves and muscle atonia. The PPT sends projections to the lateral geniculate nucleus (LGN) of the thalamus, and it has been hypothesized that the PPT is the site from which PGO waves are relayed from the brain stem to the thalamus (reviewed in Steriade and McCarley, 1990). Electrical stimulation of the mesencephalic reticular formation elicits PGO waves, while bilateral electrolytic lesions (Sakai et al., 1976) and cooling (Laurent et al., 1977) of pontomesencephalic areas abolishes PGO waves. In decerebrate cat, electrical stimulation of the PPT suppresses skeletal muscle tone (Lai and Siegel, 1990), and in intact, anesthetized cat elemetrical stimulation of the LDT/PPT decreases tone of upper airway muscles (Lydic and Baghdoyan, 1993). Together, these data support the concept that cholinergic LDT/PPT projections to the mPRF comprise key components of an anatomically distributed network generating REM sleep. Although numerous molecules have been implicated to participate in REM sleep regulation, experimental focus on defining a limited portion of this cholinergic network has helped identify the presynaptic and postsynaptic mechanisms generating REM sleep.

The Cholinergic Model of REM Sleep

As noted elsewhere (Lydic, 1997; Lydic and Baghdoyan, 1992), a key advantage of the cholinergic model of REM sleep is that it promotes the testing of mechanistic, causal hypotheses. The cholinergic induction of a REM sleep-like state allows one to make causal inferences about the brain region microinjected (Baghdoyan et al., 1984; Calvo et al., 1996; Datta et al., 1991; Semba, 1993), the signal transduction mechanisms activated (Capece et al., 1998; Capece and Lydic, 1997; Leonard and Lydic, 1997; Shuman et al., 1995), and the gene products altered (Shiromani et al., 1992; Yamuy et al., 1993).

The REM sleep-like state induced by carbachol exhibits behavioral, electrographic, neurotransmitter, and molecular features associated with spontaneous REM sleep (Fig. 1). The carbachol-induced REM sleep-like state differs from natural REM sleep in its temporal organization. Following microinjection of carbachol, cats enter REM sleep directly from wakefulness, whereas spontaneously occurring REM sleep always is preceded by non-REM (NREM) sleep. Cholinergic REM sleep and natural REM also differ in the latency to REM sleep onset and in the duration of REM sleep. Cholinergic REM sleep in cat is characterized by a very short (< 10 min) latency to onset and a long duration (30–60 min) of individual episodes. Spontaneous REM sleep has a longer onset latency (> 15 min), and in cat the individual epochs are rarely greater than 12 min in duration.

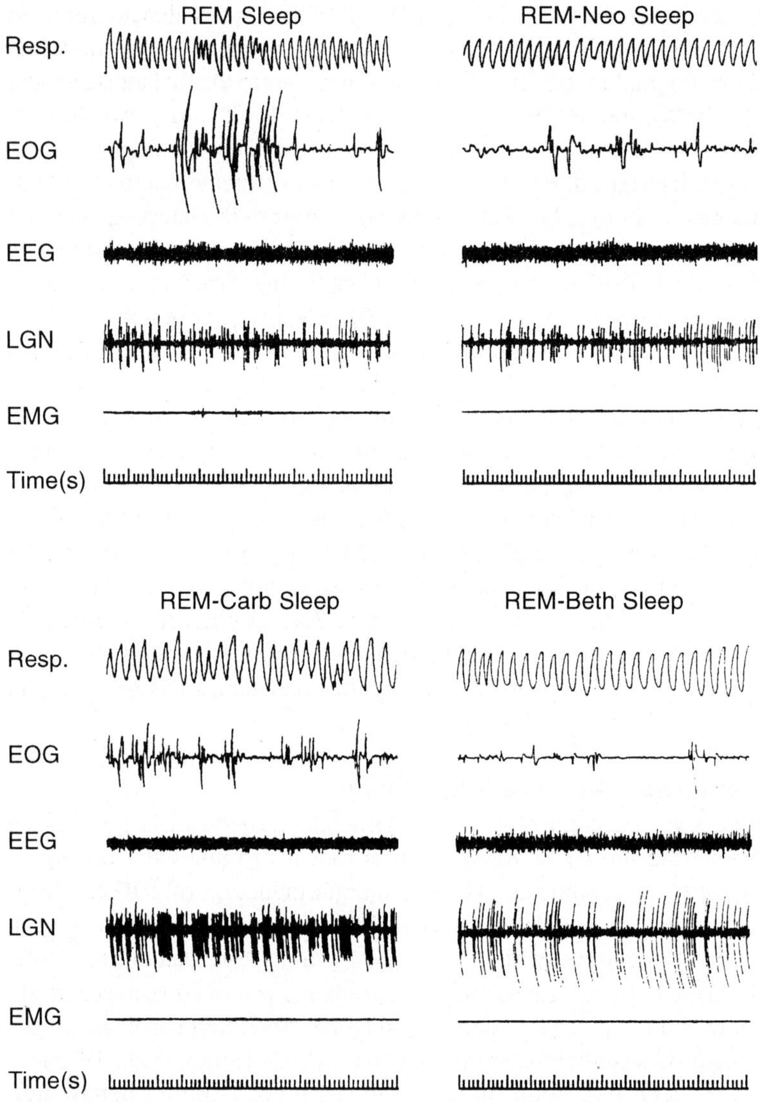

Fig. 1. Polygraphic Similarities Between Natural REM Sleep and Cholinergically Induced REM Sleep. Each column denotes a 1 min polygraphic recording following mPRF microinjection of either saline as a vehicle control (REM sleep), the acetylcholinesterase inhibitor neostigmine (REM-Neo), the mixed cholinergic agonist carbachol (REM-Carb), or the muscarinic agonist bethanechol (REM-Beth). Comparison of each of these recordings shows that natural REM sleep and the cholinergically induced REM sleep-like states are polygraphically similar for the pattern of respiratory airflow (Resp.), presence of rapid eye movements (electrooculogram; EOG), cortical activity (electroencephalogram; EEG), presence of ponto-geniculo-occipital (PGO) waves recorded from the lateral geniculate nucleus of the thalamus (LGN), and muscle atonia (electromyogram; EMG).

There are significant species differences in the amount of REM sleep that can be produced by pontine administration of cholinomimetics in cat compared to rat (Deurveilher et al., 1997). The first dose-response data for cholinergic REM sleep in cat revealed REM sleep enhancement of >300% (Baghdoyan et al., 1989). In contrast, long-standing (Shiromani and Fishbeïn, 1986) and recent (Bourgin et al., 1995; Imeri et al., 1994) studies in rat demonstrated much smaller, but statistically significant, REM sleep enhancement following pontine administration of carbachol. During continuous infusion of carbachol in the rat, REM sleep increased about 50% (Shiromani and Fishbein, 1986) and microinjection of carbachol into the caudal PnO increased the number, but not the duration, of REM sleep epochs (Bourgin et al., 1995). A systematic effort to understand these species-specific differences may help to elucidate cholinergic mechanisms contributing to REM sleep generation.

Recent reviews make it clear that natural REM sleep and the carbachol-induced REM sleep-like state are remarkably similar in their effects on multiple physiological systems (Baghdoyan, 1997a; Lydic, 1997). Statistically significant similarities during the two states have been reported for measures of blood pressure (Shiromani et al., 1986), heart rate (Garzón et al., 1995), EEG frequency (Garzón et al., 1997), and motor atonia (Morales et al., 1987). Cellular level analyses document a resemblance between REM sleep and carbachol-induced REM sleep for measures of neuronal excitability. To date, similarities have been reported for discharge rate of reticular (Shiromani and McGinty, 1986), parabrachial (Gilbert and Lydic, 1990), and medullary raphe (Woch et al., 1996) neurons. Intracellular studies revealed consistent effects of carbachol on membrane potential for *in vivo* and *in vitro* recordings (Greene et al., 1989; Ito and McCarley, 1984; Kohlmeier et al., 1996). Additional studies report REM sleep-related increases in brain energy utilization (Lydic et al., 1991a) and increased immediate early gene expression during the cholinergically induced REM sleep-like state (Shiromani et al., 1992; Shiromani et al., 1995; Yamuy et al., 1993).

It was not until end of 1980s that the cholinergic model of REM sleep first began to be applied systematically to the study of breathing during sleep (Lydic and Baghdoyan, 1989; Lydic et al., 1989). These respiratory studies found that in intact, unanesthetized cat, the REM sleep-like state caused by carbachol is characterized by significant reductions in ventilation (Lydic and Baghdoyan, 1989), upper airway muscle tone (Lydic et al., 1989), the ventilatory response to hyperoxic hypercapnia (Lydic et al., 1991c), and the discharge of pontine respiratory neurons (Gilbert and Lydic, 1994). Even when normal sleep and waking states are eliminated by decerebration, pontine administration of carbachol causes significant depression of respiratory control (Kimura et al., 1990; Kubin et al., 1992a; Kubin et al., 1994; Kubin et al., 1992b; Kubin et al., 1996; Pack, 1995). Thus, considerable data show that cholinoceptive mechanisms in the mPRF contribute to state-dependent changes in respiratory control even though the mPRF does not contain respiratory neurons.

Despite the differences between REM sleep and the cholinergically induced REM sleep-like state, pontine levels of endogenous ACh significantly increase during both cholinergically evoked (Lydic et al., 1991b) and spontaneous (Kodama et al., 1990; Leonard and Lydic, 1995) REM sleep. In addition, increasing levels of electrical stimulation in the LDT/PPT cause a monotonic increase in ACh release within the mPRF (Lydic and Baghdoyan, 1993) and enhance REM sleep (Thakkar et al., 1996). These findings are consistent with the view that the cholinergic LDT/PPT and cholinoceptive mPRF comprise portions of a network regulating REM sleep. The next section discusses presynaptic cholinergic mechanisms within the mPRF involved in modulating REM sleep.

Presynaptic Cholinergic Mechanisms and REM Sleep Generation

Similar to other cholinergic neurons, LDT/PPT neurons can be defined phenotypically based on the presence of the ACh synthetic enzyme choline acetyltransferase (ChAT) and by the presence of a vesicular ACh transporter (VAChT). As a presynaptic protein, the VAChT is located on the vesicle membrane, interacts with the cytoplasm of the presynaptic terminal, and functions to actively transport ACh into presynaptic vesicles (Parsons et al., 1993a; Parsons et al., 1993b). This ACh transporter also contains a binding site referred to as the vesamicol receptor. The compound 2-(4-phenylpiperidinyl) cyclohexanol (vesamicol) has been shown to bind to this site, block presynaptic packaging of ACh, and decrease the amount of ACh released from presynaptic terminals (Marshall and Parsons, 1987; Prior et al., 1992). As an interesting aside, the endogenous ligand for the vesamicol receptor remains unknown. What is known—and what has provided a useful tool for neurobiology—is the fact that vesamicol blocks central cholinergic neurotransmission (Buccafusco et al., 1991). Of direct relevance to this chapter are the recent findings that intracerebroventricular injections of vesamicol in rat inhibit REM sleep (Salin-Pascual and Jimenez-Anguiano, 1995) and mPRF microinjection of (±)-4-aminobenzovesamicol (ABV) in cat significantly blocks natural REM sleep and the neostigmine-induced REM sleep-like state (Capece et al., 1997). Such data are consistent with the conclusion that normal REM sleep generation requires normal vesicular ACh transport in presynaptic terminals of LDT/PPT neurons known to regulate ACh release in the mPRF (Capece et al., 1997; Lydic and Baghdoyan, 1993).

The 1991 finding that unilateral mPRF administration of carbachol caused a significant increase in ACh release in the contralateral mPRF (Lydic et al., 1991b) raised an interesting question. By what mechanisms do LDT/PPT terminals regulate ACh release? Although the answer to this question remains incomplete, it recently has become clear that muscarinic autoreceptors modulate ACh release in the mPRF (Roth et al., 1996). Muscarinic autoreceptors located on presynaptic cholinergic terminals have been shown to regulate ACh release in many brain regions (reviewed in Roth et al., 1996). Agonist occupation of autoreceptors provides a negative feedback control signal informing the releasing terminal that appropriate levels of the neurotransmitter are present in the synaptic junction. When the autoreceptor is blocked by an antagonist, transmembrane signaling systems are activated to increase neurotransmitter release. With this basic knowledge, the muscarinic cholinergic antagonist scopolamine was administered by microdialysis to the cat mPRF while obtaining simultaneous measures of ACh relase. The results revealed that, in every case, scopolamine administration to the mPRF caused increased ACh release (Roth et al., 1996). Since mPRF neurons are not cholinergic, and since LDT/PPT neurons regulate ACh release in the mPRF (Lydic and Baghdoyan, 1993), the scopolamine-induced increase in ACh release is consistent with the possibility that these autoreceptors reside on LDT/PPT terminals in the mPRF (Roth et al., 1996). Electrophysiological studies of LDT/PPT neurons have shown that cell excitability is modulated by muscarinic autoreceptors (Luebke et al., 1993). Electrophysiological and microdialysis data also suggest that noradrenaline may presynaptically inhibit ACh release in the pontine reticular formation (Semba et al., 1997). In addition, GABAergic interneurons may modulate ACh release from presynaptic LDT/PPT terminals within the mPRF. Blocking GABAergic synaptic transmission by mPRF microinjection of the $GABA_A$ antagonist bicuculline has been shown to induce a REM sleep-like state in cat (Xi et al., 1997). Recent studies demonstrated that after sleep deprivation, LDT neurons positively labeled for the GABA synthetic enzyme (GAD) showed an increase in Fos protein (Maloney and Jones, 1997). Since Fos expression has been linked to cell activity, these data are coinsistent with the view that GABAergic cells may contribute to REM sleep generation. Together, these recent data demonstrate that an exciting future direction will include efforts to specify the presynaptic mechanisms contributing to REM sleep control.

REM Sleep Generation by Postsynaptic Cholinergic Receptors

Two major classes of cholinergic receptors are nicotinic and muscarinic. Nicotinic receptors are multi-subunit, ligand-gated ion channels (ionotropic). mAChRs are single-subunit, G protein-coupled (metabatropic) receptors. While there is uniform agreement that mAChRs modulate REM sleep (Baghdoyan, 1997a; Gillin et al., 1993), the role of nicotinic receptors in REM sleep control is not presently clear. Some data suggest that nicotinic receptors contribute to REM sleep generation, yet contradictory data suggest no role for nicotinic mechanisms in the regulation of REM sleep. Pontine microinjections of nicotine have been reported to have no effect on REM sleep (George et al., 1964), or to increase REM sleep at the expense of waking and NREM sleep (Velazquez-Moctezuma et al., 1990). PGO waves, one of the electrographic signs of REM sleep, have been reported to be blocked by nicotinic antagonists (Hu et al.,

1988) and to be abolished by patch-delivery of nicotine (Vazquez et al., 1996). Subcutaneous (Jewett and Norton, 1966) and intravenous (Domino and Yamamoto, 1965) injections of nicotine have been reported to induce REM sleep, and this increase was prevented by the nicotinic blocker mecamylamine (Domino and Yamamoto, 1965). More recently, transdermal nicotine patches have been reported to increase REM sleep in normal volunteers and to improve mood and increase REM sleep in depressed patients (Salin-Pascual and Drucker-Colin, 1998). Obviously, the precise roles of nicotinic cholinergic receptors in REM sleep generation remains to be clarified.

Specifying Muscarinic Receptor Subtypes Contributing to REM Sleep Generation

A large body of *in vivo* pharmacological evidence shows that muscarinic agonists reliably induce REM sleep when injected into the mPRF, and that this REM sleep enhancement is dose-dependent and blocked by muscarinic antagonists (reviewed in Baghdoyan, 1997a). The muscarinic agonist bethanechol can produce REM sleep when microinjected into the mPRF of cat, and pontine microinjection of atropine blocks both natural REM sleep and REM sleep enhancement by bethanechol (Lee et al., 1995). Pontine administration of carbachol enhances REM sleep in rat (Shiromani and Fishbein, 1986), and REM sleep is decreased by pontine (Shiromani and Fishbein, 1986) or systemic (Zoltoski et al., 1993) administration of the muscarinic antagonist scopolamine.

In vitro data confirm that the pons contains mRNA for mAChRs (Buckley et al., 1988; Vilaro et al., 1992), 1992), and it has been shown that mAChRs are expressed in the mPRF (Baghdoyan, 1997b; Baghdoyan et al., 1994; Levey, 1993). In fact, there are a greater number of mAChRs present in rat pons during REM sleep compared to the number present during NREM sleep (Pompeiano and Tononi, 1990). Taken together, these findings support the view that mAChRs within the mPRF modulate REM sleep.

Molecular cloning studies have revealed five subtypes of mAChRs in mammalian brain (m1-m5) (reviewed in Buckley, 1990). Four mAChR subtypes have been defined pharmacologically (M1-M4) based on binding affinities for a series of muscarinic antagonists (Caulfield et al., 1993). In general, there is good correspondence between the pharmacologically defined M1-M3 and molecularly identified m1-m3 mAChR subtypes. Currently, there is considerable interest in identifying the functional roles of mAChR subtypes within the mPRF with respect to REM sleep generation. These research efforts must contend with the fact that, at present, there are no agonists which are purely selective for a single subtype of mAChR (Jones et al., 1992). Therefore, ongoing research efforts are attempting to gain information about mAChR subtype modulation of REM sleep by quantifying the ability of relatively selective mAChR antagonists to inhibit REM sleep. These studies have shown that in the mPRF of cat (Baghdoyan, 1997a;

Velazquez-Moctezuma et al., 1991), and in the PnO of rat (Imeri et al., 1994), M2 antagonists are more effective than M1 or M3 antagonists in blocking REM sleep. In pontine regions of cat brain rostral and dorsal to the mPRF, microdialysis application of cholinergic antagonists led to the suggestion that M3 mAChRs play a critical role in REM sleep generation (Sakai and Onoe, 1997). Autoradiographic data show that the M2 subtype comprises the predominant mAChR binding site in the mPRF of cat (Baghdoyan et al., 1994) and rat (Baghdoyan, 1997b) (Fig. 2A). The use of subtype selective mAChR antibodies also shows that m2 receptors are the predominant subtype in rat brain stem although additional subtypes are present (Levey et al., 1991). Considered together, these data demonstrate the importance of non-M1 mAChRs for REM sleep generation.

Signal Transduction Pathways Linked to Muscarinic Receptors

All five mAChR subtypes interact with guanine nucleotide binding proteins (G proteins), and each subtype is associated with one of two possible signal transduction pathways (Caulfield et al., 1993). The m1/m3/m5 subtypes are coupled to stimulatory G proteins (G_s) which stimulate phospholipase C (PLC). When activated, PLC hydrolyzes phosphatidyl inositol bisphosphate (PIP2) into diacylglycerol (DAG) and inositol triphosphate (IP3). DAG then stimulates protein kinase C (PKC), and IP3 releases intracellular Ca^{2+} pools (Felder, 1995). The m2/m4 subtypes are linked to inhibitory G proteins (G_i), the major type of G protein present in mammalian brain (Spiegel, 1992). The signal transduction cascade modulated by m2/m4 subtypes involves inhibition of adenylate cyclase (AC), a decrease in cyclic adenosine monophosphate (cAMP) production, and subsequent inhibition of protein kinase A (PKA) (Felder, 1995). Stimulation of PKA leads to phosphorylation of serine and threonine residues of various intracellular proteins such as ion channels, receptors, G proteins, and synthetic enzymes. Phosphorylation of these proteins leads to the alteration of neuronal excitability, the synthesis and release of neurotransmitters, and changes in receptor sensitivity.

Multiple lines of evidence (Fig. 3) suggest that within the pontine reticular formation the signal transduction pathway activated by m2/m4 receptors contributes to cholinergic REM sleep generation. It is highly probable that components of the m1/m3/m5 signal transduction pathway also are involved in REM sleep generation. Characterizing the role of these effector and second CRLF messenger proteins offers an exciting future direction for sleep neurobiology.

Carbachol Activates G Proteins in Brain Stem Regions Regulating REM Sleep

G proteins are membrane-bound heterotrimeric proteins consisting of α, β, and γ subunits that transduce a chemical signal from receptors into an

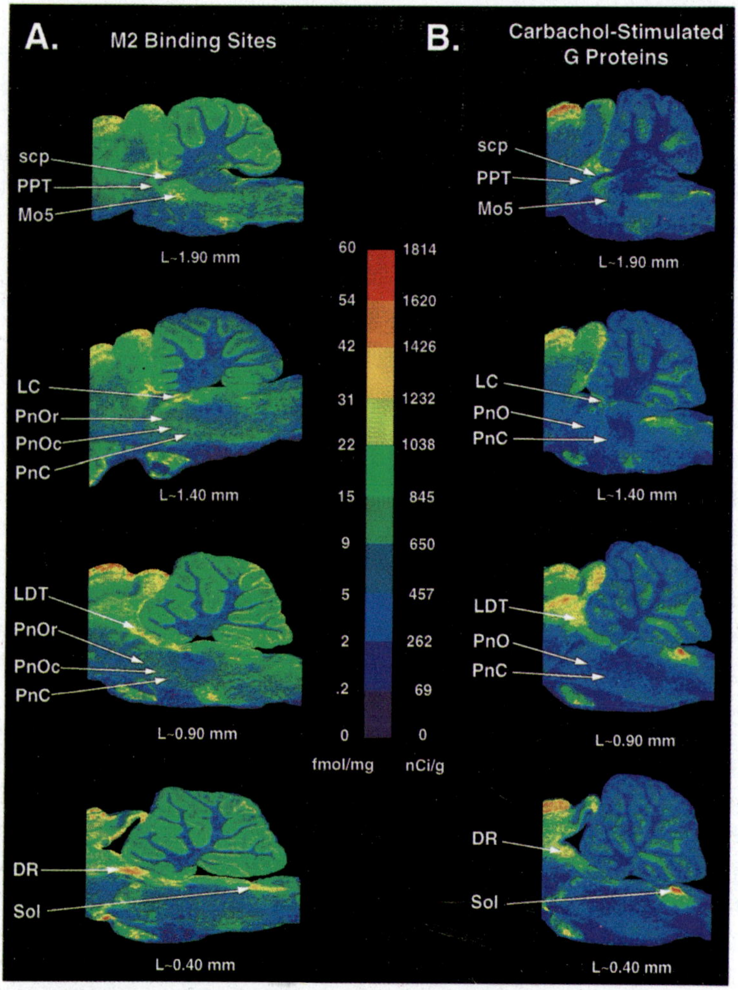

Fig. 2. **Localization of Muscarinic Receptors and Muscarinically Activated G Proteins in Rat Brain Stem. Color-coded autoradiograms of sagittal sections show the distribution of M2 binding (A) and carbachol-stimulated [^{35}S] GTP$_\gamma$S binding (B) across the rostro-caudal extent of the rat brain stem at four different lateralities. The color bar codes for levels of M2 and [^{35}S] GTP$_\gamma$S binding. Total binding is represented as fmol/mg tissue equivalent for M2 binding (values left of color bar) and *n*Ci/g for G protein binding (values right of color bar). The ability of carbachol to induce [^{35}S] GTP$_\gamma$S binding was dose dependent and antagonized by atropine. Note the similar localization of M2 and carbachol-stimulated G protein binding sites. Nuclei were identified according to the rat atlas of Paxinos and Watson (1986), and these data are described in detail elsewhere (Baghdoyan, 1997b; Capece et al., 1998). Abbreviations: DR, dorsal raphe nucleus; LC, locus coeruleus; LDT, laterodorsal tegmental nucleus; Mo5, trigeminal motor nucleus; PnC, nucleus pontis caudalis; PnO, nucleus pontis oralis; PPT, pedunculopontine tegmental nucleus; scp, superior cerebellar peduncle; Sol, nucleus of the solitary tract.**

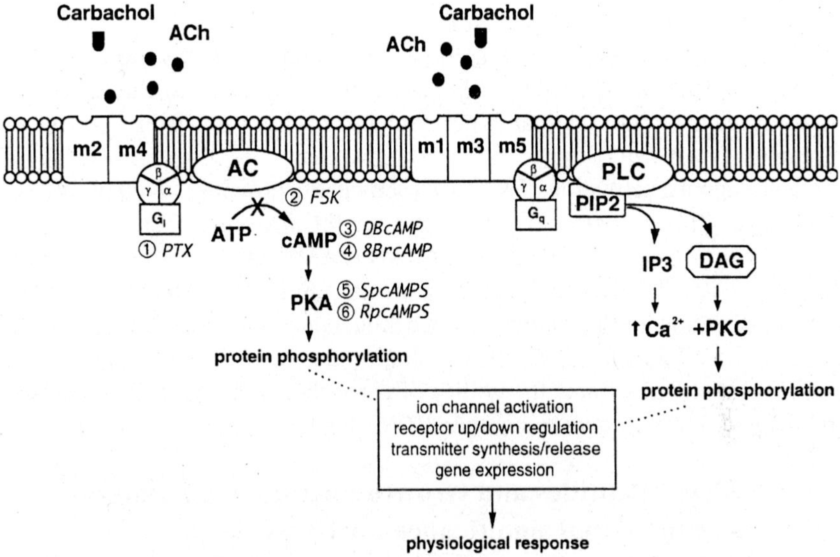

Fig. 3. Signal Transduction Pathways Linked to the Muscarinic Receptors. Pharmacological data suggest that the M2 receptor is responsible for regulating REM sleep, and multiple lines of evidence confirm that the signal transduction components (G_i, AC, cAMP, and PKA) linked to m2/m4 receptor subtypes play a key role in modulating REM sleep generation. Pontine microinjection of both pertussis toxin (PTX; 1), which inhibits G_i, and forskolin (FSK; 2), which stimulates AC, significantly reduce the ability of carbachol to induce the REM sleep-like state. Data also show that pontine administration of cAMP analogs dibutyryl cAMP (DB-cAMP; 3) and 8-bromo-cAMP (8Br-cAMP; 4) decrease cholinergic REM sleep. In addition, mPRF injection of the PKA stimulator Sp-cAMPS (5) decreases cholinergic REM sleep, whereas PKA inhibitor Rp-cAMPS (6) has no effect on REM sleep. Thus, six lines of evidence support a role for cAMP and PKA as modulators of cholinergic REM sleep generation. From (Capece and Lydic, 1997).

intracellular signal that can modulate second messenger systems (Aronstam and Dennison, 1989). To date, over 20 different G protein subtypes have been identified (Neer, 1995). The majority of G proteins are linked to receptors that contain seven transmembrane spanning domains, but some G proteins are associated with ion channels (Brann et al., 1993; Caulfield et al., 1993). In an inactive state, mAChRs are coupled to G proteins. The α subunit of inactive G protein is bound to guanosine diphosphate (GDP). Upon agonist activation of the mAChR, the receptor undergoes a conformational change which facilitates activation of the G protein (Strader et al., 1994). Activation of the G protein causes a conformational change in the structure of the α subunit, which results in a decreased affinity for GDP. Since guanosine triphosphate (GTP) is present in the cytoplasm in a greater concentration than GDP, the α subunit replaces the lost GDP with GTP. GTP binding uncouples the G protein from the receptor (Aronstam and Dennison, 1989) and reduces the receptor's affinity for the agonist. Soon after the α subunit binds GTP, the α subunit dissociates from the $\beta\gamma$ complex and is free to interact with an effector enzyme (AC or PLC, for example). Inactivation of the G protein is accomplished when GTPase activity intrinsic to the α subunit hydrolyzes GTP into GDP and the α subunit resumes its affinity for the $\beta\gamma$ comples (Neer, 1995).

Thus, agonist activation of mAChRs activates G proteins, which transduces the extracellular chemical signal into an intracellular chemical signal. Stimulation of second messenger systems (cAMP, DAG, Ca^{2+}) by primary effectors (AC, PLC) fosters a multitude of intracellular events including protein phosphorylation and alterations in ion conductance (Wess et al., 1995). These long-term cellular processes clearly contribute to REM sleep generataion, and Figure 2B illustrates the ability of carbachol to activate G proteins in brain stem nuclei known to regulate REM sleep.

Co-Localized Peptides and Growth Factors in Cholinergic Neurons: Functional significance for REM sleep

This article provides a selective overview of pontine cholinergic mechanisms contributing to REM sleep regulation. It should be clear, however, that no single neurotransmitter can account for the generation of any complex behavioral state. The complexity of the neurochemical control of REM sleep is further illustrated by the observation that peptide molecules within cholinergic neurons also can modulate REM sleep. Long-standing (Drucker-Colin et al., 1984; Obál et al., 1986) and recent (Jimenez-Anguiano et al., 1996; Obál et al., 1994) data from rat suggest that vasoactive intestinal peptide (VIP) participates in REM sleep regulation. VIP is localized in the Cholinergic LDT (Sutin and Jacobwitz, 1988), and microinjection of VIP into the caudal PnO enhances REM sleep in rat (Bourgin et al., 1997). Recently, muscarinic receptors have been shown to modulate the ability of pontine VIP to cause REM sleep enhancement (Bourgin et al., 1998).

The neurotrophin nerve growth factor (NGF) also is co-localized in some cholinergic neurons, and in cat mPRF microinjections of NGF have been reported to increase REM sleep (Yamuy et al., 1995). The mechanisms by which NGF enhances REM sleep remain unclear, but the results demonstrate the importance of future studies aiming to understand the role of endogenous protein molecules contributing to the generation of REM sleep.

Conclusions

This article discussed the cholinergic model of REM sleep and provided an overview of some of the pre- and postsynaptic mechanisms modulating REM sleep. Within the brain stem, the neurochemical network comprised of cholinergic LDT/PPT neurons and noncholinergic, cholinoceptive pontine reticular formation neurons is critical for regulating REM sleep. Decreasing ACh release from presynaptic LDT/PPT terminals or manipulating the signal transduction pathway modulated by m2/m4 mAChRs disrupts REM sleep.

Recent data have localized both mAChRs and muscarinically coupled G proteins to the mPRF. These exciting findings encourage future investigation into the precise role of specific intracellular molecules in REM sleep generation.

Acknowledgements

This work supported by the Departments of Anesthesia and Neuroscience & Anatomy, and by grants MH45361 (HAB) and HL40881 (RL). We thank P. Myers for expert secretarial assistance.

References

Aronstam, R.S. and Dennison, R.L. Anesthetic effects on muscarinic signal transfuction, International Anesthesiology Clinics, 27 (1989) 265–272.

Baghdoyan, H.A., Cholinergic mechanisms regulating REM sleep. In: W.J. Schwartz (Ed.), *Sleep Science: Integrating Basic Research and Clinical Practice. Monographs in Clinical Neuroscience, Vol. 15*, Karger, Basel, 1997a, pp. 88–116.

Baghdoyan, H.A. Location and quantification of muscarinic receptor subtypes in rat pons: Implications for REM sleep generation, American Journal of Physiology, 273 (1997b) R896–R904.

Baghdoyan, H.A., Lydic, R., Callaway, C.W. and Hobson, J.A. The carbachol-induced enhancement of desynchronized sleep signs is dose dependent and antagonized by centrally administered atropine, Neuropsychopharmacology, 2 (1989) 67–79.

Baghdoyan, H.A., Mallios, V.J., Duckrow, R.B. and Mash, D.C. Localization of muscarinic receptor subtypes in brain stem areas regulating sleep, NeuroReport, 5 (1994) 1631–1634.

Baghdoyan, H.A., Rodrigo-Angulo, M.L., McCarley, R.W. and Hobson, J.A. Site-specific enhancement and suppression of desynchronized sleep signs following cholinergic stimulation of three brain stem regions, Brain Research, 306 (1984) 39–52.

Baghdoyan, H.A., Spotts, J.L. and Snyder, S.G. Simultaneous pontine and basal forebrain microinjections of carbachol suppress REM sleep, Journal of Neuroscience, 13 (1993) 229–242.

Baxter, B.L. Induction of both emotional behavior and a novel form of REM sleep by chemical stimulation applied to cat mesencephalon, Experimental Neurology, 23 (1969) 220–229.

Berman, A.L., The Brain Stem of the Cat. University of Wisconsin Press, Madison, 1968.

Bourgin, P., Ahnaou, A., Laporte, A.M., Hamon, M. and Adrien, J. REM sleep induction by VIP infused into the oral pontine tegmentum of the rat may involve muscarinic receptors, Neuroscience (1998), In Press.

Bourgin, P., Escourrou, P., Gaultier, C. and Adrien, J. Induction of rapid eye movement sleep by carbachol infusion into the pontine reticular formation in the rat, NeuroReport, 6 (1995) 532–563.

Bourgin, P., Lebrand, C., Escourrou, P., Gaultier, C., Franc, B., Hamon, M. and Adrien, J. Vasoactive intestinal polypeptide microinjections into the oral pontine tegmentum enhance rapid eye movement sleep in the rat, Neuroscience, 77 (1997) 351–360.

Brann, M.R., Ellis, J., Jorgensen, H., Hill-Eubanks, D. and Jones, S.V.P. Muscarinic acetylcholine receptor subtypes: localization and structure/function, Progress in Brain Research, 98 (1993) 121–127.

Buccafusco, J.L., Wei, J. and Kraft, K.L. The effect of the acetylcholine transport blocker vesamicol on central cholinergic pressor neurons, Synapse, 8 (1991) 301–306.

Buckley, N.J., Molecular pharmacology of cloned muscarinic receptors. In: S.R. Nahorski (Ed.), *Transmembrane Signalling, Intracellular Messengers and Implications for Drug Development,* John Wiley & Sons Ltd., Chichester, 1990, pp. 11–30.

Buckley, N.J., Bonner, T.I. and Brann, M.R. Localization of a family of muscarinic receptor mRNAs in rat brain, Journal of Neuroscience, 8 (1988) 4646–4652.

Calvo, J.M., Simón-Arceo, K. and Fernández-Mas, R. Prolonged enhancement of REM sleep produced by carbachol microinjection into the amygdala, NeuroReport, 7 (1996) 577–580.

Capece, M.L., Baghdoyan, H.A. and Lydic, R. Carbachol-stimulated [^{35}S]Guanylyl 5′-(γ-Thio) triphosphate binding in sleep-related brain stem nuclei of rat, Journal of Neuroscience, 18 (1999) 3779–3785.

Capece, M.L., Efange, S.M.N. and Lydic, R. Vesicular acetylcholine transport inhibitor suppresses REM sleep, Neuro Report, 8 (1997) 481–484.

Capece, M.L. and Lydic, R. Cyclic AMP and protein kinase A modulate cholinergic rapid eye movement (REM) sleep generation, American Journal of Physiology, 273 (1997) R1430–R1440.

Caulfield, M.P., Robbins, J. Higashida, H. and Brown, D.A. Postsynaptic actions of acetylcholine: the coupling of muscarinic receptor subtypes to neuronal ion channels, Progress in Brain Research, 98 (1993) 293–301.

Cordeau, J.P., Moreau, A., Beaulnes, A. and Laurin, C. EEG and behavioral changes following microinjections of ACh and adrenaline in the brain stem of cats, Archives of Italian Biology, 101 (1963) 30–47.

Cuello, A.C. Cholinergic function and dysfunction, Progress in Brain Research, 98 (1993) 1–462.

Datta, S. Calva, J.M., Quattrochi, J. and Hobson, J.A. Long-term enhancement of REM sleep following cholinergic stimulation, NeuroReport, 2 (1991) 619–622.

Deurveilher, S., Hars, B. and Hennevin, E. Pontine microinjection of carbachol does not reliably enhance paradoxical sleep in rats, Sleep, 20 (1997) 593–607.

Domino, E. and Yamamoto, K. Nicotine: effect on the sleep cycle of cat, Science, 150 (1965) 637–638.

Drucker-Colin, R., Bernal-Pedraza, J., Fernandez-Cancino, F. and Oksenberg, A. Is vasoactive intestinal polypeptide (VIP) a sleep factor?, Peptides, 5 (1984) 837–840.

El Mansari, M., Sakai, K. and Jouvet, M. Unitary characteristics of presumptive cholinergic tegmental neurons during the sleep-waking cycle in freely moving cats, Experimental Brain Research, 76 (1989) 519–529.

Felder, C.C., Muscarinic acetylcholine receptors: signal transduction through multiple effectors, FASEB Journal, 9 (1995) 619–625.

Garzón, M., De Andrés, I. and Reinoso-Suárez, F. Dose-related effects on heart rate during cholinergic-induced paradoxical sleep after carbachol microinjections in the ventral oral pontine tegmentum, Sleep Research, 24A (1995) 83.

Garzón, M., De Andrés, I. and Reinoso-Suárez, F. Neocortical and hippocampal electrical activities are similar in spontaneous and cholinergic-induced REM sleep, Brain Research, 766 (1997) 266–270.

George, R., Haslett, W.L. and Jenden, D.J. A cholinergic mechanism in the brainstem reticular formation: induction of paradoxical sleep, International Journal of Neuropharmacology, 3 (1964) 541–552.

Gilbert, K.A. and Lydic, R. Parabrachial neuron discharge in the cat is altered during carbachol-induced REM sleep-like state (DCarb), Neuroscience Letters, 120 (1990) 241–244.

Gilbert, K.A. and Lydic, R. Pontine cholinergic reticular mechanisms cause state-dependent changes in the discharge of parabrachial neurons, American Journal of Physiology, 266 (1994) R136–R150.

Gillin, J.C., Salin-Pascual, R., Velazquez-Moctezuma, J., Shiromani, P.J. and Zoltoski, R. Cholinergic receptor subtypes and REM sleep in animals and normal controls, Progress in Brain Research, 98 (1993) 379–387.

Greene, R.W., Hass, H.L., Gerber, U. and McCarley, R.W., Cholinergic activation of medial pontine reticular formation neurons in vitro. In: M. Frotscher and U. Misgeld (Eds.), *Central Cholinergic Neurotransmission,* Birkhauser Verlag, Boston, 1989, pp. 123–137.

Hernandez-Peon, R., Chavez-Ibarra, G., Morgane, P.J. and Timo-Iaria, C. Limbic cholinergic pathways involved in sleep and emotional behavior, Experimental Neurology, 8 (1963) 93–111.

Hobson, J.A. and Steriade, M. Neuronal basis of behavioral state control. In: V.B. Mountcastel and F.E. Bloom (Eds.), *Handbook of Physiology, Section I, Vol. IV*, American Physiological Society, Bethesda, MD, 1986, pp. 701–823.

Hu, B., Bouhassira, D., Steriade, M. and Deschenes, M. The blockage of ponto-geniculo-occipital waves in the cat lateral geniculate nucleus by nicotinic antagonist, Brain Research, 473 (1988) 394-397.

Imeri, L., Bianchi, S., Angeli, P. and Mancia, M. Selective blockade of different brain stem muscarinic receptor subtypes: effects on the sleep-wake cycle, Brain Research, 636 (1994) 68–72.

Ito, K. and McCarley, R.W. Alterations in membrane potential and excitability of cat medial pontine reticular formation neurons during naturally occurring sleep-waking states, Brain Research, 292 (1984) 169–175.

Jewett, R.E. and Norton, S. Effects of some stimulant and depressant drugs on sleep cycles of cats, Experimental Neurology, 15 (1966) 463–474.

Jimenez-Anguiano, A., Gracia-Garcia, F., Mendoza-Ramirez, J.L., Duran-Vazquez, A. and Drucker-Colin, R. Brain Distribution of Vasoactive intestinal peptide receptors following REM sleep deprivation, Brain Research, 728 (1996) 37–46.

Jones, B. Immunohistochemical study of choline acetyltransferase-immunoreactive processes and cells innervating the pontomedullary reticular formation in the rat, Journal of Comparative Neurology, 295 (1990) 485–514.

Jones, S.V.P., Levey, A.I., Weiner, D.M., Ellis, J., Novotny, E., Yu, S.-H., Dörje, F., Wess, J. and Brann, M.R. Muscarinic acetylcholine receptors. In: M.R. Brann (Ed.), *Molecular Biology of G-Protein-Coupled Receptors*, Birkhäuser, Boston, 1992, pp. 170–197.

Jouvet, M. Recherches sur les structures nerveuses et les mechanisms responsables des differentes phases du sommeil physiologique, Archives italiennes de Biologie, 100 (1962) 125–206.

Jouvet, M. The role of monoamines and acetylcholine containing neurons in the regulation of the sleep waking cycle, Reviews in Physiology, 64 (1972) 225.

Kayama, Y., Ohta, M. and Jodo, E. Firing of 'possibly' cholinergic neurons in the rat laterodorsal tegmental nucleus during sleep and wakefulness, Brain Research, 569 (1992) 210–220.

Kimura, H., Kubin, L., Davies, R.O. and Pack, A.I. Cholinergic stimulation of the pons depresses respiration in decerebrate cat, Journal of Applied Physiology, 69 (1990) 2280–2289.

Kodama, T., Takahashi, Y. and Honda, Y. Enhancement of acetylcholine release during paradoxical sleep in the dorsal tegmental field of the cat brain stem, Neuroscience Letters, 114 (1990) 277–282.

Kohlmeier, K.A., Lopez-Rodriguez, F., Lui, R.-H., Morales, F.R. and Chase, M.H. State-dependent phenomena in cat masseter motoneurons, Brain Research, 722 (1996) 30–48.

Kryger, M.H., Roth, T. and Dement, W.C. (Eds.), Principles and Practice of Sleep Medicine. W.B. Saunders Co., Philadelphia, 1989.

Kubin, L., Kimura, H., Tojima, H., Pack, I. and Davies, R.O. Behavior of VRG neurons during the atonia of REM sleep induced by pontine carbachol in decerebrate cats, Brain Research, 592 (1992a) 91–100.

Kubin, L., Reignier, C., Tojima, H., Taguchi, O., Pack, A.I. and Davies, R.O. Changes in serotonin level in the hypoglossal nucleus region during carbachol-induced atonia, Brain Research, 645 (1994) 291–302.

Kubin, L., Tojima, H., Davies, R.O. and Pack, A.I. Changes in serotonin level in the hypoglossal nucleus region during carbachol-induced atonia, Neuroscience Letters, 139 (1992b) 243–248.

Kubin, L., Tojima, H., Reignier, C., Pack, A.I. and Davies, R.O. Interaction of serotonergic excitatory drive to hypoglossal motoneurons with carbachol-induced, REM sleep-like atonia, Sleep, 19 (1996) 187–195.

Lai, Y.Y. and Siegel, J.M. Muscle tone suppression and stepping produced by stimulation of midbrain and rostral pontine reticular formation, Journal of Neuroscience, 10 (1990) 2727–2734.

Laurent, J.P., Rondouin, G., Benita, M. and Jouvet, M. Reversible blockade of PGO waves and concomitant modifications of thalamic unit activity in chronic cats, Brain Research, 137 (1977) 305–322.

Lee, L.H., Friedman, D.B. and Lydic, R. Respiratory nuclei share synaptic connectivity with pontine reticular regions regulating REM sleep, American Journal of Physiology, 268 (1995) L251–262.

Leonard, T.O. and Lydic, R. Nitric oxide synthase inhibition decreases pontine acetylcholine release, NeuroReport, 6 (1995) 1525–1529.

Leonard, T.O. and Lydic, R. Pontine nitric oxide modulates acetylcholine release, rapid eye movement sleep generation, and respiratory rate, Journal of Neuroscience, 17 (1997) 774–785.

Levey, A.I., Immunological localization of m1-m5 muscarinic acetylcholine receptors in peripheral tissues and brain, Life Sciences, 52 (1993) 441–448.

Levey, A.I., Kitt, C.A., Simonds, W.F., Price, D.L. and Brann, M.R. Identification and localization of muscarinic acetylcholine receptor proteins in brain with subtype-specific antibodies, Journal of Neuroscience, 11 (1991) 3218–3226.

Luebke, J.L., McCarley, R.W. and Greene, R.W. Inhibitory action of muscarinic agonsits on neurons in the rat laterodorsal tegmental nucleus in vitro, Journal of Neurophysiology, 70 (1993) 2128–2135.

Lydic, R., Respiratory modulation by non-respiratory neurons. In: W.J. Schwartz (Ed.), *Sleep Science: Integrating Basic Research and Clinical Practice. Monographs in Clinical Neuroscience, Vol. 15*, Karger, Basel, 1997, pp. 177–142.

Lydic, R. and Baghdoyan, H.A. Cholinoceptive pontine reticular mechanisms cause state-dependent respiratory changes in the cat, Neuroscience Letters, 102 (1989) 211–216.

Lydic, R. and Baghdoyan, H.A. Cholinergic pontine mechanisms causing state-dependent respiratory depression, News in Physiological Sciences, 7 (1992) 220–224.

Lydic, R. and Baghdoyan, H.A. Pedunculopontine stimulation alters respiration and increases ACh release in the pontine reticular formation, American Journal of physiology, 264 (1993) R544–R554.

Lydic, R. and Baghdoyan, H.A. The neurobiology of rapid-eye-movement sleep. In: N.A. Saunders and C.E. Sullivan (Eds.), *Sleep and Breathing, 2nd Edition*, Marcel Dekker, New York, 1994, pp. 47–77.

Lydic, R. and Baghdoyan, H.A. Cholinergic contributions to the control of consciousness. In: T.L. Yaksh, C. Lynch II, W.M. Zapol, M. Maze, J.F. Biebuyck and L.J. Saidman (Eds.), *Anesthesia: Biologic Foundations, Vol. 2: Integrated Systems*, Raven Press, New York, 1998, pp. 433–450.

Lydic, R., Baghdoyan, H.A., Hibbard, L., Bonyak, E.V., DeJoseph, M.R. and Hawkins, R.A. Regional brain glucose metabolism is altered during rapid eye movement sleep in the cat: a preliminary study, Journal of Comparative Neurology, 304 (1991a) 517–529.

Lydic, R., Baghdoyan, H.A. and Lorinc, Z. Microdialysis of cat pons reveals enhanced acetylcholine release during state-dependent respiratory depression, American Journal of Physiology, 261 (1991b) R766–R770.

Lydic, R., Baghdoyan, H.A., Wertz, R. and White, D.P. Cholinergic reticular mechanisms influence state-dependent ventilatory response to hypercapnia, American Journal of Physiology, 261 (1991c) R738–R746.

Lydic, R., Baghdoyan, H.A. and Zwillich, C.W. State-dependent hypotonia in the posterior cricoarytenoid muscles of the larynx caused by cholinoceptive reticular mechanisms, FASEB Journal, 3 (1989) 1625–1631.

Lydic, R. and Biebuyck, J.F. (Eds.), Clinical Physiology of Sleep. American Physiological Society, Bethesda, MD, 1988.

Maloney, K.J. and Jones, B.E. C-FOS expression in cholinergic, GABAergic, and monoaminergic cell groups during paradoxical sleep deprivation and recovery, Society for Neuroscience Abstract, 23 (1997) 2131.

Marshall, I.G. and Parsons, S.M. The vesicular acetylcholine transport system, Trends in Neurosciences, 10 (1987) 174–177.

McCarley, R.W., Greene, R.W., Rainnie, D. and Portas, C.M. Brainstem neuromodulation and REM sleep, Seminars in Neuroscience, 7 (1995) 341–354.

Mitani, A., Ito, K., Hallanger, A.E., Wainer, B.H., Kataoka, K. and McCarley, R.W. Cholinergic projections from the laterodorsal and pedunculopontine tegmental nuclei to the pontine gigantocellular tegmental field in the cat, Brain Research, 451 (1988) 397–402.

Morales, F.R., Engelhardt, J.K., Soja, P.A., Perreda, E.A. and Chase, M.H. Motoneuron properties during motor inhibition produced by microinjection of carbachol into the pontine reticular formation of the decerebrate cat, Journal of Neurophysiology, 57 (1987) 1118–1129.

Neer, E.J., Heterotrimeric G proteins: organizers of transmembrane signals, Cell, 80 (1995) 249–257.

Obal, F. Payne, L., Kacsoh, B., Opp, M., Kapas, L., Grosvenor, C.E. and Krueger, J.M. Involvement of prolactin in the REM sleep-promoting activity of systemic vasoactive intestinal peptide (VIP), Brain Research, 645 (1994) 143–149.

Obal, F., Sary, G., Alfoldi, P., Rubicsek, G. and Obal, F. Vasoactive intestinal polypeptide promotes sleep without effects on brain temperature in rats at night, Neuroscience Letters, 64 (1986) 236–240.

Orem, J.M. and Barnes, C.D. (Eds.), Physiology in Sleep. Academic Press, New York, 1980.

Pack, A.I., Changes in respiratory motor activity during rapid eye movement sleep. In: J.A. Dempsey and A.I. Pack (Eds.), *Regulation of Breathing, 2nd Edition*, Marcel Dekker, New York, 1995, pp. 983–1010.

Parsons, S.M., Bahr, B.A., Rogers, G.A., Clarkson, E.D., Noremberg, K. and Hicks, B.W. Acetylcholine transporter—vesamicol receptor pharmacology and structure, Progress in Brain Research, 98 (1993a) 175–181.

Parsons, S.M., Prior, C. and Marshall, I.G. Acetylcholine transport, storage, and release, International Review of Neurobiology, 35 (1993b) 279–390.

Paxinos, G. and Watson, C. The Rat Brain in Stereotaxic Coordinates, 2nd Edition. Academic Press, New York, 1986.

Pompeiano, M. and Tononi, G. Changes in pontine muscarinic receptor binding during sleep-waking states in the rat, Neuroscience Letters, 109 (1990) 347–352.

Prior, C., Marshall, I.G. and Parsons, S.M. The pharmacology of vesamicol: an inhibitor of the vesicular acetylcholine transporter, General Pharmacology, 23 (1992) 1017–1022.

Roth, M.T., Fleegal, M.A., Lydic, R. and Baghdoyan, H.A. Pontine acetylcholine release is regulated by muscarinic autoreceptors, NeuroReport, 7 (1996) 3069–3072.

Sakai, K. and Onoe, H. Critical role for M_3 muscarinic receptors in paradoxical sleep generation in the cat, European Journal of Neuroscience, 9 (1997) 414–423.

Sakai, K., Petitjean, F. and Jouvet, M. Effects of ponto-mesencephalic lesions and electrical stimulation upon PGO waves and EMPs in unanesthetized cats, Electroencephalography and Clinical Neurophysiology, 41 (1976) 49–63.

Salin-Pascual, R.J. and Drucker-Colin, R. A novel effect of nicotine on mood and sleep in major depression, NeuroReport, 9 (1998) 57–60.

Salin-Pascual, R.J. and Jimenez-Anguiano, A. Vesamicol, an acetylcholine uptake blocker in presynaptic vesicles, suppresses rapid eye movement (REM) sleep in the rat, Psychopharmacology, 121 (1995) 485–487.

Semba, K., Aminergic and cholinergic afferents to REM sleep induction regions of the pontine reticular formation in the rat, Journal of Comparative Neurology, 330 (1993) 543–556.

Semba, K., Greene, R.W., Rasmusson, D.D., McCarley, R.W. and J.W. Noradrenergic presynaptic inhibition of acetylcholine release in the rat pontine reticular formation: an in vitro electrophysiological and in vivo microdialysis study, Society for Neuroscience Abstract, 23 (1997) 1065.

Semba, K., Reiner, P.B. and Fibiger, H.C. Single cholinergic mesopontine tegmental neurons project to both the pontine reticular formation and the thalamus in rat, Neuroscience, 38 (1990) 643–654.

Shiromani, P.J., Armstrong, D.M. and Gillin, J.C. Cholinergic neurons from the dorsolateral pons project to the medial pons: a WGA-HRP and choline acetyltransferase immunohistochemical study, Neuroscience Letters, 95 (1988) 19–23.

Shiromani, P.J. and Fishbein, W. Continuous pontine cholinergic microinfusion via mini-pump induces sustained alterations in rapid eye movement (REM) sleep, Pharmacology Biochemistry and Behavior, 25 (1986) 1253–1261.

Shiromani, P.J., Kilduff, T.S., Bloom, F.E. and McCarley, R.W. Cholinergically induced REM sleep triggers Fos-like immunoreactivity in dorsolateral pontine regions associated with REM sleep, Brain Research, 580 (1922) 351–357.

Shiromani, P.J., Malik, M., Winston, S. and McCarley, R.W. Time course of Fos-like immunoreactivity associated with cholinergically induced REM sleep, Journal of Neuroscience, 15 (1995) 3500–3508.

Shiromani, P.J. and McGinty, D.J. Pontine neuronal response to local cholinergic infusion: relation to REM sleep, Brain Research, 386 (1986) 20–31.

Shiromani, P.J., Siegel, J.M., Tomaszewski, K.S. and McGinty, D.J. Alternations in blood pressure and REM sleep after pontine carbachol microinjections, Experimental Neurology, 91 (1986) 285–292.

Shuman, S.L., Capece, M.L., Baghdoyan, H.A. and Lydic, R. Pertussis toxin-sensitive G proteins mediate carbachol-induced REM sleep and respiratory depression, American Journal of Physiology, 269 (1995) R308–R317.

Spiegel, A.M., Receptor-effector coupling by G proteins: implications for normal and abnormal signal transduction, Endocrine Reviews, 13 (1992) 536–565.

Steriade, M. and McCarley, R.W. Brainstem Control of Wakefulness and Sleep. Plenum, New York, 1990.

Strader, C.D., Fong, T.M., Tota, M.R. and Underwood, D. Structure and function of G protein-coupled receptors, Annual Review of Biochemistry, 63 (1994) 101–132.

Sutin, E.L. and Jacobwitz, D.M. Immunocytochemical localization of peptides and other neurochemicals in rat laterodorsal tegmental nucleus and adjacent areas, Journal of Comparative Neurology, 270 (1988) 243–270.

Thakkar, M., Portas, C. and McCarley, R.W. Chronic low-amplitude electrical stimulation of the laterodorsal tegmental nucleus of freely moving cats increases REM sleep, Brain Research, 723 (1996) 223–227.

Vazquez, J., Guzman-Martin, R., Salin-Pasqual, R.J. and Drucker-Colin, R. Transdermal nicotine on sleep and PGO spikes, Brain Research, 737 (1996) 317–320.

Velazquez-Moctezuma, J., Shalauta, M., Gillin, J.C. and Shiromani, P.J. Cholinergic antagonists and REM sleep generation, Brain Research, 543 (1991) 175–179.

Velazquez-Moctezuma, J., Shalauta, M.D., Gillin, J.C. and Shiromani, P.J. Microinjections of nicotine in the medial pontine reticular formation elicits REM sleep, Neuroscience Letters, 115 (1990) 265–268.

Vilaro, M.T., Wiederhold, K.-H., Palacios, J.M. and Mengod, G. Muscarinic M2 receptor mRNA expression and receptor binding in cholinergic and non-cholinergic cells in the rat brain: a correlative study using in situ hybridization histochemistry and receptor autoradiography, neuroscience, 47 (1992) 367–393.

Wainer, B.H., Steininger, T.L., Roback, J.D., Burke-Watson, M.A., Mufson, E.J. and Kordower, J. Ascending cholinergic pathways: functional organization and implications for disease models, Progress in Brain Research, 98 (1993) 9–30.

Webster, H.H. and Jones, B.E. Neurotoxic lesions of the dorsolateral pontomesencephalic tegmentum-cholinergic cell area in the cat. II. Effects upon sleep-waking states, Brain Research, 458 (1998) 285–302.

Wess, J., Blin, N., Mutschler, E. and Bluml, K. Muscarinic acetylcholine receptors: structural basis of ligand binding and protein coupling, Life Sciences, 56 (1995) 915–922.

Woch, G., Davies, R.O., Pack, A. and Kubin, L. Behavior of raphe cells projecting to the dorsomedial medulla during carbachol-induced atonia in the cat, Journal of Physiology, 490 (1996) 745–758.

Xi, M.-C., Morales, F.R. and Chase, M.H. GABAergic transmission in the nucleus pontis oralis: a mechanism controlling the generation of active sleep, Society for Neuroscience Abstract, 23 (1997) 1066.

Yamuy, J., Mancillas, J.R., Morales, F.R. and Chase, M.H. C-fos expression in the pons and medulla of the cat during carbachol-induced active sleep, Journal of Neuroscience, 13 (1993) 2703–2718.

Yamuy, J. Morales, F.R. and Chase, M.H. Induction of rapid eye movement sleep by the microinjection of nerve growth factor into the pontine reticular formation of cat, Neuroscience, 66 (1995) 9–13.

Zoltoski, R.K., Velazquez-Moctezuma, J., Shiromani, P.J. and Gillin, J.C. The relative effects of selective M_1 muscarinic antagonists on rapid eye movement sleep, Brain Research, 608 (1993) 186–190.

Rapid Eye Movement Sleep
B.N. Mallick and S. Inoué (Eds)

9. Functional Role of Serotonin 5-HT$_1$ and 5-HT$_2$ Receptor in the Regulation of REM Sleep

Jaime M. Monti[1] and Daniel Monti[2]

[1]Department of Pharmacology and Therapeutics, Clinics Hospital, Montevideo, Uruguay

[2]Department of Psychiatry, The University of Utah at Salt Lake City, USA

SUMMARY

Cholinergic neurons in the laterodorsal (LDT) and pedunculopontine (PPT) tegmental nuclei act to promote REM sleep (REMS). These neurons are in turn inhibited by serotonergic (5-HT), noradrenergic (NE) and presumably histaminergic and dopaminergic cells. 5-Hydroxytryptamine neurons virtually cease firing when the animal starts REMS, consequently decreasing the release of 5-HT from dorsal raphe nuclei (DRN) during this state. The activation of GABA$_A$ receptors is apparently responsible for this phenomena. Accordingly, direct microinjection of the GABA$_A$ agonist muscimol into the DRN increases REMS.

Systemic administration of the selective 5-HT$_{1A}$ receptor agonist 8-OHDPAT induced dose-dependent effects; i.e. low doses increased slow wave sleep (SWS) and reduced waking (W), while large doses caused opposite effects. When injected directly into the LDT, 8-OHDPAT selectively inhibited REMS. On the other hand, microdialysis perfusion of 8-OHDPAT in the DRN decreased 5-HT levels and significantly increased REMS. Systemic administration of the azapirones buspirone, gepirone and ipsapirone increased W and reduced SWS and REMS in laboratory animals, normal volunteers and insomniac patients. The β-adrenoceptor and 5-HT$_{1A/B}$ antagonist (–) pindolol prevented the effect of the 5-HT$_{1A}$ agonists on SWS and W. However, (–) pindolol failed to attenuate REMS suppression. Moreover, the drug by itself produced a dose-related reduction in REMS, which is tentatively ascribed to an increased release of 5-HT at postsynaptic sites.

Preliminary data show that CP-94, 253, a selective agonist of the 5-HT$_{1B}$ receptor increases W and reduces SWS and REMS.

Administration of 5-HT$_2$ receptor agonists DOI or DOM promotes the increase of W and decrease of SWS and REMS. In contrast, the nonselective 5-HT$_2$ antagonist ritanserin increases SWS and suppresses REMS. To date, the mechanism of action by which 5-HT$_{1B}$ and 5-HT$_2$ agonists or ritanserin suppress REMS has not been established.

Key Words: serotonin, REM sleep, dorsal raphe nuclei, 5-HT$_{1A}$ receptor, 5-HT$_{1B}$ receptor, 5-HT$_{2A/C}$ receptor, 8-OHDPAT, azapirones, pindolol.

The original proposal by Jouvet (1972) associating serotonin-containing neurons to sleep mechanisms, and catecholamine-containing neurons (mainly noradrenergic and dopaminergic neurons) to "executive mechanisms" of REM sleep (REMS) as well as behavioral and electrographic waking (W) is no longer tenable (Jones, 1990). Moreover, there is no firm evidence in support of the alternative hypothesis (Jouvet, 1984), suggesting that 5-HT released during W might act as a neurohormone inducing the synthesis and/or release of hypnogenic factors, secondarily responsible for slow wave sleep (SWS) and REMS.

On the other hand, the proposal that identifies cholinergic neurons in the laterodorsal (LDT) and pedunculopontine (PPT) tegmental nuclei with promoting REMS, and their inhibition by 5-HT afferents from the dorsal raphe nuclei (DRN) and noradrenergic afferents from the locus coenuleus (LC) (McCarley, 1983) is one that best contemplates the experimental evidence to elucidate REMS.

Current lesion, electrophysiological (recording of neuronal activity) neurochemical (measurement of extracelluar 5-HT levels) and pharmacological (local and systemic administration of drugs) data strongly support the hypothesis that serotonin can inhibit REMS. The characterization of several subtypes of central 5-HT receptors allowed a new approach in the study of the role of 5-HT in REMS regulation. In addition, availability of selective agonists and antagonists improved our capability to determine serotonergic effects on REMS through specific 5-HT receptors.

Structures Responsible for Activation and Inhibition of REMS

REM sleep is identified by the presence of a desynchronized cortical EEG (which results from the asynchronous activity of thalamocortical proyection neurons), a highly synchronized hippocampal EEG (theta or rhythmical slow activity, generated in the pyramidal cells of CA1, dentate gyrus and medial entorhinal cortex), ponto-geniculo-occipital (PGO) spikes (generated in the pons), and muscle atonia (which results from a cessation of discharge in the motoneurons supplying these muscles) (Siegel,1994).

Cholinergic neurons most important for REMS are located in the pontine reticular formation (Mitani et al., 1988) and project to the midbrain and bulbar reticular formation, limbic system, thalamus and basal forebrain, where effector neurons responsible for the occurrence of tonic or phasic events of REMS are present. These cholinergic neurons discharge selectively during REMS, which correlates with an increased release of acetylcholine (ACh) (McCarley et al., 1978; Kodama et al., 1990). Although cholinergic neurons are controlled mainly by noradrenergic and serotonergic neurons, preliminary evidence indicates that they receive also epinephrinergic, dopaminergic and histaminergic projections (Shammah-Lagnado et al., 1987; Semba, 1993; Iwakiri et al., 1994).

The activity of DRN serotonergic neurons and LC noradrenergic neurons is at its highest during W, it diminishes during SWS and is virtually suppressed when the animal starts REMS (Trulson and Jacobs, 1979), which argues against the original proposal that these neurotransmitters maintain REMS. A similar pattern has been reported with presumed histaminergic neurons in the posterior hypothalamus.

As a result, the release of 5-HT, norepinephrine (NE), histamine (HA) and ACh is high during W, whereas during REMS only ACh is released at a significant rate. Release of all four neurotransmitters is reduced during SWS.

It was initially proposed that virtual silence on DRN and LC neurons during REMS could be secondary to the activation of inhibitory 5-HT_{1A} receptors in DRN and α_2 receptors in LC through a process of feedback inhibition, dependent on recurrent inhibitory collaterals (McCarley et al., 1995). However, the concentration of 5-HT in the DRN is reduced during REMS, and the somatodendritic receptor-dependent inhibition of DRN neurons is mainly present during W (Portas and McCarley, 1994; Fornal et al., 1994). This argues against the activation of inhibitory 5-HT_{1A} receptors as the main mechanism responsible for the decreased activity of serotonergic cells during REMS. It has been also suggested that GABA input from the reticular formation could inhibit DRN and LC cells during REMS. In this respect, Wang et al. (1992) found GABAergic axon terminals making synaptic connections with DRN cell bodies and dendrites, and Stamp and Semba (1995) determined the existence of neurons containing both serotonin and GABA in the brainstem raphe nuclei. Moreover, Nitz and Siegel (1993, 1997b) described higher GABA levels during REMS compared with W at both LC and DRN sites. Reports indicating that systemic administration or direct microinjection of the $GABA_A$ agonist muscimol into the DRN increases REMS, whereas the GABA antagonists bicuculline and picrotoxin induce opposite effects further supports GABA participation in the inhibition of 5-HT neurons and facilitation of REMS production (Lancel et al., 1996; Nitz and Siegel, 1997a). Interestingly enough, although infusion of either $GABA_A$ agonist muscimol or $GABA_B$ agonist baclofen induce a decrease in DRN 5-HT, tonic inhibition of 5-HT release is mainly dependent on the activation of $GABA_A$ receptor (Tao et al., 1996).

Active inhibition of LC neurons secondary to the activation of α_2 receptors could explain their reduced firing rate during REMS (Ashton Jones and Bloom, 1981). Nevertheless, GABAergic inhibition has been also recently found to participate in the cessation discharge in LC neurons during REMS (Nitz and Siegel, 1997b).

The principal projections of the rostral raphe nuclei extend to forebrain structures. Input coming from the DRN predominate in the brain stem, ventral hippocampus, amygdala, striatum, accumbens nuclei and frontal cortex, while the dorsal hippocampus, dentate gyrus and septal nuclei receive mainly axons from median raphe nuclei (MRN). Shared projection sites from the DRN and MRN in the forebrain include the thalamus (intralaminar and mediodorsal nuclei) and hypothalamus (Hjorth, 1992; Baker and Halliday, 1995). It should be emphasized that DRN neurons not only project to neural structures linked to physiological events of REMS, but also to mesopontine and pedunculopontine cholinergic neurons that participate in the production of REMS (Honda and Semba, 1994). In addition, it has been established that 5-HT levels at the DRN and distant projection areas show a progressive decline from W to SWS and to REMS (Auerbach et al., 1989; Wilkinson et al., 1991; Portas and McCarley, 1994). Hence, if is noteworthy to consider whether 5-HT, either directly or indirectly (through inhibition of LDT/PPT neurons), influences effector cells.

Operational Characteristics of 5-HT receptors

The 5-HT receptors can be classified into at least seven classes designated 5-HT$_{1-7}$. The 5-HT$_1$, 5-HT$_2$ and 5HT$_5$ classes consist of five (5-HT$_{1A\text{-}B\text{-}D\text{-}E\text{-}F}$), three (5-HT $_{2A\text{-}B\text{-}C}$) and two (5HT5$_{A\text{-}B}$) subtypes, respectively, while the 5-HT$_3$, 5-HT$_4$, 5-HT$_6$ and 5-HT$_7$ classes have at present one subtype each (Hoyer et al., 1994; Hoyer and Martin, 1996; Baez et al., 1995). Except for the 5-HT$_3$ receptor, all other 5-HT receptors are structurally related to the superfamily of G-protein-coupled receptors.

The 5-HT$_{1A}$ Receptor

The 5-HT$_{1A}$ receptor is coupled to adenylate cyclase. Its activation inhibits the activity of the enzyme, which could be related to the coupling of 5-HT$_{1A}$ receptor to a Gi protein. The 5-HT$_{1A}$ receptor is located on the soma and dendrites of 5-HT neurons within the DRN and at postsynaptic sites. Stimulation of the somatodendritic 5-HT$_{1A}$ receptor inhibits the firing rate of serotonergic neurons, while activation of the postsynaptic receptor reproduces the action of endogenous 5-HT on target structurs. Brain areas rich in 5-HT$_{1A}$ receptors include the neocortex, hippocampus, septal nuclei, hypothalamus, and some amygdaloid and raphe nuclei, particularly the DRN. Currently, there are a number of ligands which show high selectivity for the 5-HT$_{1A}$ binding site in rat and cat brain. They include 8-OHDPAT, a full agonist at pre- and postsynaptic sites, and the azapirones buspirone, gepirone and ipsapirone, which behave as full agonists at the somatodendritic autoreceptor but only as partial agonists at the postsynaptic level. Mixed β-adrenoceptor and 5-HT$_{1A/B}$ receptor antagonist (–) pindolol acts both at pre- and postsynaptic sites, while WAY 100635 is a selective high affinity silent antagonist.

8-OHDPAT, when given systemically or applied directly into the DRN, reduces firing rate in 5-HT neurons, and the synthesis and release of 5-HT in neocortex, nucleus accumbens, dorso-lateral neostriatum and septum, an effect blocked by 5-HT$_{1A}$ antagonists.

Effect of 5-HT$_{1A}$ receptor agonists on REMS

8-OHDPAT and REMS

Initially, it was shown that systemic injection of relatively high doses of 8-

OHDPAT (0.125–2.0 mg/kg, s.c. or i.p.) consistently reduced SWS and REMS and increased W in the rat (Dzoljic et al., 1987; St. Hilaire Kafi et al., 1987; Dugovic and Wauquier, 1987; Depoortere, 1988; Monti et al., 1990). Moreover, when administered directly into the LDT, 8-OHDPAT selectively inhibited REMS in the cat (Sanford et al., 1994). In a subsequent study where 8-OHDPAT was given over a wide range of doses, biphasic effects on W and sleep became apparent. Accordingly, a 10 μg/kg dose of the 5-HT$_{1A}$ agonist significantly increased SWS and reduced W; REM sleep was slightly but not significantly increased. In contrast, a higher dose (375 μg/kg) induced opposite effects. Pretreatment with (–) pindolol reversed the effects of both small and large doses of 8-OHDPAT on SWS and W (Monti and Jantos, 1992). It was suggested that the sleep increase after a low dose of 8-OHDPAT could depend on the activation of inhibitory 5-HT$_{1A}$ autoreceptors which have been characterized in the DRN (Lucki and Wieland, 1990). In support of this proposal, injection of a 10 μg/kg dose of 8-OHDPAT in 5,7- dihydroxytryptamine-treated animals did not result in significant changes in sleep or W (Monti et al., 1994). Reduction of SWS and REMS after a large dose of 8-OHDPAT could be related to the activation of postsynaptic 5-HT$_{1A}$ receptors. REM sleep suppression induced by 8-OHDPAT could be due to an inhibitory action of the drug on REM-on neurons, similar to the physiological 5-HT effect.

Portas et al. (1996) provided direct evidence that suppression of DRN serotonergic activity increases REMS. To this purpose, the authors measured simultaneously extracellular 5-HT in the DRN and behavioral state changes by simultaneous polygraphic recordings. Microdialysis perfusion of 8-OHDPAT in the DRN decreased 5-HT levels by 50% during W and significantly increased REMS. In agreement with the reciprocal interaction model proposed by McCarley and Hobson (1975), inhibition of DRN activity following stimulation of somatodendritic 5-HT$_{1A}$ receptors by 8-OHDPAT suppressed 5-HT inhibition of mesopontine cholinergic neurons and increased REMS.

The Azapirones and REMS

The azapirones buspirone, ipsapirone and gepirone function as full agonists at 5-HT$_{1A}$ autoreceptors, and as partial agonists at postsynaptic 5-HT$_{1A}$ receptors (Glaser et al., 1992). All three compounds have been shown to increase total wake time and to reduce SWS and REMS when given i.p. or s.c in rats (Lerman et al., 1986; Monti et al., 1995a). Subcutaneous injection of the azapirones to 5,7-dihydroxytryptamine-pretreated rats led also to an increase of W and a reduction of sleep, which further supports the dependence of these effects on postsynaptic 5-HT$_{1A}$ receptors (Monti et al., 1995a). Similar sleep-disrupting effects for buspirone and ipsapirone have been described in normal volunteers and insomniac patients. Thus, buspirone administration increased W, prolonged REM latency and decreased REMS time in insomniac patients (Seidel et al., 1985; Manfredi et al., 1991), while ipsapirone prolonged REM latency and reduced REMS and SWS time in healthy subjects (Gillin et al., 1994; Driver et al., 1995). Thus, both 8-OHDPAT and azapirones in laboratory animals and man are capable of reducing SWS and REMS. The decrease in REMS after buspirone, gepirone or ipsapirone administration is compatible with the reciprocal interaction hypothesis for the regulation of desynchronized sleep. Therefore, azapirones could be acting on postsynaptic 5-HT$_{1A}$ receptors in the pontine reticular formation to inhibit REM-on neurons.

5-HT$_{1A}$ receptor antagonists and REMS

Several apparent antagonists have been characterised. The β-adrenoceptor antagonist pindolol shows also 5-HT$_{1A/1B}$ antagonistic activity. Compounds NAN 190, BMY 7378, WAY 100, 135 and SDZ 216525 have all demonstrated partial agonist properties in studies of somatodendritic autoreceptor function. The only selective high affinity silent antagonist at this receptor is the recently developed compound WAY 100,635 (Forster et al., 1995). However, to date pindolol has been the most thoroughly studied. Pindolol produces a dose-related suppression of REMS in laboratory animals and healthy subjects. At relatively high doses the compound increases W and reduces SWS in the rat, and decreases EEG spectral power during non-REM sleep in man (Monti and Jantos, 1994; Seifritz et al, 1997). Pretreatment with (-) pindolol has been shown to reverse the effect of both small and large doses of 8-OHDPAT on SWS and W. Nevertheless, 8-OHDPAT-related suppression of REMS remained unchanged (Monti and Jantos, 1994). Thus, irrespective of their opposed mechanism of action, 5-HT$_{1A}$ agonists and (–) pindolol tend to suppress REMS.

Seifritz et al. (1997) have proposed that REMS reduction after (–) pindolol administration could be mediated through inactivation of inhibitory somatodendritic 5-HT$_{1A}$ receptors. Their proposal is based on the finding that systemic administration of pindolol increases 5-HT release at postsynaptic sites in the rat. This in turn would inhibit REM-on neurons in the pontine reticular formation.

The 5-HT$_{1B}$ Receptor

The 5-HT$_{1B}$ receptor is specific to rats and some other rodents. The 5-HT$_{1D\beta}$ receptor is a close homologue (93% identical overall) found in higher species including man. The 5-HT$_{1B}$ receptor is linked to the inhibition of adenylyl cyclase. The receptor is located at presynaptic sites (5-HT axon terminals) and postsynaptically in the substantia nigra, dorsal subiculum, corpus striatum, cerebellum, nucleus accumbens and cerebral cortex. Preliminary evidence suggests that a small population of 5-HT$_{1B}$ receptors may also act as somatodendritic autoreceptors in the DRN (Davidson and Stamford, 1995). Moreover, 5-HT$_{1B}$ receptor mRNA expression has been reported in raphe nuclei cells (Voigt et al., 1991).

Several agonists show selectivity for the 5-HT$_{1B}$ receptor including CP-93, 129, CP-94,253 and CGS 12066B. Recently, selective antagonists for the

$5\text{-}HT_{1B}$ receptor also became available. These include compounds GR 127935, SB-224289 and GR 55562.

CP-94, 253 and REMS
In the study by Monti et al. (1995b), the $5\text{-}HT_{1B}$ agonist CP-94,253 was injected s.c. over a wide range of doses in rats. The compound significantly increased W and reduced SWS and REMS. Pindolol reversed the effect of CP-94,253 on W and SWS, while REMS remained unchanged. To date, no $5\text{-}HT_{1B}$ receptors have been found on cholinergic REM-on cells in the pontine reticular formation. Moreover, no attempts have been made to determine the effect of local administration of $5\text{-}HT_{1B}$ agonists on the laterodorsal and pedunculopontine tegmental nuclei, remaining the localization of brain areas responsible for CP-94,253 REMS suppressant properties still unknown.

5-HT$_2$ Receptor
Three $5\text{-}HT_2$ receptor subtypes are currently recognized and termed $5\text{-}HT_{2A}$, $5\text{-}HT_{2B}$ and $5\text{-}HT_{2C}$. All three are thought to stimulate phospholipase C to hydrolyse phosphatidylinositol 4,5-biphosphate with the formation of inositol triphosphate and diacylglycerol. They are thus linked to the phosphoinositide metabolism signal transduction system. The $5\text{-}HT_2$ receptor subtypes are located at postsynaptic sites.

The $5\text{-}HT_{2A}$ receptor is primarily found in neocortex, some components of the limbic system, olfactory nuclei, claustrum, basal ganglia and raphe nuclei (Roberts and Davies, 1989).

$5\text{-}HT_{2B}$ mRNA has been detected to some extent in rat brain (Foguet et al., 1992).

$5\text{-}HT_{2C}$ sites are present in the choroid plexus and at lower densities in the limbic system, basal ganglia, olfactory and subthalamic nucleus.

Most $5\text{-}HT_{2A}$ agonists such as the phenalkylamine derivatives DOI and DOM have no selectivity over the other $5\text{-}HT_2$ receptor subtypes. α-Methyl-5-HT is a full agonist with high affinity for the $5HT_{2B}$ site. Despite the large number of antagonist compounds with high affinity for the $5\text{-}HT_2$ receptor, few are truly selective. Ketanserin is 70 fold selective for the $5\text{-}HT_{2A}$ over the $5\text{-}HT_{2B}$ and $5\text{-}HT_{2C}$ receptors but is also a potent α_1 antagonist. Ritanserin has high affinity for $5\text{-}HT_{2B}$, $5\text{-}HT_{2C}$ and H_1 receptors. SB204741 has been reported as the first selective $5\text{-}HT_{2B}$ receptor antagonist with approximately 100 fold selectivity over the $5\text{-}HT_{2A}$ and $5\text{-}HT_{2C}$ sites.

Phenylalkylamines and REMS

Systemic administration of the $5\text{-}HT_2$ receptor agonists DOI or DOM increased W and reduced SWS and REMS in the rat. Pretreatment with ritanserin prevented the effect of the agonists on W and SWS, while REMS remained suppressed (Dugovic et al., 1989; Monti et al., 1990).

Ritanserin and REMS

Intraperitoneal injection of ritanserin in rats during the light period increased SWS and reduced REMS. Meanwhile, waking remained unchanged or showed a moderate decrease (Monti et al., 1990; Dugovic, 1992).

Thus, as already described for the 5-HT$_{1A}$ and 5-HT$_{1B}$ receptor, compounds with opposite effects suppressed REMS. *In vitro* physiological activation of 5-HT$_2$ receptors has been shown to hyperpolarize and depolarize separate populations of pontine reticular formation neurons. However, no further steps have been taken to elucidate the mechanism by which 5-HT$_2$ agonists and antagonists suppress REMS.

References

Ashton-Jones, G. and Bloom, F.E. Activity of norepinephrine-containing locus coeruleus neurons in behaving rats anticipates fluctuations in the sleep-waking cycle, J. Neurosci., 1 (1981) 876–886.

Auerbach, S.B., Minzenberg, M.J. and Wilkinson, L.O. Extracelluar serotonin and 5-hydroxyindolacetic acid in hypothalamus of the unanesthetized rat measured by in vivo dialysis coupled to high performance liquid chromatography with electrochemical detection: dialysate serotonin reflects neuronal release, Brain Res., 499 (1989) 281–290.

Baez, M., Kursar, J.D., Helton, L.A., Wainscott, D.B. and Nelson, D.L.G. Molecular biology of serotonin receptors, Obesity Res., 3 (suppl.4) (1995) 441S–447S.

Baker, K.G. and Halliday, G.M. Ascending noradrenergic and serotonergic systems in the human brainstem. In: D.J. Tracey, G. Paxinos and J. Stone (Eds.) *Neurotransmitters in the Human Brain,* Plenum Press, New York, 1995, pp. 155–171.

Davidson, C. and Stamford, J.A. Evidence that 5-hydroxytryptamine release in rat dorsal raphe nucleus is controlled by 5-HT$_{1A}$, 5-HT$_{1B}$ and 5-HT$_{1D}$ autoreceptors, Br. J. Pharmacol., 114 (1995) 1107–1109.

Depoortere, H., Effects of 5-HT$_{1A}$ agonists on the sleep-wakefulness cycle in rats. in W.P. Koella, F. Obál, H. Schulz and P. Visser (Eds.), *Sleep '86'* Gustav Fischer Verlag, Sttutgart, 1988, pp. 346–348.

Driver, H.S., Flanigan, M.J., Bentley, A.J., Luus, H.G., Shapiro, C.M. and Mitchell, D. The influence of ipsapirone, a 5-HT$_{1A}$ agonist, on sleep patterns in healthy subjects, Psychopharmacology, 117 (1995) 186–192.

Dugovic, Ch. Functional activity of 5-HT$_2$ receptors in the modulation of the sleep/wakefulness states, J. Sleep Res., 1 (1992) 163–168.

Dugovic, Ch. and Wauquier, A. 5-HT$_2$ receptors could be primarily involved in the regulation of slow-wave sleep in the rat, Eur. J. Pharmacol., 137 (1987) 145–146.

Dugovic, Ch., Wauquier, A., Leysen, J.E., Marrannes, R. and Janssen, P.A.J. Functional role of 5-HT$_2$ receptors in the regulation of sleep and wakefulness in the rat, Psychopharmacology, 97: 436–442, 1989.

Dzoljic, M.R., Saxena, P.R. and Ukponmwan, O.E. 5-HT$_1$ receptor agonists enhance alertness, sleep Res., 16 (1987) 88.

Forster, E.A., Cliffe, I.A., Bill, D.J., Dover, G.M., Jones, D., Reilly, Y. and Fletcher, A. A pharmacological profile of the selective silent 5-HT$_{1A}$ receptor antagonist, WAY-100635, Eur. J. Pharmacol., 281 (1995) 81–88.

Foguet, M., Hoyer, D., Pardo, L.A., Parekh, A., Kluxen, F.W., Kalkman, H.O., Stuhmer,

W. and Lubbert, H. Cloning and functional expression of the rat stomach fundus serotonin receptor, EMBO J., 11 (1992) 3481–3487.

Fornal, C.A., Litto, W.J., Metzler, C.W. and Marrosu, F. Single unit responses of dorsal raphe neurons to 5-HT_{1A} agonist and antagonist drug administration in behaving cats, J. Pharmacol, Exp. Ther., 270 (1994) 1345–1358.

Gillin, J.C., Jernajcyk, W., Valladares-Neto, D.D.C., Golshan, S., Lardon, M. and Stahl, S.M. Inhibition of REM sleep by ipsapirone, a 5-HT_{1A} agonist in normal volunteers, Psychopharmacology, 116 (1994) 433–436.

Glaser, T. and De Vry, J. Neurobiology of 5-HT_{1A} receptors. In: M. Stahl, J.M. Gastpar, J.M. Keppel Hesselink, J. Traber (Eds.), *Serotonin$_{1A}$ Receptors in Depression and Anxiety,* Raven Press, New York, 1992, pp. 35–53.

Hjorth, S., Functional differences between ascending 5-HT systems. In: P.B. Bradley, S.L. Handley, S.J. Cooper, B.J. Key, N.M. Barnes and J.H. Coote (Eds.) *Serotonin, CNS Receptors and Brain Function,* Pergamon Press, Oxford, 1992, pp. 203–218.

Honda, T. and Semba, K. Serotonergic synaptic input to cholinergic neurons in the rat mesopontine tegmentum, Brain Res., 47 (1994) 299–306.

Hoyer, D. and Martin, G.R. Classification and nomenclature of 5-HT receptors: a comment on current issues, Behav. Brain Res., 73 (1996) 263–268.

Hoyer, D., Clarke, D.E., Fozard, J.R., Hartig, P.R., Martin, G.R., Mylecharane, E.J., Saxena, P.R. and Humphrey, P.P.A. VII. International Union of Pharmacology Classification of receptors for 5-Hydroxytryptamine (Serotonin), Pharmacol. Rev., 46 (1994) 157–203

Iwakiri, H., Takekusaki, K., Nonaka, S. and Mori,S. Extracelluar levels of serotonin in the medial pontine reticular formation in acute decerebrate cats with a microdialysis technique, Neurosc. Lett., 177 (1994) 19–22.

Jones, B.E., Influence of the brainstem reticular formation, including intrinsic monoaminergic and cholinergic neurons, on forebrain mechanisms of sleep and waking. In: M. Mancia and G. Marini (Eds.), *The Diencephalon and sleep,* Raven Press, New York, 1990, pp. 31–48.

Jouvet, M. The role of monoamines and acetylcholine-containing neurons in the regulation of the sleep-waking cycle, Ergebn. Physiol.. 64 (1972) 166–307.

Jouvet, M. Neuromediateurs et facteurs hypnogénes, Rev. Neurol., 140 (1984) 389–400.

Kodama, T., Takahashi, Y. and Honda, Y. Enhancement of acetylcholine release during paradoxical sleep in the dorsal tegmental field of the cat brain stem, Neurosci. Lett., 114. (1990) 277.

Lancel, M., Crönlein, Faulhaber, J. Role of $GABA_A$ receptors in sleep regulation, Neuropsychopharmacology, 15: (1996) 63–74.

Lerman, J.A., Kaitin, K.I., Dement, D.C. and Peroutka, S.J. The effects of buspirone on sleep in the rat, Neurosci. Lett., 72 (1986) 64–68.

Lucki, I. and Wieland, S. 5-Hydroxytryptamine 1A receptors and behavioral responses, Neuropsychopharmacology, 3 (1990) 481–493.

Manfredi, R.L., Kales, A., Vgontzas, A.N., Bixler, E.O., Isaac, M.A. and Falcone, Ch.M. Buspirone: sedative or stimulant effects?, Am. J. Psychiatry, 148 (1991) 1213–1217.

McCarley, R.W, REM dreams, REM sleep and their isomorphisms, In: M. Chase and E.D. Weitzman. (Eds.), *Sleep Disorders: Basic and Clinical Research,* Spectrum Publications, New York, 1983, pp. 363–392.

McCarley, R.W. and Hobson, J.A. Neuronal excitability modulation over the sleep cycle: a structural and mathematical model, Science, 189: (1975) 58–60.

McCarley, R.W.., Nelson, J.P. and Hobson, J.A. PGO burst neurons: correlative evidence for neuronal generators of PGO waves, Science, 201 (1978) 209–262.

McCarley, R.W., Greene, R.W., Rainnie, D. and Portas, C.M. Brainstem neuromodulation and REM sleep, Seminars in The Neurosciences, 7 (1995) 341–354.

Mitani, A., Ito, K., Hallanger, A.E., Wainer, B.H., Kataoka, K. and McCarley, R.W. Cholinergic projections from the laterodorsal and pedunculopontine tegmental nuclei to the pontine gigantocellular tegmental field in the cat, Brain Res., 451 (1988) 397–402.

Monti, J.M. and Jantos, H. Dose-dependent effects of the 5-HT$_{1A}$ receptor agonist 8-OH-DPAT on sleep and wakefulness in the rat, J. Sleep Res., 1 (1992) 169–175)

Monti, J.M. and Jantos H. Stereoselective antagonism by the pindolol enantiomers of 8-OH-DPAT-induced changes of sleep and wakefulness, Neuropharmacology, 33 (1994) 705–708.

Monti, J.M., Orellana, C., Boussard, M., Jantos, H., Labraga, P., Olivera, S. and Alvariño, F. 5-HT receptor agonists 1-(2,5-dimethoxy-4-iodophenyl)-2-aminopropane (DOI) and 8-OH-DPAT increase wakefulness in the rat, Biogen. Amines, 7 (1990) 145–151.

Monti, J.M. Jantos, H., Silveira, R., Reyes-Parada, M., Scorza, C. and Prunell, C. Depletion of brain serotonin by 5,7-DHT: Effects on the 8-OH-DPAT-induced changes of sleep and waking in the rat, Psychopharmacology 115 (1994) 273–277.

Monti, J.M., Jantos, H., Silveira, R., Reyes-Parada, M. and Scorza, C. Sleep and waking in 5,7,-DHT-lesioned or (–)-pindolol-pretreated rats after administration of buspirone, ipsapirone, or gepirone, Pharmacol. Biochem. Behav., 52 (1995a) 305–312.

Monti, J.M., Monti, D., Jantos, H. and Ponzoni, A. Effects of selective activation of the 5-HT$_{1B}$ receptor with CP-94,253 on sleep and wakefulness in the rat, Neuropharmacology, 12 (1995b) 1647–1651.

Nitz, D. and Siegel, J. Inhibitory amino acid neurotransmission in the dorsal raphe nucleus during sleep-wake states, Soc. Neurosci. Abstr., 19 (1993) 742.

Nitz, D. and Siegel, J. GABA release in the dorsal raphe nucleus: role in the control of REM sleep, Am. J. Physiol., 273 (1997a) R451–R455.

Nitz, D. and Siegel, J. GABA release in the locus coeruleus as a function of sleep/wake state, Neuroscience, 78 (1997b) 795–801.

Portas, C.M. and McCarley, R.W. Behavioral state related changes of extracelluar serotonin concentration in the dorsal raphe nucleus: a microdialysis study in the freely moving cat, Brain Res., 648 (1994) 306–312.

Portas, C.M., Thakkar, M., Rainnie, D. and McCarley, R.W. Microdialysis perfusion of 8-hydroxy-2-(di-n-propylamino) tetralin (8-OH-DPAT) in the dorsal raphe nucleus decreases serotonin release and increases rapid eye movement sleep in the freely moving cat, J. Neurosci., 16 (1996) 2820–2828.

Roberts, M.H.T. and Davies, M. In vivo electrophysiology of receptors mediating the central nervous system actions of 5-hydroxytryptamine. In: E.J. Mylecharane, J.A. Angus, I.S. de la Lande and P.P.A. Humphrey (Eds.), *Serotonin: Actions, Receptors and Pathophysiology,* MacMillan, London, 1989, pp. 70–76.

Sanford, L.D., Ross, R.J., Seggos, A.E., Morrison, A.R., Ball, W.A. and Mann, G.L. Central administration of two 5-HT receptor agonists: effect on REM sleep initiation and PGO waves, Pharmacol. Biochem. Behav., 49 (1994) 93–100.

Shammah-Lagnado, S.J., Negrao, N., Silva, B.A. and Ricardo, J.A. Afferent connections of the nuclei reticularis pontis oralis and caudalis: A horseradish peroxidase study in the rat Neuroscience, 20 (1987) 961–989.

Seidel, W.F., Cohen, S.A., Bliwise, N.G. and Dement, W.C. Buspirone: an anxiolytic without sedative effect, Psychopharmacology, 87 (1985) 371–373.

Seifritz, E., Stahl, S.M. and Gillin, J.C. Human sleep EEG following the 5-HT$_{1A}$ antagonist pindolol: possible disinhibition of raphe neuron activity, Brain Res., 759 (1997) 84–91.

Semba, K. Aminergic and cholinergic afferents to REM sleep induction regions of the pontine reticular formation in the rat, J. Comp. neurol., 330 (1993) 543–556.

Siegel, J.M., Brainstem mechanisms generating REM sleep, In: M.H. Kryger, T. Roth and W.C. Dement (Eds.), *Principles and Practice of Sleep Medicine,* Saunders, Philadelphia, 1994, pp. 125–144.

St. Hilaire-Kafi, S., Hjorth, S. and Gaillard, J.M. Effect of 8-OH-DPAT on the sleep-waking cycle in the rat. In: C.T. Dourish, S. Ahlenius and P.H. Huston (Eds.), *Brain 5-HT$_{1A}$* Receptors: Behavioral and Neurochemical Pharmacology, Ellis Horwood, Chichester, 1987, 135–139.

Stamp, J.A. and Semba, K. Extent of colocalization of serotonin and GABA in the neurons of the rat raphe nuclei, Brain Res., 677 (1995) 39–49.

Tao, R., Ma, Z. and Auerbach, S.B. Differential regulation of 5-hydroxytryptamine release by $GABA_A$ and $GABA_B$ receptors in midbrain raphe nuclei and forebrain of rats, Br. J. Pharmacol., 119 (1996) 1375–1384.

Trulson, M.E. and Jacobs, B.L. Raphe unit activity in freely moving cats: correlation with level of behavioral arousal, Brain Res., 163 (1979) 135–150.

Voigt, M.M., Laurie, D.J., Seeburg, P.H. and Bach, A. Molecular cloning and characterisation of a rat brian cDNA encoding a 5-hydroxytryptamine$_{1B}$ receptor, EMBO J., 10 (1991) 4017–4023.

Wang, Q.P., Ochiai, H. and Nakai, Y. GABAergic innervation of serotonergic neurons in the dorsal raphe nucleus of the rat studied by electron microscopy double immunostaining, Brain Res. Bull., 29 (1992) 943–948.

Wilkinson, L.O., Auerbach, S.B. and Jacobs, B.L. Extracellular serotonin levels change with behavioral state but not with pyrogen-induced hyperthermia, J. Neurosci., 11 (1991) 2732–2741.

Rapid Eye Movement Sleep
B.N. Mallick and S. Inoué (Eds)

10. Possible Role of GABA in the Regulation of REM Sleep with Special Reference to REM-OFF Neurons

Birendra N. Mallick, Satvinder Kaur, Sushil K. Jha and Jerome M. Siegel[1]

School of Life Sciences, Jawaharlal Nehru University, New Delhi 110 067, India
[1]Neurobiology Research (151A3), VA Medical Center, North Hills, CA 91343, Department of Psychiatry and Brain Research Institute, University of California at Los Angeles, CA 90024, USA

SUMMARY

GABA is the most abundant inhibitory neurotransmitter in the central nervous system. Its possible role in regulation of REM sleep by inhibiting the REM-OFF neurons has been reviewed in this chapter. Some of the neurons located in the locus coeruleus (noradrenergic), in the raphe (serotonergic) and in the ventrolateral posterior hypothalamus (histaminergic) have been found to be REM-OFF type. The cessation of these neurons during REM sleep is possibly mediated by GABA coming from interneurons and/or GABA-ergic inputs from areas which have REM-ON like neuronal activity. The REM-ON neurons also might possibly mediate inhibition of REM-OFF neurons through the GABA-ergic neurons as they have often been found to project to areas with GABA-ergic neurons.

Key Words: GABA, histamine, inhibition of REM-OFF neurons, locus coeruleus, norepinephrine, raphe, serotonin, ventrolateral posterior hypothalamus.

Introduction

Rapid eye movement (REM) sleep is electrophysiologically characterized by the presence of tonic desynchronization of the cortical electroencephalogram (EEG), somatic muscular atonia, phasic bursts of eye movement and presence of pontogeniculo-occipital (PGO) waves. It is known to be present in all placental and marsupial mammals. Based on earlier studies (Allison et al., 1972) where REM sleep could not be identified in echidna, a monotreme, it was suggested that REM sleep is of relatively recent origin. However, recent studies (Siegel et al., 1996, 1997) showed that signs resembling REM sleep are present in echidna, the spiny ant eater and in platypus, both egg laying mammals. This suggests that the REM sleep is likely to be relatively primitive. Transection and lesion studies were conducted in an attempt to identify the brain areas responsible for generation and maintenance of REM sleep (Sakai, 1985; Siegel, 1985). These studies suggested that structures in the pontine region are both necessary and sufficient to generate the basic phenomenon of REM sleep. The importance of the pontine structures in REM sleep has further been supported by studying the behaviour of single neurons in this area. It was found that there are neurons in the brain stem whose firing rate showed state specific changes in relation to REM sleep. Some neurons decreased firing, while others increased firing during REM sleep. These neurons have been termed as REM-OFF and REM-ON respectively. The norepinephrinergic neurons in locus coeruleus (LC) (Aston-Jones and Bloom, 1981, Jacobs, 1986), the serotonergic neurons in nucleus raphe (McGinty and Harper, 1976) and the histaminergic neurons in posterior hypothalamus (Vanni Mercier et al., 1984; Lin et al., 1986) have been found to be of REM-OFF type. The REM-ON neurons are presumably glutamatergic, GABA-ergic and cholinergic, and are primarily located in the laterodorsal tegmental and pedunculo pontine (LDT/PPT) areas (El Mansari et al., 1989; Kayama et al., 1992). Besides the Brainstem region, the preoptico-anterior hypothalamic area (POAH) also has neurons which increase firing in anticipation of REM sleep (Koyama and Hayaishi, 1994).

A reciprocal interaction between the cholinergic REM-ON neurons in the LDT/PPT and the nor-epinephrinergic REM-OFF neurons present in the LC for the generation of REM sleep has been proposed (Hobson et al., 1975; Sakai, 1988). This was supported by to and fro projections between those two areas (Jones, 1990). The presence of muscarinic receptors on the LC neurons (Baghdoyan et al., 1994) also support the existence of a functional link between cholinergic REM-ON and aminergic REM-OFF neurons. Further anatomical studies showed projections from the cholinergic LDT/PPT area to the serotonergic nucleus raphe (Gang et al., 1990; Hermann et al., 1997) and to the histaminergic posterior hypothalamic area (Satoh and Fibiger 1986; Woolf and Butcher, 1986) where REM-OFF neurons are located as mentioned above. In order to investigate if the cessation of REM-OFF neurons located in the LC, is required for the generation of REM sleep, attempts were made not to allow the REM OFF neurons to stop firing by continuous application of low frequency and low intensity of stimulation at the LC (Singh and Mallick, 1996). Such stimulation resulted in a significant decrease in REM sleep during stimulation and a rebound increase in REM sleep after the stoppage of stimulation. A rebound increase in REM sleep is also observed during recovery period after REM sleep deprivation (Dement, 1960). The decrease in REM sleep was due to reduction in frequency of generation of REM sleep while duration of REM sleep per episode remained unaffected. This suggested that activation of neurons in the LC region, where REM OFF neurons are located, might not have allowed initiation of REM sleep. However, the mechanism of inhibition of the neurons is unknown. An active inhibitory mechanism is likely to be involved for cessation of the REM-OFF neurons and generation of REM sleep. Glycine and GABA are the two common inhibitory neurotransmitters widely distributed in the CNS.

Fort et al. (1993) found the presence of glycine immunoreactive boutons in the LC. Although glycine tonically inhibits LC neurons, it does not mediate changes in the sleep state dependent firing of noradrenergic LC neurons (Darracq et al., 1996). Further, it has been shown by microdialysis studies that the concentration of glycine in LC remained unchanged during REM sleep (Nitz and Siegel, 1997). These reports suggest that glycine is unlikely to play a significant role in triggering of REM sleep. GABA is the other possible inhibitory neurotransmitter involved in the generation of REM sleep which will be reviewed in this chapter.

Evidence for Involvement of GABA in REM Sleep

A possibility of involvement of GABA in REM sleep may be supported by the fact that chemicals which affect GABA receptors have been found to modulate REM sleep. The GABA interacts with two receptor subtypes, the GABA-A and GABA-B. The former is a macromolecular protein composed of a receptor gated chloride channel, a GABA recognition site and modulatory sites for benzodiazepines, picrotoxin, barbiturates and anaesthetic steroids (Deutsch et al., 1992; Macdonald and Olsen, 1994). The GABA-A receptors are generally postsynaptic, where as GABA-B receptors are located on presynaptic neurons. The activation of GABA-A complex by GABA increases the membrane conductance for chloride ions resulting in hyperpolarization of neurons. The GABA-A may act as presynaptic heteroreceptor and can modulate the release of the neurotransmitter (Fillenz, 1990). The GABA-B receptor works by decreasing the inward flux of Ca^{++} into the presynaptic nerve terminal, thereby reducing the release of the neurotransmitter. This decrease of GABA level ultimately produces facilitation due to disinhibition.

Benzodiazepines (BNZ) alter sleep by modulating GABA-ergic transmission by acting on BNZ—GABA-A receptor—chloride channel supramolecular complex in the CNS. Gaillard et al. (1973) studied the effect of benzodiazepines on sleep in humans. Flunitrazepam (FNZ), a BNZ-receptor agonist affected the sleep by reducing the number of awakenings, stage shifts and body movements. Slow activities in EEG, namely delta, theta and *K* potentials were markedly increased in amplitude whereas alpha activity and spindles were reduced. It also prolonged the latency of REM sleep. However, higher doses of FNZ suppressed REM sleep. Clonazepam, a partial antagonist of BNZ receptors with comparable dosage did not change sleep-waking balance but inhibited REM sleep to a limited extent. Chronic administration of Ro 15-1788 (Flumazenil) (3.6 mg/kg/day in drinking water), a BNZ-receptor antagonist significantly enhanced both REM sleep and total sleep duration (O'Connor et al., 1991). In humans, Ro 15-1788 delayed the onset of REM sleep, but did not affect overall REM sleep time (Ziegler et al., 1986). Polc et al. (1982) had shown that BNZ augmented non-REM sleep and reduced REM sleep but, Scherschlicht and Marias (1983) had shown that low doses of BNZ caused proportional augmentation of REM sleep.

Barbiturates, which although do not bind to the same site as the BNZ's on the supramolecular complex, still affect Cl^- conductance, exerting an inhibitory effect on the neurons. These are also been known to have hypnogenic effect and inhibit REM sleep mainly by decreasing latency and frequency of occurrence (Gaillard, 1989). They also influenced slow wave sleep, reduced K potentials and affected spindles. Schneider et al. (1977), studied the effect of Di-N-propylacetic acid, which acts by enhancement of GABA levels in humans brain. After short term application, it resulted in shortening of time to fall asleep as well as of the waking time. Long term administration caused a decrease in deep synchronous sleep. As opposed to the results known from animal studies, no marked influence on REM sleep was observed.

Progesterone, an allosteric modulator of GABA-A receptor, has also been shown to 'affect sleep-wakefulness including REM sleep. Intraperitoneal administration of progesterone (30, 90 and 180 mg/kg) at the light onset of sleep in rats affected sleep-wakefulness (Lancel et al., 1996). It was observed that progesterone dose dependently shortened non REM latency, lengthened REM latency, and decreased the amount of wakefulness and REM sleep. It also enhanced the EEG activity during non REM sleep in higher frequency range >7 Hz).

Localization of GABA Neurons and Terminals in Brain Areas Involved in REM Sleep

GABA is likely to be involved in REM sleep because anatomically GABA-ergic neurons and terminals have been localized in areas of the brain which are known to influence REM sleep. In the brain stem, GABA-ergic neurons and terminals have been reported in the LC (Iijima and Ohtomo,1988; Ford et al., 1995; Shipley et al., 1996) and also in the medullary reticular formation projecting to dorsal raphe (Holmes et al. 1994). The medullary nuclei, paragigantocellularis and prepositus hypoglossi, have strong efferents to the LC, the latter being predominently GABA-ergic (Ennis and Aston-Jones, 1989). The medial preoptic area, which plays a critical role in the regulation of sleep wakefulness, also has been shown to contain high levels of GABA as well as its synthesizing enzyme, glutamic acid decarboxylase (Tappaz et al., 1976; Gritti et al., 1993) containing neurons.

GABA in LC and REM Sleep

Both tyrosine hydroxylase (TH) and glutamic acid decarboxylase (GAD) immunoreactivities have been found in LC cells (Ford et al., 1995; Shipley et al., 1996). They have been colocalized in the same neuron as well (Ijima and Ohtomo, 1988). Similarly, autoradiographic studies demonstrated the presence of GABA-A receptors in LC (Wamsley and Palacios, 1984; Luque et al., 1994). Electrophysiologically GABA has been shown to inhibit firing in LC neurons (Cedarbaum and Aghajanian, 1977). This inhibition predominently has been shown to be mediated through GABA-A receptors (Cherubini et al., 1988). Microinjection of picrotoxin, a GABA-A antagonist, into LC induced a significant reduction in duration of REM sleep (Kaur et al., 1997a). It was found in this study that the effect on REM sleep was primarily due to a reduction in mean duration of REM sleep per episode and not due to a reduction in the frequency of generation of REM sleep. Thus, these observations suggest that GABA in LC is responsible for maintenance and tonic regulation of REM sleep by acting on the GABA-A receptor. This is supported by the recent finding that the concentration of GABA increased in LC during REM sleep (Nitz and Siegel, 1997).

Acetylcholine in the pontine area including LC was found to increase during REM sleep (Kodama et al., 1990) and cholinergic stimulation of dorsolateral pontine area increased REM sleep (Baghdoyan et al., 1984). Similarly, cholinergic induction of REM sleep has been found to trigger Fos-like immunoreactivity in dorsolateral pontine neurons (Shiromani et al., 1992), where REM sleep related and unrelated neurons are present. However, the REM OFF neurons in the LC cease firing during REM sleep, but acetylcholine does not inhibit the LC neurons (Egan and North, 1986). Hence, it was hypothesized that acetylcholine action in LC may be mediated through GABA (Kaur et al., 1997a). Further it has recently been shown that the carbachol action in LC in increasing REM sleep was likely to be predominantly, although not exclusively mediated through GABA (Kaur et al., 1997b). The GABA may be coming from local interneurons or from other areas. The LC has been shown to receive strong input from nucleus prepositus hypoglossi (PrH) and paragigantocellularis (PGi), where atleast the latter has been reported to have GABA-ergic neurons (Mugnaini and Oertel, 1985). Although the chemical nature of the neurons are not known, some of the neurons in and around PrH are the REM ON type (Sakai, 1985). The PrH provides a direct inbibitory synaptic input to LC, which is likely to be mediated by GABA (Ennis and Aston Jones, 1989). Neurons in PrH were found to be immunoreactive for c-fos, an immediate early gene, after induction of REM sleep by microinjection of the muscarine receptor agonist, carbachol, into the cat pontine reticular formation (Yamuy et al., 1993). C-fos expression has been correlated with neuronal activity (Sagar et al.,1987). Thus, a subgroup of neurons in PrH must be relatively active during REM sleep. Some of these neurons may be GABA-ergic and possibly mediate the inhibition of the LC noradrenergic neurons during REM sleep.

GABA in Dorsal Raphe Nucleus and REM Sleep

It has been reported that neurons with a REM-OFF discharge profile are present in raphe nucleus (Mcginty and Harper, 1976; Cespuglio et al., 1981).

Lydic et al. (1987) have quantified the cyclic progressive decrease in dorsal raphae (DR) discharge in slow waves sleep (SWS) and suggested that the DR cells participate in a cyclical process controling SWS-REM sleep.

Pharmacological studies demonstrated a possible interaction between serotonergic and presumed REM-ON neurons, since it was only after depletion of monoamines with reserpine that the state of REM sleep could be induced with ACh-esterase inhibitor eserine (Karczmar et al., 1970). Similarly, following specific depletion of serotonin, a cholinergic agonist enhanced the phasic phenomenon of REM sleep (Jacobs and Fornal, 1991) revealing the potential interaction of serotonergic and cholinergic neurons in generation of REM sleep and its component variables. Cytotoxic lesion studies, in the medial medullary reticular formation (MMRF), further clarifies the role of cholinergic, GABAergic and serotonergic interaction across the sleep-awake states. The destruction of a portion of MMRF had little effect on wakefulness and slow wave sleep but produced a reduction in REM sleep, loss of muscle atonia and motor inhibition. The variable amount of post lesion REM sleep was correlated positively with the number of the cholinergic cells surviving, negatively with the number of surviving serotonergic cells and positivelty with the ratio of the surviving cholinergic or GABA-ergic cells to serotonergic cells (Holmes and Jones, 1994).

An inhibitory influence of serotonergic neurons has been supported by pharmacological evidence that enhanced levels of serotonin prevent the occurrence of REM sleep (Pastel et al., 1987) and depletion of serotonin facilitated the appearance of REM sleep or its phasic activity (Delrome et al., 1966; Karczmar et al., 1970). Anatomically, a large population of GABA-ergic neurons within the medullary reticular formation innervate the medullary reticular formation and raphe nucleus (Holmes et al., 1994). On other hand, serotonin-containing neurons of the midbrain raphe nuclei are under inhibitory control by GABA. Injection of the GABA antagonists picrotoxin and bicuculline and the GABA agonist muscimol into the median raphe nucleus increased and decreased, respectively, the serotonin turnover and the steady-state content of 5-hydroxyindoleacetic acid. These results provide evidence of a tonic inhibition of serotonergic neuronal activity in the median raphe nucleus (Forchetti and Meek, 1981). This inhibitory effect is likely to be potentiated by benzodiazepines (Mendelson et al., 1987). Thus, GABA may possibly act by suppressing the activity of serotonergic neurons, which could be important for the appearance of REM sleep. They could thus serve as intermediary between the cholinergic, putative REM-ON cells and serotonergic, putative REM-OFF cells.

GABA in Hypothalamus and REM Sleep

The role of hypothalamus in regulating REM sleep in particular, is not very clear. However, some evidence suggests that it may play a modulatory role. Lesion (McGinty and Sterman, 1968; Szymusiak et al., 1986) and local reversible inactivation (Alam and Mallick, 1990; Mallick and Alam, 1991) of the anterior part of the hypothalamus, including the preoptic area, caused loss of sleep and altered thermoregulation. Alteration in both, sleep and body temperature might affect REM sleep. Warming of the anterior hypothalamus during REM sleep was shown to increase the duration of REM sleep episodes (Parmeggiani et al., 1974). An increase in REM sleep was observed after ibotenic acid lesion of neurons in the ventrolateral part of the posterior hypothalamus (Sallanon et al., 1988). Some REM sleep specific neurons also have been identified in the hypothalamic area. Koyama and Hayaishi (1994) reported some neurons in the POAH whose firing rate increased during and prior to REM sleep.These neurons behaved as those of REM-ON neurons and their properties were similar to LDT neurons. On the other hand, neurons have been found in the ventrolateral part of the posterior hypothalamus whose firing rate is reduced during REM and non-REM sleep (Vanni-Mercier et al., 1984). These REM-OFF neurons are within the histaminergic group of neurons and are located in the tubero-mammillary region of the posterior hypothalamus (Lin et al., 1986).

Although the hypothalamus may modulate REM sleep, the specific neurotransmitter(s) involved in this mechanism are not known. There are large number of GABA-ergic neurons, GABA-ceptive neurons and GABA-terminals present in this region (Brashear et al., 1986; Gritti et al., 1993). Severe insomnia and reduction in REM sleep were observed when POAH was chemically lesioned (Szymusiak and McGinty, 1986) or reversibly inactivated (Alam and Mallick, 1990; Mallick and Alam, 1991). In another study, it was observed that microinjection of muscimol into the posterior hypothalamus of POAH lesioned cats there was hypersomnia and increased REM sleep (Sallanon et al., 1989). On the other hand, muscimol microinjection into the ventrolateral portion of the posterior hypothalamus induced an increase in slow wave as well as REM sleep (Lin et al., 1989). Recently microinjection of a GABA antagonist, picrotoxin, into the POAH was found to reduce REM sleep (Ali and Mallick, 1996). Thus, there may be reciprocal connection between the anterior and the posterior hypothalamic neurons for the modulation of REM sleep similar to that of regulation of sleep. However, in the above mentioned studies sleep was also reduced significantly. Thus, whether the effect on REM sleep was secondary to the effect on sleep is unknown.

The suppression of REM sleep may be due to inactivation of some population of hypothalamic neurons located in the ventrolateral posterior hypothalamus. This might exert a tonic inhibitory influence on REM sleep executive neurons located in the mediodorsal pontine tegmentum (Sakai, 1985). The withdrawl of tonic inhibition on the hypothalamic neurons might be responsible for induction of REM sleep. This is supported by the finding that there are direct projections of some ventrolateral posterior hypothalamic neurons to the mediodorsal pontine tegmentum (Sakai et al., 1977). Neural signals to induce REM sleep may be produced in the brain stem and then conveyed to the LDT neurons projecting towards basal forebrain (Ford et al., 1995; Satoh and Fibiger, 1986; Woolf and Butcher, 1986; Kayama and Ogawa, 1987) and relayed directly or indirectly to the anterior hypothalamic REM sleep related neurons. The latter may be a part of descending inhibitory projection to the posterior hypothalamus REM-OFF neurons. In that case it is reasonable to observe a reduction of REM sleep with microinjection of picrotoxin (a GABA-A antagonist) into the POAH area (Ali and Mallick, 1996) possibly by inhibiting the neurons located in the former. Therefore, it can be hypothesized that the REM sleep related neurons in the POAH may be a part of inhibitory projection to the posterior hypothalamus which may be acting through GABA.

There are GABA-ergic projections as well as interneurons in the preoptic area. GABA has been colocalized within the cholinergic neurons in the basal forebrain including POAH (Gritti et al., 1993). A GABA-ergic projection from the POAH to the posterior hypothalamus has also been demonstrated recently (Gritti et al., 1994). A continuous supply of GABA there might be responsible for inhibiting, while its withdrawal may be resposible for induction of REM sleep. The above mechanism may be supported by the finding that during REM sleep the GABA concentration decreases in the posterior hypothalamus (Nitz and Seigel, 1996). In addition, the septal nucleus is rich in GABA-ergic neurons (Gritti et al., 1994), which have a strong reciprocal interconnections with the hippocampus (Alonso and Kohler, 1982; Freund and Antal, 1988; Toth et al., 1993; Linke et al., 1994). These neurons may have a significant role to play in REM sleep particularly in the control of the hippocampal theta rhythm that characterizes this state. This is supported by the recent finding that GABA concentration in the septal region of basal forebrain was highest in REM sleep with significantly lower levels in quiet waking, non REM sleep and lowest levels in active waking (Mallick et al., 1997).

Conclusion

So far three groups of REM-OFF neurons have been identified in the brain viz. the NE-ergic neurons in the LC, the serotonergic neurons in the raphe and the histaminergic neurons in the ventrolateral posterior hypothalamic area. All these brain areas (containing REM-OFF neurons) receive GABA-ergic projections and /or contain GABA-ergic interneurons and often the REM-ON neurons project to areas having GABA-ergic neurons. This information have been schematically represented in Fig. 1. Although there are some indirect evidences that the REM-OFF neurons may possibly be inhibited by GABA during REM sleep, further studies are needed for obtaining direct evidence.

Acknowledgement

Fundings from CSIR, DBT, DST and ICMR are acknowledged.

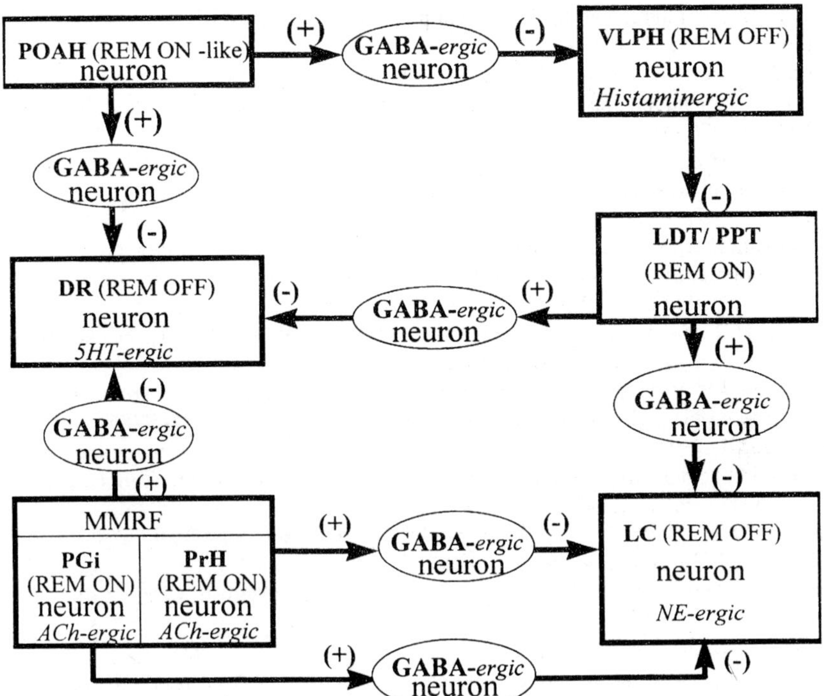

Fig.1. Schematic representation of GABA-ergic inputs and/or interneurons to the brain areas where REM-OFF neurons have been identified. As may be seen that the brain areas having REM-ON neurons have projections to the areas having GABA-ergic neurons which in turn project to the area having REM-OFF neurons. These GABA-ergic neurons are likely to play a significant role during REM sleep. Details of the possibility of these neurons inhibiting the REM-OFF neurons during REM sleep have been discussed in the text. 5HT—5-hydroxytprytamine (serotonin); ACh—acetylcholine; DR—dorsal raphe; GABA—gamma amino butyric acid; LC—locus coeruleus area; LDT/PPT—dorsolateral/pedunculopontine tegmental area; MMRF—medial medullary reticular formation; NE—norepinepherine; PGi—paragigantocellularis area; POAH—preoptico-anterior hypothalamic area; PrH—prepositus hypoglossi area; VLPH—ventrolateral posterior hypothalamic area. (+) and (–) show possible excitation and inhibition, respectively.

References

Alam, M.N. and Mallick, B.N. Differential acute influence of medial and lateral preoptic areas on sleep wakefulness in freely moving animals, Brain Res., 525 (1990) 242–248.

Ali, M. and Mallick, B.N. Involvement of medial preoptic area GABA-A receptor in sleep-wakefulness and thermoregulation, XIth National Symposium of the Indian Society for Chronobiology, (1996) 25.

Allison, T., Van Twyver, G. and Goff, W.R. Electrophysiological studies of the echidna, Tachyglossus aculeatus, Arch. Ital. Biol., 110 (1972) 145–184.

Alonso, A. and Kohler, C. Evidence for separate projections of hippocampal pyramidal and non-pyramidal neurons to different parts of the septum in the rat brain, Neurosci. Lett., 31 (1982) 209–214.

Aston-Jones, G. and Bloom, F.E. Activity of norepinephrine containing locus coeruleus neurons in behaving rats anticipates fluctuations in the sleep-waking cycle, J. Neurosci., 1 (1981) 876–886.

Baghdoyan, H.A., Mallios, V.J., Duckrow, R.B. and Mash, D.C. Localization of muscarinic receptor subtypes in brain stem areas regulating sleep, Neuroreport, 5 (1994) 1631–1634.

Baghdoyan, H.A., Rodrigo-Anglio, M.L. and McCarley, R.W. Site-specific enhancement and suppression of desynchronised sleep signs following cholinergic stimulation of three brainstem areas, Brain Res., 306 (1984) 39-52.

Brashear, H.R., Zaborszky, L. and Heimer, L. Distribution of GABA-ergic and cholinergic neurons in the rat diagonal band, Neuroscience, 17 (1986) 439-451.

Cedarbaum, J.M. and Aghajanian, G.K. Catecholamine receptors on locus coeruleus neurons: Pharmacological characterization, Eur. J. Pharmacol., 44 (1977) 375-385.

Cespuglio, R., Faradji, H. and Gomez, M.E., Jouvet, M. Single unit recordings in the nuclei raphe dorsalis and magnus during sleep-waking cycle of semi chronic prepared cats, Neurosci. Lett., 24 (1981) 133-138.

Cherubini, J.M., North, R.A. and Williams, J.T. Synaptic potentials in rat locus coeruleus neurons, J. Physiol., 406 (1988) 431-442.

Darracq, L., Gervasoni, D., Soulier, F., Lin, J.S., Fort, P., Chouvet, G. and Luppi, P.H. Effect of strychnine on rat locus coeruleus neurons during sleep and wakefulness, Neuroreport, 8 (1996) 351–355.

Delrome, F., Froment, J.L. and Jouvet, M. Suppression du sommeil par la p. chloromethamphetamine et la p. chloroalanine, C. R. Seanc. Soc. Biol., 160 (1966) 2347.

Dement, W.C. The effect of dream deprivation, Science, 131 (1960) 1705–1707.

Deutsch, S.I., Mastropaolo, J. and Hitri, A. GABA-active steriods: endogenous modulators of GABA-gated chloride ion conductance, Clin. Neuropharmacol., 15 (1992) 352-364.

Egan, T.M. and North, R.A. Action of acetylcholine and nicotine on rat locus coeruleus neurons in vitro, Neuroscience, 19 (1986) 565-571.

El Mansari, M., Sakai, M. and Jouvet, M. Unitary characteristics of presumptive cholinergic tegmental neurons during sleep-waking cycle in freely moving cats, Exptl. Brain. Res., 76 (1989) 519-529.

Ennis, M. and Aston-Jones, G. GABA mediated inhibition of locus coeruleus from the dorsomedial rostral medulla, J. Neurosci., 9 (1989) 2973-2981.

Fillenz, M., Regulation of synthesis. In: M. Fillenz (Ed.), Noradrenergic neurons, Cambridge university press, Cambridge, 1990, pp. 109-124.

Forchetti, C.M. and Meek, J.L. Evidence for a tonic GABA-ergic control of sertonon neurons in the median raphae nucleus, Brain Res., 206 (1981)108-212.

Ford, B., Holmes, C.J., Mainville, L. and Jones. B.E. GABA-ergic neurons in the rat pontomesencephalic tegmentum: Codistribution with cholinergic and other tegmental neurons projecting to the posterior lateral hypothalamus, J. Comp. Neurol., 363 (1995) 177-196.

Fort, P., Luppi, P. and Jouvet, M. Glycine immunoreactive neurons in the cat brainstem reticular formation, Neuroreport, 49 (1993) 1123-1126.

Freund, T.F. and Antal, M. GABA-containing neurons in the septum control inhibitory interneurons in the hippocampus, Nature, 336 (1988) 170-173.

Gaillard, J.M., Benzodiazepines and GABA-ergic transmission. In: M.H. Kryger, T. Roth and W. Dement (Eds.), Principles and practice of sleep medicine, W.B. Saunders Company, Philadelphia, 1989, pp. 213–218.

Gaillard, J.M., Schulz, P. and Tissot, R. Effect of three benzodiazepines (Nitrazepam, flunitrazepam and bromazepam) on sleep of normal subject, studied with an automatic sleep scoring system, Pharmacopsychiatry, 6 (1973) 207–217.

Gang, S., Mizuguch, A., Kobayashi, N. and Aoki, M. Descending axonal projections from the medial parabrachial and kolliker fuse nucleur complex to the nucleus raphae magnus in cats, Neurosci. Lett., 118 (1990) 273–275.

Gritti, I., Mainville, I. and Jones, B.E. Codistribution of GABA with acetylecholine synthesizing neurons in the basal forebrain of the rat, J. Comp. Neurol., 329 (1993) 438–457.

Gritti, I., Mainville, L. and Jones, B.E. Projections of GABA-ergic and Cholinergic basal forebrain and GABA-ergic preoptic anterior hypothalamic neurons to the posterior lateral hypothalamus of the rat, J. Comp. Neurol., 339 (1994) 251–268.

Hermann, D.M., Luppi, P.H., Peyron, C. and Jouvet, M. Afferent projections to the rat nuclei raphae magnus, raphae pallidus and reticularis gigantocelluaris pars alpha demonstrated by iontophoretic application of choleratoxin (subunit B), J. Chem. Neuroanat., 13 (1997) 1–21.

Hobson, J.A., McCarley, R.W. and Wyzinski, P.W. Sleep cycle oscillation: recirocal discharge by two brain stem neuronal groups, Science, 189 (1975) 55–58.

Holmes, C.J. and Jones, B. Importance of cholinergic, GABAergic, serotonergic and other neurons in the medial reticular formation for sleep-wake state studied by cytotoxic lesions in the cat, Neuroscience, 62 (1994) 1199–1200.

Holmes, C.J., Mainville, L. and Jones, B. Distribution of cholinergic and serotonergic neurons in the medullary reticular formation and their projections studied by cytotoxic lesions in the cat, Neuroscience, 62 (1994) 1155–1178.

Iijima, K. and Ohtomo, K. Immunohitochemical study using GABA antiserum for demonstration of inhibitory neurons in the rat locus coeruleus, Am. J. Anat., 181 (1988) 183–194.

Jacobs, B.L., Single unit activity of locus coeruleus neurons in behaving animals, Prog. Neurobiol., 27 (1986) 183–194.

Jacobs, G.L. and Fornal, C.A. Activity of brain serotonergic neurons in the behaving animals, Pharmacol. Rev., 43 (1991) 563–578.

Jones, B. Immunohistochemical study of choline acetyl transferase-immunoreactive processes and cells innervating the pontomedullary reticular formation in the rat, J. Comp. Neurol. 295 (1990) 485–514.

Karczmar, A.G., Longo, V.G. and Scotti de Carolis, A. A pharmacological model of paradoxical sleep: the role of cholinergic and monoamine systems, Physiol. Behav., 5 (1970) 175–182.

Kaur, S., Saxena, R.N. and Mallick, B.N. GABA in locus coeruleus regulates spontaneous rapid eye movement sleep by acting on GABA-A recptors in freely moving rats, Neurosci. Lett., 223 (1997a) 105–108.

Kaur, S., Saxena, R.N. and Mallick, B.N. Role of GABA in locus coeruleus mediated cholinergic induction of rapid eye movement sleep, Second Congress of Asian Sleep Research Society, (1997b) 64.

Kayama, Y. and Ogawa, T. Electrophysiology of ascending, possibly cholinergic neurons in the rat laterodorsal tegmental nucleus: comparison with monoaminergic neurons, Neurosci. Lett., 77 (1987) 277–282.

Kayama, Y., Ohta, M. and Jodo, E. Firing of possibly cholinergic neurons in the rat laterodorsal tegmental nucleus during sleep-wakefulness, Brain Res., 569 (1992) 20–21.

Kodama, T., Takahashi, Y. and Honda, Y. Enhancement of acetylcholine release during paradoxical sleep in the rat laterodorsal tegmental field of the cat brain stem, Neurosci. lett., 114 (1990) 277–282.

Koyama, Y. and Hayaishi, O. Firing of neurons in the preoptic/anterior hypothalamic areas in rat: its possible involvement in slow wave sleep and paradoxical sleep, Neuroscience Res., 19 (1994) 31–38.

Lancel, M., Faulhaber, J., Holsboer, F. and Rupprecht, R. Progesteron induces changes in sleep comparable to those of agonistic GABA-A receptor modulators, Am. J. Physiol., 271 (1996) E763–E772.

Lin, J.S., Luppi, P.H., Salvert, D., Sakai, K. and Jouvet, M. Histamine containing neurons in the cat hypothalamus, C.R. Acad. Sci., 303 (1986) 371–376.

Lin, J.S., Sakai, K., Vanni-Mercier, G. and Jouvet, M. A critical role of the posterior hypothalamus in the mechanisms of wakefulness determined by microinjection of muscimol in freely moving cats, Brain Res., 479 (1989) 225–240.

Linke, R., Schwegler, H. and Boldyreva, M. Cholinergic and GABA-ergic septo-hippocampal projection neurons in mice: a retrograde tracing study combined with double immunocytochemistry for choline acetyltransferase and parvalbumin, Brain Res., 653 (1994) 73–80.

Luque, J.M., Malherbe, P. and Rrichards, J.G. Localization of GABA-A receptor subunit mRNAs in the rat locus coeruleus, Mol. Brain Res., 24 (1994) 219–226.

Lydic, R., McCarley, R.W. and Hobson, J. Serotonin neurons and sleep: II time course of dorsal raphae discharge, PGO waves and behavioral states, Arch. Ital. Biol., 126 (1987) 1–28.

Macdonald, R.L. and Olsen, R.W. GABA-A receptor Channels, Ann. Rev. Neurosci., 17 (1994) 569–602.

Mallick, B.N. and Alam, M.N. Medial preoptic area affects sleep-wakefulness independent of associated body temperature change in free moving rats, Brain Res. Bull., 26 (1991) 215–218.

Mallick, B.N., Nitz, D., Fahringer, H.M. and Seigel, J.M. GABA release in basal forebrain/medial septal region across sleep cycle, Sleep Research, 26 (1997) 26.

McGinty, D.J. and Harper, R.M Dorsal raphae neurons: depression of firing during sleep in cats, Brain Res., 101 (1976) 569–575.

McGinty, D.J. and Sterman, M.B. Sleep suppression after basal forebrain lesions in the cat, Science, 160 (1968) 1253–1255.

Mendelson, W.B., Martin, J.V., Perlis, M. and Wagner, R. Arousal induced by injection of trizolam in to dorsal raphae, Neuropsychopharmacol., 1 (1987) 85–88.

Mugnaini, E. and Oertel, W.H. The distribution of GABAergic neurons and terminals in rat CNS as revealed by GAD immunohistochemistry. In: A. Bjorkland and T. Hokfelt (Eds.), Handbook of Chemical Neuroanatomy, Elsevier, amsterdam, 1985, pp. 436–608.

Nitz, D. and Siegel, J.M. GABA release in posterior hypothalamus across sleep-wake cycle, Am. J. Physiol., 271 (1996) R1707–R1712.

Nitz, D. and Siegel, J.M. GABA release in the locus coeruleus as a function of sleep/wake state, Neuroscience, 78 (1997) 795–801.

O'Connor, S.D., Urbancic, M., Marczynski, T.J. and Radulowacki, M. Chronic RO 15-1788 treatment increases REM sleep in rats, Brain Res. Bull., 27 (1991) 251–253.

Parmeggiani, P.L., Zamboni, G., Cianci, T., Agnati, L.F. and Ricci, C. Influence of anterior hypothalamic heating on the duration of fazt wave sleep episodes, Electroencephalogr. Clinc. Neurophysiol., 36 (1974) 465–470.

Pastel, R.H. and Fernstrom, J.D. Short term effects of fluxetine and trifluoromethlyl-phenylpiperazine on electroencephalographic sleep in rat, Brain Res., 436 (1987) 92–102.

Polc, P., Bonetti, E.P., Schaffner, R. and Haefely, W. A three state model of benzodiazepine receptor explains the interactions between benzodiazepine antagonist RO 15-1788,

Benzodiazepine tranquilizers, beta- carbolines and phenobarbitone, Naunyan-schmied. Arch. Pharmacol., 321 (1982) 260–264.

Sagar, S.M., Sharp. F.R. and Curran, T. Expression of c-fos protein in brain: metabolic mapping at cellular level, Science, 240 (1987) 1328–1331.

Sakai, K., Anatomical and physiological basis of paradoxical sleep. In: D.J. McGinty, R. Drucker-Colin, A. Morrison and P.L. Parmeggiani (Eds.), Brain Mechanism of Sleep, Raven Press New York, 1985, pp. 111–137.

Sakai, K. Executive mechanisms of paradoxical sleep, Arch. Ital. Biol. 126 (1988) 239–257.

Sakai, K. Touret, M., Salvert, D., Legger, L. and Jouvet, M. Afferent projection to the cat locus coeruleus as visulized by the horseradish peroxidase technique. Brain Res., 119 (1977) 21–41.

Sallanon, M., Denoyer, M., Kitahama, K., Aubert, C., Gay, N. and Jouvet, M. Long lasting insomnia induced by preoptic neurons lesion and its transient reversal by muscimol injection in to the posterior hypothalamus in the cat, Neuroscience, 32 (1989) 669–683.

Sallanon, M., Sakai, K., Buda, C., Puymartin, M. and Jouvet, M. Increase of paradoxical sleep induced by injection of ibotenic acid into the ventrolateral part of the posterior hypothalamus in the cat, Archieves Italiennes de Biologie, 126 (1988) 87–97.

Satoh, K. and Fibiger, H.C. Cholinergic neurons of the laterodorsal tegmental nucleus: efferent and afferent connections, J. Comp. Neurol., 253 (1986) 277-302.

Scherschlicht, R. and Marias, J. Effect of oral and intravenous midazolam, triazolam and flunitrazepam on the sleep wakefulness cycle of rabbits, Br. J. Clin. Pharmacol., 16 (1983) 29s–35s.

Schneider, E., Ziegler, B. and Maxion, H. Gamma-aminobutyric acid (GABA) and sleep: The influence of di-n- propylacetic acid on sleep in man, Eur. J. Pharmacol., 15 (1977) 146–152.

Shipley, M.T., Fu, L., Ennis, M., Liu, W.L. and Aston-Jones, G. Dendrites of locus coeruleus neurons extend preferentially in to two pericoerulear zones, J. Comp. Neurol., 365 (1996) 56–68.

Shiromani, P.J., Kilduff, T.S., Bloom, F.E. and McCarley, R.W. Cholinergically induced REM sleep triggers Fos-like immunoreactivity in dorsolateral pontine regions associated with REM sleep, Brain Res., 580 (1992) 351–357.

Siegel, J.M. Pontomedullary interactions in the generation of REM sleep. In: D.J. McGinty, R. Drucker-Colin, A. Morrison and P.L. Parmeggiani (Eds.), Brain mechanism of sleep, Raven Press, New York, 1985, pp. 157–174.

Siegel, J.M., Manger, P., Nienhuis, R., Fahringer, H.M. and Pettigrew, J. The echidna Tachyglossus aculeatus combines REM and non-REM aspects in a single sleep states: implications for the evolution of sleep, J. Neurosci., 16 (1996) 3500–3506.

Siegel, J.M., Manger, P., Nienhuis, R., Fahringer, H.M. and Pettigrew, J. The platypus has REM sleep, Sleep Res. 26 (1997) 177.

Singh, S. and Mallick, B.N. Mild electrical stimulation of pontine tegmentum around locus coeruleus reduces rapid eye movement sleep in rats, Neurosci. Res., 24 (1996) 227–235.

Szymusiak, R. and McGinty, D.J. Sleep suppression following kainic acid induced lesions of the basal forebrain, Exp. Neurol., 94 (1986) 598–614.

Tappaz, M.L., Brownstem, M.J. and Palkovits, M. Distribution of glutamate decarboxylase in discrete brain nuclei, Brain Res., 108 (1976) 371–379.

Toth, K., Borhegyi, Z. and Freund, T.F. Postsynaptic targets of GABA-ergic hippocampal neurons in the medial septum-diagonal band of Broca complex, J. Neurosci., 13 (1993) 3712–3724.

Vanni Mercier, G., Sakai, K. and Jouvet, M. Waking state specific neurons in the caudal hypothalamus of the cat, C.R. Acad. Sci., 298 (1984) 195–220.
Wamsley, J.K. and Palacios, J.M. Aminoacid and benzodiazepine receptors. In: A. Bjorkland, T. Hokfelt and M.J. Kuhar (Eds.), Handbook of Chemical Neuroanatomy, Elsevier, Amsterdam, 1984, pp. 352–385.
Woolf, N.J. and Butcher, L.L. Cholinergic systems in the rat brain; III. Projections from the pontomesencephalic tegmentum to the thalamus, tectum, basal ganglia, and basal forebrain, Brain Res. Bull., 16 (1986) 603–637.
Yamuy, J., Mancillas, J.R., Monales, F.R. and Chase, M.H. C-fos expression in pons and medulla of cat during carbachol induced active sleep, J. Neurosci., 13 (1993) 2703–2718.
Ziegler, G., Ludwig, L. and Fritz, G. Effects of specific benzodiazepine antagonist RO 15-1788 on sleep, Pharmacopsychiatry, 19 (1986) 200–201.

Rapid Eye Movement Sleep
B.N. Mallick and S. Inoué (Eds)

11. Nitric Oxide: A diffusible modulator of physiological traits and behavioral states

Timothy O. Leonard and R. Lydic

Department of Anesthesia, The Pennsylvania State University, College of Medicine, Hershey, PA 17033 USA

SUMMARY

The data outlined in this chapter indicate a compelling role for brain nitric oxide (NO) in modulating levels of behavioral arousal. Nitric oxide participates in the regulation of sleep/wake state and cortical electrical activity by altering neurotransmitter release in key areas of the brain—such as the cholinergic mesopontine tegmentum, the cholinoceptive pontine reticular formation, and the thalamus—which are known to contribute to REM sleep regulation. The possibility that NO can alter directly the activity of neuronal circuits mediating states of consciousness—through regulation of gene transcription and/or protein modulation—cannot be overlooked. Nitric oxide now is appreciated to be a ubiquitous molecule modulating diverse physiologic functions. With respect to modulation of physiologic traits and behavioral states such as REM sleep, NO does not act through any single transduction pathway. NO may serve a unifying role in the regulation of arousal states by functioning as an intercellular messenger modulating neuronal excitability. Continuing efforts to elucidate the role of NO as a transmembrane signal transduction molecule offers an exciting opportunity for mechanistic studies of REM sleep generation.

Key Words: REM Sleep, Anesthesia, Arousal, Cholinergic, Signal Transduction

Introduction

"Perhaps the last frontier of science—its ultimate challenge—is to understand the biological basis of consciousness...". Thus begins *Principles of Neural Science* (Kandel et. al., 1991) a leading textbook of neuroscience used in medical and graduate education. Indeed, many aspects of consciousness and unconsciousness have come to be understood in terms of the physiological traits that comprise behavioral states. This chapter provides a brief and selective overview of nitric oxide's diverse roles in modulating physiological traits and behavioral states. The chapter concludes with an overview of available evidence suggesting a role for nitric oxide in the regulation of REM sleep.

Interested readers are referred elsewhere for comprehensive reviews on nitric oxide (Chiueh et. al., 1994; Vincent, 1995).

Discovery of the Biological Functions of Nitric Oxide

In the mid-1980s, the free radical gas nitric oxide (NO) was considered to be just another environmental pollutant, the unwanted waste-product of automobile and industrial combustion processes, jet planes, lightning, cigarette smoke, and microorganisms such as *Alcaligenes faecalis* and *Bacillus halodentrificans* (Culotta and Koshland, 1992; Nathan, 1992). Nitric oxide was suspected to be a carcinogen, to destroy ozone, and to produce acid rain (Star, 1993). Anesthesiologists may recall a series of fatalities that occurred in the 1960s resulting from the inadvertent contamination of gas cylinders of nitrous oxide with nitric oxide (Clutton, 1967). Now known to be produced in mammals as well as microorganisms, NO is appreciated as an important regulatory molecule (Nathan, 1992; Radomski, 1995).

Current understanding of the biological functions of NO developed from two independent lines of research. Experiments investigating the mechanisms of immune-mediated cytotoxicity demonstrated that activated macrophages explanted from mice secreted nitrite and nitrate (Stuehr and Marletta, 1985). Subsequently, the production of nitrite and nitrate was shown to result from the oxidation of a guanidino nitrogen of L-arginine which yielded L-citrulline and a compound able to produce nitrosamines upon interaction with amines (Iyengar et al., 1987; Miwa et al., 1987). Furthermore, L-arginine was found to be a necessary metabolite for maintaining macrophage-mediated cytotoxicity (Hibbs et al., 1990). The cytotoxic effect of activated macrophages was mimicked by dissolving NO gas in the cell culture media and blocked by addition of NO scavengers such as hemoglobin (Marletta et al., 1988). Therefore, these studies of immune function demonstrated that NO is produced in macrophages and mediates some of their cytotoxic effects.

A second, independent line of investigation discovered that NO causes endothelium-dependent vasodilation. A fortuitous experimental error led to the discovery in 1978 that the vasodilation elicited by acetylcholine (ACh) in rat aorta is dependent upon the presence of the endothelium (Furchgott, 1995). It was found later that ACh stimulated the secretion of a substance from the endothelium (Furchgott and Zawadski, 1980) that mediated vasodilation through increases in cyclic guanine monophosphate (cGMP) levels in the muscle (reviewed in Furchgott, 1995). At a meeting on Mechanisms of Vasodilation in Rochester, MN, in 1986, it was proposed independently by Louis Ignarro and Richard Furchgott that endothelial derived relaxing factor (EDRF) is NO (reviewed in Furchgott, 1995). The declaration a decade ago that NO possesses important biological activity stimulated many studies of NO function. The discovery of NO as a unique neurochemical messenger has challenged the conventional way of looking at chemical neurotransmission. The contrast between the properties of conventional chemical neurotransmission and the neurochemical NO is shown in Table 1.

Table 1 Comparison between the properties of classical neurotransmitters and nitric oxide (NO). Nitric oxide is the smallest and lightest known neurotransmitter-like substance, and NO is unique in its properties as an intercellular messenger in the nervous system. Unlike conventional neurotransmitters such as ACh, NO is not stored in synaptic vesicles. Furthermore, the release of NO is not regulated. Once it is synthesized, lipophilic NO passes through cellular membranes and interacts with iron or cysteine residues on intracellular proteins in neighboring cells. NO acts more like a paracrine autacoid than a classical neurotransmitter (Paakkari and Lindberg, 1995). NO has a large sphere of influence over intracellular proteins and is well suited for a role in synchronizing and recruiting synaptic activity (Bach-Y-Rita, 1993; Schuman and Madison, 1994b; Wood and Garthwaite, 1994)

Table 1 Properties of Neurochemical Messengers

Property	Classical Neurotransmitter	Nitric Oxide
Synthesis/storage	Synthesis is largely constitutive, molecules are stored in membrane synaptic vesicles	Synthesis is not constitutive but occurs rapidly upon stimulation by Ca^{2+}; as a lipid-soluble gas, NO is not stored
Release	Ca^{2+}-dependent; localized to specialized region of synapse; molecules released in discrete quanta	Not regulated: NO freely diffuses across cellular membranes once synthesized
Action	Reversibly interacts with surface membrane protein receptor; transmitter-receptor specificity	Forms covalent linkage with several potential intracellular protein targets
Sphere of influence	Immediate synaptic region	200 μm diameter (estimated 2 million synapses possibly affected) (Abu-Soudt et al., 1994)
Biological half-life	Milliseconds	3-6 seconds
Termination of Action	Presynaptic reuptake and recycling; enzymatic degradation (often with recycling)	Diffusion away from targets; spontaneous oxidation; covalent linkage with scavenger compound (superoxide anion, hemoglobin)

The L-Arginine/Nitric Oxide/cGMP Pathway

NO is generated by the conversion of the amino acid L-arginine and molecular oxygen to the amino acid L-citrulline by the nitric oxide synthase (NOS) enzyme complex (Fig. 1A) (Marletta et al., 1988; McCall and Vallance, 1992; Moncada and Higgs, 1993; Nathan, 1992). Experiments using activated macrophages (Hibbs et al., 1990; Iyengar et al., 1987; Miwa et al., 1987) and endothelial cells (Palmer et al., 1988; Schmidt et al., 1988) demonstrated that L-arginine is the precursor to NO synthesis. The NOS enzyme complexes that synthesize NO from the substrate L-arginine are some of the most highly regulated enzymes in the human body (Nathan, 1992). Mechanisms that

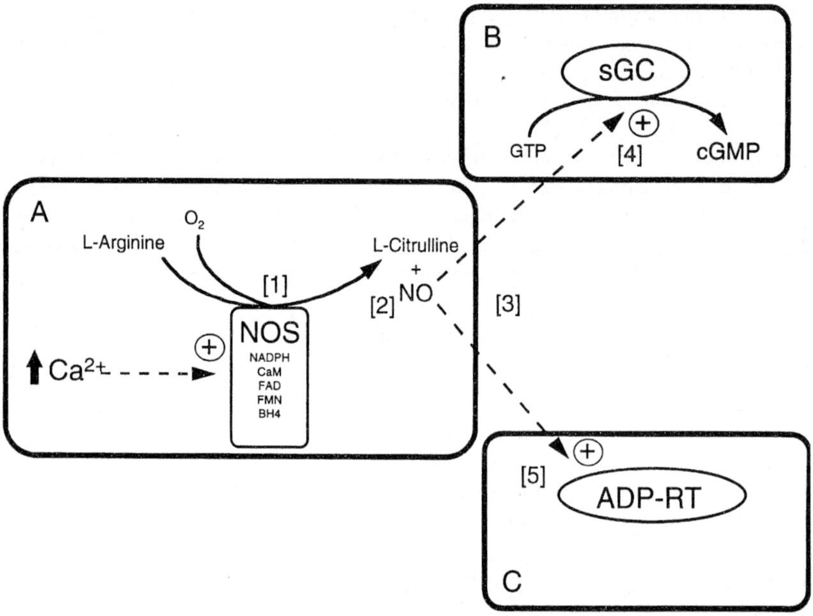

Fig. 1 **Schematic illustration of L-arginine/nitric oxide/cGMP pathway. This depicts molecular and cellular participants in production and action of the intercellular messenger nitric oxide (NO). An increase in intracellular free calcium levels in a cell (A) containing nitric oxide synthase (NOS) results in the activation of NOS. The enzymatic substrate L-arginine is converted in the presence of molecular oxygen to L-citrulline and the free radical gas NO. Due to its lipid solubility and small size, NO exits the cell in which it was produced (A) and enters a neighboring cell (B). In the target cell (B), NO can activate the soluble guanylate cyclase (sGC) enzyme. Guanylate cyclase converts intracellular GTP to cGMP. cGMP then can act as a second messenger to alter protein phosphorylation or activity of membrane ion channels (Garthwaite, 1991). Another intercellular target for NO is ADP-ribosyltransferase (ADP-RT) (C). The activation of presynaptic ADP-RT by NO may lead to mono-ADP-ribosylation of GTP-binding proteins which have been shown to modulate exocytosis (reviewed by Schuman et al., 1994).**

In experimental systems, L-arginine/NO/cGMP pathway can be manipulated at a number of steps (numbered 1 through 5) to examine the role of NO in physiologic processes. L-arginine analogs, such as N^G-nitro-L-arginine (NLA) compete with L-arginine at the active site on NOS [1] and reduce NO production. Exogenous sources of NO [2], such as NO gas or an NO donor (e.g., nitroprusside) dissolved in a perfusate, increase NO levels and can enhance the biological effects of NO. Macromolecules, such as hemoglobin, which contain heme or iron moieties and are unable to pass through cellular membranes, can be administered to the extracellular compartment. These macromolecules bind and scavenge extracellular NO and prevent the entrance of NO into the target cell [3] thereby reducing the biologic effects attributable to NO. Agents such as methylene blue inhibit guanylate cyclase [4] and reduce production of cGMP, the downstream effector of NO production. Finally, inhibitors of ADP-RT [5] such as phylloquinone have been used to demonstrate the role of ADP-RT stimulation in cellular activity.

regulate NOS and the production of NO are as complicated as the roles of NO are diverse.

There are at least three different isoenzymes responsible for the production of NO. Table 2 lists the three conventional isoforms of NOS and some of the characteristics of each form. The isoforms of NOS reflect the diversity of function and tissue location for NO activity (Katusik and Cosentino, 1994). The three principal forms of NOS are the neuronal, the macrophage or inducible,

Table 2 Characteristics of the nitric oxide synthase (NOS) enzymes. There are at least three different types of NOS enzymes: neuronal NOS (nNOS or NOS-1), macrophage NOS (NOS-2), and endothelial NOS (eNOS or NOS-3). The major characteristics of NOS isoforms are listed in the column at the far left. NOS enzymes have been classified according to anatomical location of the enzyme, gene location and nucleotide sequence, and regulation of gene expression and enzymatic activity. Both neuronal NOS and endothelial NOS are expressed constitutively and are activated by elevations in intracellular free calcium to NO and L-citrulline from L-arginine. Expression of macrophage NOS, however, requires gene induction, and enzyme activity is not dependent upon free calcium levels. All forms of NOS require as cofactors NADPH, calmodulin, tetrahydrobiopterin, flavin adenine dinucleotide (FAD), and flavin mononucleotide (FMN).

Characteristic	Neuronal NOS	Macrophage NOS	Endothelial NOS
Conventional Terminology	Type 1/nNOS/cNOS	Type 2/iNOS	Type 3/eNOS/cNOS
First Tissue Source Described	cerebellum	activated macrophage	vascular endothelium
Tissue Locations Discovered to Date	neurons; macula densa cells; bronchial epithelium; gastric epithelium; mast cells; human skeletal muscle; photoreceptor inner segments	macrophages cardiomyocytes hepatocytes megakaryocytes keratinocytes chondrocytes	endothelium neurons
Chromosomal Location (Rat)	$12_q24.2$ - 24.31	17_{cen} - $17_q11.2$	7_q35 - 36
Gene Expression	constitutive (cNOS)	inducible (iNOS)	constitutive (cNOS)
Molecular Weight (Daltons)	160	130	133
Activation Requirements			
rise in intracellular calcium	yes/causes calmodulin binding	no	yes/causes calmodulin binding
calmodulin binding	yes/reversible binding	yes/irreversible binding	yes/reversible binding
tetrahydrobiopterin	yes	yes	yes
FAD	yes	yes	yes
FMN	yes	yes	yes
NADPH	yes	yes	yes
Activation Period	short-term (sec-min)	long-term (hr-day)	short-term (sec-min)

and the endothelial (Snyder, 1995). The NOS enzymes are a group of multifunctional oxidoreductases that share certain important properties. All isoforms catalyze the formation of NO and L-citrulline from L-arginine via a transfer of 5 electrons from molecular oxygen and nicotinamide adenine dinucleotide phosphate (NADPH) to a guanidino nitrogen of L-arginine (Abu-Soudt et al., 1994). Similarly, all require the cofactors calmodulin (CaM), tetrahydrobiopterin (BH4), flavin adenine dinucleotide (FAD), and flavin mononucleotide (FMN) in addition to NADPH (Schmidt et al., 1994). The formation of NO from L-arginine by NOS is among the most highly regulated enzymatic reactions known (Schmidt et al., 1994). The complexity of the reaction forming NO requires the simultaneous cooperation of multiple cofactors and provides a number of methods for manipulating NO formation (reviewed in (Fukuto and Chaudhuri, 1995; Griffith and Stuehr, 1995).

The distinction between the activity of cNOS (the endothelial and neuronal forms) and iNOS conveys important differences in the cellular roles of NO. Normally, macrophages and other cells of the reticuloendothelial system do not contain NOS (Moilanen and Vapaatalo, 1995). Upon induction of NOS gene expression by mediators of inflammation or infection, such as interleukin-1, interferon-gamma, or lipopolysaccharide of gram negative bacteria, NOS protein is produced in cells of the nonspecific immune system (Schmidt et al., 1994). While it may require 6-24 h for the formation of iNOS enzyme following gene induction (Kitamura, 1994), NO is produced continuously by iNOS at an estimated rate of 1600 nmol/mg cell weight/min (Schmidt et al., 1994). The rapid and prolonged production of NO leads to accumulation of large amounts of NO that diffuse over a distance possibly as great as 1 mm (Moilanen and Vapaatalo, 1995). In such large quantities, NO functions as a potent oxidant that damages cellular membranes and DNA of bacteria and eukaryotic cells (Moilanen and Vapaatalo, 1995; Schmidt et al., 1994). The NOS present in macrophages is well-suited for the cytotoxic role of NO.

In contrast to iNOS, cNOS present in endothelium and in neurons functions well as the generator of an intercellular messenger molecule. In neurons and endothelial cells containing cNOS genes, the NOS protein is continually present. The NOS enzyme is quiescent until stimulated by increases in intracellular levels of free calcium (Knowles, 1994). NO is produced by cNOS in response to stimuli which increase calcium concentrations from resting levels of less than 100 nM to levels greater than 500 nM (Schmidt et al., 1994). The production of NO from cNOS can occur within seconds to minutes following the rise in intracellular calcium (Kitamura and Nomura, 1994). The production of NO by cNOS occurs at an estimated rate of 200 nmol/mg cell weight/min, eight-fold slower than the production of NO by iNOS (Schmidt et al., 1994). The estimated radius of NO diffusion in the brain is less than 200 μm from its source of production (Wood and Garthwaite, 1994). NO production is regulated on a minute-by-minute basis by cNOS. Given its lability and lipid solubility, NO can diffuse from its cell of origin to a cellular target within a 200 μm radius. NO in the brain thus can serve as a discrete intercellular mediator of cellular activity. Furthermore, localized inhibition of cNOS and reduction of NO production in experimental systems likely would have an effect on NO function within only a 500 μm radius (Wood and Grathwaite, 1994). The proposed limited region of effect of NOS inhibition on NO function is important when considering the role of NOS inhibition on any neuronal network contributing to arousal state control.

Cellular Targets of Nitric Oxide

As an intercellular messenger, NO leaves the cell in which it was synthesized and enters a neighboring cell where it interacts with many possible intracellular targets. The cytotoxic functions of NO are mediated, in part, through the ability of the free radical nitrogen monoxide gas to act as a strong oxidant, and thereby cause membrane lipid peroxidation and cell lysis (Dawson et al., 1994). The unpaired electron of NO is involved in electrophilic reactions with phospholipids, nucleotides, and nucleic acids (Reiter, 1995). Either directly, or by conversion into another molecular species with strong oxidant properties (peroxynitrite, hydroxyl radical), NO interacts indiscriminately with DNA or cellular enzymes to mediate cytotoxic effects (Gutteridge, 1994).

NO also can react with specific intracellular targets, thus causing the production of NO in one cell to directly affect the function of another cell in a paracrine fashion (Sato and Murota, 1995). Nitric oxide has a strong affinity for heme or ferrous (Fe^{2+})-containing proteins (Butler, 1995), and the most relevant physiological activator of soluble guanylate cyclase (sGC) appears to be nitric oxide (Mayer et al., 1993). Activated guanylate cyclase converts GTP into cGMP (Fig. 1B). Increases in cGMP then can affect protein phosphorylation through cGMP-dependent protein kinase, directly alter membrane ion channel permeability, or influence the activity of phosphodiesterase enzymes (Garthwaite and Boulton, 1995). An additional, albeit little-studied, target and cellular effect of NO independent of guanylate cyclase includes stimulation of ADP-ribosyltransferase activity (Fig. 1C) (Schuman and Madison, 1994a).

Experimental Approaches to Studying Nitric Oxide Function

The first research tools used to assess the role of NO in biological systems were analogs of L-arginine. Prior to the discovery that NO was crucial for macrophage-mediated cytotoxicity (Hibbs et al., 1990), it was determined that L-arginine analogs substituted at one or both of the guanidino nitrogens blocked cytotoxicity in a stereospecific manner (Hibbs et al., 1987). Reversible binding of substituted L-arginine compounds inhibits NO production and allows for testing whether a cellular phenomenon is dependent upon NO (Fukuto and Chaudhuri, 1995). Biological phenomena have been considered to be mediated by NO or to be NOS-dependent if they are attenuated stereoselectively by L-arginine NOS inhibitors (Griffith and Stuehr, 1995).

The use of substituted L-arginine analogs has become the cornerstone of NO research (Griffith and Stuehr, 1995; Nathan, 1992). Figure 1 depicts the pathway of NO production from L-arginine and the interaction of NO with intracellular tragets within neighboring cells.

Physiologic Roles of Nitric Oxide

Thus far, two major physiologic roles for NO have been defined: cytotoxicity and intercellular communication (McCall and Vallance, 1992). Many studies show that NO participates in a wide variety of physiologic and pathologic processes. NO is involved in cardiovascular regulation, immunological function, endocrine regulation, and the functioning of the peripheral and central nervous systems (McCall and Vallance, 1992). The relative ubiquity of NO means that the diversity of NO function derives not from the biomolecular properties of NO, but from the specific site of NO production and interaction with its cellular targets. Where possible, this section highlights the role of NO in relation to state-dependent changes in autonomic control and sleep disorders medicine (Dement, 1994).

Nitric Oxide and Immunologic Function

Considerable data generated from preclinical and clinical studies demonstrate that sleep deprivation can impair the function of the immune system (Benca et al., 1989; Bergman et al., 1996; Everson, 1993; Irwin et al., 1996; Moldofsky et al., 1989). Nitric oxide has been identified as a mediator of inflammation and nonspecific immunity. Endotoxin, lipopolysaccharide, and interleukin-1—stimuli which cause induction of the iNOS gene and production of NO—have been shown to produce sleep in humans and experimental animals. In humans, intravenous injections of purified endotoxin (*Salmonella abortus equi*) resulted in an increase in non-rapid eye movement (NREM) sleep time and rapid eye movement (REM) sleep was minimally affected during a 12 h polygraphic recording period post injection (Pollmacher et al., 1993). Intracerebroventricular (icv) and intraperitoneal injections of interleukin-1 increased NREM sleep time and had no statistically significant effect on REM sleep during a 3 h recording period following interleukin-1 administration (Kapas et al., 1994).

In addition to these results linking NO production with the generation of NREM sleep, there also is evidence that NO generated through infectious, inflammatory responses may facilitate REM sleep. Lipopolysaccharide (LPS, *S. abortus equi*) injections in rats did not alter NREM sleep time over a 24 h period post intraperitoneal injection (Lancel et al., 1995). REM sleep, however, was decreased during the first half of the recording and was significantly increased during the second 12 h following injection of LPS (Lancel et al., 1995). The findings of Lancel et al., are interesting in light of the evidence showing that induction of iNOS by LPS and other cytokines requires from several to 24 h (Kitamura and Nomura, 1994). Therefore, the

12 h delay to increases in REM sleep time seen in rats following LPS injection (Lancel et al., 1995) would indicate that REM sleep enhancement could have been mediated by iNOS induction and production of NO in response to LPS.

Nitric Oxide and Cardiopulmonary Function

The balance between factors mediating vasodilation or vasoconstriction determines vascular tone on a moment-by-moment basis (Star, 1993). NO can contribute to cardiovascular regulation in the following way. NO is released from endothelial cells by shear stress to the vessel wall or by ACh, bradykinin, angiotensin II, serotonin, adenine nucleotides, and thrombin (Star, 1993). NO passes out of endothelial cells into neighboring smooth muscle cells where it enhances cGMP levels and mediates smooth muscle relaxation (Star, 1993). Continuous synthesis of NO is important in determining basal vascular tone and systemic vascular resistance (Vallance and Collier, 1994). The importance of NO in normal cardiovascular function is underscored by the fact that use of nitrovasodilators (nitroglycerin and nitroprusside) in the treatment of angina and arterial vasospasm relies upon the exogenous delivery of NO (Radomski, 1995).

NO released from neurons and astrocytes surrounding cerebral vessels results in vasodilation (Paakkari and Lindberg, 1995; Tomimoto et al., 1994) and may be important in coupling neuronal activity and local cerebral blood flow (Paakkari and Lindberg, 1995). In some brain regions cerebral glucose utilization (Lydic et al., 1991) and cerebral blood flow increase during REM sleep compared to NREM sleep and wakefulness (Abrams et al., 1991; Asenbum et al., 1995; Madsen et al., 1991; Maquet et al., 1996, 1997).

Few studies have addressed the role of NO in regulating respiratory rhythmogenesis. NOS inhibition in the pontine respiratory group of decerebrate, paralyzed, and artificially ventilated cat resulted in prolonged periods of inspiration, or apneusis (Ling et al., 1992). These data suggested that NO production in the neurons of the pontine respiratory group is important for regulating the pneumotaxic mechanism: the transition from inspiration to expiration (Ling et al., 1992). Thus, NO levels within the rostral ventrolateral pons may be important in modulating respiratory timing.

Brain NO levels also have been shown to participate in the physiologic response to hypoxia. The nucleus tractus solitarius (NTS) is the first central nervous system relay in the chemoreceptor afferents related to hypoxic ventilatory changes. The NTS stained positively for NADPH-diaphorase (Gozal et al., 1997), and intraperitoneal administration of the NOS inhibitor N^G-nitro-L-arginine methyl ester (L-NAME) enhanced initial minute ventilation in response to hypoxia (Gozal et al., 1997). Therefore, NO produced within the NTS appears to modulate the hypoxic ventilatory response.

Obstructive sleep apnea results from the repetitive obstruction of a hypotonic upper airway during sleep (McNamara et al., 1994). While the precise mechanisms of airway obstruction during sleep apnea remain unknown

(McNamara et al., 1994), it has been shown that injection of cholinergic agonist into the medial pontine reticular formation (mPRF) of cat causes a REM sleep-like state and upper airway muscle hypotonia (reviewed in Lydic, 1997). A direct relationship between upper airway muscle tone and NO levels is purely speculative, but a possible relationship is suggested by two lines of evidence. First, stimulation of cholinergic neurons causing increased ACh release in the mPRF also caused apneas (Lydic and Baghdoyan, 1993). Second, mPRF administration of NOS inhibitors decreases mPRF levels of ACh while reversing cholinergically-evoked respiratory rate depression (Leonard and Lydic, 1997). Nitric oxide as a modulator of cardiopulmonary interaction also is suggested by clinical data. For example, there is an association between sleep apnea, atherogenesis, and hypertension (Ferguson and Fleetham, 1995). Impaired vascular NO production may contribute to the formation of atheromatous plaques (Nava et al., 1995) and to the etiology of both primary and secondary hypertension (Tikkanen and Fyhrquist, 1995).

Nitric Oxide and Endocrine Regulation

For many years, endocrinologists have used L-arginine to stimulate the release of growth hormone, insulin, pancreatic polypeptide, and other hormones (reviewed in Vallance and Collier, 1994). The hormone-releasing effect of L-arginine may be related to the production of NO (reviewed in Vallance and Collier, 1994). NO has been shown to mediate the release of insulin from pancreatic b cells in the presence of D-glucose and L-arginine (Schmidt et al., 1992). These results suggest a possible link between endocrine regulation, sleep and NO.

Drowsiness following food consumption is common. Evidence suggests that sleep propensity following a meal is due in part to the secretion of insulin. Systemic (Danguir and Nicolaidis, 1980) or icv (Danguir and Nicolaidis, 1984) administration of insulin enhances NREM sleep, whereas immunoneutralization of endogenous insulin reduces NREM sleep (reviewed by Krueger and Obál, 1994).

Recently, the potential role of melatonin in the regulation of sleep states has gained scientific and popular attention. The synthesis and secretion of melatonin by the pineal body exhibits a clear circadian modulation, with increasing serum melatonin levels after the onset of darkness and peaks in levels between 2 and 4 a.m. Melatonin levels gradually taper throughout the latter portion of the night (reviewed in Brzezinski, 1997). Several studies in humans have demonstrated that oral administration of melatonin in physiologic levels has a sleep-inducing effect. In elderly people who are found to have abnormally low levels of serum melatonin and complaints of poor sleep, controlled-release administration of melatonin had a beneficial effect on sleep quality (Garfinkel et al., 1995). In healthy, young adults, ingestion of melatonin prior to bedtime significantly decreased sleep onset latency, latency to onset of stage II sleep (Zhdanova et al., 1995), and duration of REM sleep (Tzischinsky and Lavie, 1994). Melatonin may participate in the regulation of sleep states by altering neurotransmitter levels and other brain stem neuronal networks known to be involved in the generation of sleep (reviewed in Rotenberg, 1992), and it is relevant that the circadian modulating effects of melatonin on suprachiasmatic neuron activity are mediated through the generation of nitric oxide (Starkey, 1996).

Nitric Oxide Function in the Nervous System

Anatomical Localization of NOS in the Brain

Data concerning the location of NOS within the brain is important for studies aiming to characterize the functional roles of NO. Therefore, histochemical mapping of brain NOS has played a key role in studies of brain NO function. In 1961, Thomas and Pearse noted the ability of a discrete population of neurons to convert a soluble tetrazolium salt into an insoluble and visible formazan reaction product by a process dependent upon reduced NADPH (Thomas and Pearse, 1961). A single enzyme, NADPH-diaphorase, later was shown to be responsible for the formation of this dye product in aldehyde-fixed brain sections (Scherer-Singer et al., 1983). Although the function of NADPH-diaphorase remained uncertain, the differential distribution of formazan staining throughout the brain suggested that special functional properties were conveyed to the NADPH-diaphorase-positive neurons (Mizukawa et al., 1989).

The identity of the neuronal NADPH-diaphorase enzyme finally was discovered independently by two different research groups in 1991. Hope et al. (1991) showed that NADPH-diaphorase and NOS purified from rat brain neurons are identical. NOS and NADPH-diaphorase shared a molecular weight of approximately 150 kD, were immunoprecipitated from crude supernatants with NADPH- diaphorase-specific antiserum, and shared the enzyme substrate nitro blue tetrazolium (Hope et al., 1991). Additionally, it was shown that throughout the brain and peripheral nervous system all NO-synthase-staining cells also reacted positively for the presence of NADPH-diaphorase (Dawson et al., 1991). Similarly, nearly all the cells, including those cells of the mesopontine tegmentum, that are NADPH-diaphorase-positive also are NOS-immunoreactive (Dawson et al., 1991).

The discharge rate of laterodorsal and pedunculopontine (LDT/PPT) neurons repeatedly has been found to vary with levels of sleep and wakefulness (Datta, 1995; El Mansari et al., 1989, 1990; Sakai and Koyama, 1996). To date, no published data from cat have localized NOS mRNA to cholinergic terminals of the LDT/PPT neurons. The finding that NADPH-diaphorase staining is a reliable and valid histochemical technique for localizing nNOS is essential in providing anatomical evidence for the presence of NOS in cholinergic LDT/PPT neurons. Based upon NADPH-diaphorase histochemistry using 4% paraformaldehyde brain fixation (Bickford et al., 1993; Mizukawa et al.,

1989; Vincent et al., 1983), and choline acetyltransferase immunoreactivity (Bickford et al., 1993), cholinergic neurons of the cat LDT/PPT have been shown to contain NOS in their cell bodies, processes, and axon terminals. Additional immunohistochemical *in situ* hybridization analyses of the rat brain have revealed that, in addition to NADPH-diaphorase staining, the LDT/PPT nuclei contain both NOS protein and mRNA (Bredt et al., 1990). Finally, autoradiographic localization using tritiated L-arginine and L-arginine analog competitive inhibitors of NOS has shown the presence of NOS in the LDT of the rat (Buzarin and Gundlach, 1995). Thus, at least in the rat, NADPH-diaphorase positive LDT/PPT neurons have been shown to transcribe nNOS genes and to translate the nNOS mRNA into NOS protein.

Nitric Oxide and Synaptic Function

The current appreciation that NO modulates synaptic function and brain excitability were advanced by studies of N-methyl-D-aspartate (NMDA) receptors. In 1987, it was shown that cells from the rat cerebellum responded to NMDA receptor agonists with an increase in cellular cGMP levels (Garthwaite and Garthwaite, 1987). Subsequently, Garthwaite et al. (1988) demonstrated that the increase in cGMP levels in cerebellar cell suspensions was mediated by stimulating NMDA receptors and the production of a labile factor resembling EDRF. Effects of NMDA receptor stimulation were mimicked by the NO donor sodium nitroprusside and attenuated by the NO scavenger hemoglobin and the guanylate cyclase inhibitor methylene blue (Garthwaite et al., 1988). These findings led to the understanding that the EDRF-like factor evoked by NMDA agonists was nitric oxide.

The initial concept of how NO functioned in the brain focused on changes in NO synthesis in response to NMDA receptor stimulation and the ability of NO to enhance cGMP levels (Garthwaite, 1991). Binding of glutamate or other excitatory amino acid agonists to the inotropic NMDA receptor opens the voltage-sensitive calcium channels in the cell membrane. Calcium then enters the cell and binds to calmodulin which activates neuronal NOS (Snyder, 1995). Activated NOS converts intracellular L-arginine to L-citrulline and NO. NO freely diffuses out of the cell of origin and enters a neighboring neuron where it binds to cytosolic guanylate cyclase and enhances production of cGMP (Garthwaite, 1991). The ways in which NO exerts its downstream effects has not been delineated completely. It is clear, however, that NO produced in one neuron can alter the synaptic function of neighboring axons (Schuman and Madison, 1994b).

One of the most widely studied functions of NO in the brain is its role in potentiating local synaptic activity (Holscher, 1997). Models of learning, memory, and synaptic reorganization within the brain rely on alterations in the synaptic interaction between circuits of neurons (Schuman and Madison, 1994b). Included in these models are processes whereby multiple synapses performing similar functions are co-activated and tend to function as a group, thus strengthening the functional output of this local circuit of neurons (Schuman and Madison, 1994a). Long-term potentiation (LTP) of synaptic activity by enhanced hippocampal transmission is believed to play a role in memory consolidation (Hawkins et al., 1994).

The circuit of neurons in the hippocampus involved in the initial retention of declarative memory (persons, places, and things) has been studied extensively (Hawkins et al., 1994; Schuman and Madison, 1994b). The LTP function in this system of neurons consists of pyramidal cells of the CA3 region of the hippocampus which project axonal processes (Schaffer collaterals) to the pyramidal neurons in the CA1 hippocampal region (Hawkins et al., 1994). Action potentials transmitted down the length of the Schaffer collateral result in excitatory postsynaptic potential (EPSP) activity in the innervated CA1 neuron.

The role of NO in mediating hippocampal LTP was demonstrated by Schuman and Madison (1994a). By recording the activity of neighboring CA1 neurons while simultaneously stimulating the presynaptic Schaffer collaterals, Schuman and Madison showed that induction of LTP was dependent upon the presence of excitatory potentials in both the presynaptic and postsynaptic neurons (Schuman and Madison, 1994a). Furthermore, the depolarization of one CA1 neuron produced a labile mediator that elicited LTP at Schaffer collateral terminals on neighboring CA1 neurons. The finding that addition of the NOS inhibitor L-NMMA eliminated LTP in the neighboring CA1 neuron suggested that NO was the labile mediator of LTP (Schuman and Madison, 1994a). Postsynaptic injections of calcium chelators demonstrated that the labile mediator of LTP required increased intracellular calcium, presumably for the stimulation of NO production by NOS (Schuman and Madison, 1994a). From these experiments, it was shown that NO produced in one neuron is capable of eliciting increased synaptic neurotransmission at a neighboring neuron. This line of research demonstrated that NO can alter the activity of postsynaptic neurons by modulating synaptic activity.

A large body of data indicate that sleep, and particularly the rapid eye movement (REM) phase of sleep, is important in the process of memory formation. In fact, evidence from a number of studies suggests that memory consolidation and learning may occur during sleep (reviewed in Hennevin et al., 1989; Hobson and Stickgold, 1995; Rotenberg, 1992; Smith, 1995; Wilson and McNaughton, 1994. In a study involving young men and women, Karni and colleagues (1994) demonstrated that postlearning REM sleep deprivation significantly impaired retention of material memorized prior to sleep onset. Animal studies have shown a similar association between REM sleep and learning. During location-specific activities such as maze running, multi-electrode recordings of pairs of rat hippocampal CA1 neurons exhibit a characteristic spatial firing pattern, referred to as "place fields" (Wilson and McNaughton, 1994). These place field patterns of firing are specific to a particular activity and are not exhibited during behavioral inactivity or during deep or slow wave sleep. During REM sleep, however, hippocampal cells are engaged in the same place field pattern of firing as seen during previous periods of waking, task-specific activity (Wilson and McNaughton, 1994). Such evidence suggests that during REM sleep, there is a kind of replaying of information acquired during wakefulness. This replaying of information during REM sleep may contribute to the process of memory consolidation (reviewed in Guiditta et al., 1995). Furthermore, this replaying of information through a repeated neuronal firing pattern is consistent with the function of long-term potentiation. NO contributes not only to long-term potentiation within the hippocampus, but, as will be discussed later, NO participates in the maintenance of REM sleep episodes (Leonard and Lydic, 1997).

Nitric Oxide Modulation of Neurotransmitter Release

The specific mechanisms whereby NO alters synaptic activity remain incompletely understood. One of the intracellular targets of NO is ADP-ribosyltransferase (Schuman et al., 1994). The activation of presynaptic ADP-ribosyltransferase by NO may lead to mono-ADP-ribosylation of GTP-binding proteins which have been shown to modulate exocytosis (reviewed in Schuman et al., 1994). Therefore, the regulation of exocytosis by NO-stimulated ADP-ribosyltransferase (Schuman et al., 1994) or soluble guanylate cyclase (Garthwaite, 1991) could increase presynaptic neurotransmitter release. More commonly, the effects of NO are thought to be mediated through cGMP (Garthwaite, 1991). The specific downstream effects of NO mediated by increased cGMP levels remain to be determined, but cGMP levels have been found to influence neurotransmitter release. Dialysis of the rat striatum with membrane-permeable analogs of cGMP causes a dose-dependent increase in ACh release (Guevara-Guzman et al., 1994).

Numerous *in vitro* and *in vivo* studies have demonstrated the ability of NO to alter neurotransmitter release. Perfusion of the rat hippocampus with an NO donor increased release of both GABA and glutamate in a dose-dependent manner dependent upon the dose of the NO liberating compound (Segovia et al., 1994). Within the rat striatum, NO was shown to facilitate release of GABA, glutamate (Guevara-Guzman et al., 1994), and dopamine (Guevara-Guzman et al., 1994; Sandor et al., 1995). A number of studies using *in vivo* brain microdialysis or push-pull cannula perfusion have demonstrated that NO can modulate neurotransmitter release. Delivery of L-arginine analog NOS inhibitors was shown to reduce ACh release in the rat ventral striatum (Guevara-Guzman et al., 1994; Prast et al., 1995) and basal forebrain (Prast et al., 1992). Perfusion of the rat striatum with NO releasing compounds (Prast et al., 1995) or with NO gas dissolved in artificial cerebrospinal fluid (Guevara-Guzman et al., 1994) resulted in an increase in striatal ACh release. Similarly, NO was shown to facilitate ACh release in cerebral cortical neurons (Ohkuma et al., 1995; Ohkuma and Kuriyama, 1994).

Nitric Oxide and the Regulation of Arousal States

As researchers continue to investigate the many biological functions of NO, it becomes increasingly clear that NO possesses a vast repertoire of physiologic roles. States of consciousness and state-dependent physiologic processes are regulated by an elusively complex array of molecules. Given the ubiquitous nature of NO, it is not surprising that nitric oxide has been found to participate in the regulation of nociception and arousal states such as sleep and anesthesia.

Efforts aiming to elucidate the contribution of NO to nociceptive processing and arousal state control have focused on neuronal networks known to play a regulatory role. Transmission of pain information clearly involves both spinal and supraspinal levels of the neuraxis. With regard to anesthetic states, no single brain region has been identified which is primarily responsible for mediating the drug induced loss of waking consciousness. In contrast, specific brain regions, neuronal networks, neurotransmitters, and signal transduction cascades have been identified as modulators of sleep and wakefulness. One neuronal network known to be involved in REM sleep regulation includes cholinergic and non-cholinergic, cholinoceptive neurons within the mesopontine brain stem (Baghdoyan, 1997). In particular, two brain regions—(1) the LDT/PPT nuclei which contain cholinergic neurons and project to the (2) medial pontine reticular formation (mPRF)—have been shown to mediate the cholinergic regulation of REM sleep. The role of this cholinergic pathway as well as other neurotransmitter systems has been reviewed extensively elsewhere (Kryger et al., 1994; Schwartz, 1997). Indeed, a REM sleep-like state can be induced reliably by discrete administration of cholinomimetic agents to the mPRF (reviewed in Baghdoyan, 1997). The remainder of this chapter notes some of the salient features of neuronal NO function related to nociceptive processing, anesthesia, and REM sleep.

Nitric Oxide and Nociception

Evidence from studies of the thermal withdrawal response in rats indicates that stimulation of NMDA receptors, production of NO, and elevation of cGMP levels in dorsal root ganglion neurons mediate synaptic transmission and spinal nociceptive processing (Meller and Gebhart, 1993). Spinal NO mechanisms involving both muscarinic neurotransmission and NO have been shown to mediate analgesia (Iwamoto and Marion, 1994b; Zhuo et al., 1993). Eisenach and colleagues have shown that muscarinic receptor stimulation by ACh causes NO production and that ACh mediates a spinal cholinomimetic nociceptive effect through the action of NO (Xu et al., 1996a, 1996b). Supraspinal cholinergic antinociception recently has been shown to be evoked from the same pontine reticular regions of cat known to contribute to REM sleep generation (Kshatri et al., 1997).

Cholinergic neurons of the LDT/PPT project to the rostroventral medulla (RVLM) (Yasui et al., 1990). Microinjection of the muscarinic cholinergic receptor agonist (+)-*cis*-methyldioxolane (CD) into the RVLM elicited antinociception in the rat as measured by hot plate and tail flick tests (Iwamoto and Marion, 1994a). An injection of the NOS inhibitor N^G-nitro-L-arginine (NLA) into the RVLM prior to CD administration blocked the antinociceptive effects of muscarinic stimulation. These results suggest that NO participates in the cholinergic mediation of antinociception in the medulla.

Nitric Oxide and General Anesthetic States

The biological actions of general anesthetic agents such as halothane and isoflurane, although heavily investigated, remain incompletely understood. It is unlikely that a single unifying mechanism can adequately explain the effects of these agents in rendering the loss of waking consciousness and a decreased perception of noxious stimuli (Lydic and Biebuyck, 1994). A number of studies do demonstrate, however, that NO is one participant in the mechanisms of action of general volatile anesthetics (Pelligrino et al., 1996). Halothane has been shown to attenuate NO-mediated vascular relaxation (Blaise et al., 1994). Halothane likely does not directly alter the production of NO by inhibiting NO synthase (Rengsamy et al., 1995), but interferes in some way with the downstream effects of NO once it has been produced (Blaise et al., 1994). Indeed, halothane has been shown to decrease the production of cGMP in numerous brain regions (Pelligrino et al., 1996).

Just as volatile anesthetics interfere with the function of NO, the manipulation of NO levels alters the ability of these anesthetics to produce their effects. NOS inhibitors cause a dose-dependent reduction in anesthetic threshold or the minimum alveolar concentration (MAC) for halothane and isoflurane in rats (Johns et al., 1992) and in mice (Ichinose et al., 1995). Similarly, targeted disruption of the NOS gene with subsequent malfunction in NO production, resulted in diminished MAC for halothane anesthesia. Two different investigational approaches provide evidence: (1) that volatile anesthetics diminish NO activity; and (2) that diminished NO activity potentiates the anesthetic effects of halothane and other inhalational agents. The questions of how NO levels in the brain influence the ability of volatile anesthetics to render a patient unconscious and less sensitive to noxious stimulation remain to be answered.

One approach to studying the mechanisms of anesthetic action has involved efforts to understand how anesthetics diminish cellular excitability while depressing levels of behavioral arousal (Johns, 1996). General anesthesia reduces ACh release in a number of brain regions (Lydic and Biebuyck, 1994). Microdialysis experiments in cat brain have demonstrated that halothane anesthesia (1.2% in O_2) decreases pontine ACh release (Keifer et al., 1996) and evokes cortical EEG spindles characteristic of NREM sleep (Keifer, et al., 1994). With termination of halothane, ACh release increased in the mPRF during emergence and resumption of wakefulness. Halothane anesthesia both decreases NO activity (Blaise et al., 1994; Rengsamy et al., 1995) and decreases ACh release in the mPRF (Keifer et al., 1996). The next section reviews the experimental evidence that NO participates in ACh release within the mPRF and influences REM sleep regulation.

Nitric Oxide and REM Sleep

Within the past two years, a number of studies have focused on the role of NO in the control of sleep and wakefulness (Burlet and Cespuglio, 1997; Datta et al., 1997; Dugovic et al., 1995; Dzoljic and de Vries, 1994, 1995; Kapas et al., 1994a; 1994b; Leonard and Lydic, 1997; Matsumara et al., 1995; Okabe et al., 1995). The effects of NOS inhibitors on sleep/wake timing reported by these different studies have been conflicting. The overall findings of these studies are reviewed below.

Intravenous or intraperitoneal (i.p.) injections of the NOS inhibitor N^G-nitro-monomethyl L-arginine (L-NMMA) in rats was shown to increase both NREM sleep and REM sleep and to decrease wakefulness (Dzoljic and de Vries, 1994). The same investigators, using the same animal model, found that the NOS inhibitor 7-nitro-indazole (7-NI), while producing a general state of sedation, diminished electrographic and behavioral signs of wakefulness as well as NREM sleep and REM sleep (Dzoljic and de Veries, 1995). Intracerebroventricular (i.c.v.) or i.p. administration of the L-arginine analog N^G-nitro-L-arginine methyl ester (L-NAME) in rabbits (Kapas et al., 1994b) and rats (Kapas et al., 1994a) reduced both NREM sleep and REM sleep. L-NAME infusion in the rat, however, was reported to augment REM sleep (Matsumara et al., 1995). L-NAME, when microinjected into the mesopontine tegmentum of three rats, elicited an increased REM sleep epoch duration, decreased frequency of REM sleep epochs, and no overall effect on percent of time spent in REM sleep (Okabe et al., 1995). Finally, i.p. injections of L-NAME in rat has been reported to produce both sleep (NREM and REM) and arousal, depending upon the dose administered (Dugovic et al., 1995).

The picture of the involvement of NO in modulating arousal states which has emerged from the aforementioned work is complex if not confusing. The diversity of effects attributed to NO with respect to sleep/wake control may be due, in part, to the use of different NOS inhibitors, varying doses of drugs, different species of animals, and the chosen route of drug administration. In six out of seven of these recent experiments, NOS inhibitors were given either systemically or into the brain cerebrospinal fluid. Multiple organ systems and brain regions were exposed to NOS inhibitor drugs. As noted elsewhere (Burlet and Cespuglio, 1995) widespread inhibition of NOS or difficulties for NOS inhibitors to cross the blood brain barrier likely contribute to confounding results. These initial studies have provided some insight into the potential role of NO in affecting states of arousal. An important advance toward achieving the goal of clarifying the role of NO in sleep/wake control was the discovery that NO modulates the release of numerous neurotransmitters (reviewed in Garthwaite and Boulton, 1995).

Recent work has demonstrated the ability of NO to alter the release of ACh in multiple brain regions. Of particular interest to the understanding of REM sleep generation are findings related to ACh release within the mPRF terminal field of the LDT/PPT cholinergic neurons. Leonard and Lydic (1995) showed that ACh release in the mPRF is highest during REM sleep, a finding consistent with observations of increased ACh release during the carbachol-induced REM sleep state (Lydic et al., 1991a). Administration of nerve growth factor (NGF) into the mPRF also has been shown to cause an increase in REM sleep (Yamuy et al., 1995), and data suggest that the REM sleep enhancing effects of NGF may be modulated by nitric oxide and ACh. In various brain regions NGF has been shown to increase both NOS (Peunova and Enikolopov, 1995) and ACh (Knipper et al., 1994). It will be important for future studies to determine by direct measurement whether NGF causes increased ACh release in the mPRF.

Leonard and Lydic (1997) discovered that stereospecific inhibition of NOS in the mPRF decreased release of ACh in the mPRF. Inhibition of NOS within the mPRF attenuated naturally occurring REM sleep, and reduced the ability of mPRF cholinergic agonist injection to elicit a REM sleep-like state. This same study quantified the effects of mPRF NOS inhibitor administration on 3 variables of REM sleep: the duration of individual REM sleep epochs, the frequency with which REM sleep epochs occurred, and the latency to onset of the first REM sleep epoch. NLA microinjection had a significant effect on REM sleep epoch duration but not on REM sleep epoch frequency or latency to onset (Leonard and Lydic, 1997). The ability of mPRF NOS inhibition to inhibit REM sleep resulted from a reduction in the duration of each REM sleep epoch. REM sleep epochs occurred at a normal frequency, and the

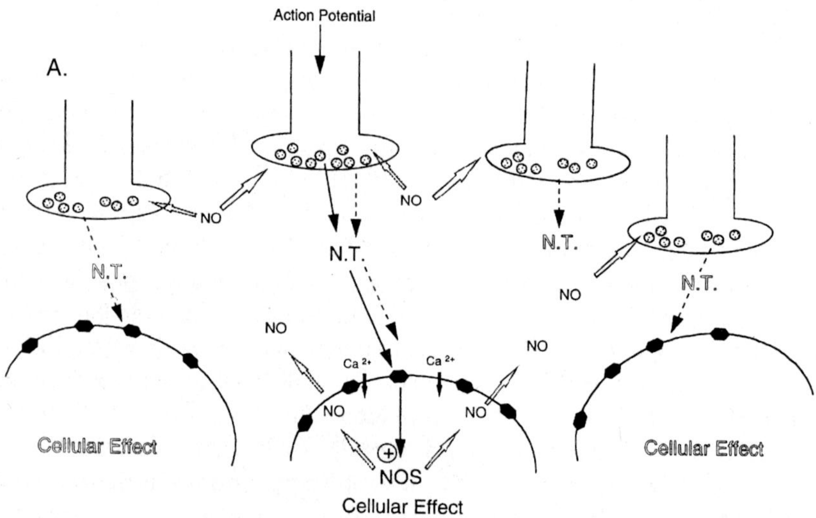

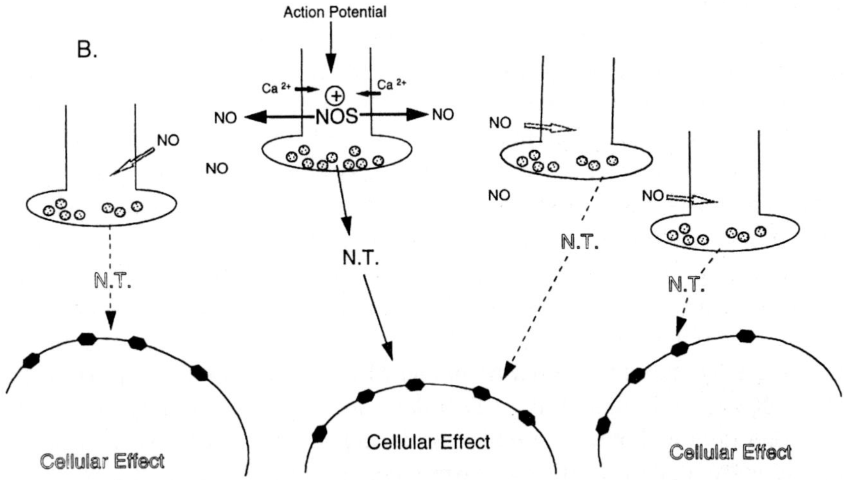

Fig. 2. **Proposed model for the role of NO in modulating physiologic, behavioral states through the recruitment of local synaptic circuits. A and B show four axon terminals and three postsynaptic neurons. Circles in the axon terminals represent synaptic vesicles. In Fig. 2A, an action potential propagated down one axon results in neurotransmitter (N.T.) release. The ligand-receptor interaction of this neurotransmitter with a postsynaptic neuron results in an influx of calcium intracellularly which activates constitutive nitric oxide synthase (NOS) and generates nitric oxide (NO). The NO produced in this postsynaptic neuron diffuses across the cellular membrane and enters nearby axon terminals. Within these presynaptic sites, NO interacts with intracellular targets to effect an increase in neurotransmitter release. The resultant increase in local synaptic neurotransmitter release causes increased receptor binding of postsynaptic neurons and therefore enhances the effect of ligand-receptor interaction in this local circuit of neurons. B schematizes NOS to be present in an axon terminal. Propagation of an axon potential results in an influx of calcium ions into the axon terminal. In addition to causing synaptic release of neurotransmitter (N.T.), increased intracellular calcium levels activate NOS leading to NO production. NO diffuses to neighboring axon terminals where it interacts with intracellular targets to cause increased neurotransmitter release from these nearby axons. Thus, presynaptic production of NO can also effect increased synaptic release of neurotransmitter within a local circuit of synapses and enhance downstream effects of receptor-ligand interaction. Produced in either axon terminals or within postsynaptic sights, NO can function as a transsynaptic neurochemical which may prolong or enhance the cellular effcts of synaptic activity on postsynaptic neurons. Such likely cellular effects include functions of memory formation and consolidation, the mediation of sleep states, and the modulation of other circadian behavioral processes.**

initiation of REM sleep was not prolonged. These data suggest that mPRF NO levels, by altering pontine cholinergic neurotransmission, can modulate the maintenance of REM sleep but not the initiation of REM sleep.

Recent experiments also have looked at the ability of NO-generating compounds to alter sleep/wake states. In rats, the drugs SIN-1 and SNAP which serve as exogenous sources of NO increased NREM sleep 9 h after intracerebroventricular injection (Kapas and Krueger, 1996). These NREM sleep-promoting effects lasted for 24–48 h after injection, suggesting that the effects of SIN-1 and SNAP may have been mediated through transcriptional regulation of unidentified somnogenic factors (Kapas and Krueger, 1996). Injection of SIN-1 and SNAP did not significantly alter REM sleep time, and EEG power spectra were increased in waking and sleeping states. Additional studies in cats have shown the NO levels are linked to behavioral states. Local injection of SNAP into the PPT produced a concentration-dependent increase in both NREM and REM sleep whereas injection of the NOS inhibitor L-NAME decreased both NREM and REM sleep compared to control (Datta et al., 1997). Williams et al. (1997) have shown by microdialysis that NO levels within the thalamus are highest during waking and REM sleep states. Furthermore, thalamic NO levels were increased by electrical stimulation of LDT (Williams et al., 1997).

Figure 2 depicts one possible synaptic arrangement whereby NO modulates diverse physiologic traits comprising complex behavioral states, such as REM sleep. Thus, while the functional roles of NO are diverse, the means of accomplishing these various tasks may be unified through NO's ability to potentiate multiple neuronal circuits and prolong intracellular events regulating neuronal excitability.

Acknowledgments

This work supported by the Departments of Anesthesia and Neuroscience & Anatomy and by grant HL-40881 (RL).We thank P. Myers for expert secretarial assistance.

References

Abrams, R.M., Gerhardt, K.J. and Burchfield, D.J. Behavioral state transition and local cerebral blood flow in fetal sheep, Journal of Developmental Physiology, 15, (1991) 283–288.

Abu-Soudt, H.M., Feldman, P.L., Clark, P. and Stuerh, D.J. Electron transfer in the nitric-oxide synthases: characterization of L-arginine analogs that block heme iron reduction, The Journal of Biological Chemistry, 269, (1994) 32318–32326.

Asenbaum, S., Zeithofer, J., Saletu, B., Frey, R., Brucke, T., Podreka, I. and Deecke, L. Technetium-99m-HMPAO SPECT imaging of cerebral blood flow during REM sleep in narcoleptics, Journal of Nuclear Medicine, 36, (1995) 1150–1155.

Bach-Y-Rita, P. Nonsynaptic diffusion neurotransmission (NDN) in the brain, Neurochemistry International, 23, (1993) 297–318.

Baghdoyan, H.A., Cholinergic mechanisms regulating REM sleep: Relevance for clinical neurology. In: *Sleep Science: Monographs in Clinical Neurology*, W.J. Schwartz, (Ed.), Karger, Basel, 1997, 88–116.

Benca, R.M., Kushida, C.A., Everson, C.A., Kalski, R., Bergmann, B.M. and Rechtschaffen, A. Sleep deprivation in the rat: VII. Immune function, Sleep, 12, (1989) 47–52.

Bergman, B.M., Rechtschaffen, A., Gilliland, M.A. and Quintans, J. Effect of extended sleep deprivation on tumor growth in rats, Am. J. Physiol., 271 (1996) R1460–R1464.

Bickford, M.E. Gunluk, A.E., Guido, W. and Sherman, S.M. Evidence that cholinergic axons from the parabrachial region of the brainstem are the exclusive source of nitric oxide in the lateral geniculate nucleus of the cat, The Journal of Comparative Neurology, 334, (1993) 410–430.

Blaise, G., Quy, T., Parent, M., Lagarde, B., Asenjo, F. and Sauve, R. Does halothane interfere with the release, action or stability of endothelium-derived relaxing factor/ nitric oxide?, Anesthesiology, 80, (1994) 417–426.

Bredt, D.S., Hwang, P.M. and Snyder, S.H. Localization of nitric oxide synthase indicating a neural role for nitric oxide, Nature, 347, (1990) 768–770.

Brzezinski, A. Melatonin in humans, The New England Journal of Medicine, 336, (1997) 186–195.

Burlet, S. and Cespuglio, R. Voltammetric detection of nitric oxide (NO) in the rat brain: its variations throughout the sleep-wake cycle, Neurosci. Letts., 226. (1997) 131–135.

Butler, A.R., Flitney, F.W. and Williams, D.L.H. NO, nitrosium ions, nitroxide ions, nitrosothiols and iron-nitrosyls in biology: a chemists perspective, Trends in Physiological Science, 16, (1995) 18–22.

Buzarin, T.C.D. and Gundlach, A.L. Localization of NO synthase in rat brain by [^{3}H]L–N^G-nitro-arginine autoradiography, NeuroReport, 6, (1995) 1842–1844.

Chiueh, C.C., Gilbert, D.L. and Colton, C.A. (Ed.), *The Neurobiology of NO and OH*, the New York Academy of Sciences, New York, 1994, 1–471.

Clutton, B.J. Two cases of poisoning by contamination of nitrous oxide with higher oxides of nitrogen during anaesthesia, British Journal of Anaesthesia, 39, (1967) 388–392.

Culotta, E. and Koshland, D.E. NO news is good news, Science, 258, (1992) 1862–1864.

Danguir, J. and Nicolaidis, S. Circadian sleep and feeding patterns in the rat: possible dependence on lipogenesis and lipolysis, American Journal of Physiology, 208, (1980) E223–E230.

Danguir, J. and Nicolaidis, S. Chronic intracerebroventricular infusion of insulin causes selective increases in slow wave sleep in rats, Brain Research, 306, (1984) 97–103.

Datta, S. Neuronal activity in the peribrachial area: relationship to behavioral state control, Neuroscience and Biobehavioral Reviews, 19, (1995) 67–84.

Datta, S., Patterson, E.H. and Siwek, D.F. Endogenous and exogenous nitric oxide in the pedunculopontine tegmentum induces sleep, Synapse, 27, (1997) 69–78.

Dawson, T.M., Bredt, D.S., Fotuhi, M., Hwang, P.M. and Snyder, S.H. Nitric oxide synthase and neuronal NADPH diaphorase are identical in brain and peripheral tissues, Proc. Natl. Acad. Sci. USA 88, (1991) 7797–7801.

Dawson, T.M., Dawson, V.L. and Snyder, S.H. Molecular mechanisms of nitric oxide actions in the brain. In: *The Neurobiology of NO and OH,* C.C. Chiueh, D.L. Gilbert, and C.A. Colton, (Eds.), New York Academy of Sciences, New York, 1994, 77–85.

Dement, W.C., History of sleep physiology and medicine. In: *Principles and Practice of Sleep Medicine*, 2nd, M.H. Kryger, T. Roth and W.C. Dement, (Eds.), W.B. Saunders Company, Philadelphia, 1994, 3–15.

Dugovic, C., Van den Broeck, W.A.E., De Ryck, M. and Clincke, G.H.C. Inhibition of nitric oxide synthesis induces opposite effects on slow wave sleep and paradoxical sleep, Sleep Research, 24A, (1995) 119.

Dzoljic, M.R. and de Vries, R. Nitric oxide synthase inhibition reduces wakefulness, Neuropharmacology, 33, (1994) 1505–1509.

Dzoljic, M.R. and de Vries, R. Nitric oxide (NO) a modulator of vigilance: effects of 7-nitro indazole, an inhibitor of central NO synthase, Sleep Research, 24A, (1995) 120.

El Mansari, M., Sakai, K. and Jouvet, M. Unitary characteristics of presumptive cholinergic tegmental neurons during the sleep-waking cycle in freely moving cats, Experimental Brain Research, 76, (1989) 519–529.

El Mansari, M., Saki, K. and Jouvet, M. Responses of presumed cholinergic mesopontine tegmental neurons to carbachol microinjections in freely moving cats, Experimental Brain Research, 83, (1990) 115–123.

Everson, C.A. Sustained sleep deprivation impairs host defense, American Journal of Physiology, 265, (1993) R1148–R1154.

Ferguson, K.A. and Fleetham, J.A. Consequences of sleep disordered breathing, Thorax, 50 (1995) 998–1004.

Fukuto, J.M. and Chaudhuri, G. Inhibition of constitutive and inducible nitric oxide synthase: potential selective inhibition, Annual Review of Pharmacology and Toxicology, 35, (1995) 165–194.

Furchgott, R.F., A research trail over half a century, Annual Review of Pharmacology and Toxicology, 35, (1995) 1–27.

Furchgott, R.F. and Zawadski, J.V. The obligatory role of endothelial cells in the relaxation of arterial smooth muscle by acetylcholine, Nature, 288, (1980) 373–376.

Garfinkel, D., Laudon, M., Nof, D. and Zisapel, N. Improvement of sleep quality in elderly people by controlled-release melatonin, Lancet, 346, (1995) 542–544.

Garthwaite, J. Glutamate, nitric oxide and cell-cell signalling in the nervous systems, Trends in Neuroscience, 14, (1991) 60–67.

Garthwaite, J. and Boulton, C.L. Nitric oxide signaling in the central nervous system, Annual Review of Physiology, 57, (1995) 683–706.

Garthwaite, J., Charles, S.L. and Chess-Williams, R. Endothelium-derived relaxing factor release on activation of NMDA receptors suggests role as intercellular messenger in the brain, Nature, 336 (1988) 385–387.

Garthwaite, J. and Garthwaite, G. Cellular origins of cyclic GMP responses to excitatory amino acid receptor agonists in rat cerebellum in vitro, Journal of Neurochemistry, 48, (1987) 29–39.

Gozal, D., Gozal, E., Tores, J.E., Gozal, Y.M., Nuckton, T.J. and Hornby, P.J. Nitric oxide modulates ventilatory responses to hypoxia in the developing rat, American Journal of Respiratory and Critical Care Medicine, 155 (1997).

Griffith, O.W. and Stuehr, D.J. Nitric oxide synthases: properties and catalytic mechanism, Annual Review of Physiology, 57, (1995) 707–736.

Guevara-Guzman, R., Emson, P.C. and Kedrick, K.M. Modulation of in vivo striatal transmitter release by nitric oxide and cyclic GMP, Journal of Neurochemistry, 62, (1994) 807–810.

Guiditta, A., Ambrosini, M.V., Montagnese, P., Mandile, P., Cotugno, M., Zucconi, G.G. and Vescia, S. The sequential hypothesis of the function of sleep, Behavioral Brain Research, 69, (1995) 157–166.

Gutteridge, J.M.C. Biological origin of free radicals, and mechanisms of antioxidant protection, Chemico-Biological Interactions, 91, (1994) 133–140.

Hawkins, R.D., Zhuo, M. and Arancio, O. Nitric oxide and carbon monoxide as possible retrograde messengers in hippocampal longer-term potentiation. In: *Nitric Oxide:Roles in Neuronal Communication and Neurotoxicity,* H. Tagagi, N. Toda and R.D. Hawkins, (Eds.), Japan Scientific Societies Press, Tokyo, Japan, 1994, 129–147.

Hennevin, E., Hars, B. and Bloch, V. Improvement of learning by mesencephalic reticular stimulation during postlearning paradoxical sleep, Behav. Neural Biol., 51, (1989) 291–306.

Hibbs, J.B., Taintor, R.R., Vavrin, V., Granger, D.L., Drapier, J.-C., Amber, I.J. and Lancaster, J.R. Synthesis of nitric oxide from a guanidino nitrogen of L-arginine: a molecular mechanism that targets intracellular iron, In: *Nitric Oxide from L-Arginine: A*

Bioregulatory System, S. Moncada, and E.A. Higgs, (Eds.), Elsevier, Amsterdam, 1990, 189–223.

Hibbs, J.B., Taintor, R.R. and Vavrin, Z. Macrophage cytotoxicity: role for L-arginine deaminase and imino nitrogen oxidation to nitrite, Science, 235, (1987) 473–476.

Hobson, J.A. and Stickgold, R. Sleep the beloved teacher?, Current Biology, 5, (1995) 35–36.

Holscher, C. Nitric oxide, the enigmatic neuronal messenger: its role in synaptic plasticity, Trends in Neuroscience. 22, (1997) 298–303

Hope, B.T., Michael, G.J., Knigge, K.M. and Vincent, S.R. Neuronal NADPH diaphorase is a nitric oxide synthase, Proceedings of the National Academy of Science USA, 88, (1991) 2811–2814.

Ichinose, F., Huang, P.L. and Zapol, W.M. Effects of targeted neuronal nitric oxide synthase gene disruption and nitro G L-arginine methylester on the threshold for isoflurane anesthesia, Anesthesiology, 83, (1995) 101–108.

Irwin, M., McClintick, J., Costlow, C., Fortner, M., White, J. and Gillin, J.C. Partial night sleep deprivation reduces natural killer and cellular immune responses in humans FASEB J., 10, (1996) 643–653.

Iwamoto, E.T. and Marion, L. Pharmacological evidence that nitric oxide mediates the antinociception produced by muscarinic agonists in the rostral ventral medulla of rats, The Journal of Pharmacology and Experimental Therapeutics, 269, (1994a) 699–708.

Iwamoto, E.T. and Marion, L. Pharmacological evidence that spinal muscarinic analgesia is mediated by an L-arginine/nitric oxide/cyclic GMP cascade in rats, The Journal of Pharmacology and Experimental Therapeutics, 271, (1994b) 601–608.

Iyengar, R., Stuehr, D.J. and Marletta, M.A. Macrophage synthesis of nitrite, nitrate, and N-nitrosamines: precursors and role of respiratory burst, Proceedings of the National Academy of Sciences USA, 84, (1987) 6369–6373.

Johns, R.A., Nitric oxide, cyclic guanosine monophosphate, and the anesthetic state, Anesthesiology, 85, (1996) 457–459.

Johns, R.A., Moscicki, J.C. and DiFazio, C.A. Nitric oxide synthase inhibitor dose-dependently and reversibly reduces the threshold for halothane anesthesia, Anesthesiology, 77, (1992) 779–784.

Kandel, E.R., Schwartz, J.H. and Jessel, T.M. (Ed.), *Principles of Neural Science,* 3rd, Elsevier, New York, 1991.

Kapas, L., Fang, J. and Krueger, J.M. Inhibition of nitric oxide synthesis inhibits rat sleep, Brain Research, 664, (1994a) 189–196.

Kapas, L. and Krueger, J.M. Nitric oxide donors SIN-1 and SNAP promote nonrapid-eye-movement sleep in rats, Brain Research Bulletin, 41, (1996) 293–298.

Kapas, L., Shibita, M., Kimura, M. and Krueger, J.M. Inhibition of nitric oxide synthesis suppresses sleep in rabbits, American Journal of Physiology, 266, (1994b) R151–R157.

Karni, A., Tanne, D., Rubenstein, B.S., Askenasy, J.J.M. and Sagi, D. Dependence on REM sleep of overnight improvement of perceptual skill, Science, 265, (1994) 679–682.

Katusik, Z.S. and Cosentino, F. Nitric oxide synthase: from molecular biology to cerebrovascular physiology, News in Physiological Science, 9, (1994) 64–67.

Keifer, J.C., Baghdoyan, H.A., Becker, L. and Lydic, R. Halothane decreases pontine acetylcholine release and increases EEG spindles, NeuroReport, 5, (1994) 577–580.

Keifer, J.C., Baghdoyan, H.A. and Lydic, R. Pontine cholinergic mechanisms modulate the cortical electroencephalographic spindles of halothane anesthesia, Anesthesiology, 84, (1996) 945–954.

Kitamura, Y. and Nomura, Y. Inducible nitric oxide synthase (i-NOS) in rat glia cells. In:

Nitric Oxide: Roles in Neuronal Communication and Neurotoxicity, H. Takagi, N. Toda, and R.D. Hawkins, (Eds.), Japan Scientific Societies Press, Tokyo, 1994, 1994, 41–51.

Knipper, M., Berzaghi, M.P., Blochl, A., Breer, H., Thoenen, H. and Lindholm, D. Positive feedback between acetylcholine and the neurotrophins nerve growth factor and brain-derived neurotrophic factor in the rat hippocampus, Eur. J. Neurosci., 6, (1994) 668–671.

Knowles, R.G., Properties and regulation of brain nitric oxide synthase, In: *Nitric Oxide: Roles in Neuronal Communication and Neurotoxicity,* H. Takagi, N. Toda, and R.D. Hawkins, (Eds.), Japan Scientific Societies Press, Tokyo, 1994, 29–39.

Krueger, J.M. and Obal, F. Sleep Factors, In: *Sleep and Breathing,* 2nd, N.A. Saunders and C.E. Sullivan, (Eds.), Marcel Dekker, New York, 1994, 79–112.

Kryger, M.H., Roth, T. and Dement, W.C. *Principles and Practice of Sleep Medicine,* 2nd, W.B. Saunders Company, Philadelphia, 1994, 1–1067.

Kshatri, A.M., Fleegal, M.A., Baghdoyan, H.A. and Lydic, R. Cholinomimetics, but not morphine, evoke antinociception from pontine reticular regions generating REM sleep Soc. Neurosci. Abst., 23, (1997) 2132.

Lancel, M., Cronlein, J., Muller-Preuss, P. and Holsboer, F. Lipopolysaccharide increases EEG delta activity within non-REM sleep and disrupts sleep continuity in rats, American Journal of Physiology, 268, (1995) R1310–1318.

Leonard, T.O. and Lydic, R. Nitric oxide synthase inhibition decreases pontine acetylcholine release, NeuroReport, 6, (1995) 1525–1529.

Leonard, T.O. and Lydic, R. Pontine nitric oxide modulates acetylcholine release, rapid eye movement sleep generation, and respiratory rate, J. Neurosci., 17, (1997) 774–785.

Ling, L., Karius, D.R., Fiscus, R.R. and Speck, D.F. Endogenous nitric oxide required for an integrative respiratory function in the cat brain, Journal of Neurophysiology, 68. (1992) 1910–1912.

Lydic, R., Respiratory modulation by non-respiratory neurons: Localization of function varies with arousal state. In: *Sleep: Karger Monograph in Clinical Neuroscience,* W. Schwartz, (Ed.), Karger, Bern, 1997, 117–143.

Lydic, R. and Baghdoyan, H.A. Pedunculopontine stimulation alters respiration and increases ACh release in the pontine reticular formation, American Journal of Physiology, 264, (1993) R544–R554.

Lydic, R., Baghdoyan, H.A., Hibbard, L., Bonyak, E.V. and DeJoseph, M.R. Regional brain glucose metabolism is altered during rapid eye movement sleep in the cat: a preliminary study, The Journal of Comparative Neurology, 304, (1991b) 517–529.

Lydic, R., Baghdoyan, H.A. and Lorinc, Z. Microdialysis of cat pons reveals enhanced acetylcholine release during state-dependent respiratory depression, American Journal of Physiology, 261, (1991a) R766–R770.

Lydic, R. and Biebuyck, J.F. Sleep neurobiology: relevance for mechanistic studies of Anaesthesia, British Journal of Anaesthesia, 72, (1994) 506–508.

Madsen, P.L., Holm, S., Vorstrup, S., Friberg, L. and Lassen, N.A. Human regional cerebral blood flow during rapid-eye-movement sleep, Journal of Cerebral Blood Flow and Metabolism, 11 (1991) 502–507.

Maquet, P., Degueldre, C., Delfiore, G., Aerts, J., Peters, J.-M., Luxen, A. and Franck, G. Functional neuroanatomy of human slow wave sleep, Journal of Neuroscience, 17, (1997) 2807–2812.

Maquet, P., Peters, J.–M., Aerts, J., Delfiore, G., Degueldre, C., Luxen, A. and Franck, G. Functional neuroanatomy of human rapid-eye-movement sleep and dreaming, Nature, 383, (1996) 163–166.

Marletta, M.A., Yoon, P.S., Iyengar, R. and Leaf, C.D. Macrophage oxidation of L-arginine to nitrite and nitrate: nitric oxide is an intermediate, Biochemistry, 21, (1988) 8706–8711.

Matsumara, H., Gerashchenko, D., Satoh, S., Nakajima, T., Kasahara, K. and Hayaishi, O. Prostaglandin D_2 and nitric oxide: key factors underlying the regulation of slow-wave sleep and paradoxical sleep in the rat brain, Sleep Research, 24A, (1995) 134.

Mayer, B., Koesling, D. and Bohme, E. Characterization of nitric oxide synthase, soluble guanylyl cyclase, and Ca2+/calmodulin-stimulated cGMP phosphodiesterase as components of neuronal signal transduction, In: *Advances in Second Messenger and Phosphoprotein Research*, B.L. Brown and P.R.M. Dobson, (Eds.), Raven Press, New York, 1993, 111–119.

McCall, T. and Vallance, P. Nitric oxide takes center stage with newly defined roles, Trends in Physiological Science, 13, 1–6, 1992.

McNamara, S.G., Cistulli, P.A., Sullivan, C.E. and Strohl, K.P. Clinical aspects of sleep apnea, In: *Sleep and Breathing,* 2nd, N.A. Saunders, and C.E. Sullivan, (Eds.), Marcel Dekker, New York, 1994, 493–528

Meller, S.T. and Gebhart, G.F. Nitric oxide (NO) and nociceptive processing in the spinal cord, Pain, 52, (1993) 127–136.

Miwa, M., Stuehr, D.J., Marletta, M.A., Wishnok, J.S. and Tannebaum, S.R. Nitrosation of amines by stimulated macrophages, Carcinogenesis, 8, (1987) 955–958.

Mizukawa, K., Vincent, S.R., McGeer, P.L. and McGeer, E.G. Distribution of reduced-nicotinamide-adenine-dinucleotide-phosphate diaphorase-positive cells and fibers in the cat central nervous system, The Journal of Comparative Neurology, 279, (1989) 281–311.

Moilanen, E. and Vapaatalo, H. Nitric oxide in inflammation and immune response, Annals of Medicine, 27, (1995) 359–367.

Moldofsky, H., Lue, F.A., Davidson, J.R. and Gorczynski, R. Effects of sleep deprivation on human immune function, FASEB J., 3, (1989) 1972–1977.

Moncada, S. and Higgs, A. The L-arginine-nitric oxide pathway, New England Journal of Medicine, 329, (1993) 2002–2012.

Nathan, C. Nitric oxide as a secretory product of mammalian cells, FASEB Journal, 6, (1992) 3051–3064.

Nava, E., Noll, G. and Luscher, T.F. Nitric oxide in cardiovascular disease, Annals of Medicine, 27, (1995) 343–351.

Ohkuma, S., Katsura, M., Chen, D.–Z., Narihara, H. and Kuriyama, K. Facilitation of N-methyl-D-aspartate-evoked acetylcholine release by hydroxyl radical scavengers, NeuroReport, 6, (1995) 2033–2036.

Ohkuma, S. and Kuriyama, K. Roles of nitric oxide in neurotransmitter release: analysis using primary cultured CNS neurons, In: *Nitric Oxide: Roles in Neuronal Communication and Neurotoxicity,* H. Takagi, N. Toda, and R.D. Hawkins, (Eds.), Japan Scientific Societies Press, Tokyo, 1994, 171–180.

Okabe, S., Sanford, L.D., Veasey, S.C. and Kubin, L. Effect of nitric oxide synthase inhibitor microinjections into the rat mesopontine tegmentum on sleep, Sleep Research, 24A, (1995) 50.

Paakkari, I. and Lindberg, P. Nitric oxide in the central nervous system, Annals of Medicine, 27, (1995) 369–377.

Palmer, R.M.J., Ashton, D.S. and Moncada, S. Vascular endothelial cells synthesize nitric oxide from L-arginine, Nature (London), 333, (1988) 664–666.

Pelligrino, D.A., Baughman, V.L. and Koenig, H.M. Nitric oxide and the brain, International Anesthesiology Clinics, 34, (1996) 113–132.

Peunova, N. and Enikolopov, G. Nitric oxide triggers a switch to growth arrest during differentiation of neuronal cells, Nature, 375, (1995) 68–73.

Pollmacher, T., Schreiber, W., Gudewill, S., Vedder, H., Fasbender, K., Wiedemann, K., Trachsel, L., Galanos, C. and Holsboer, F. Influence of endotoxin on nocturnal sleep in humans, American Journal of Physiology, 264, (1993) R1077–R1083.

Prast, H., Fischer, H., Werner, E., Werner-Felmayer, G. and Philippu, A. Nitric oxide modulates the release of acetylcholine in the ventral striatum of the freely moving rat, Naunyn-Schmiedeberg's Arch Pharmacol, 352, (1995) 67–73.

Prast, H. and Philippu, A. Nitric oxide releases acetylcholine in the basal forebrain, European Journal of Pharmacology, 216, (1992) 139–140.

Radomski, M.W. Nitric oxide: biological mediator, modulator and effector, Annals of Medicine, 27, (1995) 321–329.

Reiter, R.J. Oxidative processes and antioxidative defense mechanisms in the aging brain, FASEB Journal, 9, (1995) 526–533.

Rengsamy, A., Ravichandran, L.V., Reikerdorfer, C.G. and Johns, R.A. Inhalational anesthetics do not alter nitric oxide synthase activity, The Journal of Pharmacology and Experimental Therapeutics, 273, (1995) 599–604.

Rotenberg, V.S. Sleep and memory I: The influence of different sleep stages on memory, Neurosci. Biobehav. Rev., 16, (1992) 497–502.

Sakai, K. and Koyama, Y. Are there cholinergic and non-cholingeric paradoxical sleep-on neurons in the pons? NeuroReport, 7, (1996) 2449–2453.

Sandor, N.T., Brassai, A., Puskas, A. and Lendai, B. Role of nitric oxide in modulating neurotransmitter release from rat striatum, Brain Research Bulletin, 36, (1995) 483–486.

Sato, I. and Murota, S.–I., Paracrine functions of endothelium-derived nitric oxide, Life Sciences, 56, (1995) 1079–1087.

Scherer-Singer, U., Vincent, S.R., Kimura, H. and McGeer, E.G. Demonstration of a unique population of neurons with NADPH-diaphorase histochemistry, Journal of Neuroscience Methods, 9, (1983) 229–234.

Schmidt, H., Warner, T.D., Sheng, H. and Murad, F. Insulin secretion from pancreatic β cells caused by L-arginine-derived nitrogen oxides, Science, 255, (1992) 721–723.

Schmidt, H.H.H.W., Hofman, H., Olgilvie, P., Sennefelder, H. and Meinberg, R.J. Biochemistry and regulation of nitric oxide synthase, In: *Nitric Oxide: Roles in Neuronal Communication of Neurotoxicity,* H. Takagi, N. Toda, and R.D. Hawkins, (Eds.), Japan Scientific Societies Press, Tokyo, 1994, 3–18.

Schmidt, H.H.H.W., Nau, H., Wittfoth, W., Gerlach, J., Prescher, K.-E., Klein, M.M., Niroomand, F. and Bohme, E. Arginine is a physiological precursor of the endothelium-derived nitric oxide, European Journal of Pharmacology, 154, (1988) 213–216.

Schuman, E.M. and Madison, D.V. Locally distributed synaptic potentiation in the hippocampus, Science, 263, (1994a) 532–536.

Schuman, E.M. and Madison, D.V. Nitric oxide and synaptic function, *Annual Review of* Neuroscience, 17, (1994b) 153–183.

Schuman, E.M., Meffert, M.K., Schulman, H. and Madison, D.V. An ADP-ribosyltransferase as a potential target for nitric oxide action in hippocampal long-term potentiation, Proceedings of the National Academy of Sciences USA, 91 (1994) 11958–11962.

Schwartz, W.J. (Ed.), *Sleep Science: Integrating Basic Research and Clinical Practice*, Karger, Basel, 1997, 1–256.

Segovia, G., Porras, A. and Mora, F. Effects of a nitric oxide donor on glutamate and GABA release in striatum and hippocampus of the conscious rat, NeuroReport, 5, (1994) 1937–1940.

Smith, C. Sleep states and memory processes, Behavioral Brain Research, 69, (1995) 137–145.

Snyder, S.H. No endothelial NO, Science, 377, (1995) 196–197.

Star, R.A. Southwestern International Medicine Conference: Nitric oxide, The American Journal of the Medical Sciences, 306, (1993) 348–358.

Starkey, S.J., Melatonin and 5-hydroxytryptamine phase-advance the rat circadian clock by activation of nitric oxide synthesis, Neuroscience Letters, 211, (1996) 199–202.

Stuehr, D.J. and Marletta, M.A. Mammalian nitrate biosynthesis: mouse macrophages produce nitrite and nitrate in response to *Escherichia coli* lipopolysaccharide, Proceedings of the National Academy of the Sciences USA, 82, (1985) 7738–7742.

Thomas, E. and Pearse, A.G.E. The fine localization of dehydrogenases in the nervous system, Histochemie, 2, (1961) 266–282.

Tikkanen, I. and Fyhrquist, F. Nitric oxide in hypertension and renal disease, Annals of Medicine, 27, (1995) 353–357.

Tomimoto, H., Nishimura, M., Suenaga, T., Nakamura, S., Akiguchi, I., Wakita, H., Kimura, J. and Mayer, B. Distribution of nitric oxide synthase in human cerebral blood vessels and brain tissue, Journal of Cerebral Blood Flow and Metabolism, 14, (1994) 930–938.

Tzischinsky, O. and Lavie, P. Melatonin possesses time-dependent hypnotic effects, Sleep, 17, (1994) 638–645.

Vallance, P. and Collier, J. Biology and clinical relevance of nitric oxide, British Medical Journal, 309, (1994) 453–457.

Vincent, S. (Ed.), *Nitric Oxide in the Nervous System,* Academic Press, San Diego, 1995, 1–317.

Vincent, S.R., Satoh, K., Armstrong, D.M. and Fibiger, H.C. NADPH-diaphorase: a selective histochemical marker for the cholinergic neurons of the pontine reticular formation, Neuroscience Letters, 43, (1983a) 31–36.

Williams, J.A., Vincent, S.R. and Reiner, P.B. Nitric oxide production in rat thalamus changes with behavioral state, local depolarization, and brainstem stimulation, The Journal of Neuroscience, 17, (1997) 420–427.

Wilson, M.A. and McNaughton, B.L. Reactivation of hippocampal ensemble memories during sleep, Science, 265, (1994) 676–679.

Wood, J. and Garthwaite, J. Models of the diffusional spread of nitric oxide: implications for neural nitric oxide signalling and its pharmacological properties, Neuropharmacology, 33, (1994) 1235–1244.

Xu, Z., Chuanyao, T. and Eisenach, J.C. Acetylcholine stimulates the release of nitric oxide from rat spinal cord, Anesthesiology, 85, (1996a) 107–111.

Xu, Z., Chuanyao, T., Figueroa, J., Tobin, J.R. and Eisenach, J.R. Location and characteristics of nitric oxide synthase in sheep spinal cord and its interaction with alpha2-adrenergic and cholinergic antinociception, Anesthesiology, 84, (1996b) 890–899.

Yamuy, J., Morales, F.R. and Chase, M.H. Induction of rapid eye movement sleep by the microinjection of nerve growth factor into the pontine reticular formation of the cat, Neuroscience, 66, (1995) 9–13.

Yasui, Y., Cechetto, D.F. and Saper, C.B. Evidence for a cholinergic projection from the pedunculopontine tegmental nucleus to the rostral ventrolateral medulla in the rat, Brain Research, 517, (1990) 19–24.

Zhdanova, I.V., Wurtman, R.J., Lynch, H.J., Ives, J.R., Dollins, A.B., Morabito, C., Matheson, J.K. and Schomer, D.L. Sleep-inducing effects of low doses of melatonin ingested in the evening, Clinical Pharmacology and Therapeutics, 57, (1995) 552–558.

Zhuo, M., Meller, S.T. and Gebhart, G.F. Endogenous nitric oxide is required for tonic cholinergic inhibition of spinal mechanical transmission, Pain, 54, (1993) 71–78.

Rapid Eye Movement Sleep
B.N. Mallick and S. Inoué (Eds)

12. Neurotransmitters Changes and REM Sleep

Tohru Kodama

Dept. Psychology, Tokyo Metropolitan Institute for Neuroscience, 2–6 Musashidai, Fuchu, Tokyo 183-0041, Japan

SUMMARY

REM-on neurons in the pons are known to be responsible for the REM sleep generation but, not much is known about what and how the discharges of the REM-on neurons are regulated. The first step in answering these questions is to measure the neurotransmitter releases in the pontine regions across the sleep-wake cycle. Positive reports on the sleep related change of neurotransmitters are increasing in this decade, but, not many reports are available for the clarification of the REM sleep specific changes in neurotransmitters. ACh in the pontine reticular formation (PIA) which is a potent REM sleep generating center, and nucleus paramedianus in the medulla (NPM), glutamate in PIA and nucleus magnocellularis in the medulla(NMC), serotonin in PIA and nucleus raphe (DRN), and GABA in locus coeruleus(LC) and DRN are the major factors included in the positive results. However, while most of the neurotransmitters across sleep stages are supposed to change as a result of the generated REM sleep period, only PIA is known to indicate a change prior to the REM sleep generation. In order to investigate the source of the REM sleep generating mechanisms, the cholinergic and glutamatergic inputs into PIA from: NMC; the other side of PIA(cPIA); the pedunculopontine tegmentum(PPN); and the retrorubral nucleus(RRN), were investigated. The results indicate that ACh and glutamate releases from these nuclei, especially from NMC and cPIA, affect the REM generating mechanisms in PIA. Further, ACh-glutamate interaction in PIA was observed to indicate the importance of the interaction among the neurotransmitters.

Key Words: Microdialysis, ACh, glutamate, pontine inhibitory area(PIA), nucleus magnocellularis of the medulla (NMC), neurotransmitter interaction, REM sleep.

This chapter summarizes some recently obtained evidence concerning which neurotransmitters show the sleep related changes and what kind of neurotransmitters are responsible for the REM sleep generating mechanism. The field of the synaptic gap is the place where the information of the neuronal inputs are summed and interact with each other, contributing to a much complicated transition of information. As described later, the neurotransmitters work synergistically, changing the receptor sensitivity and regulating neurotransmitter release. In this micro-environment the neurotransmitters are the stars in a play. This is what is to be emphasized in this chapter.

In recent years, sleep studies have prospered and accumulated considerable knowledge with the development of the electrophysiological procedures: The methodology of recording the electroencephalogram(EEG) has lead us to the finding of REM sleep (Aserinsky and Kleitman, 1953), and the usage of microwires for recording neuronal discharges lead us to conclude that neurons have their own properties of firing across sleep-wake cycles (McGinty et al., 1974). Other procedures, such as anatomical and pharmacological procedure, supplemented the electrophysiological findings, and we now know that pontine REM-on neurons are responsible for the generation of the characteristic phenomena during REM sleep. However, sleep was originally discussed in relation to the substances related: In ancient Greece, Hippokrates and Empedoklees described sleep in relationship to the fundamental elements (Borbely, 1984). This was not actually substance but only a certain concept. Early in the 20th century, Pieron and Ishimori, independently, discussed sleep generation mechanisms in relation to the sleep substances. Subsequently, the work of Jouvet's group, followed by that of Hobson and McCalrey or Sakai, recognized the importance of the roles of neurotransmitters. Finally, with the latest technological advancements, the measurement of neurotransmitters allows us to discuss the micro-environment of the neurons.

In the CNS, the neurons utilize the chemical synapses instead of the electrical synapses. This way of transmission is somewhat of a primitive mode and supposedly not very efficient, but it has some advantages as compared with that with electrical signal transmission only. First, let us recall how the information is transferred from one neuron to another. When electrical signals reach the axonal end, neurotransmitters are released into the synaptic gap by that electrical signal and diffused to bind the receptor on the postsynaptic (partially presynaptic) membrane, and then depolarize the postsynaptic membrane to make an output signal. In electrophysiological research, we record these output signals to discuss the neuronal function. For example, in the field of higher cognition research, the unit discharge pattern by itself may have a significant meaning. The unit discharge of the "Grandmother neuron", which discharges when the subject animal watches a special pattern like the face of a grandmother, has a reason for its independent treatment. Similarly, in the research of motor function, there is first an intention to make a movement. That followed by muscle changes with the reflection of the intention to move. The output signal which starts discharging prior to the muscle movement, is more important than the input signal which is represented by neurotransmitters.

In the case of sleep study, REM-on neurons were also found by electrophysiological procedures. This type of neurons start discharging prior to REM sleep and keep discharging during the REM sleep period, indicating REM-on neurons have an important role in generating REM sleep. However, there is no intention to start the REM sleep period unlike that in the case of the start of the motor function. Animals can not intend to start REM sleep. The timing of REM sleep is decided, as it were, semiautomatically by a couple of specific conditions. Therefore, the interest of investigation of the sleep mechanism is bound to the research of, how and by what does REM sleep start. What kind of input would be responsible for the discharge of the REM-on neurons? There are a couple of possibilities. The input to the REM-on neurons might be simply the signal from other REM-on neuron. Another possibility which is a little more complicated is that the excitatory input from type I neurons and inhibitory input from REM-off neurons may work synergistically,generating a REM-on discharge pattern.

Furthermore, as the changes across sleep-wakefulness are much slower than unit discharges and have more general effects on the entire brain, the contributions of relatively slow changing factors, such as receptor changes, second messenger contribution or sleep inducing factors need to be considered when researchers attempt to explain sleep-wake transition mechanisms. Any kind of these slow factor changes are mainly caused by the neurotransmitter release. Accordingly, further studies of neurotransmitters are needed to discuss the sleep-wake cycle and consequently the REM generation mechanisms.

The first step in solving the question of how and by what REM sleep starts is to measure the neurotransmitter release in the targeted regions across the sleep-wake cycle.

In Vivo Measurement of Local Neurotransmitter Release

From the studies of the recording of neuronal discharges, we now have much knowledge about the REM-on neurons which are related to the REM generation in the brain. If such REM-on cells in the brain regions are normally involved in the REM sleep, there should be a specific increase of neurotransmitter release at neuronal terminals during or prior to REM sleep. This is the general expectation in investigating neurotransmitter release across the sleep-wake cycle. However, it was very difficult to measure the local release of transmitters associated with REM sleep in a small brain region. In the early experiments, the cortical cup method (MacIntosh and Oborin, 1953) and a push-pull cannulation method (Gaddum, 1961) were used to measure ACh release at the cortical surface of cats. Gadea-Ciria et al. (1973) measured ACh release in the caudate nucleus of cats during sleep and wakefulness with a push-pull cannula and a radioenzyme assay, indicating ACh increase in the wake state. However, push-pull cannulation causes substantial damage to the brain tissues around the tip of the cannulae. The brain dialysis method was developed as an alternative method of push-pull cannulation, and later refined to a microdialysis method with the use of semipermeable membrane (Ungerstedt and Pycock, 1974).

The two approaches that have contributed most to the understanding of neurotransmitters responsible for or related to REM generation are the *in vivo* microdialysis and the *in vivo* voltametry methods. In the following, advantage and disadvantage of each approach will be discussed.

Principle and Necessary Caution of Microdialysis

The microdialysis method uses the principle of dialysis. The semipermeable membrane separates brain interstitial fluid from the perfusing solution. If there is a difference in the concentration of any substance to which the membrane is permeable, the concentration gradient causes diffusion of the substance from interstitial space into dialysis membrane.

Perfusion rate, temperature of perfusate, dialysis-tube length, composition of perfusate (Ca^{2+}, pH etc.), time after probe insertion and the draining-effect are factors that affect the collection. There are three criteria to be kept in mind when we measure neuronal release of a substance from the terminals; TTX (tetrodotoxin) sensitivity of the change, Ca^{2+} dependency and the increase of release produced by high K^+ concentration Ringer solution (Westerink, 1987, 1988, 1989). The inserted probe damages the tissue, and prolonged insertion of a probe will cause a change of the microenvironment by tissue gliosis. We have found that the most reliable results are obtained if samples are collected for 24 h, starting at 12 to 18 h after insertion.

The smallest probe now utilized is 0.22 mm in diameter and 0.5 mm in length. Taking the perfusion area of 0.5 mm into consideration, the research target must be bigger than $1 mm^3$. Good detection and fine time resolution are incompatible. Dectection limit is 0.05 picomole in ACh assay and ca one hundred femtogram in amine assay. Accordingly the time resolution of the sampling is limited approximately to the order of 1 min.

Microdialysis Methods in the Sleep Study

Advantages of microdialysis method in sleep studies are: (1) damage to the tissue is relatively small compared to the push-pull cannulation method, as the perfusing liquid does not contact the tissue directly. This allows a more prolonged collection of samples and (2) relation between the release of neurotransmitters and the animal behavior can be investigated, as samples can be collected serially from moving animals without disturbing the sleep.

Together with the improvement of the detection system for neurotransmitters, such as high performance liquid chromatography with electro-chemical detection, the present microdialysis method has enabled us to measure various neurotransmitters in a small quantity, *in vivo* and *in situ.* Measurement of neurotransmitters, such as ACh, dopamine, norepinephrine, serotonin and glutamate, by the microdialysis method under pharmacological stimulation has been reported since 1980's (Fisone, et al., 1987; Hernandez, et al., 1986; Ungerstedt and Hallstorm, 1987; Drejer et al., 1985). Recently, the number of such brain neurotransmitter studies has been increasing. However, there are only several reports on the relationships between endogenous brain neurotransmitters and sleep-wake stages, because the physiological changes in their concentration during the sleep-wake cycle are quite small compared to those produced by electrical and pharmacological stimulation. On the other hand, much knowledge has been accumulated by pharmacological stimulation experiments, using the advantage of microdialysis methods that, when the substance is added to perfusion liquid in a higher concentration than in the intersticial fluid, the substance diffuses out of the membrane and into the tissue around the membrane.

In Vivo Voltametry and Enzyme Electrodes

There are two methods that are capable of being alternatives to the microdialysis methods, the *in vivo* voltametry and enzyme electrode methods. Voltametry, which is based on the voltametric measurement of extracellular fluid in the brain, has an advantage, in some circumstances, over the microdialysis: the high temporal resolution of up to ten seconds and the better spatial resolution of up to a hundred micrometers. However, because of several disadvantages, such as poor separation of neurotransmitters, low sensitivity and limitation of detectable substances (electrically active molecules only), the voltametry method does not play the leading role now.

Another possible alternative, the enzyme electrode method is a modified method of the voltametry, measuring extracellular fluid directly in contact with the electrode. On the surface of the electrode the enzyme generates H_2O_2 from the aimed molecule and the amount of generated H_2O_2 is measured quantitatively by the voltametric procedure. We utilized the good time resolution of the enzyme methods (SYCOPEL, UK) and clarified that the perfusion area of the microdialysis method is only within 0.5 mm from the probe surface and the delay caused by membrane transition and by the distance form the probe was only 2-3 mim. This method settled the problem of poor separation of the neurotransmitters in the voltametry, but still requires some modification for use in sleep study.

Neurotransmitter Change Across Sleep Stage

Many neuronal transmitters and neuropeptides, such as dopamine (Ungerstedt and Pycock, 1974), norepinephrine (L'Heureux et al., 1986), serotonin (5HT) (Kato et al., 1986), amino acids (Lehmann et al., 1983), melatonin (Azekawa et al., 1990) and adenosine (Ballarin et al., 1987), have been measured with microdialysis technique in the past decade. However, there are only about 50 reports of neurotransmitters changes related to sleep or circadian rhythms. Here, the major neurotransmitters related to sleep stages are summarized briefly (Table 1).

Acetylcholine

The number of microdialysis studies dealing with changes in ACh release related to sleep-wake stages and circadian rhythm are now gradually increasing

Table 1 Neurotransmitters release across sleep-wake cycle and circadian rhythms

Acetylcholine	**PIA**	S=W < R	Kodama, 1990
	Hippocampas (HIP)	S < W=R	Kametani, 1990
	Hippocampas	S < R	Marrosu, 1995
	Thalamus	S < = R	WIlliams, 1994
	Caudate nucleus	S < W = R	Kodama, 1990
	mPRF	S = W < R	Leonard and Lydic, 1997
	Cortex	S < R < W	Marrosu, 1995
	Cortex	S < W	Jimenez-Capdeville, 1996
	PPN	S < W = R	Kodama, unpublished data
	NPM	S < W < R	Kodama, 1992
	NMC	S < W = R	Kodama, 1992
	Caudate nucleus	S < W	Gadea-Ciria, 1973
	Frontal Cortex (Fr-Cx)	L < D	Mitushima, 1996
	Hippocampas	L < D	Mizuno, 1994
	Sensory Cortex	L < D	Jimenez-Capdeville, 1993
	Hippocampas	L < D	Mizuno, 1991
Serotonin	**mPRF**	R < S < W	Iwakiri, 1993
	Raphe	R < S < W	Portas, 1994
	Hypothalmus	S < W	Auerbach, 1989
	Hypothalmus/Pre Optic Area	S < W	Wilkinson, 1991
	Hypothalmus	S < W	Imeri, 1994
	HIP, STR, Amygdara, Fr-Cx	L = D < D-Onset	Rueter, 1996
NE	Pineal gland.	L < D	Drijfhout, 1996
	Para ventricular n. (PVN)	L < D	Morien, 1995
	PVN	L < D	Mitome, 1994
	PVN	L < D	Stanley, 1989
Dopamine	Striatum (STR)	L < D	Paulson, 1994
	Striatum	L < D	Smith, 1992
Histamine	Hypothalamus	L < W	Mochizuki, 1992
glutamate	PIA	S = W = R	Kodama, 1997
	Basal Forebrain	S = W = R	Azuma, 1997
	NMC	W < R	Kodama, 1997
	PPN	S = R < W	Kodama, unpublished data
	NPM	S = W = R	Kodama, 1997
	Ventroposterolateral n. (VPL)	R = W < S	Kekesi, 1997
	mPRF	S < W < R	Hasegawa, 1997 (abstr)
	Laterodorsal tegmental n. (LDT)	S < W	Kodama, 1997
	Suprachiasmatic n. (SCN)	L < D	Glass, 1993
	SCN	L < D	Honma, 1996
GABA	**LC**	S = W < R	Nitz, 1997
	Raphe	W < S < R	Nitz, 1997
	Hypothalmus	W = R < S	Nitz, 1996

(Contd)

	VPL	R < W < S	Kekesi, 1997
Gln	**VPL**	W < R	Kekesi, 1997
Thr	VPL	W = R < S	Kekesi, 1997
Asn	VPL	W = R < S	Kekesi, 1997
Ser	VPL	W = R < S	Kekesi, 1997
Tau	VPL	W = R < S	Kekesi, 1997
Gly	VPL	W = R < S	Kekesi, 1997
Asp	VPL	W = R < S	Kekesi, 1997
	SCN	L < D	Honma, 1996
Adenosine	Basal Forebrain	S < W	Porkka, 1997
	Neostratum, Hippocampas	L < D	Huston, 1996
NO	Frontal Cortex	S = R < W	S. Burlet, 1997
	Thalamus	W = R < S	J. A. Williams, 1997
cGMP	Hippocampas, midbrain, pons medulla, cerebellum	S < R < W	S. Ogasawara, 1981
	Striatum	S < W < R	S. Ogasawara, 1981
	LDT	S < R < W	Kodama, 1997

*Bold indicates the areas where a REM sleep specific changes are observed.
**S: Slow wave sleep, R: REM sleep, W: wake, L: light phase, D: dark phase.

(Table 1). In the cortex (Cx) and hippocampus (HIP), ACh release increases during dark period (D) compared to light period (L) (Mizuno et al., 1994; Mitushima et al., 1996; Capdeville, et al., 1993). Changes in ACh release across sleep-wake cycle in HIP, the cerebral cortex and the caudate nucleus (CA) are also reported (Kametani and Kawamura, 1990, 1991; Williams et al., 1994) to indicate that ACh release is higher during wakefulness and REM sleep than during SWS. Because of the absence of REM sleep-specific increase of ACh release, these areas may not be directly involved in REM sleep generation.

Acetylcholine release in the brain is basically high during both wake and REM sleep as described above. This might correspond to the fact that most of the cholinergic neurons discharge in the manner of high discharge rate during both wake and REM sleep (Type I neurons, classified by Sakai). But, in some regions where acetylcholine plays an important role in generating the REM phenomena, for example, the pontine reticular formation including the n. reticularis pontis oralis and the rostral portion of n. reticularis pontis caudalis (pontine inhibitory area: PIA) or the nucleus paramedianus (NPM), enhanced ACh releases during REM sleep were observed. Recently, Leonard and Lydic also reported an increase of ACh release in the dorsal pons of cats during REM sleep (1997) (Table 1).

ACh release was observed in our previous study of cats both in PIA, a possible REM-generating locus, and in CA, which is a subcortical region rich in ACh but not directly related to the REM-generation system (Kodama et al., 1990). ACh release during REM sleep in the dorsolateral pons was about two times higher ($p<0.001$) than those during slow-wave sleep (SWS) and wakefulness (Table 1). ACh release was consistently the highest during REM sleep in each session and each animal. ACh release in the caudate nucleus was not significantly different between REM sleep and wakefulness. ACh release in the caudate was significantly higher during REM sleep and wakefulness than during SWS (Table 1). The overall mean amount of ACh release was 10 times higher in the caudate nucleus than in the dorsolateral pons.

The studies in decerebrated animals have shown that cholinergic stimulation, without glutamate stimulation of the caudal portion of medulla, corresponding to NPM, triggers atonia (Lai and Siegel, 1988). In contrast, glutamate, stimulation is effective in triggering muscle atonia in the rostral area, corresponding to the nucleus magnocellularis (NMC). In NPM and NMC, ACh releases during sleep-waking cycle were measured (Kodama et al., 1992) to indicate that ACh release in NPM is significantly higher during REM sleep than during SWS and wakefulness. By contrast, ACh release in NMC was not significantly different between REM sleep and wakefulness. These results indicate that the cholinoceptive neurons in NPM are related to REM sleep, especially to muscle atonia.

Amine

Serotonin

Serotonin (5HT) plays an important role in REM sleep generation. Many researchers tried to measure the concentration of 5HT, but, low 5HT concentration during REM sleep and data variation caused by blood contamination made it difficult to obtain constant results. Recently, Iwakiri et al. (1993) reported 5HT changes in mPRF and Portas and McCarley in raphe (1994). Kubin et al. also reported that 5HT decreased in XII nucleus in relation to the atonia during REM sleep (1994).

Others (NE, Dopamine, Histamine)

Though there are a couple of reports about circadian changes of norepinephrine (NE) (Stanley et al., 1989) dopamine (Smith, 1992) and of Histamine (Mochizuki et al., 1992), there is no report of sleep specific change of those neurotransmitter releases. These neurotransmitters should also have important function on sleep mechanisms even if there is no sleep specific change in the release. We reported on the result of glutamatergic inputs into cholinergic neurons in LDT generated NO. NO works as a retrograde transmitters to generate cGMP and affect presynaps of NE neurons to regulate NE release (Kodama, et al., 1997). We also observed atonia related NE changes in XII (data not shown).

Amino Acids

Glutamate

Glutamate and ACh are the major excitatory inputs to the neurons. However, in many cases glutamate release in the brain does not show significant changes across sleep stages. For example, the changes in levels of glutamate in PIA detected by dialysis were smaller than those of ACh and showed no significant change across sleep stages, in spite of the facts that glutamate is involved in REM sleep. The most likely explanation for this is that synaptically released glutamate in this area is obscured by the large glial release.

In NMC, where our group found the evidence which showed that glutamate might be involved in REM sleep control, we detected a REM sleep related change (11%) of glutamate release. We reported the sleep stage related increases of glutamate in LDT and basal forebrain (Kodama et al., 1997; Azuma et al., 1997), but they were also small as in NMC. Tetrodotoxin (TTX) studies have shown that the neuronal release of glutamate amounts to only 10% (our observation) or less (Westerink, 1989) of the ambient level as measured by HPLC. The major source of basal glutamate is presumably glia cells surrounding neurons (Villablanca, 1966) and metabolic sources, rather than synaptic releases, which makes it difficult to determine the significant glutamate changes. Under TTX perfusion, we observed little glutamate change across sleep stages, which suggested most of the changes of glutamate content came from neuronal release (1998). From these aspects, the 11% increase observed in NMC is equal to almost twice the basal neuronal release, which could be interpreted as a sufficiently large change.

GABA

GABA is another important neurotransmitter functioning as inhibitory inputs. Recently, very important findings were reported by Nitz and Siegel that GABA increases in DRN and LC during REM sleep (1997). GABA is discussed to be a strong source to inhibit serotonergic neurons in dorsal raphe nucleus during REM sleep. As some of REM-on neurons are considered to contain GABA as a transmitter, further study of GABA release and interaction between GABA and other transmitters is desired.

Purine

Adenosine

Adenosine is known as an potent sleep inducing substance. Important finding that adenosine concentration in the basal forebrain increased during wakefulness to inhibit cholinergic neuron in the basal forebrain and decreased in sleep phase was reported by Porkka et al. (1997).

Others

Nitric Oxide

Nitric oxide (NO) is now considered to play an important retrograde messenger. Our data suggested NO regulated NE release in LDT area to coordinate ACh-

NE release balance across sleepwake cycle (Kodama et al., 1997). Burlet and Cespuglio reported NO increase during wakefulness in frontal cortex (1997). Williams's report indicated the NO increases in the thalamus during SWS (1997).

cGMP

cGMP also change its concentration across sleep-waking cycle in LDT (Kodama et al., 1997), hippocampus, pons, medulla and cerebellum (Ogasawara et al., 1981). Although the direct affect on sleep mechanisms is not known now, cGMP plays a definitely important role as a modulator to regulate neuronal release in the presynaptic field.

In this decade, positive reports of sleep related change of neurotransmitters are increasing as described above. But, there are not many reports clarifying the REM sleep specific change of neurotransmitters. Furthermore, when the contribution of the neurotransmitter on sleep stages is discussed, the point is whether the change comes out as a result or as a cause. For example, we observed a REM sleep specific increase of ACh release in PIA, but, it remains unclear whether this increase of ACh triggers REM or an increase in discharge of cholinergic neurons during REM sleep enhances ACh release in dorsal pons. Our previous study using 2.5 min sampling period suggested that ACh release in PIA already began to increase during the transition period from SWS to REM sleep. The ACh release during 2.5 min sampling including the transient period amounted to over 80% of that of the subsequent full-blown REM period (Kodama et al., 1990). That means the cholinergic input into cholinoceptive neurons in PIA is involved in the REM sleep generating mechanism. However, most of the neurotransmitters across sleep stages are supposed to change as a result of the sleep-wake transition. To determine exactly the temporal relationship between neurotransmitters release and the onset of REM sleep, a dialysate sampling period much shorter than 1 min is needed, which is not possible with the present sensitivity of the HPLC method.

Neurotransmitters Changes in PIA and REM Sleep Generation

Recent anatomical and neurophysiological studies suggest that, as excitatory inputs, both ACh and glutamate in the dorsolateral pontine area are involved in generating REM sleep. (Although there are reports that the basal forebrain and the hypothalamus affect the REM sleep, transection studies indicate that the inputs from these areas are not required for basic REM phenomena.) Here, the cholinergic and glutamatergic inputs and their interaction in PIA will be discussed from the standpoint of REM generating mechanisms, and the interaction between excitatory and inhibitory input which affect REM sleep starting and ending mechanisms is also discussed.

Cholinergic and glutamatergic inputs into PIA

The interaction between the medulla and the pons is thought to be important

(Siegel, 1985). Sakai et al. reported that there are cholinergic projections from NMC to the pons (1988) and that these projections are necessary for REM sleep. Our report examining the effect of electrical or chemical stimulation of NMC on ACh release in the pons indicated that glutamate stimulation, but not ACh stimulation in NMC, enhanced ACh release in PIA and changed the REM sleep latency and the amount. This is consistent with our findings that there is significant change not in ACh release but in glutamate release in NMC during REM sleep. Therefore dialysis and stimulation data indicate that ACh release in NMC may relate to arousal system, rather than REM sleep generation. These results suggest that a non-cholinoceptive, glutamate-receptive system in NMC is involved in the REM sleep-specific enhancement of ACh release in the dorsal pons (Fig. 1).

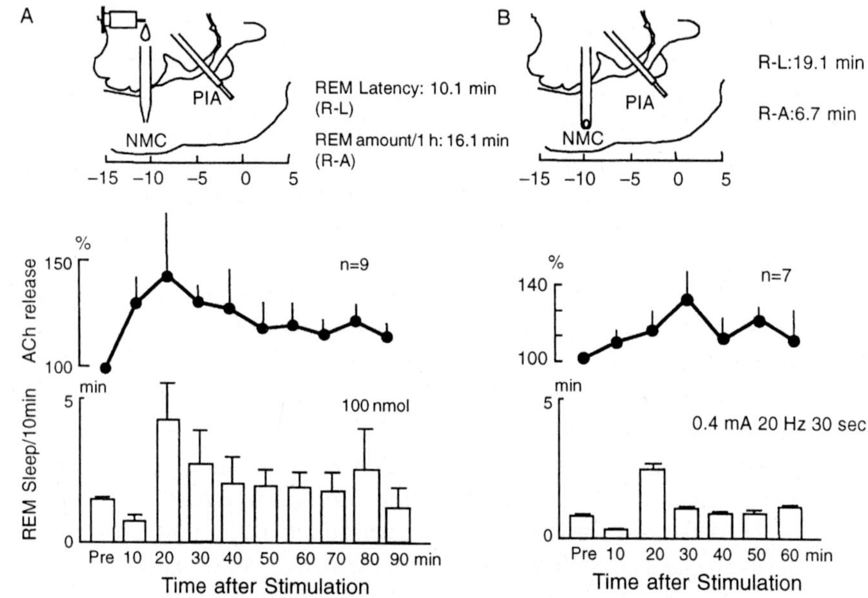

Fig. 1. **Microdialysis probe was placed into the dorsolateral pons. We activated NMC with electrical stimulation (30 sec trains of 0.1 msec pulses at 20 Hz), carbachol injection (40 nmol/0.4 μl) or glutamate injection (100 nmol/0.5 μl): (A) Microinjection of glutamate into NMC during wakefulness enhanced ACh release. This microinjection significantly shortened REM latency (R-L) to 10.1 min and also significantly enhanced the amount of REM during one hour (R-A) to 16.1 min compared with 9.6 min in the control. (B) Electrical stimulation significantly enhanced ACh release in the dorsolateral pons during the first, third and fifth 5 min periods after stimulation. This stimulation significantly shortened REM latency to 19.1 min in comparison to 33.0 min in the control, but did not significantly affect the amount of REM sleep during each one hour period following stimulation. In contrast, microinjection of carbachol into NMC during wakefulness failed to affect ACh release, REM latency and the amount of REM sleep. Injection of carbachol into NMC tended to arouse the cats.**

Other candidates as the generators of the change of ACh and glutamate release in PIA are the pedunculopontine nucleus (PPN) and the retrorubral nucleus (RRN). A UCLA group reported that electrical stimulation into PPN and RRN (midbrain) produced muscle atonia (Lai and Siegel, 1988). The anatomical studies indicated both cholinergic and glutamatergic projections from both PPN and RRN to PIA. The ACh and glutamate release in PIA were measured with electrical stimulations in PPN and RRN (Fig. 2). There were both ACh and glutamate increases after the electrical stimulation in PPN. Similarly, in RRN, the enhancement of both glutamate and ACh were observed in the first 5 min and the first 10 min periods, respectively. However, the data indicated that the relationship between PPN/RRN and ACh/ glutamate was not intimate. Furthermore, in the experiment with intact cats, there were no muscle atonia observed in the cases of both PPN and RRN stimulation, suggesting these pathways are not active when there are the regulations from the forebrain. Therefore, it seems that there exists not a direct regulation but modification of cholinergic and glutamatergic input from PPN and RRN to PIA.

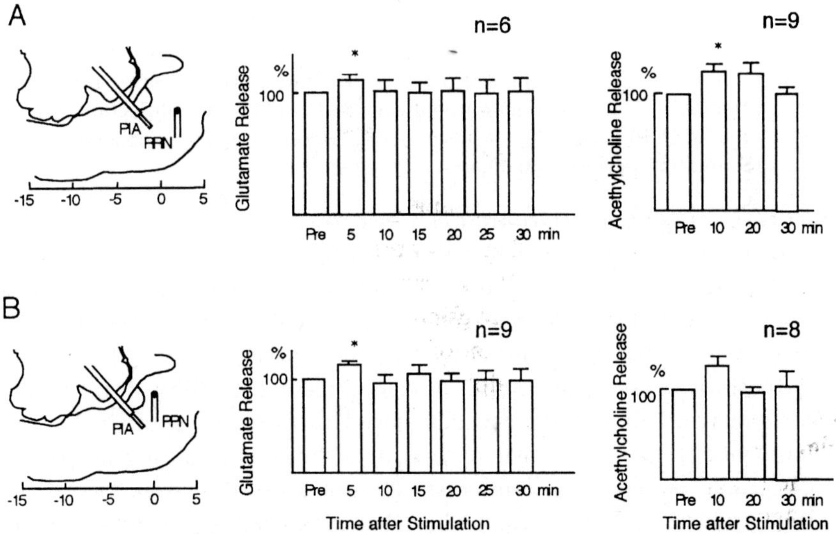

Fig. 2. (A) ACh and glutamate release after the electrical stimulation in RRN. The enhancement of both glutamate and ACh were observed in the first 5 min and the first 10 min periods, respectively. (B) There are both ACh and glutamate increases after the electrical stimulation in PPN. The increase of glutamate and ACh release were observed in the first 5 min and the first 10 min, respectively. ACh samples were collected every 10 min. and glutamate samples were collected every 5 min. Stimulation is 30 sec trains of 0.1 mV rectangular pulses at 100 Hz.

There are anatomical connections between bilateral PIA. Lydic et al. reported ACh increase in one side PIA after the carbachol injection into the other side

PIA (1991). In our experiment, the increase of glutamate was observed in short latency with ACh microinjection (100 μg/0.5 μl) (Fig. 3). From these findings, it may be concluded that the interaction between bilateral PIA is also important to regulate ACh and glutamate concentration in PIA in generating the REM sleep phenomena.

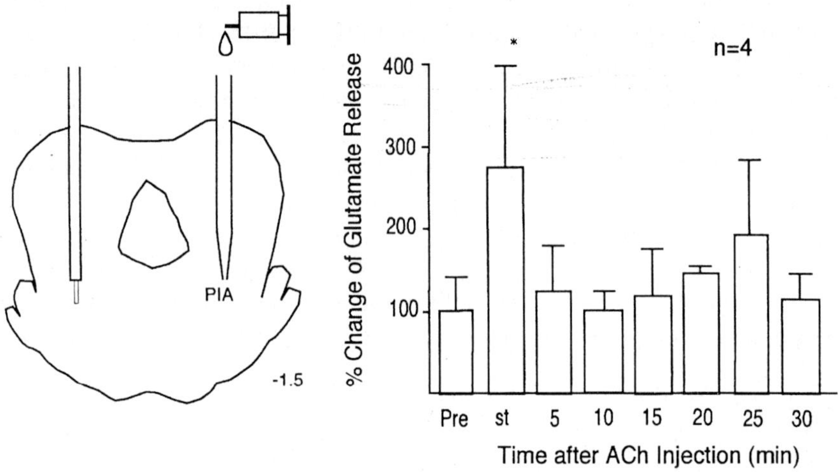

Fig. 3. Glutamate release in the one side PIA in 5 min period of pre-, during and after ACh (100 μg/0.5 μ*l*) microinjection after ACh microinjection into the other side PIA. Significant increase in glutamate release is observed in the stimulating period (P<0.05, n=4).

Interaction among neurotransmitters in PIA

Lai and Siegel first reported that activation of non NMDA glutamate receptors in PIA triggered atonia (1991). Elazar et al. (1990) found that injection of glutamate antagonists in this region disrupted REM sleep. Onoe and Sakai (1995) showed that application of kinate through a microdialysis probe in PIA (peri-locus coeruleus alpha) produced increased REM sleep without contribution of cholinergic mechanisms. However, we could not detect significant change in glutamate release in the dorsal pons during sleep-wake cycle (Table 1). Two possibilities are discussed as the reason. One is that synaptic release in the pons cannot be detected because it is overwhelmed by non-synaptic release. Another possibility is that it is not the specifc increase of glutamate release during REM sleep but the glutamate-ACh interaction that is important for REM generation. We stimulated the dorsolateral portion of pons with glutamate/ACh through a microdialysis probe and simultaneously measured ACh/glutamate release with the same probe. During the continuous glutamate stimulation, ACh release increased initially, but later decreased significantly. ACh release, then, returned to the control level after glutamate stimulation (Fig. 4). During glutamate stimulation, the REM sleep amount was only 1.8 min/ 60 min and during the withdrawal phase, the REM amount significantly increased to 12.2 min. The REM latency after stimulation was

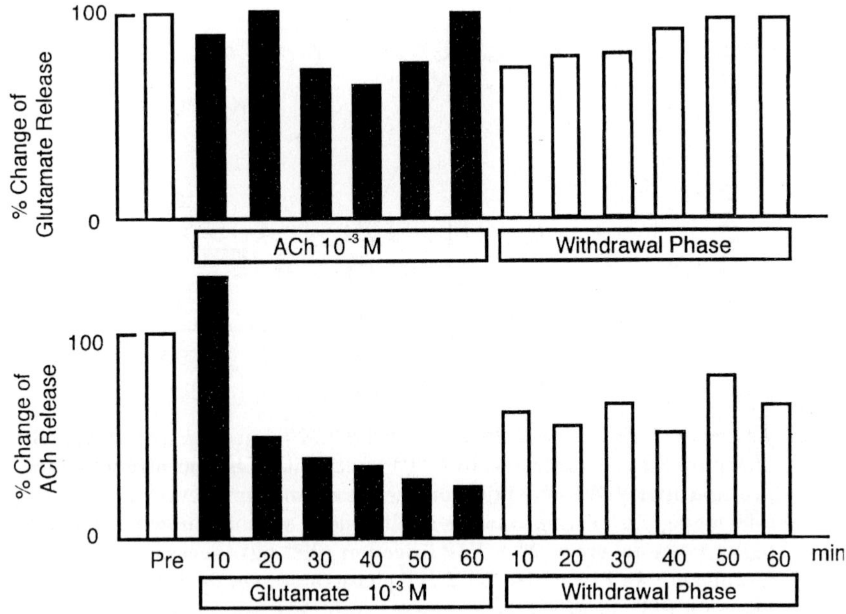

Fig. 4. ACh-glutamate interaction in the PIA. In the PIA, ACh effect on glutamate release and glutamate effect on ACh release were investigated. During the ACh application, glutamate release decreased and returned to the control level after stop application (top). During glutamate application, ACh release initially increase but then decreased gradually and returned to control level in the withdrawal phase (bottom).

15.3 min and significantly shorter than the control. Similarly, during ACh stimulation, glutamate release decreased and then returned to the control level after stimulation. It has been reported that the function of glutamate receptors is changed by ACh release (Akaike et al., 1991) and our data showed that ACh and glutamate release were modulated by each other. We hypothesize that the ACh increase prior to REM sleep inhibits glutamate release and the low concentration of glutamate works on mainly in kinate receptor to work cooperatively with ACh to maintain REM sleep, because kinate is more sensitive than other glutamatergic receptor. On the other hand, in latter part of REM sleep, we hypothesize that glutamate increase by the change of serotonergic affinity not only decrease ACh release but also affect on NMDA receptor to make the animal wake up. In this way, the interaction between ACh and glutamate could contribute to the REM generation system. Furthermore, there is evidence that neurotransmitters and neuropeptides (Fig. 5) interact and modulate ACh release. Clarification of the interaction not only between neurotransmitters but also between these substances is critical for the understanding of the REM sleep control.

Neurotransmitters change and REM sleep

It is clear that cholinoceptive REM-on neurons in the dorsolateral pontine

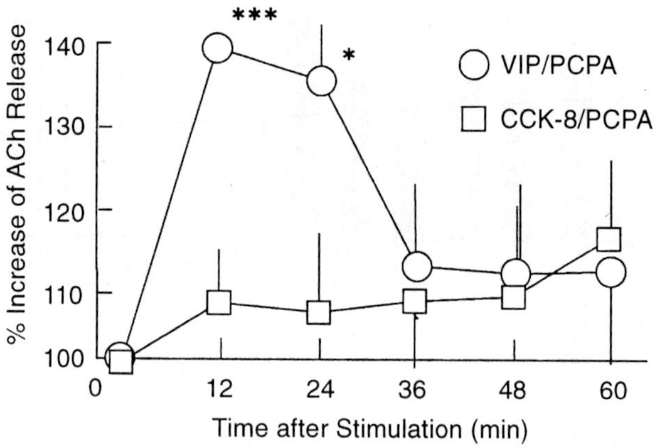

Fig. 5. Change in ACh release in PIA after VIP/CCK 8 injection into IV ventricle under low 5HT concentration by PCPA treatment. ACh concentrations during pre- and after VIP/CCK8 injection periods are measured every 12 min PCPA (500 mg/kg, i.p.) was injected as a initial dose, with additional doses of 200 ng/kg twice in every 24 h VIP injection (200 ng) increase ACh release significantly in PIA under low 5 HT phase (*n*=4). On the other hand, CCK injection did not affect ACh release (*n*=5). VIP was suggested to have a presynaptic effect on cholinergic terminal to increase ACh release under low concentration of 5HT.

reticular formation are important in generating REM sleep. REM sleep starts with the increase of cholinergic or partially glutamatergic excitatory input from other structures in the pons and the medulla. ACh release in PIA is known to increase prior to REM sleep, but the source of this ACh increase is not clear. Thakkar et al. explained the increase by $5HT_{1A}$ sensitivity in LDT neurons (1997) and Semba et al. explained it by NE inhibition on the presynaps of LDT neurons (1997). Lydic's group (Roth et al., 1996; Baghdoyan et al., 1997) explained the ACh increase in PIA by autoregulation through mACh receptors (M_2). The characteristic common to these explanations is that ACh release of cholinergic neurons (LDT/PPN) is somehow regulated by inhibitory input, such as 5HT, NE or autoreceptors, to make a REM sleep specific increase of ACh release in PIA. Serotonin and NE, which are the typical inhibitory inputs and stop firing during REM sleep, are in turn regulated by GABA release (Nitz and Siegel, 1997; Bijorkum et al., 1997) or by adenosine accumulated during wakefulness (Strecker et al., 1997). So what regulates GABA release?

There are few findings about what is responsible for the end of REM sleep. Akaike reported serotonin changes its function from an inhibitory to excitatory action accompanied by the change of its concentration. In Hasegawa's report, they observed a glutamate surge during REM sleep (1997). From these observations and the importance of interaction of neurotransmitters, we hypothesize that when 5HT concentration decreases to the critical point of inhibitory/excitatory function, the glia neurons are excited by 5HT and release glutamate promptly, and that this surge of glutamate depolarizes the arousal neurons in the rostral part of PIA to end the REM period. That is the reason why we supposed that no wake-starting neurons are reported.

In this decade, much knowledge has been accumulated about sleep related neurotransmitters, but further investigation is needed for complete explanation of the starting and stopping mechanisms of REM sleep.

References

Akaike, N., Harata, N. and Tateishi, N. Modulatory action of cholinergic drugs on N-methyl-D-asparatate response in dissociated rat nucleus basalis of Meynert neurons, Neurosci., 130 (1991) 243–247.

Aserinsky, E. and Kleitman, N. Regularly occurring periods of eye motility, and concomitant phenomena, during sleep, Science, 118 (1953) 273–274.

Auerbach, S.B., Minzenberg, M. J. and Wilkinson, L.O. Extracellular serotonin and 5-hydroxyindola cetic acid in hypothalmus of the unanesthetized rat measured by in vivo dialysis coupled to high performance liquid chromatography with electrochemical detection: dialysate serotonin refrects neeuronal release, Brain Res., 499 (1989) 281–290.

Azekawa, T., Sano A., Aoi, K., Sei, H. and Morita, Y. Concurrent on-line sampling of melatonin in pineal microdialysates from conscious rat and its analysis by high-performance liquid chromatography with electrochemical detection, J. Chromatgr., 530 (1990) 47–55.

Azuma, A., Kodama, T., Honda, K. and Inoue, S. State-dependent changes of extracellular glutamate in the medial pre-optic area in freely behaving rats, Neurosci. Lett., 214 (1996) 179–82.

Baghdoyan, H.A., Fleegal, M.A. and Lydic, R. Acetylcholine(ACh) release in the medial pontine reticular formation (mPRF) is regulated by M2 muscarinic autoreceptor, Neurosci. Abstr., 23 (1997) 2131.

Ballarin, M., Herrera-Marschitz, M., Casas, M. and Ungerstedt, U. Striatal adenosine levels measured 'in vivo' by microdialysis in rats with unilateral dopamine denervation, Neurosci. Lett., 83 (1987) 338–344.

Bjorkum, A.A., Strecker, R.E., Porkka-Heiskanen, T., Stenberg, D. and McCarley, R.W. GABA modulates the dorsal raphe and alters the pattern of sleep-waking, *Neurosci. Abstr.,* 23 (1997) 1065.

Borbely, A.A., *Das Geheimnis des Schlafe: Neue Wege und Erkenntnisse der Forshung.* Deutsche Verlags-Anstalt GmbH.

Burlet, S. and Cespuglio, R. Voltametric detection of nitric oxide (NO) In the rat brain: its variations throughout the sleep-wake cycle, Neurosci. Lett., 226 (1997) 131–135.

Drejer, J., Benveniste, H., Diemer, N.H. and Schousboe, A. Cellular origin of ischemia-induced glutamate release from brain tissue in vivo and in vitro, J. Neurochem., 45 (1985) 145–51.

Drijfhout, W.J., van-derLinde, A.G., Kooi, S.E., Grol, C.J. and Westerink, B.H. Norepinephrine release in the rat pineal glang: the input from the biological clock measured by in vivo microdialysis, J. Neurochem., 66 (1996) 748–55.

Elazar, Z., Fahringer, H.M., Baldwin, C., Camelleri, C. and Siegel, J.M. Effect of pontine injection of glutamate antagonists on REM sleep, Sleep Res., 19 (1990) 12.

Fisone, G., Wu, C.F., Consolo, S., Nordstrom, O., Brynne, N., Bartfai, T., Melander, T. and Hokfelt, T. Galanin inhibits acetylcholine release in the ventral hippocampus of the rat:

histochemical, autoradiographic, in vivo, and in vitro studies, Proc. Natl. Acad. Sci., 84 (1987) 7339–43.

Gaddum, J.H. Push-pull cannulae, J. Physiol., 155 (1961) 1.

Gadea-Ciria., M., Stadler, H., Lloyd, K.G. and Batholini, G. Acetylcholine release within the cat striatum during the sleep-wakefulness cycle, Nature, 243 (1973) 518–519.

Glass, J.D., Hauser, U.E., Blank, J.L., Selim, M. and Rea, M.A. Differential timing of aminoacid and 5-HIAA rhythms in suprachiasmatic hypothalmus, Am. J. Physiol., 265 (1993) R504–511.

Hasegawa, T., Azuma, S., Kimura, M. and Inoue, S. State dependent changes of amino acids in the rat oral pontine reticular nuclei measured by microdialysis, Neurosci. Abstr., 23 (1997) 2132.

Hernandez, L., Stanley, B.G. and Hoebel, B.G. A small, removable microdialysis probe, Life Sci., 39 (1986) 2629–37

Honma, S., Katsuno, Y., Shimohara, K., Abe, H. and Honma, K. Circadian rhythm and response to light of extracellular glutamate and asparatate in rat suprochiasmatic nucleus, Am. J. Physiol., 271 (1996) R579–85.

Huston, J.P., Haas, H.L., Boix, F., Pfister, M., Decking, U., Schrader, J. and Schwarting, R.K. Extracellular adenosine levels in neostriatum and hippocampus during rest and activity periods of rats, Neurosci., 73 (1996) 99–107.

Imeri, L., Simoni, M.G., Giglio, R., Clavenna, A. and Mancia, M. Changes in the serotonergic system during the sleep-wake cycle: simultaneous polygraphic and voltammetric recordings in hypothalamus using a telemetry system, Neurosci., 58 (1994) 353–358.

Iwakiri, H., Matsuyama, K. and Mori, S. Extracellular levels of serotonin in the medial pontine reticular formation in relation to sleep-wake cycle in cats: a microdialysis study, Neurosci. Res., 18 (1993) 157–170.

Jimenez-Capdeville, M.E. and Dykes, R.W. Daily changes in the release of acetylcholine from rat primary somatosensory cortex, Brain Res., 625 (1993) 152–8.

Jimenez-Capdeville, M.E. and Dykes, R.W. Changes in cortical acetylcholine release in the rat during day and night: differences between motor and sensory areas, Neurosci., 71 (1996) 567–79.

Kametani, H. and Kawamura, H. Alterations in acetylcholine release in the rat hippocampus during sleepwakefulness detected by intracerebral dialysis, Life Sci., 47 (1990) 421–426.

Kametani, H. Kawamura, H. Circadian rhythm of cortical acetylcholine release as measured by in vivo microdialysis in freely moving rats, Neurosci Lett., 132 (1991) 263–266.

Kato, T., Dong, B., Ishii, K. and Kinemuchi, H. Brain dialysis: In vivo metabolism of dopamine and serotonin by monoamine oxidase A but B in the striatum of unrestrained rats, J. Neurochem., 46 (1986) 1277–1282.

Kekesi, K.A., Dobolyi, A., Salfay, O., Nyitrai, G. and Juhasz, G. Slow wave sleep is accompanied by release of certain aminoacids in the thalmus of cats, NeuroReport, 8 (1997) 1183–1186

Kodama, T., Takahashi, Y. and Honda, Y. Enhancement of acetylcholine release during paradoxical sleep in the dorsal tegmental field of the cat brain stem, Neurosci. Lett., 114 (1990) 277–282.

Kodama, T., Lai, Y.Y. and Siegel, J.M. Enhancement of acetylcholine release during REM sleep in the caudomedial medulla as measured by in vivo microdialysis, Brain Res., 580 (1992) 348–350.

Kodama, T., Lai, Y.Y. and Siegel, J. M. Enhanced glutamate release during REM sleep in the rostromedial medulla as measured by in vivo microdialysis, Brain Res., 780 (1998) 178–181.

Kodama, T., Honda, Y. and Koyama, Y. Glutamate-noradrenergic interaction in the laterodorsal tegmentum neurons mediated by nitric oxide, Psychiatry Clin. Neurosci., 51 (1997) S45.

Kubin, L., Reignier, C., Tojima, H., Taguchi, O., Pack, A.I. and Davies, R.O. Changes in serotonin level in the hypoglossal nucleus region during carbachol-induced atonia, Brain Res., 645 (1994) 291–302.

Lai, Y.Y. and Siegel, J.M. Medullary regions mediating atonia, J. Neurosci., 8 (1988) 4790–4796.

Lai, Y.Y. and Siegel, J.M. Muscle tone suppression and stepping produced by stimulation of midbrain and rostral pontine reticular formation, J. Neurosci., 10 (1990) 2727–2734.

Lai, Y.Y. and Siegel, J.M. Ponto-medullary glutamate receptors mediating locomotion and muscle tone suppression, J. Neurosci., 11 (1991) 2931–2937.

Lehmann, A. and Hamberger, A. Dihydrokainic acid affects extracellular taurine and phosphoethnolamine levels in the hippocampus, Neurosci. Lett., 38 (1983) 67–72.

Leonard, T.O. and Lydic, R. Pontine nitric oxide modulates acetylcholine release, rapid eye movement sleep generation, and respiratory rate, J. Neurosci., 17 (1997) 774–785.

L′ Heureux, R., Dennis, T., Curet, O. and Scatton, B. Measurement of endogenous noradrenaline release in the rat cerebral cortex in vivo by transcortical dialysis: Effects of drugs affecting noradrenergic transmission, J. Neurochem., 46 (1986) 1794–1801.

Lydic, R., Baghdoyan, H.A. and Lorinc, Z. Microdialysis of cat pons reveals enhanced acetylcholine release during state-dependent respiratory depression, Am. J. Physiol., 261 (1991) R766–70.

MacIntosh, F.C. and Oborin, P.E. Release of acetylcholine from intact cerebral cortex. In: XIX International Physiol. Congress, (1953) 580–581.

Marrosu, F., Portas, C., Mascia, M.S., Casu, M.A., Fa, M., Giagheddu, M., Imperato, A. and Gessa, G.L. Microdialysis measurement of cortical and hippocampal c-acetylcholine release during sleepwake cycle in freely moving cats, Brain Res., 671 (1995) 329–332.

McGinty, D.J., Harper, R.M. and Fairbanks, M.K. Neuronal unit activity and the control of sleep states, Advances in Sleep Research, 1 (1974) 217–250.

Mitome, M., Honma, S., Yoshihara, T. and Honma, K. Prefeeding increase in paraventricular NE release is regulated by a feeding -associated rythm in rats, Am. J. Physiol., 266 (1994) E606–11.

Mitsushima, D., Mizuno, T. and Kimura, F. Age-related changes in diurnal acetylcholine release in the prefrontal cortex of male rats as measured by microdialysis, Neurosci., 72 (1996) 429–34.

Mizuno, T., Arita, J. and Kimura, F. Spontaneous acetylcholine release in the hippocampus exhibits a diurnal variation in both young and old rats, Neurosci. Lett., 178 (1994) 271–4.

Mizuno, T., Endo, Y., Arita, J. and Kimura, F. Acetylcholine release in the rat hippocampus as measured by the microdialysis method correlates with motor activities and exhibits a diurnal variation, Neurosci., 44 (1991) 607–12.

Mochizuki, T., Yamatodani. A., Okakura, K., Horii, A., Inagaki, N. and Wada, H. Circadian rhythm of histamine release from the hypothalamus of freely moving rats, Physiol. Behavior, 51 (1992) 391–394.

Morien, A., Wellman, P.J. and Fojt, J. Diurnal rhythms of paraventricular hypothalamic norepinephrine and food intake in rats, Pharmacol. Biochem. Behav., 52 (1995) 169–74.

Nitz, D. and Siegel, J.M. GABA release in posterior hypothalamus across sleep-wake cycle, Am. J. Physiol., 271 (1996) R 1707–12.

Nitz, D. and Siegel, J.M. GABA release in the dorsal raphe nucleus: role in the control of REM sleep, Am. J. Physiol., 273 (1997) R451– R455.

Nitz, D. and Siegel, J.M. GABA release in the locus coeruleus as a function of sleep/wake state, Neurosci., 78 (1997) 795–801.
Ogasahara, S., Taniguchi, Y. and Wada, H. Changes in the levels of cyclic nucleotides in rat brain during the sleep-wakefulness cycle, Brain Res., 213 (1981) 163–171.
Onoe, H. and Sakai, K. Kinate receptors: a novel mechanism in paradoxical (REM) sleep generation, Neuroreport, 6 (1995) 353–356.
Paulson, P.E. and Robinson, T.E. Relationship between circadian changes in spontaneous motor activity and dorsal versus ventral striatal dopamine neurotransmission assessed with on-line microdialysis, Behav. Neurosci., 108 (1994) 624–35.
Porkka-Heiskanen, T., Strecker, R.E., Thakkar, M., Bjorkum, A.A., Greene, R.W. and McCarley, R.W. Adenosine: A mediator of the sleep-inducing effects of prolonged wakefulness, Science, 276 (1997) 1265–1268.
Portas, C.M. and McCarley, R.W. Behavioral state-related changes of extracelullar serotonin concentration in the dorsal raphe nucleus: a microdialysis study in the freely moving cat.
Roth, M.T., Fleegal, M.A., Lydic, R. and Baghodyan, H.A. Pontine acetylcholine release is regulated by muscarinic autoreceptors, Neuroreport, 7 (1996) 306–972.
Rueter, L.E. and Jacobs, B.L.Changes in forebrain serotonin at the light-dark transition: correlation with behaviour , NeuroReport, 7 (1996) 1107–1111.
Sakai, K., Executive mechanisms of paradoxical sleep. Arch. Ital. Biol., 126 (1988) 239–257.
Semba, K., Greene, R.W., Rasmusson, D.D., McCarley, R.W. and Weider, J. Noradrenergic presynaptic inhibition of acetylcholine release in the rat pontine reticular formation: an in vitro electrophysiological and in vivo microdialysis study, Neurosci. Abstr., 23 (1997) 1065.
Shimizu, N., Duan, S.M., Hori, T. and Oomura, Y. Glutamate modulate dopamine release in the striatum as measured by brain microdialysis, Brain. Res. Bull., 25 (1990) 99–102.
Siegel, J.M., Pontomedullary interactions in the generation of REM sleep. In: D.J. McGinty, R. Drucker-Colin, A.R. Morrison, P.L. Parmeggiani (Eds.), Brain Mechanisms of Sleep, Raven Press, New York, 1985, pp. 157–174.
Smith, A.D., Olson, R.J. and Justice, J.B., Jr. Quantitative microdialysis of dopamine in the striatum: effect of circadian variation, J. Neurosci. Methods, 44 (1992) 33–41.
Stanley, B.G., Schwartz, D.H., Hernandez, L., Hoebel, B.G. and Leibowitz, S.F. Patterns of extracellular norepinephrine in paraventricular hypothalamus: relationship to circadian rhythm and deprivation-induced eating behavior, Life Sci., 45 (1989) 275–282.
Strecker, R.E., Porkka-Heiskanen, T., Bjorkum, A.A. and McCarley, R.W. Adenosine actions of the dorsal raphe nucleus: Altered sleep-waking pattern, Neurosci. Abstr., 23 (1997) 1065.
Thakkar, M.M., Strecker, R.E. and McCarley, R.W. A 5-HT1A agonist in the laterodorsal tegmental nucleus (LDT) inhibits REM-on neurons but has no effect on wake/REM-on neurons as revealed by combined unit recording and microdialysis, Neurosci. Abstr., 23 (1997) 1065.
Ungerstedt, U. and Hallstrom, A. In vivo microdialysis—a new approach to the analysis of neurotransmitters in the brain, Life Sci., 41 (1987) 861–4
Ungerstedt, U. and Pycock, C. Functional correlates of dopamine neurotransmission, Bulletin der Schweizerishen Akademie der Medizinishen Wissenschaften, 127 (1974) 44–55.
Villablanca, J., Behavioral and polygraphic study of "sleep" and "wakefulness" in chronic decerebrate cats, Electroenceph. Clin. Neurophysiol., 21 (1966) 562–577.
Westerink, B.H.C., Damsma, G., Rollema, H., De Vries, J.B. Horn, A.S. Scope and limitations

of in vivo brain dialysis: A comparison of its application to various neurotransmitter systems, Life Sci., 41 (1987) 1763–1776.

Westerink, B.H.C., Hofsteede, H.M., Damsma, G. and De Vries, J.B. The significance of extracellular calcium for the release of dopamine, acetylcholine and amino acids in conscious rats, evaluated by brain microdialysis, Naunyn-Schmiedeberg's Archives of Pharmacol., 337 (1988) 373–378.

Westerink, B.H.C., Rollema, H., De Vries, J.B. and Damsma, G. Levels of neurotransmitters and precursors in microdialysates: criteria for neurogenic origin, Current Separations, 9 (1989) 76.

Wilkinson, L.O., Auerbach, S.B. and Jacobs, B.L. Extracellular serotonin levels change with behavioral state but not with pyrogen-induced hyperthermia, J. Neurosci., 11 (1991) 2732–2741.

Williams, J.A., Comisarow, J., Day, J., Fibiger, H.C. and Reiner, P.B. State-dependent release of acetylcholine in rat thalamus measured by in vivo microdialysis, J. Neurosci., 14 (1994) 5236–42.

Williams, J.A., Vincent, S.R. and Reiner, P.B. Nitric oxide production in rat thalmus changes with behavioral states, local depolarization, and brainstem stimulation, J. Neurosci., 17 (1997) 420–427.

Rapid Eye Movement Sleep
B.N. Mallick and S. Inoué (Eds)

13. Spatio-Temporal Distribution of Neuronal Activities and REM Sleep

Shojiro Inoué, Utpal Kumar Saha[1] and Toshimitsu Musha[1]

Institute for Medical and Dental Engineering, Tokyo Medical and Dental University, Kanda-Surugadai 2-3-10, Chiyoda-ku, Tokyo 101-0062, Japan

[1]Brain Functions Laboratory, Sakato, Takatsu-ku, Kawasaki 213-0012, Japan

SUMMARY

The dipole tracing method, a non-invasive technique to estimate spatial and temporal locations of the most active neuron groups in the brain, was first applied to sleeping and dreaming humans. Stages II-IV non-rapid-eye-movement (NREM) sleep were characterized by dipole positions which were stably localized in restricted regions of the diencephalon. Vector moments of dipoles remained small and unidirectional. In contrast, stage I NREM sleep and REM sleep were characterized by dipole positions which were unstably distributed in the wide brainstem regions including the thalamus, midbrain and pons and also in the hippocampus and the visual cortex. Vector moments of dipoles were large and bidirectional or multidirectional. Frequent translocations of the dipole positions indicated the existence of ponto-geniculo-occipital wave-like activity during dreaming. In conclusion, different neuronal groups in the subcortical regions were differentially involved in the generation and maintenance of different vigilance states in the human brain, and the simultaneous activation of cortical and hippocampal neurons were specific to REM sleep.

Key Words: Brainstem, dipole tracing method, dreaming, human, NREM sleep, pontogeniculo-occipital wave-like activity.

Introduction

The dipole tracing method estimates with a high spatial accuracy the source of electrical brain potentials (He et al., 1987; Musha and Honma, 1990; Nakajima et al., 1990; Musha et al., 1994). In this novel method, activated or depolarized neurons in the brain are replaced by a small current dipole and, if these neurons are limited within a small volume, the localized bioelectric activities are approximated by equivalent current dipoles. Position and vector moment of the dipoles are estimated from scalp potentials using a head model of the scalp, skull and brain tissue with uniform electric conductivities (Musha et al., 1994). The DT method has been successfully applied to human brain in finding pathways of somatosensory and cognitive signals or identifying the location of epileptic focus. Positions of activated neurons detected by the DT method are in good agreement with those detected by other methods like magnetoencephalography and positron emission tomography (PET).

Scalp potentials are routinely and easily measured for study of sleep and wakefulness. However, depth brain potentials which can be routinely measured by invasive methods in animals are hardly available in humans. For example, pontogeniculo-occipital (PGO) waves are widely known to occur in various mammalian species (Callaway et al., 1987) but not usually detectable in humans (Carskadon and Dement, 1994). Hence the non-invasive DT method seems to be of special value for human sleep study. We for the first time applied this method to an analysis of spatio-temporal patterns of brain bioelectric activities during sleep in healthy humans. Preliminary results were partially published in abstracts (Saha et al., 1994; Inoué et al., 1995).

Application of the DT Method to Sleeping Humans

Seven-night sleep records were obtained from five healthy males aged 23-30 yr, who had previous experience of sleeping in laboratory as subjects. They were attached with 21-channel scalp electroencephalographic (EEG), 2-channel electromyographic (EMG) and 2-channel electro-oculographic (EOG) electrodes. An exact head geometry of the subjects was measured with a special instrument system (HS-1, NF Electronics Instruments), and stored in a computer for correcting the effect of the skull on dipole location. Bioelectric data were processed and stored by an on-line real-time analyzer system (BFA-T1, NF Electronics Instruments), which enabled continuously to observe the time-course of brain mapping in combination with spectral analysis throughout a 6 to 8 h bed rest period starting at 23.00–24.00 h. High-sensitive television camera systems were simultaneously used to watch the movements of the body and eyes. Upon spontaneous awaking, the subjects were asked as to whether or not they experienced dreaming.

The stored EEG records were used for off-line analyses by the DT method. The location and vector moment of dipoles in the brain were searched in successive 20–1340 msec EEG fractions, taken from typical episodes of non-rapid-eye-movement (NREM) sleep and REM sleep. Stages of sleep were determined by visual scoring of the stored EEG, EMG and EOG records according to the Rechtschaffen-Kales criteria. Original EEG was sometimes filtered to isolate components of the delta, theta and alpha waves. Dipolarity was calculated usually with a threshold level of higher than 98%. Anatomical loci corresponding to each detected dipole position were identified by three-dimensional correlates of the head geometry, which were matched with horizontal, sagittal and frontal cross sections of nuclear magnetic resonance images (MRI) of the subject's brain.

DT Activities in the Sleeping Brain

The state of sleep in general was characterized with centripetal dipole positions, which mainly located in the brainstem regions including the thalamus, hypothalamus, midbrain and pons. Most of the dipole positions asymmetrically appeared in a unilateral side of the brain within a limited period. During stages II-IV NREM sleep, dipole positions were stably localized inside restricted regions of the brainstem, whereas they showed relatively wide and unstable distributions during stage I NREM sleep and REM sleep. Vector moments of the dipoles were usually small, stable and unidirectional during stages II-IV NREM sleep, while they were large, bidirectional, sometimes multidirectional, and dynamically translocating during stage I NREM sleep and REM sleep.

Figure 1 shows a typical example for the locations and vector moments of dipoles during a period of 120 msec in stage III NREM sleep. Synchronized delta waves were observed in 16 out of 21 EEG channels. The activated neuron groups were restricted in the left medial preoptic area exhibiting small unidirectional vector moments. Figure 2 shows another example during a period of 30 msec in stage III NREM sleep. Synchronized delta waves

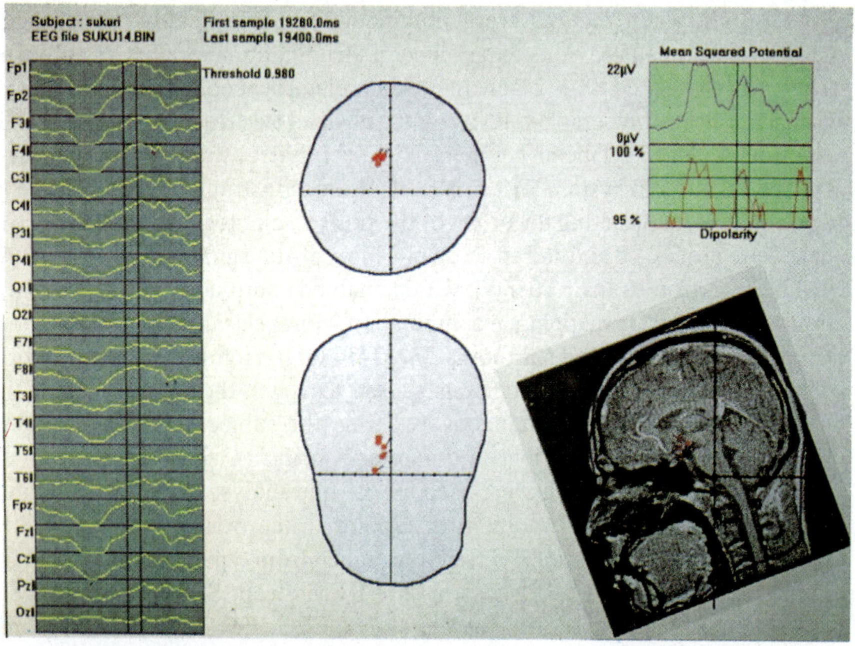

Fig. 1 **Locations with vector moments of dipoles during stage III NREM sleep. Red points and lines indicate the location and vector moments, respectively, of dipoles. Left: A 120 msec fraction (within two vertical lines) in the original 1500 msec scalp EEG was examined for dipolarity. Middle upper: Horizontal plane of the subject's head, in which the upper side is frontal. Middle lower: Vertical plane of the subject's head. Right upper: Time course of dipolarity corresponding to that of EEG in the left panel. Dipoles appearing at the threshold of higher than 98% were shown in the head planes. Right lower: Lateral plane of the subject's head superimposed with MRI photography. For further explanations, see text.**

appeared in all EEG channels. The activated neuron groups with unidirectional vector moments were restricted in the left thalamic area.

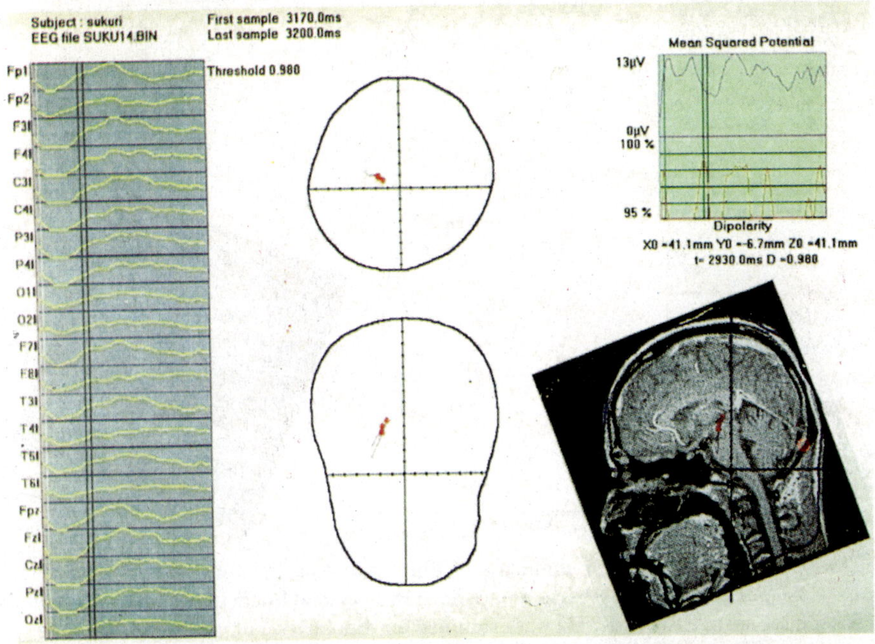

Fig. 2 Locations with vector moments of dipoles during stage III NREM sleep. Left: A 30 msec fraction (within two vertical lines) in the original 3000 msec scalp EEG was examined for dipolarity. For other explanations, see legend of Fig. 1 and text.

Figure 3 shows a typical example for the locations and vector moments of dipoles during a period of 320 msec in the tonic stage of REM sleep. Theta components of 6-7 Hz were filtered from original EEG. Dipoles with large upward and downward vector moments were detectable in the left thalamic and hippocampal regions. Figure 4 shows a typical example for the locations and vector moments of dipoles during a period of 330 msec in the phasic stage of REM sleep. Desynchronized beta waves appeared in all EEG channels. Dipoles with large bidirectional vector moments were detectable in the right thalamic and hippocampal regions as well as in the bilateral occipital cortex (visual area). Frequent translocations of dipole positions were observed from the thalamus to the hippocampus and/or the cortex or vice versa in synchrony with rapid eye movements. The subject recollected the experience of dreaming upon awaking.

Possible Neuronal Activities During Different Sleep Stages

By means of the DT method we first found that, in the sleeping human brain, activated neuronal groups were stably located in restricted areas of the subcortical regions during NREM sleep. We also found that activated neuronal groups were unstably located not only in the subcortical regions but in the

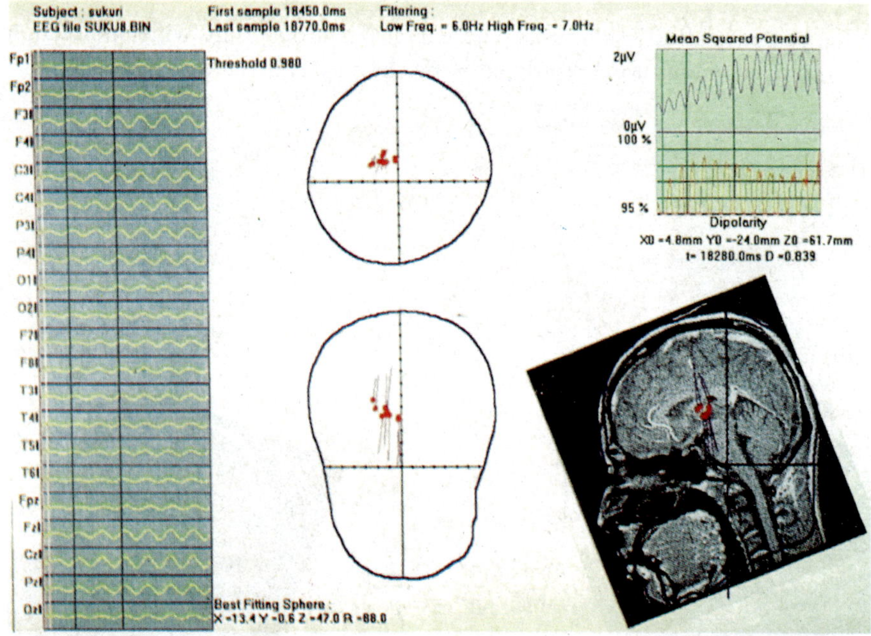

Fig. 3. Locations with vector moments of dipoles during the tonic stage of REM sleep. Left: A 320 msec fraction (within two vertical lines) in the filtered 1000 msec scalp EEG of 6-7 Hz was examined for dipolarity. For other explanations, see legend of Fig. 1 and text.

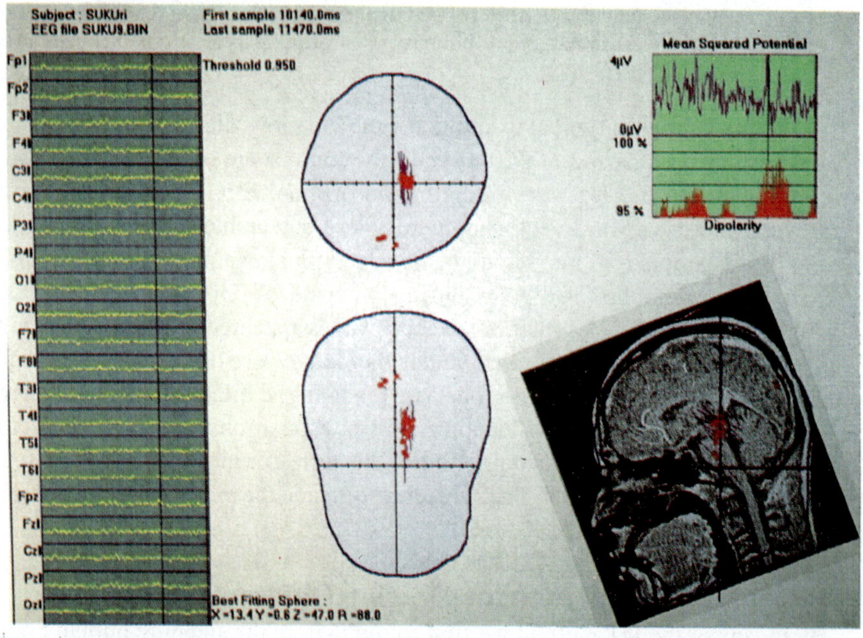

Fig. 4 Locations with vector moments of dipoles during the phasic stage of REM sleep. Left: A 330 msec fraction (within two vertical lines) in the original 1200 msec scalp EEG was examined for dipolarity. For other explanations, see legend of Fig. 1 and text.

occipital cortex during REM sleep. These findings make a sharp contrast to distributions of dipole positions in the cerebral cortex during wakefulness (He et al., 1987). Thus the present study demonstrated differential spatio-temporal bioelectric/neuronal activity corresponding to different sleep-waking states in the human brain.

In animal studies, it is well known that the electric activity of the thalamus crucially depends on the states of sleep and wakefulness (Steriade, 1994). It is also known that the hypothalamic preoptic area and the midbrain/pontine regions plays an essential role in the generation of slow-wave sleep (Jones, 1994) and REM sleep (Siegel, 1994), respectively. Our findings by the non-invasive method were largely consistent with those investigated in animals by various invasive methods.

The findings on REM sleep is of special interest in terms of neuronal activity during dreaming. Dipole positions frequently translocated within a short period of less than 1 sec from the pons, thalamus and occipital cortex, indicating interactions among these brain regions. This bioelectric activity may be regarded as a direct evidence of PGO-like activity. The possible occurrence of PGO spikes in humans has been suggested on the basis of indirect PGO- or REM sleep-associated phenomena like extra-ocular potentials (Rechtschafften et al., 1970), phasic integrated potentials (Rechtschafften and Chernik, 1971), depth cortical spikes (Salzarulo et al., 1975), and eye movement-associated potentials (McCarley et al., 1983; Miyauchi et al., 1987). Recently, Maquet et al. (1996) reported that regional cerebral blood flow is positively correlated with REM sleep in the pontine tegmentum, left thalamus, both amygdaloid complexes, anterior cingulate cortex and right parietal operculum. This observation conducted by means of PET seems to be consistent with our findings.

Since our present study demonstrated that the DT method is applicable to weak subcortical source estimation during sleep, when the cortical activity is relatively inactive and easily to be filtered out, the DT method may further offer novel techniques to sleep research. Firstly, it may enable to understand the neural consequences underlying the memory storage and information processing in the sleeping, especially dreaming human brain. Secondly, it may enable to analyze sleep-disordered brains with various functional and structural damages.

Acknowledgements

This study was supported by a grant from the Ministry of Education, Science, Sports and Culture (No. 04302033) and Special Coordination Funds for Promoting Science and Technology from the Science and Technology Agency, Prime Minister's Office.

References

Callaway, C.W., Lydic, R., Baghdoyan, H.A. and Hobson, J.A. Pontogeniculooccipital waves: spontaneous visual system activity during rapid eye movement sleep, Cell. Mol. Neurobiol, 7 (1987) 105–149.

Carskadon, M.A. and Dement, W.C. Normal human sleep: an overview. In: M. Kryger, T. Roth and W. Dement (Eds.), *The Principles and Practice of Sleep Medicine,* Second Edition, W.B. Saunders, Philadelphia, 1994, pp. 16–39.

He, B., Musha, T., Okamoto, Y., Honma, S., Nakajima, Y. and Sato T. Electric dipole tracing in the brain by means of the boundary element method and its accuracy, IEEE Trans. Biomed. Eng., BME-34 (1987) 406–414.

Inoué, S., Saha, U.K. and Musha, T. Spatio-temporal patterns of bioelectric activities in a sleeping human brain analyzed by the dipole tracing method, Psychiatr. Clin. Neurosci., 49 (1995) S46.

Jones, B.E. Basic mechanisms of sleep-wake states. In: M. Kryger, T. Roth and W. Dement (Eds.), *The Principles and Practice of Sleep Medicine,* Second Edition, W.B. Saunders, Philadelphia, 1994, pp. 145–162.

Maquet, P., Péters, J.M., Aerts, J., Delfiore, G., Degueldre, C., Luxen, A. and Franck, G. Functional neuroanatomy of human rapid-eye-movement sleep and dreaming, Nature, 383 (1996) 163–166.

McCarley, R.W., Winkelman, J.W. and Duffy, F.H. Human cerebral potentials associated with REM sleep rapid eye movements: links to PGO waves and waking potentials, Brain Res., 274 (1983) 359–364.

Miyauchi, S., Takino, R., Fukuda, H. and Torii, S. Electrophysiological evidence for dreaming: human cerebral potentials associated with rapid eye movement during REM sleep, Electroenceph. Clin. Neurophysiol., 66 (1987) 383–390.

Musha, T. and Honma, S. Do optimal dipoles obtained by the dipole tracing method always suggest true source locations? Brain Topography, 3 (1990) 143–150.

Musha, T., Ivanov, V. and Konyshev, V. Source estimation in the human brain from EEG based on the SSB head model, Methods Information Medicine, 33 (1994) 120–122.

Nakajima, Y., Honma, S., Musha, T., Okamoto, Y., Ackerman, R.H., Correia, J.A. and Alpert, N.M. Dipole-tracing of abnormal slow brain potentials after cerebral stroke—EEG, PET, MRI correlations, Neurosci. Lett., 112 (1990) 59–64.

Rechtschaffen, A. and Chernik, D.A. The effect of REM derivation on periorbital spike activity in NREM sleep, Psychophysiology, 9 (1971) 128.

Rechtschaffen, A., Molinari, S., Watson, R. and Wincor, M.Z. Extra-ocular potentials: a possible indicator of PGO activity in the human, Psychophysiology, 7 (1970) 336.

Saha, U.K., Musha, T. and Inoué, S. EEG source localization during sleep, Jpn. J. Med. Electr. Biol. Eng., 32 (Suppl) (1994) 354.

Salzarulo, P., Lairy, G.C., Bancaud, J. and Munari, C. Direct depth recording of the striate cortex during REM sleep in man: are there PGO potentials?, Electroencephal. Clin. Neurophysiol., 38 (1975) 199–202.

Siegel, J.M., Brainstem mechanisms generating REM sleep, In: M. Kryger, T. Roth and W. Dement (Eds.), *The Principles and Practice of Sleep Medicine,* Second Edition, W.B. Saunders, Philadelphia, 1994, pp. 125–144.

Steriade, M., Brain electrical activity and sensory processing during waking and sleep states. In: M. Kryger, T. Roth and W. Dement (Eds.), *The Principles and Practice of Sleep Medicine,* Second Edition, W.B. Saunders, Philadelphia, 1994, pp. 105–124.

Rapid Eye Movement Sleep
B.N. Mallick and S. Inoué (Eds)

14. Different Physiological Properties of Two Populations of PS-on Neurons in the Mesopontine Tegmentum

Yoshimasa Koyama[1], Yukihiko Kayama[1] and Kazuya Sakai[2]

[1]Department of Physiology, Fukushima Medical University School of Medicine, 1 Hikari-ga-oka, Fukushima 960-1295, Japan

[2]Inserum U480, Départment de Médecine Expérimentale Université Claude Bernard, 8 avenue Rockefeller, 69373 Lyon, France

ABSTRACT

The mesopontine tegmentum has a crucial role in the regulation of paradoxical sleep (PS) and waking. It contains two populations of cholinergic neurons, laterodorsal tegmental nucleus (LDT) and pedunculopontine tegemental nucleus (PPT). Neurons in the mesopontine tegmentum show specific firing patterns during sleep-waking cycles; some are active both during waking and PS, while others are silent during waking and start to fire prior to PS (PS-on neurons). Based upon several physiological properties including their location, conduction velocity or the duration of their action potential, we have hypothesized that the mesopontine tegmental neurons with longer action potentials (broad spike) are cholinergic and those with shorter ones (brief spike) are non-cholinergic. Labeling by biotinamide (biocytin or neurobiotin) of single neurons recorded extracellularly in the rat LDT, followed by histochemical staining with NADPH-diaphorase, supports the hypothesis more confidentially.

Using unanesthetized cats, it has become clear that the PS-on neurons in the mesopontine tegmentum can be divided into two populations, broad-spike (cholinergic) and brief-spike (non-cholinergic) neurons. Iontophoretic application of several kinds of neurotransmitters has indicated that the broad-spike PS-on neurons are all inhibited by carbachol, excited by glutamate and bicuculline, while brief-spike PS-on neurons are excited by carbachol and glutamate, and inhibited by noradrenaline. These findings propose several possibilities in the two kinds of PS-on neurons concerning the regulation of PS; (i) cholinergic excitation from the cholinergic PS-on neurons to the non-cholinergic PS-on neurons, (ii) non-cholinergic excitation on the cholinergic PS-on neurons or on the medullarly neurons involved in postural atonia during PS, and (iii) noradrenergic or GABAergic inhibition on the PS-on neurons during waking.

Key Words: laterodorsal tegmental nucleus, cholinergic neurons, PS-on neurons, biocytin, iontophoresis, noradrenaline, γ-amino butyric acid.

Introduction

It has been proposed that rapid eye movement (REM) sleep, or paradoxical sleep (PS), is regulated by the mesopontine tegmentum (Hobson, Lydic et al., 1986; Steriade and McCarley, 1990; Sakai, 1984). Infusion of carbachol, a potent cholinergic agonist, in some regions in the mesopontine tegmentum induced PS (Venni-mercier, Sakai et al., 1989; Yamamoto, Mamelak et al., 1990), indicating the importance of the cholinergic and/or cholinoceptive neurons in the mesopontine tegmentum for the regulation of PS.

Anatomical studies have shown that the cholinergic neurons in the brainstem gather tightly in the laterodorsal tegmental nucleus (LDT) and scattered in the pedunuclopontine tegmental nucleus (PPT). These neurons project diffusely to the forebrain, including the thalamus, hypothalamus or frontal cortex, they also send descending axons to the spinal cord, medulla or several regions in the mesopontine tegmentum (Hallanger and Wainer, 1988; Satoh and Fibiger, 1986).

Recent studies in which single neural activity is recorded in unanesthetized animals have revealed the existence of several populations of presumptive cholinergic neurons in the mesopontine tegmentum that are characterized by their firing properties across sleep waking cycles. (El Mansari, Sakai et al., 1989; Kayama, Ohta et al., 1992; Steriade, Datta et al., 1990). A subset of them shows slow tonic discharge during both waking and PS (Type I-S neurons), while another one has tonic firing during waking and burst discharges in conjunction with PGO-wave (PGO-on neurons). Still another is a group of neurons which are completely silent during waking, show significant increase in discharge rate prior to PS and maintain tonic firing during PS; (PS-on neurons) they may be most important in the regulation of PS. The cholinergic nature of these neurons has been ascribed to several lines of evidence described in the following chapter, although, until recently, we have obtained no confident evidence.

It has been suggested that the PS-executing system in the mesopontine tegmentum is under the inhibitory influences of the monoaminergic neurons (Hobson, Lydic et al., 1986; Steriade and McCarley, 1990; Sakai, 1984). This hypothesis is supported by the anatomical studies showing that the mesopontine tegmentum receives noradrenergic and/or serotonergic afferents from the locus coeruleus (LC) and/or dorsal raphe (DR), respectively (Sakai, 1991; Semba and Fibiger, 1992). Recent *in vitro* studies have shown monoaminergic influences on the mesopontine cholinergic neurons; they are hyperpolarized by serotonin or noradrenaline (Leonard and Llinás, 1994; Luebke, Greene et al., 1992; Mühlethaler, Khateb et al., 1990; Williams and Reiner, 1993) and depolarized by noradrenaline or histamine (Kahteb, Serafin et al., 1990; Mühlethaler, Khateb et al., 1990). These studies suggest that the monoaminergic afferents to the mesopontine cholinergic neurons have not simply an inhibitory tone concerning the regulation of PS.

For further understanding of neural mechanisms regulating PS, the following questions should be solved. (1) Are the neurons showing specific firing during sleep-waking cycles really cholinergic? (or what is the transmitter phenotype of these neurons?) (2) What kind of neural input (or neurotransmitter) is involved in the regulation of activity of neurons having such firing properties across sleep-waking cycles?

Discrimination of the Cholinergic Neurons by Spike Shape

Both in cat and rat, two kinds of action potential can be recorded in and around the clusters of cholinergic neuron in the mesopontine tegmentum; one with longer duration (broad spike) and the other with shorter duration (brief spike). The broad-spike neurons in the mesopontine tegmentum are presumed to be cholinergic, since: (1) they are located exclusively in the cholinergic neural zone, especially in the rat in which the cholinergic neurons gathering tightly in the LDT are well segregated from the monaminergic neurons (Hallanger and Wainer, 1988; Satoh and Fibiger, 1986), (2) conduction velocity of the broad-spike, thalamic projecting neurons are slower than that of the brief-spike ones (El Mansari, Sakai et al., 1989, 1990), (3) great majority of the LDT neurons activated antidromically by stimulation of several forebrain sites were of the broad type, corresponding with anatomical studies in which similar percentages of the thalamic projecting neurons were cholinergic (Kayama and Ogawa, 1987; Sofroniew, Priestley et al., 1985), and (4) majority of the broad-spike neurons are inhibited by carbachol or cholinergic agonists (Koyama and Kayama, 1993; Sakai, El Mansari et al., 1990).

The presumption is supported by *in vitro* intracellular recording and staining studies in which the identified cholinergic neurons have a high-threshold calcium spike which is generated at the falling phase of the sodium spike, making the total width of the spike broader (Leonard and Llinás, 1990) and the cholinergic neurons are hyperpolarized by carbachol through autoreceptors (Leonard and Llinás, 1994; Luebke, McCarley et al., 1993). However, identification by intracellular labeling in *in vivo* animals, especially in unanesthetized ones, is virtually impossible.

Recently, Pinault has developed a method to label single neurons recorded extracellularly by biotinamide (biocytin or neurobiotin) (Pinault, D., 1996); in brief, after extracellular recording of single neurons in anesthetized rats (Figs. 1Aa, 1Ba), biotinamide was injected iontophoretically through the glass pipette electrode using anodal current pulses (200 msec on/off, 2–10 nA) for 3–10 min. The neuron labeled successfully by biotinamide could be visualized by fluorescence of Taxas red (Figs. 1Ab, 1Bb). The broad-spike neuron in Figure 1A could be identified to be cholinergic, since it was stained by NADPH-diaphorase histochemistry (Fig. 1Ac), which is a confidential marker of the cholinergic neurons in the rat brainstem (Vincent, Satoh et al., 1983). In contrast, brief-spike neuron labeled in Figure 1 Bb was negative for NADPH-diaphorase (Fig. 1 Bc), thus non-cholinergic. When broad-spike neurons were classified as those with positive component of the spike longer

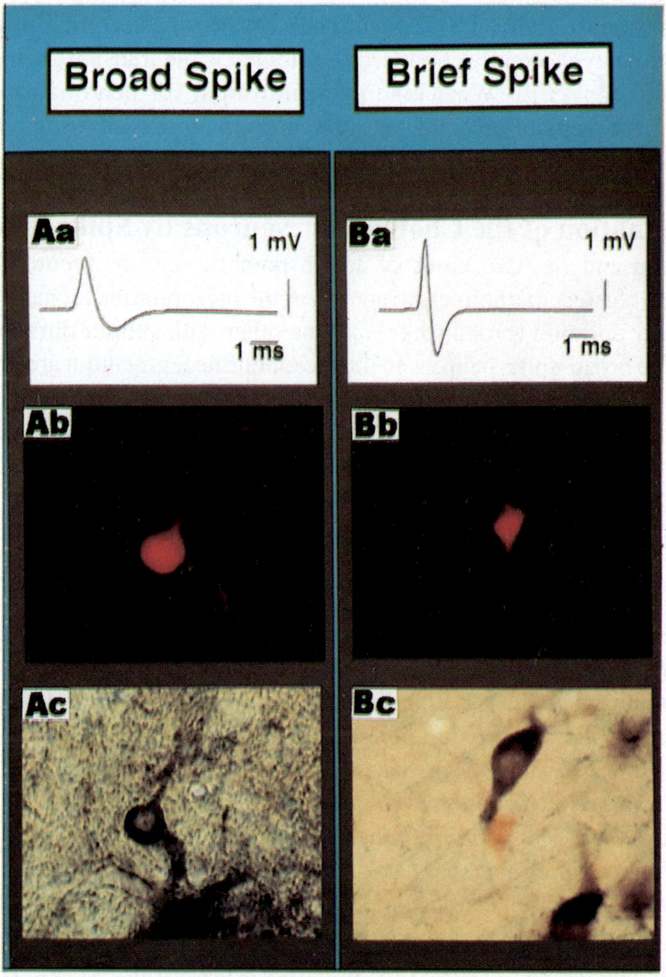

Fig. 1 Photomicrograph of neurons labeled by biocytin: An, A broad-spike neuron recorded from the laterodorsal tegmental nucleus (LDT). Ba, A brief-spike neuron recorded from outside of the LDT. b, Neurons visualized by fluorescence of Texas Red bound to biocytin. c, The same section processed for NADPH-diaphorase histochemistry, and in Bc, further for DAB reaction of biocytin. Scale bar = 20 μm (from Koyama, Honda et al., 1998).

than 0.78 msec and negative component longer than 1.0 msec, almost all (21/22) of the broad-spike neurons located within the LDT were NADPH-diaphorase positive, thus were cholinergic, while all (39/39) of the brief-spike neurons in and around the LDT were NADPH-diaphorase negative, to be non-cholinergic (Fig. 2).

Two Populations of PS-on Neurons in the Mesopontine Tegmentum

The spike width dichotomy is also observed when the metal electrodes were

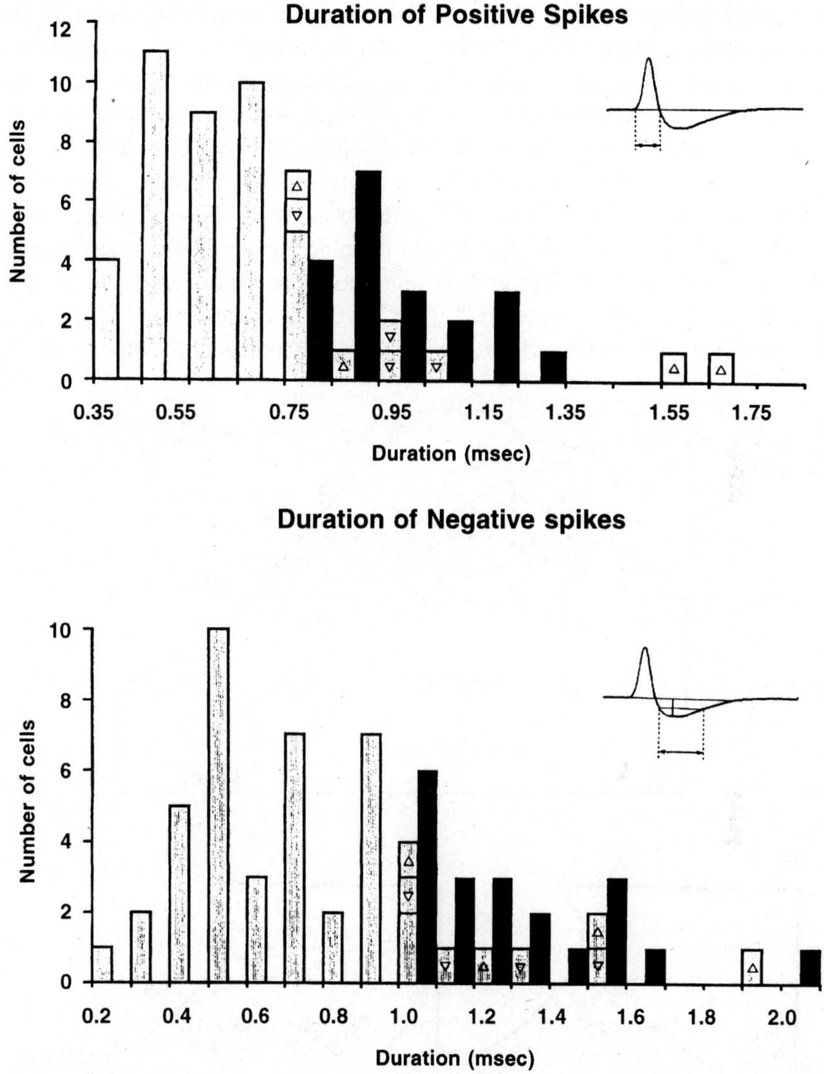

Fig. 2 **Distribution of duration of positive spikes (upper) and negative spikes (lower). Positive-spike duration is definded as from the onset to zero crossing of the spike, while negative-spike duration is defined as that at half maximum height of the negative component (see insets). Filled columns, NADPH-diaphorase positive neurons. Striped columns, NADPH-diaphorase negative neurons. Triangles, neurons presumed to be noradrenergic (upward triangles) or serotonergic (downward triangles) with their broad-spike shape and their histological location (from Koyama, Honda et al., 1998).**

used in the study in unanesthetized cats, in which two populations of PS-on neurons could be recorded in the mesopontine tegmentum; one generating broad spikes (1.16 ± 0.10 msec in duration from the onset of the first negative component to the peak of the following positive component) and the other with brief spikes (0.65 ± 0.06 msec) (Sakai and Koyama, 1996). Broad-spike PS-on neurons were located in the cholinergic neural zone including the LDT, LC α and peri-LC α, while brief-spike PS-on neurons were found, not only in the cholinergic neural zone, but in the non-cholinergic region in the mesopontine tegmentum, i.e. the caudal region of peri-LC α (Fig. 3). The broad-spike PS-on neurons were driven antidromically by stimulation of the intralaminar thalamic complex while the brief-spike PS-on neurons were mostly activated antidromically from the magnocellular nucleus of the medulla (Mc); the area receiving exclusively non-cholinergic input from the peri-LC α (Sakai, 1985) and is presumed to be involved in generation of postural atonia during PS (Sakai, Sastre et al., 1981). Conduction velocity of the

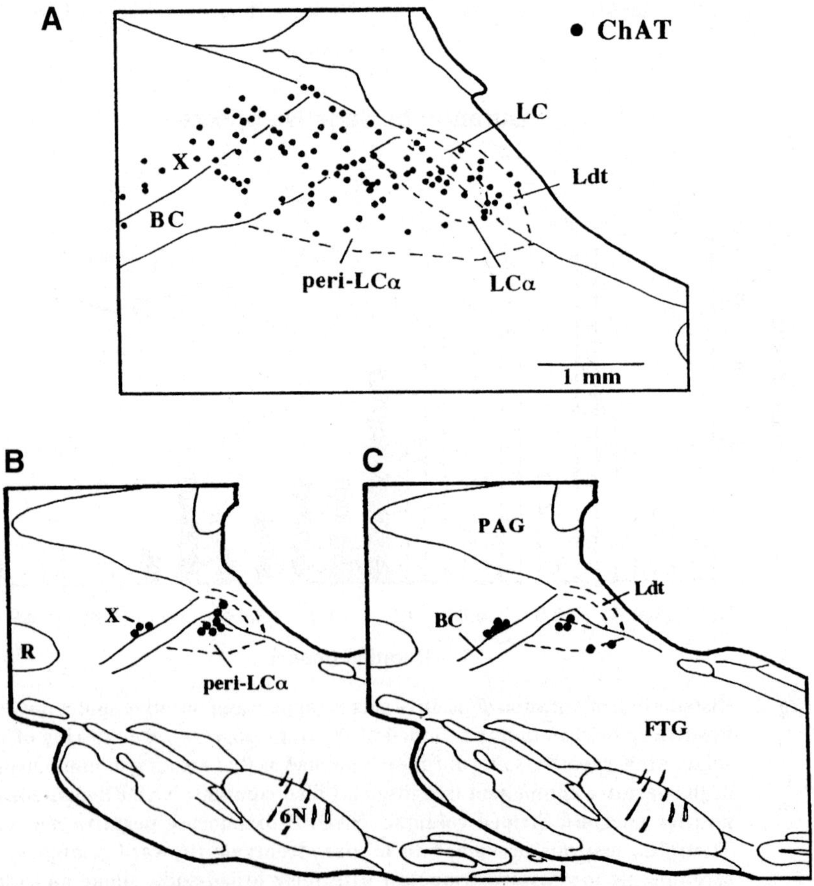

Fig. 3 **Localization of choline acetyltransferase (ChAT) immunoreactive neurons (A), broad-spike PS-on neurons (B) and brief-spike PS-on neurons (C). BC, brachium conjunctivum; FTG, gigantocellular tegmental field; LC α, locus coerulius α; Ldt, laterodorsal tegmental nucleus; PAG, periaqueductal gray; R, red nucleus; X, area X or pedunculopontine tegmental nucleus pars compacta (from Sakai and Koyama, 1996).**

former (2.3–3.3 m/sec) was slower than those of the latter (5.5–14.4 m/sec). These findings, taken together, further confirm that the broad-spike PS-on neurons are cholinergic and brief-spike ones are non-cholinergic.

Responsiveness to several neurotransmitters was quite different between the two populations of PS-on neurons. When the drugs were applied iontophoretically through the multibarreled electrodes glued to the recording pipette, activity of broad-spike PS-on neuron during PS was strongly inhibited by carbachol (arrow in Fig. 4A). During waking, that is, when the neurons were quite silent, glutamate caused excitation with very prompt onset and offset (Fig. 4A). Bicuculline (BIC), a potent GABA antagonist, also caused excitation with a time course slower than that in case of glutamate (Glu) application (Fig. 4B). Noradrenaline (NE) or serotonin (5HT) had almost no effect on the broad-spike PS-on neurons. In sharp contrast, the brief-spike PS-on neurons were all excited by carbachol (Fig. 5). Virtually all of them were inhibited by noradrenaline (Fig. 5) and excited by glutamate (not shown). Bicuculline had excitatory effect only one of four examined. None showed a significant response to serotonin.

Are There Species Difference?

In anesthetized rats, noradrenergic influence on the broad-spike LDT neurons (i.e., cholinergic neurons) are mainly inhibitory (Koyama and Kayama, 1993). *In vitro* study using neonatal rat has also shown that noradrenaline hyperpolarized all of the cholinergic LDT neurons (Williams and Reiner, 1993). In cat, however, no inhibition by noradrenaline was observed in broad-spike PS-on neurons and in other types of broad-spike neurons (Type I-S and PGO-on neurons); noradrenaline has excitatory effect on almost all of them examined (Koyama and Sakai, 1995). The findings are in agreement with those from *in vitro* study using adult guinea pig (Mühlethaler, Khateb et al., 1990).

Serotonin has an inhibitory effect on the broad-spike, or cholinergic, neurons in the LDT both in *in vivo* (anesthetized) or *in vitro* studies using rats (Koyama and Kayama, 1993; Luebke, Greene et al., 1992) or guinea pigs (Leonard and Llinás, 1994); in cats, however, no clear effect was observed on broad-spike PS-on neurons, and only slight inhibition was observed in some of the Type I-S neurons (Koyama and Sakai, 1995). Further investigation is required to clarify the discrepancy in the drug effects to be due to difference of species, preparation or age of the animals.

Implication of Two Types of PS-on Neurons for the Regulation of PS

It has been shown that the most effective site for inducing PS by carbachol infusion is the caudal peri-LC α (Vanni-mercier, Sakai et al., 1989), where some of the non-cholinergic PS-on neurons are located (Fig. 3). Since these neurons are excited by application of carbachol, it is highly probable that,

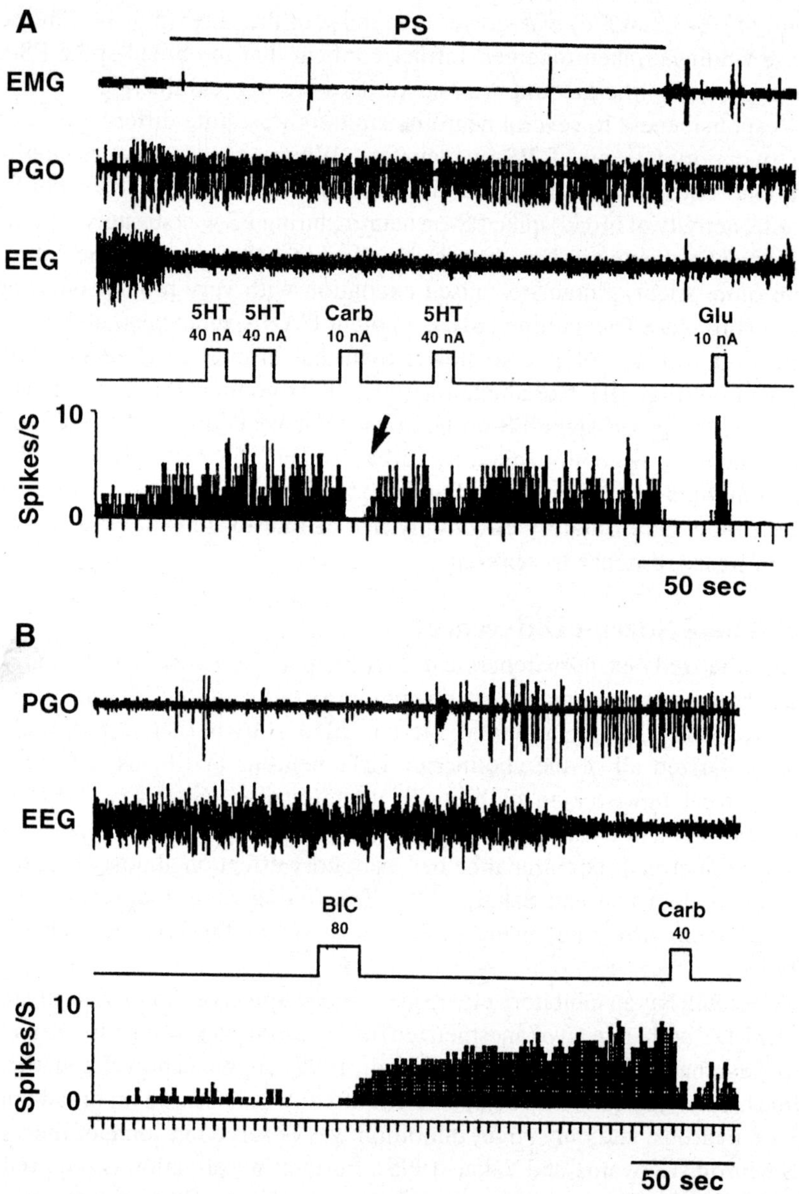

Fig. 4 (A), Effects of iontophoretically applied carbachol (Carb), serotonin (5HT) or glutamate (Glu) on a broad-spike PS-on neuron. (B), Effects of bicuculline (BIC) or carbachol (Carb) on a different broad-spike PS-on neuron. The intensity of the ejection current (nA) is indicated under the drug name. EMG, electromyogram; PGO, PGO wave from dorsal lateral geniculate nucleus; EEG, electroencephalogram (from Sakai and Koyama, 1996).

during naturally occurring PS, the brief-spike PS-on neurons in the caudal peri-LC α receive excitatory influence from the cholinergic PS-on neurons. Conversely, the broad-spike PS-on neurons, receiving cholinergic inhibition, are required to receive non-cholinergic (possibly glutamatergic) excitation from the brief-spike PS-on neurons. The brief-spike PS-on neurons descending to the Mc may execute a function to produce postural atonia during PS.

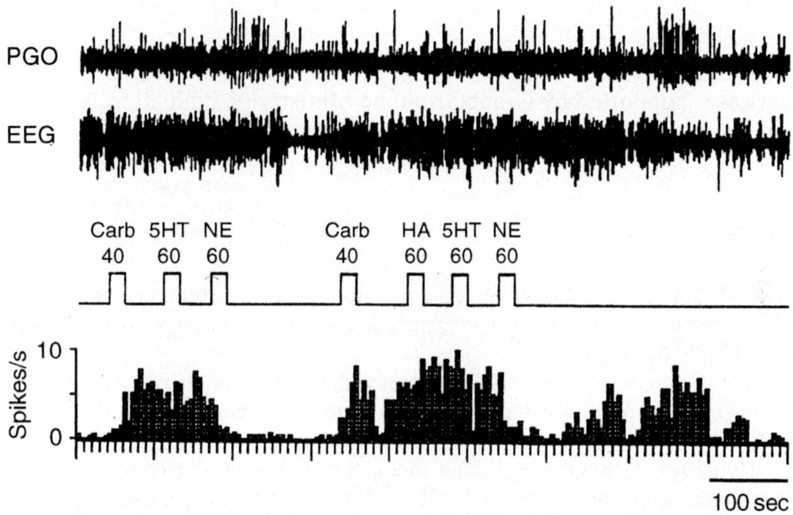

Fig. 5 Effects of iontophoretically applied drugs on a brief-spike PS-on neuron. HA, histamine; NE, noradrenaline; other abbreviations as in Fig. 4 (from Sakai and Koyama, 1996).

It is still probable that some of the brief-spike PS-on neurons are GABAergic, since large number of GABAergic neurons intermingle with the cholinergic ones in the mesopontine tegmentum (Ford, Holmes et al., 1995), c-fos expression in GABAergic neurons in the LDT increased during PS (Maloney and Jones, 1997) and GABA release in the DR and LC increased during PS (Nitz and Siegel, 1993, 1994). It should be examined whether silence of DR and LC neurons during PS may be produced by GABAergic inputs from the mesopontine tegmentum.

Two types of PS-on neurons receive different inhibitory input; GABAergic inhibition mainly on the broad-spike PS-on neurons, and noradrenergic inhibition on the brief-spike PS-on neurons. Since the noradrenergic neurons in the LC are PS-off type, these neurons are highly probable to exert inhibitory influence on the brief-spike PS-on neurons during waking. Although it has not been identified, there must be GABAergic PS-off neurons which cause inhibition on the broad-spike PS-on neurons during waking.

It has been hypothesized that the reciprocal interaction between PS-on and PS-off neurons has an important role in the regulation of PS (Hobson, Lydic et al., 1986; McGarley and Hobson, 1975; Sakai, 1984). The present article supports some of their concepts, in that, cessation of inhibition by, monoaminergic PS-off neurons (disinhibition) is required for initiation of PS, though further involvement of disinhibition from GABAergic neurons is suggested. As has been suggested, in addition to disinhibition, mutual excitatory interaction between the cholinergic and non-cholinergic (possibly glutamatergic) PS-on neurons would be crucial for the regulation of PS (McGarley, Greene et al., 1995).

Acknowledgment

This work was supported by Grants from the Ministry of Education, Science and Culture, Japan and INSERM U52 and CNRS ERS5645, France.

References

El Mansari, M., Sakai, K. and Jouvet, M. Unitary characteristics of presumptive cholinergic tegmental neurons during the sleep-waking cycle in freely moving cats, Exp. Brain Res., 76 (1989) 519–529.

El Mansari, M., Sakai, K. and Jouvet, M. Responses of presumed cholinergic mesopontine tegmental neurons to carbachol microinjection in freely moving cats, Exp. Brain Res., 83 (1990) 115–123.

Ford, B., Holmes, C.J., Mainville, L. and Jones, B.E. GABAergic neurons in the rat pontomesencephalic tegmentum: codistribution with cholinergic and other tegmental neurons projecting to the posterior lateral hypothalamus, J. Comp. Neurol., 363 (1995) 177–196.

Hallanger, A.E. and Wainer, B.H. Ascending projections from the pedunculopontine tegmental nucleus and the adjacent mesopontine tegmentum in rat, J. Comp. Neurol., 274 (1988) 483–515.

Hobson, J.A., Lydic, R. and Baghdoyan, H.A. Evolving concepts of sleep cycle generation: from brain centers to neuronal populations, Behav. Brain Sci., 9 (1986) 371–448.

Kayama, Y. and Ogawa, T. Electrophysiology of ascending, possibly cholinergic neurons in the rat laterodorsal tegmental nucleus: comparison with monoamine neurons, Neurosci. Lett., 77 (1987) 277–282.

Kayama, Y., Ohta, M. and Jodo, E. Firing of 'possibly' cholinergic neurons in the rat laterodorsal tegmental nucleus during sleep and wakefulness, Brain Res., 569 (1992) 210–220.

Khateb, A., Serafin, M. and Mühlethaler, M. Histamine excites pedunculopontine neurons in guinea pig brainstem slices, Neurosci. Lett., 112 (1990) 57–62.

Koyama, Y. and Kayama, Y. Mutual interactions among cholinergic, noradrenergic and serotonergic neurons studied by ionophoresis of these transmitters in rat brainstem nuclei, Neuroscience, 55 (1993) 1117–1126.

Koyama, Y., Honda, T., Kusakabe, M., Kayama, Y. and Sugiura, Y. *In vivo* electrophysiological distinction of histochemically-identified cholinergic neurons using extracellular recording and labeling in rat laterodorsal tegmental nucleus, Neuroscience 83 (1998) 1105–1112.

Koyama, Y. and Sakai, K. Cholinergic inhibition and monoaminergic excitation of presumed cholinergic mesopontine tegmental neurons: a microiontophoretic study in unanesthetized cats, Sleep Res., 24A (1995) 35.

Leonard, C.S. and Llinás, R.R. Electrophysiology of mammalian pedunculopontine and laterodorsal tegmental neurons in vitro: implication for the control of REM sleep. In:

M. Steriade and D. Biesold (Eds.), *Brain Cholinergic Systems*. Oxford University Press, New York, 1990, pp. 205–223.

Leonard, C.S. and Llinás, R.R. Serotonergic and cholinergic inhibition of mesopontine cholinergic neurons controlling REM sleep: an in vitro electrophysiological study, Neuroscience, 59 (1994) 309–330.

Luebke, J.I., Greene, R.W., Semba, K., Kamondi, A., McCarley, R.W. and Reiner, P.B. Serotonin hyperpolarized cholinergic low-threshold burst neurons in the rat laterodorsal tegmental nucleus in vitro, Proc. Natl. Acad. Sci., 89 (1992) 743–747.

Luebke, J.I., McCarley, R.W. and Greene, R.W. Inhibitory action of muscarinic agonists on neurons in the rat laterodorsal tegmental nucleus in vitro, J. Neurophysiol., 70 (1993) 2128–2135.

Maloney, K.J. and Jones, B.E. c-fos expression in cholinergic, GABAergic and monoaminergic cell groups during paradoxical sleep deprivation and recovery, Soc. Neurosci. Abstr., 23 (1997), 2131.

McCarley, R.W. and Hobson, J.A. Neuronal excitability modulation over the sleep cycle: a structural and mathematical model, Science, 189 (1975) 58–60.

McCarley, R.W., Greene, R.W., Rainnie, D. and Portas, C.M. Brainstem neuromodulation and REM sleep, Semin. in Neurosci., 7 (1995) 341–354.

Mühlethaler, M., Khateb, A. and Serafin, M. Effects of monoamine and opiates on pedunculopontine neurons. In: M. Mancia and G. Marini (Eds.), *The Diencephalon and Sleep*, Raven Press, New York 1990, pp. 367–378.

Nitz, D. and Siegel, J. Inhibitory amino acid neurotransmition in the dorsal raphe nucleus during sleep-wake states, Soc. Neurosci. Abstr., 19 (1993) 1815.

Nitz, D. and Siegel, J. GABA release in the locus coeruleus as a function of sleep/wake state, Soc. Neurosci. Abstr., 20 (1994) 291.

Pinault, D. A novel single-cell staining procedure performed in vivo under electrophysiological control: morpho-functional features of juxtacellularly labeled thalamic cells and other central neurons with biocytin or Neurobiotin, J. Neurosci. Meth., 65 (1996) 113–136.

Sakai, K. Central mechanisms of paradoxical sleep, Exp. Brain Res., (Suppl) 8 (1984) 3–18.

Sakai, K., Anatomical and physiological basis of paradoxical sleep. In D.J. McGinty et al. (Eds.) *Brain mechanisms of sleep. Raven* Press, New York, 1985, pp. 111–137.

Sakai, K. Physiological properties and afferent connections of the locus coeruleus and adjacent tegmental neurons involved in the generation of paradoxical sleep in the cat, Prog. Brain Res., 88 (1991) 31–45.

Sakai, K., El Mansari, M. and Jouvet M. Inhibition by carbachol microinjection of presumptive cholinergic PGO-on neurons in freely moving cats, Brain Res., 527 (1990) 213–223.

Sakai, K. and Koyama, Y. Are there cholinergic and non-cholinergic paradoxical sleep-on neurons in the pons? Neuroreport, 7 (1996) 2449–2453.

Sakai, K., Sastre, J.P., Kanamori, N. and Jouvet, M. State-specific neurons in the ponto-medullary reticular formation with special reference to the postural atonia during paradoxical sleep in the cat. In: O. Ponpeiano and C.A. Marsan (Eds.), *Brain Mechanisms and Perceptual Awareness*. Raven Press, New York, 1981, pp. 405–429.

Satoh, K. and Fibiger, H.C. Cholinergic neurons of the laterodorsal tegmental nucleus: efferent and afferent connections, J. Comp. Neurol., 253 (1986) 277–302.

Semba., K. and Fibiger, H.C. Afferent connections of the laterodorsal and the pedunculopontine tegmental nuclei in the rat: a retro- and antero-grade transport and immunohistochemical study, J. Comp. Neurol., 323 (1992) 387–410.

Sofroniew, M.V., Priestley, J.V., Consolazione, A., Eckenstein, F. and Cuello, A.C. Cholinergic projections from the midbrain and pons to the thalamus in the rat, identified by combined retrograde tracing and choline acetyltransferase immunohistochemistry, Brain Res., 329 (1985) 213–223.

Steriade, M., Datta, S., Paré, D., Oakson, G. and Curró Dossi, R. Neuronal activities in brain-stem cholinergic nuclei related to tonic activation processes in thalamocortical systems, J. Neurosci., 10 (1990) 2541–2559.

Steriade, M. and McCarley, R.W. Brainstem Control of Wakefulness and Sleep. Plenum Press, New York, 1990.

Vanni-mercier, G., Sakai, K., Lin, J.S. and Jouvet, M. Mapping of cholinoceptive brainstem structures responsible for the generation of paradoxical sleep in the cat, Archs. Ital. Biol., 127 (1989) 133–164.

Vincent, S.R., Satoh, K., Aramstrong, D.M. and Fibiger, H.C. NADPH-diaphorase: a selective histochemical marker for the cholinergic neurons of the pontine reticular formation, Neurosci. Lett., 43 (1983) 31–36.

Williams, J.A. and Reiner, P.B. Noradrenaline hyperpolarizes identified rat mesopontine cholinergic neurons in vitro, J. Neurosci., 13 (1993) 3887–3883.

Yamamoto, K., Mamelak, A.N., Quattrochi, J.J. and Hobson, J.A. A. Cholinoceptive desynchronized sleep induction zone in the anterodorsal pontine tegmentum: locus of the sensitive region, Neuroscience, 39 (1990) 279–293.

Rapid Eye Movement Sleep
B.N. Mallick and S. Inoué (Eds)

15. Hormones and REM Sleep

Ferenc Obál Jr.[1] and James M. Krueger[2]

[1]Department of Physiology, Albert Szent-Györgyi Medical University, Dóm tér 10, 6720 Szeged, Hungary

[2]Department of Veterinary and Comparative Anatomy, Pharmacology and Physiology, Washington State University, Pullman, WA 99164, USA

SUMMARY

Two aspects of the interactions between hormones and rapid eye-movement sleep (REMS) are reviewed: changes in hormone secretions during REMS and effects of hormones on REMS. *1. Hormone secretions during REMS.* The hormone, which has the best-documented relationship with the sleep cycles, is renin. Plasma renin activity (PRA) increases during non-REMS (NREMS) and decreases during REMS in both humans and rats. The variations in renin secretion may be determined by circulatory changes specific to NREMS and REMS. Secretions of most of the pituitary hormones tend to be suppressed during REMS; a decreased carotid blood flow is a possible explanation for this phenomenon. The previously proposed sleep-associated rhythm in prolactin (PRL) secretion remains controversial. PRL release seems to correlate with the intensity of NREMS, and therefore, REMS has little significance in regulating the secretory bursts of PRL. *2. Effects of hormones on REMS.* Several members of the somatotropic system exert modest REMS-promoting activity. It is suggested that growth hormone (GH) acts on REMS via metabolic actions in the brain, and this effect varies with the age and species. Stimulation of REMS by GH-releasing hormone (GHRH) is an indirect effect mediated via GH and/or somatostatin. The nature of the somatostatin-elicited enhancements in REMS is not clear; they might be mediated by neurotransmitter-like actions. One proopiomelanocortin derivative, corticotropin-like intermediate lobe peptide (CLIP), has been implicated in REMS regulation. CLIP, however, acts as a neurotransmitter not as a hormone. The hormone that has a well-documented REMS-promoting activity is PRL. Both acute and chronic rises in plasma PRL concentrations are associated with enhancements in REMS in rats. PRL may contribute to stress-induced increases in REMS. Systemic PRL in normal concentrations, however, has little impact on REMS. Under normal conditions, hypothalamic PRL-containing neurons might be more important than pituitary PRL for the regulation of REMS.

Key Words: renin, prolactin, growth hormone, somatostatin, GHRH, CLIP.

There are two aspects of the interactions between hormones and REM sleep (REMS): (1) REMS may modulate the secretions of various hormones; and

(2) hormones may influence REMS. As discussed below, the specificity of the changes is a major concern in both cases. The definition of hormones is also controversial. In contrast to neurotransmitters, autocrines, and paracrines, which act locally, hormones produced by endocrine cells and neurocrines released by nerve cells are carried by the blood stream to their action sites. There are, however, considerable overlaps among these substances. It is important to be aware of the fact that many hormones are also synthesized in nerve cells, and they function as neurotransmitters, or neuromodulators in the central nervous system. When a sleep effect of a hormone is demonstrated, it is often not clear whether the effect is a true hormonal action or if the administration of a hormone merely mimics a neurotransmitter/neuromodulator-like function of that particular substance.

REMS-associated Variations in Hormone Secretions

The concentration of a hormone in the plasma is determined by two factors: secretion and elimination. Sleep-dependent variations in the process of elimination have not been reported. Hence, we assume that changes in plasma concentrations reflect secretion. Nevertheless, knowing the kinetics of elimination, the sleep-associated variations in the secretory rate can be and have been calculated for many hormones (Van Cauter et al., 1990). Determination of the secretory rates confirmed the previously proposed correlations between sleep and hormone secretions, and as a result of the enhanced accuracy, it helped to resolve several controversies. The plasma concentrations of many hormones vary in parallel with sleep-wake activity. Correlation between sleep and hormone secretions may result from a circadian control of both events without links between them. The secretion of some hormones, however, is in fact coupled to sleep. Growth hormone (GH) is perhaps the best-documented hormone exhibiting sleep-associated secretion (reviewed in Van Cauter, 1990). The major daily surge of GH is linked to deep NREMS soon after sleep onset, and the release of GH correlates with the depth of sleep (Holl et al., 1991). If secretion of a hormone is influenced by REMS then the concentration of this hormone may vary in correlation with the ultradian rhythm of the occurrences of REMS periods. REMS tends to coincide with a decline in the secretion of the majority of the hormones of the anterior pituitary in humans (Follenius et al., 1988). Nevertheless, there are only two hormones for which REMS-associated variations have been seriously considered, prolactin (PRL) and renin; both are depressed during REMS.

Prolactin

Circulating PRL is released from the anterior pituitary. PRL was one of the first hormones whose sleep-related secretion was demonstrated in humans (Sassin et al., 1972). Rises in plasma PRL concentrations were observed after sleep onset, but the highest concentrations occurred during the morning hours. The sleep-associated secretion of PRL was confirmed although it became clear that the regulation of PRL release has a circadian component independent of sleep (see Van Cauter, 1990 for a review). Soon after the discovery of the link between sleep and secretion of PRL, it was proposed that the variations in plasma PRL concentrations correlate with the NREMS-REMS periods: plasma PRL concentrations increase and decrease during NREMS and REMS, respectively (Parker et al., 1974). Subsequent experiments, however, failed to confirm a close association between variations in PRL concentrations and sleep cycles (Van Cauter et al., 1982). Recent observations suggest that PRL secretion during sleep correlates with the slow-wave activity in the EEG during NREMS (Spiegel et al., 1995). Consequently, it is not REMS that determines the plasma concentration of PRL. Nevertheless, the general depression of pituitary hormone secretions mentioned above may contribute to decreases in PRL concentrations during REMS. The REMS-associated decline in the releases of pituitary hormones might have a circulatory explanation: carotid blood flow decreases (and vertebral perfusion of the brain increases) during REMS (Azzaroni and Parmeggiani, 1993).

Renin—(angiotensin-aldosterone)

Renin in the plasma is secreted by the kidneys. The plasma renin plays a fundamental role in the regulation of sodium and water homeostasis and blood pressure. As a proteolytic enzyme, renin releases angiotensin I from a precursor protein, angiotensinogen, which is secreted by the liver. The decapeptide angiotensin I is hydrolyzed by an endothelial cell-associated enzyme, the angiotensin-converting enzyme, yielding an octapeptide, angiotensin II, which is the active product. The various biological activities of angiotensin include contraction of the vascular smooth muscle resulting in a rise in blood pressure, stimulation of secretion of aldosterone in the adrenal cortex, and thirst; the latter is mediated by angiotensin receptive structures in some periventricular organs. Brandenberger et al. (1994b) reported that plasma renin activity (PRA) increases in subjects sleeping at night, and the period of increased PRA is shifted to the daylight hours if the subjects are awake at night and sleep during the day. In addition, PRA exhibits ultradian variations that correlate with the sleep cycles (Brandenberger et al., 1988). Rising PRA is coupled with NREMS whereas REMS episodes coincide with declining PRA. Although the general pattern of these sleep state-associated fluctuations might suggest similarities between the changes in PRA and PRL described above, they are fundamentally different. In contrast to PRL, whose sleep state-dependent fluctuations are weak if they exist at all, the robust oscillations in PRA are tightly coupled to sleep cycles in humans. Since secretion of aldosterone is controlled predominantly by the renin-angiotensin system, it is not surprising that the plasma concentration of aldosterone also tends to show an ultradian rhythm: the changes in aldosterone follow those of PRA by a time lag of approximately 20 min (Brandenberger et al., 1989).

Sleep-associated secretions of hormones have rarely been studied in experimental animals because the sleep cycles in these animals are much shorter than those in humans, and the frequent blood samplings may seriously alter sleep and other variables, including the plasma concentrations of the hormones of interest. We tried to verify if the sleep-associated ultradian rhythm of PRA also exists in rats (Obál et al., 1994). After determining the sleep cycles in individual rats, blood was withdrawn intermittently at approximately 6-min intervals while sleep-wake activity was continuously recorded. Blood sampling (200 μl each sample with the lost blood volume replete by physiological saline) was started in REMS in one group of rats, thus the next sample was taken during NREMS, then REMS, and again from NREMS. In another group of rats, the collection of blood samples was started in NREMS, followed by REMS, NREMS and REMS. The results indicated significant, sleep state-dependent variations in PRA: PRA during NREMS was higher than during REMS (Fig. 1.).

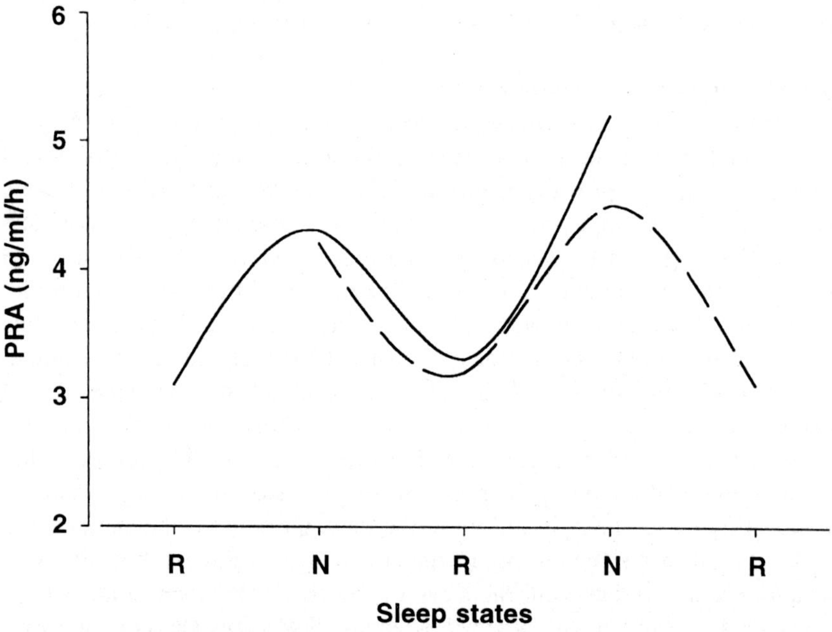

Fig. 1. **Sleep state-associated variations in plasma renin activity (PRA) in rats. Plasma samples were obtained at 6-min intervals from consecutive NREMS-REMS-NREMS-REMS periods in one group of rats (dashed line), and from consecutive REMS-NREMS-NREMS-REMS period in another group rats (continuous line). PRA is higher during NREMS episodes than during REMS episodes. In humans, where sleep episode lengths are longer, a similar relationship was described. N:NREMS, R:REMS. For details see Obál et al. (1994).**

The mechanisms and the biological significance of the sleep state-dependent changes in PRA are not known. Pharmacological manipulations remained ineffective to suppress the correlation between sleep cycles and PRA. In fact, the amplitude of the PRA oscillations is greatly enhanced in certain conditions such as heat exposure or low sodium diet (Brandenberger et al., 1989). In both humans (Sommers et al., 1993) and rats, REMS is characterized by large phasic oscillations and a tonic rise in blood pressure. It is possible, therefore, that inhibition of renin release occurring during REMS is a regulatory response to the increased blood pressure (Obál et al., 1994). Selective deprivation of REMS, however, fails to block rises in PRA during deep NREMS (stages 3 and 4) (Brandenberger et al., 1994a). The increases in PRA correlate with EEG slow-wave activity during NREMS in human subjects (Luthringer et al., 1995). These findings may suggest a common mechanism that controls both the depth of NREMS and the release of renin. To test this hypothesis, sleep and PRA were recorded after the administration of ritanserin, a 5-HT^2 receptor antagonist, that stimulates NREMS (Brandenberger et al., 1996). Ritanserin enhanced the duration of NREMS and increased EEG slow-wave activity but did not increase the oscillations of PRA (in fact slight decreases were observed that were attributed to an inhibition of serotonin-induced renin secretion by ritanserin). It seems, therefore, that mechanisms associated solely with either NREMS or REMS cannot explain the sleep state-associated ultradian rhythm in PRA. Regulatory responses to circulatory alteration in both sleep states (i.e., changes in blood pressure due to decreases in sympathetic vasomotor tone proportional to the depth of NREMS and increases in sympathetic activity during REMS), however, may result in ultradian variations in renin secretions. This proposal assumes that the importance of the renin-angiotensin system in the maintenance of blood pressure also varies with the state of sleep. Whatever is the cause of the striking variations in PRA during sleep, studying this phenomenon may provide valuable information on the physiology of both REMS and NREMS.

Modulation of REMS by Hormones

Hormones may inhibit or stimulate REMS. REMS suppression is a sensitive indicator of the discomfort caused by external or internal factors in both animals and humans. Hence, deviations in the plasma concentrations of many hormones might often be associated with decreases in REMS but these changes cannot be regarded as specific symptoms. It is, therefore, the analysis of hormone-induced enhancements in REMS that might serve as a tool to study the involvement of hormones in the regulations of REMS. Relatively few hormones exhibit REMS-promoting activities.

Hormones of the somatotropic axis

The somatotropic axis is one of the hormonal systems that can stimulate REMS (Fig. 2). Pituitary GH secretion is stimulated and inhibited by two hypothalamic neurocrines, the GH-releasing hormone (GHRH) and somatostatin, respectively. The effects of GH in the tissues (including the

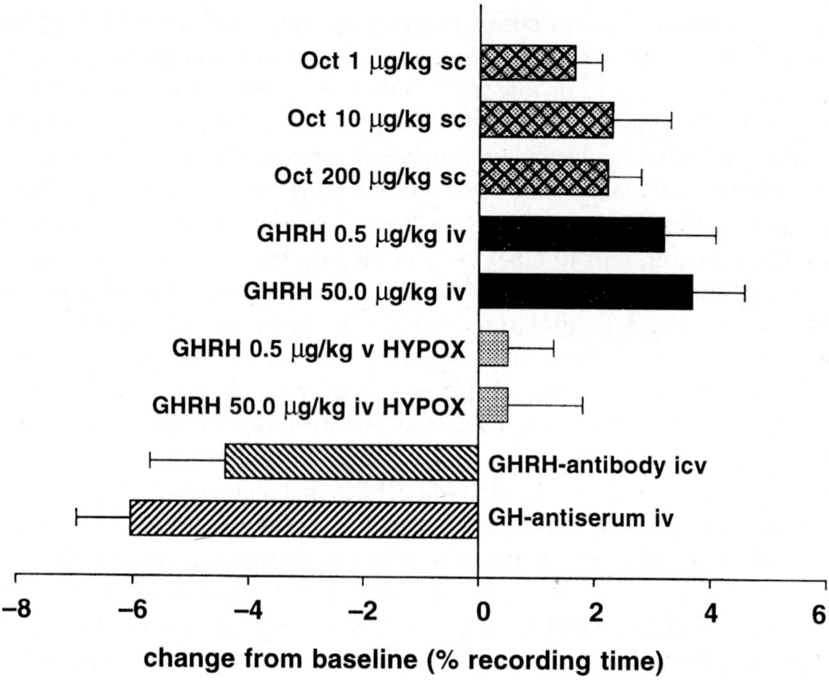

Fig. 2 **Summary of changes in REMS in response to various manipulations of the somatotropic system. Oct: significant stimulation of REMS after systemic injection of the long-acting somatostatin analog, octreotide, during the 12-h light period (Beranek et al., 1997). GHRH: significant promotion of REMS by systemic GHRH during a 6 h postinjection period (7–12 hours of the light period) in intact rats, and no effects on REMS in hypophysectomized rats (HYPOX). It is noted that the baseline REMS in HYPOX rats was less than in the intact rats (Obál et al., 1996). GHRH-antibody: significant decreases in REMS during the 12-h light period following intracerebroventricularly administered antibodies to GHRH (see reference in Krueger and Obál, 1993). GH-antiserum: significant suppression of REMS during the 12-h light period after immunoneutralization of GH (Obál et al., 1997a).**

brain) are partly direct and partly mediated by insulin-like growth factor-1 (IGF-1), which is a paracrine produced locally and is also a hormone secreted by the liver. The somatotropic system is equipped with a complicated internal regulation involving long, short, and ultrashort feedback loops resulting in inhibitory or stimulatory interactions between the individual components. GHRH is a well-documented sleep-promoting substance enhancing NREMS in rats, rabbits, and humans (reviewed in Krueger and Obál, 1993). Withdrawal of endogenous GHRH is followed by decreases in both NREMS and REMS. GHRH increases REMS. In correlation with the suppressed GH secretion, hypothalamic GHRH mRNA levels decrease and somatostatin mRNA levels tend to increase during selective REMS deprivation (Toppila et al., 1996). The REMS-promoting activity of GHRH is, however, inconsistent in rats and humans, and has a long latency in rabbits, suggesting that stimulation of REMS might be an indirect action of GHRH (reviewed in Krueger and Obál, 1993). GHRH continues to enhance NREMS in hypophysectomized rats whereas it fails to increase REMS (Obál et al., 1996); GH, therefore, might be involved in the mediation of REMS-promoting activity of GHRH. In fact, both GH and somatostatin have been implicated in the regulation of REMS. GH stimulates somatostatin; hence, GH may act on REMS via somatostatin.

A single, systemic injection of GH is followed by selective increases in REMS in rats, cats, and humans (reviewed in Krueger and Obál, 1993). Acute withdrawal of GH by means of immunoneutralization is associated with decreases in REMS, but these changes are not selective for REMS because NREMS is also suppressed (Obál et al., 1997a). Relationships between GH and sleep are controversial in conditions with chronic alterations in GH. Patients with an overproduction of GH exhibit normal REMS time, although EEG energy is enhanced (Aström and Trojaborg, 1992). Chronic GH deficiency is often accompanied by decreases in REMS when the subjects are children (Orr et al., 1977; Hayashi et al., 1992), but alterations in REMS are not observed in adults (Aström and Lindholm, 1990). Hypophysectomized rats have less REMS than do intac rats (Valatx et al., 1975). In contrast, transgenic mice that are deficient in GHRH-GH-IGF-1 have decreased NREMS but normal REMS (Zhang et al., 1996). Also, REMS is not altered in hypophysectomized cats (Sallanon et al., 1988). Finally, consistent changes in REMS were not observed after icv injections of IGF-1 in rats (unpublished observations). The effects of GH on REMS might be mediated by some GH-induced metabolic changes in the brain. This is supported by the fact that the increases in REMS occur several hours after systemic GH injection, and the changes in REMS are associated with enhanced intracerebral protein synthesis (Drucker-Colin et al., 1975). In the case of chronic alterations in GH secretions, compensatory mechanisms may maintain normal REMS. It is also likely, however, that the responsiveness to loss or excess of GH varies with the age and the species: human children and adult rats seem to be more sensitive than human adults. In conclusion, instead of a true REMS stimulating activity, GH may exert a permissive effect on REMS.

The notion that somatostatin is a physiological REMS-promoting substance stems from an impressive series of experiments published by Danguir and his collaborators. The authors reported that a systemic injection of a long-acting analog of somatostatin, octreotide (Danguir and De Saint-Hilaire-Kafi, 1989), or intracerebral infusion of somatostatin elicits selective enhancements in REMS whereas depletion of somatostatin by means of cysteamine causes selective suppressions in REMS in rats (Danguir, 1986). The somatostatin analog prevents the REMS-suppressive effect of desipramine or scopolamine whereas local infusion of an antiserum to somatostatin into the brainstem solitary tract blocks the REM enhancements that are elicited by cholinergic stimulation (Danguir and De Saint-Hilaire-Kafi, 1988). This finding was interpreted as an indication for an intracerebral, most likely a brainstem action site of somatostatin.

In our hands, systemic (Beranek et al.,1997) or intracerebral (unpublished observations) injections of a long-acting somatostatin analog, octreotide, did modestly promote REMS, but these effects were not selective and occurred after a several-hour latency. A prompt and strong inhibition of NREMS was observed that was often followed by significant rebound-like increases in NREMS. In fact, the immediate sleep-suppressive (both NREMS and REMS) activity of systemic or intracerebral somatostatin injection was one of the first behavioral responses to somatostatin described more than 30 yr ago (Havlicek et al., 1976). Systemic injection of somatostatin did not alter sleep in humans (Kupfer et al., 1992; Steiger et al., 1992), but the doses were low compared to those used in the rat. Impaired sleep was observed in elderly subjects after systemic injection of somatostatin (Frieboes et al., 1997a). Following repeated administrations of the somatostatin analog (twice a day), the REMS-promoting activity rapidly vanishes (Beranek et al., 1997). Dose-response relationships could not be detected in the effects of somatostatinergic stimulation on REMS. Finally, the cysteamine that was used by Danguir to suppress endogenous somatostatin also depletes PRL, another REMS-promoting substance. Although these findings may raise concerns about the nature of somatostatin's effects on sleep, it is clear that somatostatin promotes REMS in rats. The mechanism and the biological significance of this activity, however, remain to be determined. In addition to being a neurocrine involved in the regulation of GH secretion, somatostatin is also a widely distributed neurotransmitter in the brain. The somatostatinergic cells are interneurons in local circuits involved in the regulation of a number of functions. It is possible, therefore, that the REMS-promoting activity of somatostatin is independent from the role of this peptide in the somatotropic system. For example, somatostatin increases REMS via brainstem action sites, as suggested by Danguir and De Saint-Hilaire-Kafi (1988).

Corticotropin-like intermediate lobe hormone

The adrenocorticotrop hormone (ACTH) is the major product derived from the processing of the large prohormone, proopiomelanocortin (POMC), in the anterior pituitary. In the intermediate lobe and the arcuate nucleus, however, ACTH is cleaved into two biologically active peptides, the α-melanocyte stimulating hormone (α-MSH), and corticotropin-like intermediate lobe hormone (CLIP). Pituitary α-MSH has various functions in animals such as regulation of skin color and immune functions, but the quantity of α-MSH produced by the human pituitary is negligible. The biological significance of systemic CLIP, if any, are not known.

Intracerebroventricular administration of CLIP selectively promotes REMS (Chastrette et al., 1990). Microinjection of CLIP into the raphe also enhances REMS (El Kafi et al., 1994). The time spent in REMS increases in rats previously subjected to immobilization stress (Rampin et al., 1991). This procedure also increases POMC mRNA levels in the arcuate nucleus, and elicits enhancements in CLIP-like immunoreactivity in the dorsal raphe and the arcuate nucleus (Bonnet et al., 1997). The authors suggest that the POMC-containing neurons in the arcuate nucleus projecting to the dorsal raphe are involved in the mediation of the stress-induced increases in REMS. Neurons in the arcuate nucleus may also release CLIP into the third cerebral ventricle, and therefore, the cerebrospinal fluid is another source of CLIP for the raphe nuclei. Hence, the REMS-promoting activity of CLIP is a result of its neuromodulator or neurotransmitter role.

Prolactin

The anterior pituitary produces two hormones that belong to the somatomammotropic family: GH and PRL. The genes of these hormones have evolved from a common ancestral gene; there are significant structural homologies between GH and PRL, and they may crossreact on receptors. Human GH is known for its particularly strong lactogenic character. PRL is, however, unique for several reasons. It is involved in osmoregulation in fish. The osmoregulatory significance of PRL has been greatly reduced or lost in the course of phylogeny. There are indications that PRL plays a role in the development and function of skin accessories such as hair and glands (the mammary glands have evolved from sweat glands). Both GH and PRL have important immune regulatory functions; leukocytes may produce both hormones for local needs. Unlike IGF-1, which mediates many GH effects in the tissues, a peripheral target hormone has not been identified for PRL. The regulation of PRL secretion is also fundamentally different from the hypothalamic control of the other anterior pituitary hormones, which are stimulated and occasionally inhibited by phasically secreted releasing and inhibiting hormones. The hypothalamus tonically suppresses PRL secretion, predominantly via a non-peptidic neurocrine, dopamine. Attenuation in the dopaminergic tone results in PRL release. In addition, releasing hormones may contribute to the stimulation of PRL secretion; these neurocrines include vasoactive intestinal peptide (VIP), thyrotropin-releasing hormone (TRH), and oxytocin. The posterior pituitary also has been implicated in the simulation of PRL secretion.

Jouvet et al. (1986) noted that systemic administration of PRL increases REMS in pontine cats. Subsequently, enhancements in REMS were observed in both rabbits (Obál et al., 1989) and rats (Roky et al., 1993) in response to systemic injection of PRL. In rabbits, the time spent in REMS increased 3–6 h postinjection and NREMS was not altered. In rats, the effects of PRL on REMS were studied during both the light and the dark cycle, but REMS was enhanced only during the day. Stimulation of endogenous PRL secretion is also followed by increases in REMS. This was tested by means of systemic administration of VIP in rats (Obál et al., 1994). VIP promoted REMS. This effect was blocked when the rats were pretreated with an antiserum to PRL, thus confirming that the effects of systemic VIP on REMS were indeed mediated by PRL. Systemic VIP, in a dose that stimulated PRL release, enhanced REMS in humans, although NREMS also increased (Murck et al., 1996). A lower dose of VIP that did not elicit PRL release (in fact PRL decreased) failed to promote sleep.

REMS decreases during the light period in genetically hypoprolactinemic rats even though these rats spend more time in REMS at night (Valatx et al., 1990). We studied sleep in rats rendered hyperprolactinemic by means of pituitary grafts implanted under the capsule of the left kidney (Obál et al., 1997b). The grafts are devoid of hypothalamic control. Consequently, secretion of PRL is released from inhibition whereas the rest of the pituitary hormones are not produced due to the lack of stimulating hormones. Sleep-wake activity in the grafted rats was recorded between weeks 3 and 7 after surgery. REMS was consistently increased during the day throughout this period (Fig. 3). Sleep was not altered at night. The enhancements in REMS were particularly marked during the first portion of the light cycle: instead of the characteristic REMS peak in the afternoon, maximum REMS time was shifted to the morning hours in the hyperprolactinemic rats. Pituitary grafting also resulted in a minor, albeit statistically significant, increase in NREMS. A preliminary

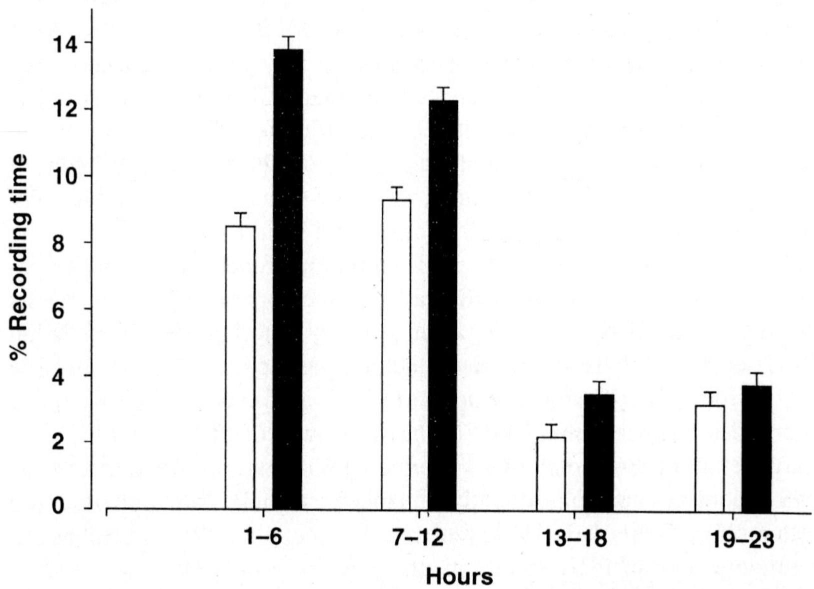

Fig. 3 **REMS in hyperprolactinemic Fisher 344 male rats (rats bearing pituitary transplants under the capsule of the left kidney, filled columns) and in sham-operated controls (open columns). Means REMS ± SE is provided for 6 h time blocks (the last time block refers to 5 h) in rats recorded between weeks 3 and 7 postinjection. REMS is significantly enhanced during the light period in the hyperprolactinemic rats. For details see Obál et al. (1997b).**

report has been published on sleep in hyperprolactinemic patients (Frieboes et al., 1997b): large increases in NREMS, particularly deep NREMS, were observed, whereas REMS did not change. To our knowledge, the effects of PRL on sleep have only been studied in male animals. There are significant differences in the quantity and the pattern of PRL secretions between males and females, and it is not clear how REMS in females reacts to PRL. Physiological hyperprolactinemia, however, occurs in females during pregnancy and lactation although the secretions of a number of hormones also are altered in these conditions. Increases in REMS were reported in both pseudopregnant (Zhang et al., 1995) and pregnant rats (Kimura et al., 1996); NREMS also increased in these conditions.

It may seem that PRL stimulates REMS, acting as a hormone. If blood-borne PRL is important for the regulation of REMS, then elimination of PRL should be followed by a suppression of REMS. Immunoneutralization of PRL, however, causes only very slight decreases in REMS while PRL drops below the detection limit (Obál et al., 1992). Therefore, in spite of the REMS-promoting activity of PRL, systemic PRL enhances REMS only when its concentration is high in the plasma, but it has little importance for REMS at normal secretions.

The action site for the REMS-promoting activity of PRL is in the brain. Intracerebral injection of PRL increases REMS, and an antiserum to PRL suppresses REMS when administered locally into the hypothalamus (Roky et al., 1994). PRL is also a neurotransmitter/neuromodulator in the central nervous system. The cell bodies of the PRL-containing neurons reside in the hypothalamus, and PRL-containing terminals and PRL receptors occur throughout the brain (reviewed in Roky et al., 1995). It is possible that under normal conditions, the central PRLergic neurons are more important for the regulation of REMS than is the pituitary PRL. There is a receptor-mediated, specific transport mechanism in the choroid plexus that carries PRL from the blood into the cerebrospinal fluid; from there PRL diffuses into the brain tissue. This way, blood-borne PRL reaches and may act on the same receptors as intracerebral PRL. The impact of circulating PRL might be significant only when high concentrations of PRL occur in the blood: PRL enhances its own transport into the brain via stimulating the production of the transport molecules in the choroid plexus (Mangurian et al., 1992). PRL secretion is greatly stimulated in stress. Systemic PRL therefore may contribute to the stress-induced enhancements in REMS.

The brain structures mediating PRL-induced increases in REMS have not been determined. PRL may act via modulating the diurnal rhythm of REMS (Roky et al., 1995). It is in fact intriguing that PRL only enhances REMS in the light period and hypoprolactinemia results in decreases and increases in REMS during the day and at night, respectively. Jouvet (1994) proposed that REMS is linked to an increased activity of pyruvate dehydrogenase resulting in an enhanced production of acetylcholine in the brainstem neurons that have a fundamental role in the generation of REMS. There is a possibility that PRL stimulates pyruvate dehydrogenase, although further experiments are required to test this hypothesis (Obál et al., 1997b).

PRL might be involved in or contribute to the mediation of the increases in REMS elicited by VIP. VIP is one of the best-documented substances with a REMS-promoting activity (Jimenez-Anguiano et al., 1993). Although VIP can be detected in the plasma, this VIP is likely to be of gastrointestinal origin and has not been implicated in sleep regulation, (i.e., VIP does not act on REMS as a hormone). VIP is a neurotransmitter/neuromodulator throughout the brain. Several lines of evidence indicate that VIP acting in the brainstem enhances REMS; i.e., promotion of REMS is a neurotransmitter-like function of VIP (Bourgin et al., 1997). VIP is, however, a hypothalamic-releasing hormone for PRL. VIP is also produced in the pituitary where it acts on PRL in an autocrine/paracrine fashion. In fact, as mentioned above, we used systemic injections of VIP to elicit PRL secretion (Obál et al., 1994). Selective deprivation of REMS causes depletion of VIP in the pituitary (Morin et al., 1992), suggesting excessive VIP release. Hypothalamic and pituitary VIP may stimulate pituitary PRL release during sleep deprivation, thus contributing to the increases in REMS after deprivation. VIP, however, may also be involved in the stimulation of intrahypothalamic PRL-containing neurons. This is indicated by the observation that intracerebral administration of VIP and pituitary adenylyl cyclase activating peptide (PACAP), in doses that enhance REMS, also elicit increases in hypothalamic PRL mRNA levels (Bredow et al., 1994).

Acknowledgments

This work was supported by ETT T-627-1996-04, OTKA T-16080 to F. Obál, and by NIH NS 27250 to J.M. Krueger.

References

Aström, C. and Lindholm, J. Growth hormone-deficient young adults have decreased deep sleep, Neuroendocrinology, 51 (1990) 82–84.

Aström, C. and Trojaborg, W. Effect of growth hormone on human sleep energy, Clin. Endocrinol., 36 (1992) 241–245.

Azzaroni, A. and Parmeggiani, P.L. Mechanisms underlying hypothalamic temperature changes during sleep in mammals, Brain Res., 1993 (632) 135–142.

Beranek, L., Obál, F., Jr., Taishi, P., Bodosi, B., Laczi, F. and Krueger, J.M. Changes in sleep after single and repeated injections of the long-acting somatostatin analog, octreotide, Am. J. Phsysiol., 273 (1997) R1484–R1491.

Bonnet, C., Léger, L., Baubet, V., Debilly, G. and Cespuglio, R. Influence of a 1 h immobilization stress on sleep states and corticotropin-like intermediate lobe peptide (CLIP or ACTH18–39, Ph-ACTH18–39) brain contents in the rat, Brain Res., 751 (1997) 54–63.

Bourgin, P., Lebrand, C., Escourrou, P., Gaultier, C., Franc, B., Jamon, M. and Adrien, J. Vasoactive intestinal polypeptide microinjections into the oral pontine tegmentum enhance rapid eye movement sleep in the rat, Neuroscience, 77 (1997) 351–360.

Brandenberger, G., Charifi, C., Muzet, A., Saini, J., Simon, C. and Follenius, M. Renin as a biological marker of the NREM-REM sleep cycle: effect of REM sleep suppression, J. Sleep Res., 3 (1994a) 30–35.

Brandenberger, G., Follenius, M., Di Nisi, J., Libert, J.P. and Simon, C. Amplification of nocturnal oscillations in PRA and aldosterone during continuous heat exposure, J. Appl. Physiol., 66 (3) (1989) 1280–1286.

Brandenberger, G., Follenius, M., Goichot, B., Saini, J., Spiegel, K., Ehrhart, J. and Simon, C. Twenty-four-hour profiles of plasma renin activity in relation to the sleep-wake cycle, J. Hypertension, 12 (1994b) 277–283.

Brandenberger, G., Follenius, M., Simon, C., Ehrhart, J. and Libert, J.P. Nocturnal oscillations in plasma renin activity and REM-NREM sleep cycles in humans: a common regulatory mechanism? Sleep, 11 (1988) 242–250.

Brandenberger, G., Luthringer, R., Muller, G., Gronfier, C., Schaltenbrand, N., Macher, J.P., Muzet, A. and Follenius, M. 5-HT2 receptors are partially involved in the relationship between renin release and delta relative power, J. Endocrinol. Invest., 19 (1996) 556–562.

Bredow, S., Kacsóh, B., Obál, F., Jr., Fang, J. and Krueger, J.M. Increase of prolactin mRNA in the rat hypothalamus after intracerebroventricular injection of VIP or PACAP, Brain Res., 660 (1994) 301–308.

Chastrette, N., Cespuglio, R. and Jouvet, M. Proopiomelanocortin (POMC)-derived peptides and sleep in the rat. Part-1, Hypnogenic properties of ACTH derivates, Neuropeptides, 15 (1990) 61–74.

Danguir, J. and De Saint-Hilaire-Kafi, S. Reversal of desipramine-induced suppression of paradoxical sleep by a long-acting somatostatin analogue (octreotide) in rats, Neurosci. Lett., 98 (1989) 154–158.

Danguir, J. and De Saint-Hilaire-Kafi, S. Somatostatin antiserum blocks carbachol-induced increases of paradoxical sleep in the rat, Brain Res. Bull., 20(1) (1988) 9–12.

Danguir, J. Intracerebroventricular infusion of somatostatin selectively increases paradoxical sleep in rats, Brain Res., 367 (1986) 26–30.

Drucker-Colin, R.R., Spanis, C.W., Hunyadi, J., Sassin, J.F. and McGaugh, J.L. Growth hormone effects on sleep and wakefulness in the rat, Neuroendocrinology, 18 (1975) 1–8.

El Kafi, B., Cespuglio, R., Leger, L., Marinesco, S. and Jouvet, M. Is the nucleus raphe dorsalis a target for the peptides possessing hypnogenic properties? Brain Res., 637 (1994) 211–221.

Follenius, M., Brandenberger, G., Simon, C. and Schlienger, J.L. REM sleep in human beings during decreased secretory activity of the anterior pituitary, Sleep, 11 (1988) 546–555.

Frieboes, R.M., Murck, H., Schier, T., Holsboer, F. and Steiger, A. Somatostatin impairs sleep in elderly human subjects, Neuropsychopharmacology, 16 (1997a) 339–345.

Frieboes, R.M., Stalla, G., Murck, H., Holsboer, F. and Steiger, A. Increased slow-wave-sleep in patients with prolactinoma, Psychoneuroendocrinology, 22 (Suppl. 2) (1997) S183.

Havlicek, V., Rezek, M. and Friesen, H. Somatostatin and thyrotropin releasing hormone: central effects on sleep and motor system, Pharmacol. Biochem. Behav., 4 (1976) 455–459.

Hayashi, M., Shimohira, M., Saisho, S., Shimozawa, K. and Iwakawa, Y. Sleep disturbance in children with growth hormone deficiency, Brain Dev., 14 (1992) 170–174.

Holl, R.W., Hartman, M.L., Veldhuis, J.D., Taylor, W.M. and Thorner M.O. Thirty-second sampling of plasma growth hormone in man: correlation with sleep stages, J. Clin. Endocrinol. Metab., 72 (1991) 854–861.

Jimenez-Anguiano, A., Baez-Saldana, A. and Drucker-Colin, R. Cerebrospinal fluid (CSF) extracted immediately after REM sleep deprivation prevents REM rebound and contains vasoactive intestinal peptide (VIP) Brain Res., 631 (1993) 345–348.

Jouvet, M. Paradoxical sleep mechanisms, Sleep, 17 (1994) S77–S83.

Jouvet, M., Buda, C., Cespuglio, R., Chastrette, N., Denoyer, M. Sallanon, M. and Sastre, J. P. Hypnogenic effects of some hypothalamo-pituitary peptides, Clin. Neuropharmacol., 9 (Suppl. 4) (1986) 465–467.

Kimura, M., Zhang, S.Q. and Inoué, S. Pregnancy-associated sleep changes in the rat, Am. J. Physiol., 271 (1996) R1063–R1069.

Krueger, J.M. and Obál, F., Jr., Growth hormone-releasing hormone and interleukin-1 in sleep regulation, FASEB J., 7 (1993) 645–652.

Kupfer, D.J., Jarrett, D.B. and Ehlers, C.L. The effect of SRIF on the EEG sleep of normal men, Psychoneuroendocrinology, 17 (1992) 37–43.

Luthringer, R., Brandenberger, G., Schaltenbrand, N., Muller, G., Spiegel, K., Macher, J.P., Muzet, A. and Follenius, M. Slow wave electroencephalographic activity parallels renin oscillations during sleep in humans, Electroenceph. Clin. Neurophysiol., 95 (1995) 318–322.

Mangurian, L.P., Walsh, R.J. and Posner, B.I. Prolactin enhancement of its own uptake at the choroid plexus, Endocrinology, 131 (1992) 698–702.

Morin, A., Deneroy, L. and Jouvet, M. Effect of paradoxical sleep deprivation on vasoactive intestinal peptide-like immunoreactivity in discrete brain areas and anterior pituitary of the rat. Brain Res. Bull., 28 (1992) 655–661.

Murck, H., Guldner, J., Colla-Muller, M., Frieboes, R.M., Schier, T., Wiedmann, K., Holsboer, F. and Steiger, A. VIP decelerates non-REM-REM cycles and modulates hormone secretion during sleep in men, Am. J. Physiol., 271 (1996) R905–R911.

Obál, F., Jr., Beranek, L. and Brandenberger G. Sleep-associated variations in plasma renin activity and blood pressure in the rat, Neurosci. Lett., 179 (1994) 83–86.

Obál, F., Jr., Bodosi, B., Szilágyi, A., Kacsóh, B. and Krueger, J.M. Antiserum to growth hormone decreases sleep in the rat, Neuroendocrinology, 66 (1997a) 9–16.

Obál, F., Jr., Floyd, R., Kapás, L., Bodosi. B. and Krueger, J.M. Effects of systemic GHRH on sleep in intact and hypophysectomized rats, Am. J. Physiol., 270 (1996) E230–E237.

Obál, F., Jr., Kacsóh, B., Alföldi, P., Payne, L., Markovic, O., Grosvenor, C. and Krueger, J.M. Antiserum to prolactin decreases rapid eye movement sleep (REM sleep) in the male rat, Physiol. Behav., 52 (1992) 1063–1968.

Obál, F., Jr., Kacsóh, B., Bredow, S., Guha-Thakurta, N. and Krueger, J.M. Sleep in rats rendered chronically hyperprolactinemic with anterior pituitary grafts, Brain Res., 755 (1997b) 130–136.

Obál, F., Jr., Opp, M., Cady, A.A., Johannsen, L. and Krueger, J.M. Prolactin, vasoactive Intestinal peptide, and peptide histidine methionine elicit selective increases in REM sleep in rabbits, Brain Res., 490 (1989) 292–300.

Obál, F., Jr., Payne, L., Kacsóh, B., Opp, M., Kapás, L., Grosvenor, C.E. and Krueger, J.M. Involvement of prolactin in the REM sleep-promoting activity of systemic vasoactive intestinal peptide (VIP), Brain Res., 645 (1994) 143–149.

Orr, W.C., Vogel, G.W., Stahl, M.L., Griffiths, W.J. and Seely, J.R. Sleep patterns in growth hormone deficient children and age-matched controls: developmental considerations, Neuroendocrinology, 24 (1977) 347–352.

Parker, D.C., Rossman, L.G. and Vanderlaan, E.F. Relation of sleep-entrained human prolactin release to REM-NonREM cycles, J. Clin. Endocrinol. Metab., 38 (1974) 646–651.

Rampin, C., Cespuglio, R., Chastrette, N. and Jouvet, M. Immobilization stress induces a paradoxical sleep rebound in rat, Neurosci. Lett., 126 (1991) 113–118.

Roky, R., Obál, F., Jr., Valatx, J.L., Bredow, S., Fang, J., Pagano, L.P. and Krueger, J.M. Prolactin and rapid eye movements sleep regulation, Sleep, 18 (1995) 536–542.

Roky, R., Valatx, J.L. and Jouvet, M. Effect of prolactin on the sleep-wake cycle in the rat. Neurosci. Lett., 156 (1993) 117–120.

Roky, R., Valatx, J.L., Paut-Pagano, L. and Jouvet, M. Hypothalamic injection of prolactin or its antibody alters the rat sleep-wake cycle, Physiol. Behav., 55 (1994) 1015–1019.

Sallanon, M., Buda, C., Puymartin, M., Denoyer, M., Kitahama, K. and Jouvet, M. Hypophysectomy does not disturb the sleep-waking cycle in the cat, Neurosci. Lett., 88 (1988) 173–178.

Sassin, J.F., Frantz, A.G., Weitzman, E. and Kapen, S. Human prolactin: 24-hour pattern with increased release during sleep, Science, 177 (1972) 1205–1207.

Sommers, V.K., Gyken, M.E., Mark, A.L. and Abboud, F.M. Sympathetic-nerve activity during sleep in normal subjects, N. Eng. J. Med., 328 (1993) 303–307.

Spiegel, K., Luthringer, R., Follenius, M., Schaltenbrand, N., Macher, J.P., Muzet, A. and Brandenberger, G. Temporal relationship between prolactin secretion and slow-wave sleep electroencephalographic activity during sleep, Sleep, 18 (1995) 543–548.

Steiger, A., Guldner, J., Hemmeter, U., Rothe, B., Wiedemann, K. and Holsboer, F. Effects of growth hormone-releasing hormone and somatostatin on sleep EEG and nocturnal hormone secretion in male controls, Neuroendocrinology, 56 (1992) 566–573.

Toppila, J., Asikainen, M., Alanko, L., Turek, F.W., Stenberg, D. and Porkka-Heiskanen, T. The effect of REM sleep deprivation on somatostatin and growth hormone-releasing hormone gene expression in the rat hypothalamus, J. Sleep Res., 5 (1966) 115–122.

Valatx, J.L., Chouvet, G. and Jouvet, M. Sleep-waking cycle of the hypophysectomized rat, Prog. Brain Res., 42 (1975) 115–120.

Valatx, J.L., Roky, R. and Paut-Pagano, L. Prolactin and sleep regulation. In: J. Horne (Ed.), *Sleep'90*, Pontenagel Press, Bochum, Germany, 1990, pp. 346–348.

Van Cauter, E., Desir, D., Refetoff, S., Spire, J.P., Noel, P., L'Hermite, M., Robyn, C. and Copinschi, G. The relationship between episodic variations of plasma prolactin and REM-Non-REM cyclicity is an artifact, J. Clin. Endocrinol. Metab., 54 (1982) 70–75.

Van Cauter, E. Diurnal and ultradian rhythms in human endocrine function: a minireview, Hormone Res., 34 (1990) 45–53.

Zhag, J., Obál, F., Jr., Fang, J., Collins, B.J. and Krueger, J.M. Non-rapid eye movement sleep is suppressed in transgenic mice with a deficiency in the somatotropic system, Neurosci. Lett., 220 (1996) 97–100.

Zhang, S.Q., Kimura, M. and Inoué, S. Sleep patterns in cyclic and pseudopregnant rats, Neurosci. Lett., 193 (1995) 125–128.

Rapid Eye Movement Sleep
B.N. Mallick and S. Inoué (Eds)

16. Endogenous Sleep Substances and REM Sleep

Shojiro Inoué, Kazuki Honda, Mayumi Kimura and Shi-Qing Zhang

Institute for Medical and Dental Engineering, Tokyo Medical and Dental University, Kanda-Surugadai 2-3-10, Chiyoda-ku, Tokyo 101-0062, Japan

SUMMARY

Among a number of endogenous sleep substances listed up to date, some have an entity of modulating both non-rapid-eye-movement (NREM) sleep and rapid-eye-movement (REM) sleep, whereas some show that of modulating selectively either NREM sleep or REM sleep. The former group includes delta-sleep-inducing peptide, uridine, oxidized glutathione and many others. The latter group includes corticotropin-like intermediate lobe peptide, prolactin, somatostatin and others. This article deal with an overview of those substances which may positively or negatively regulate REM sleep.

Key Words: corticotropin-like intermediate lobe peptide, delta-sleep-inducing peptide, oxidized glutathione, nitric oxide, prolactin, somatostatin, uridine.

Introduction

Since the classical experimental studies by Ishimori in 1909 and Legendre and Piéron in 1913, a number of candidate substances have been proposed for an endogenous sleep substance (Inoué, 1989). Interestingly, most of them were at first noted for their capability of promoting non-rapid-eye-movement (NREM) sleep but not for rapid-eye-movement (REM) sleep. A typical example is delta-sleep-inducing peptide (DSIP), which was named after its capability of promoting delta sleep, i.e. a deep stage of NREM sleep or slow wave sleep in rabbits (Schoenenberger et al., 1977). Soon later, however, it was proven that DSIP can enhance not only delta sleep but also REM sleep or paradoxical sleep in cats (Polc et al.,1978).

Today, it is established that the majority of the proposed sleep substances may modulate both states of sleep under special conditions such as recipient animal species, dosage, route of administration and time of day, and that there are only a limited number of substances which selectively promote either NREM sleep or REM sleep (Table 1). This article will deal with an overview of the REM sleep-modulatory substances.

Substances Which Affect NREM/REM Sleep

As summarized in Table 1, there are many substances which may simultaneously affect both NREM and REM sleep. Two examples, uridine and oxidized glutathione (GSSG), from our laboratory are illusrated in Figs. 1 and 2. A 10-h nocturnal intracerebroventricular (icv) infusion of each substance causes significant increases in the 12 h nocturnal amount of the two different states of sleep in freely behaving male rats. It should be noted that the uridine-induced enhancement of nocturnal sleep is based on an increase in the number of episodes of both NREM and REM sleep. In contrast, the GSSG-induced enhancement of nocturnal sleep is differentially based on a prolongation of NREM sleep episodes and an increase in the number of REM sleep episodes.

In general, however, it is not clear whether the listed NREM/REM sleep-promoting substances differentially enhance the two states of sleep. Also, it is not clear whether the enhancement of REM sleep depends on that of the preceding NREM sleep which is promoted by an actually selective NREM-enhancing substance. Since it is well known that special regions of the brain are differentially related to the generation of either NREM sleep or REM sleep and sleep substances appear to modulate neuronal and metabolic activities of the neural system as a messenger of the humoral system (Inoué, 1989), the site of action should be specified for each sleep substance. In this respect, the sleep-regulatory activity of uridine and muramyl dipeptide (MDP), a simplest synthetic form of muramyl peptides (MPs), is suggestive.

We further studied the sleep-modulatory activity of uridine alone or in combination with MDP by comparing before and after the localized bilateral lesions of the lateral preoptic area (LPOA) (Kimura-Takeuchi and Inoué, 1993). In rats with the intact LPOA, a 10-h nocturnal icv infusion of uridine alone slightly enhances both NREM and REM sleep. The administration of MDP alone markedly enhances NREM sleep and slightly suppresses REM sleep. Combined infusion of the two substances causes a marked increase in both NREM and REM sleep. The localized LPOA lesions (Fig. 3) per se exerts no remarkable alteration in the amount of nocturnal sleep (Fig. 4). Interestingly, in these lesioned rats uridine entirely fails to enhance NREM sleep but not REM sleep. In contrast, MDP still elicits the same effects on NREM sleep and REM sleep as observed under the LPOA-intact condition. Combined infusion of uridine and MDP induces the effects on NREM sleep and REM sleep similar to those induced under the LPOA-intact condition.

All these results suggest that the LPOA is required by uridine as an essential site for the enhancement of NREM sleep but not for that of REM sleep. Thus the somnogenic effects of uridine on NREM sleep and REM sleep are separately

Table 1 List of Candidate Sleep Substances Proposed up to 1997 and their Effects on Sleep (Modified after Inoué, 1995a)

Category Name (Abbreviation)	Site of presence	Effects on		Recipients	Route of administration	References
		NREM sleep	REM sleep			
Amines						
Piperidine	brain	+	+	cat	mi	#
Melatonin	pineal	+	±	cat	mi	#
		+	±	chicken	ip	#
		+	±	human	oral	#
Amino acid						
Tryptophan (Trp)		±	±	cat	oral	#
		+	+	rat	ip	#
		+/–	+/–	human	oral	#
Benzodiazepine						
Nordiazepam	brain					#
Cytokines and related substances						
Double stranded ribonucleic acid (dsRNA)	virus	+	–	rabbit	icv, iv	#
Granulocyte colony stimulating factor (G-CSF)		+	+	human	iv	Mullington et al. (1994)
Granulocyte-macrophage colony stimulating factor (GM-CSF)		+	+	rat	icv	Kimura et al. (1997)
Interferon α (INF)		+	–	rabbit	icv, iv	#
Interleukin 1 (IL-1)	blood	+	–	rabbit	icv	#
Lipid A	bacteria	+	–	rabbit	iv	#
Lipopolysaccharide (LPS)	bacteria	+	–	rabbit	iv	#
Muramyl peptides (MPs)	urine	+	–	rabbit	icv	#

Category Name (Abbreviation)	Site of presence	Effects on		Recipients	Route of administration	References
		NREM sleep	REM sleep			
Tumor necrosis factor (TNF)		+	–	rabbit	icv, iv	#
Fatty acid amide						
Oleamide (CIS-9,10)	CSF	+	–	rat	icv	Cravatt et al. (1997)*
Neuropeptides and protein hormones						
Acidic fibroblast growth factor (aFGF)	brain, CSF	+	–	rat	icv	#
Adrenocorticotropic hormone (ACTH)	pituitary	–	–	human	iv	#
(=corticotropin)		–	–	rat	icv	Chastrette et al. (1988)*
Arginine vasotocin (AVT)	pineal	+	–	cat	icv	Pavel and Adrien (1989)
		±	+	human	sc	#
Bombesin	brain, gut	+	±	rat	ip	#
Cholecystokinin (CCK)	brain, gut	+	+	rat	icv, ip	Riou et al. (1982a)
Corticotropin-like intermediate lobe peptide (CLIP) (=$ACTH_{18\text{-}39}$)	brain, pituitary	–	+	rat	icv	Chastrette and Cespuglio (1985)
Corticotropin-releasing hormone (CRH)	brain	–	±	rat	icv	Ehlers et al. (1986)
Delta-sleep-inducing peptide (DSIP)	blood, brain	+	±	rabbit	icv	Schoenenberger et al.
	etc.	+	+	cat	iv	(1977), Polc et al. (1978)
		+	±	human	iv	#
		+	±	rat	iv	#
Desacetyl-a-melanophore-stimulating hormone (des-α-MSH) (=$ACTH_{1\text{-}13}$)	brain, pituitary	–	+	rat	icv	Chastrette and Cespuglio (1985)
Growth hormone (GH)	brain, pituitary	±	+	cat	ip	#
		–	+	human	sc	#
		±	+	rat	ip	#
Growth hormone releasing hormone (GRH)	brain	+	±	rat	icv	Ehlers et al. (1986)

(Contd)

Category Name (Abbreviation)	Site of presence	Effects on NREM sleep	Effects on REM sleep	Recipients	Route of administration	References
Insulin	brain, pancreas	+	−	rat	icv	#
Oxidized glutathione (GSSG)	brain	+	+	rat	icv, md	#
		+	−	rabbit	icv	#
Prolactin (PRL)	pituitary	−	+	cat	sc	Jouvet et al. (1986)
		−	+	rabbit	sc	Obál et al. (1989)
Reduced glutathione (GSH)	brain	+	±	rat	icv	#
Somatostatin (SRIF)	brain	−	+	rat	icv	Danguir (1986)
Thyrotropin-releasing hormone (TRH)	brain	−	−	rat	icv, ip	#
Vasoactive intestinal polypeptide (VIP)	brain, gut	+	+	rat	icv	Riou et al. (1982b)
Nitrogen oxide						
Nitric oxide (NO)		+	−	rat	icv, ip	Kapás et al. (1994a)
		±	−	rabbit	icv, iv	Kapás et al. (1994b)
		−	−	rat	md	Matsumura (see text) Inoué (1989)*
Nucleosides						
Adenosine		+	±	rat	icv	#
Cytidine		±	+	rat	icv	#
Deoxycytidine		±	+	rat	icv	#
Deoxyuridine		+	±	rat	icv	#
Uridine	brain	+	+	rat	icv	#
		+	+	rat	ip	#
		+	+	mouse	ip	#
Organic bromine compound						
1-Methyl heptyl-γ-bromoacetoacetate (γ-Br, MHBAA)	CSF	−	+	cat	iv	Torii et al. (1973)

Category Name (Abbreviation)	Site of presence	Effects on NREM sleep	Effects on REM sleep	Recipients	Route of administration	References
Prostaglandins (PGs)						
Prostaglandin D_2 (PGD_2)	brain	+	+	rat	mi	#
		+	+	rhesus	icv	#
Prostaglandin E_2 (PGE_2)	brain	–	–	rat	mi	#
		–	–	rhesus	icv	#
Steroids						
$3\alpha,5\alpha$-Tetrahydrodeoxycorticosterone (THDC)	adrenal	+	–	rat	ip	#
Pregnenolone	brain	+	–	human	oral	#

CSF: cerebrospinal fluid, icv: intracerebroventricular injection or infusion, ip: intraperitoneal injection, iv: intravenous injection or infusion, md: microdialysis perfusion, mi: microinjection into brain, sc: subcutaneous injection, +: positively sleep-inducing, –: no effect or sleep-inhibiting or sometimes positively wakefulness-inducing, ±: mildly sleep-inducing or sleep-inducig under special conditions, *: review article, #: For the reference, see Inoué (1995a).

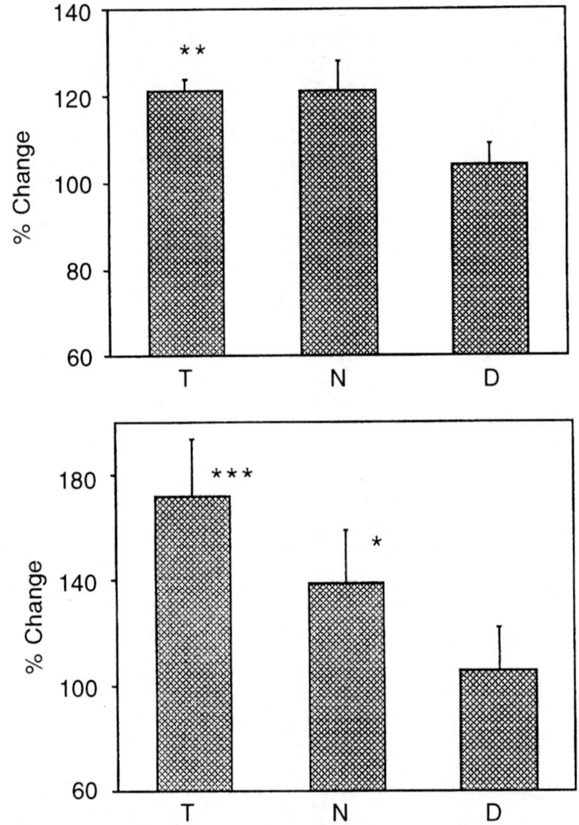

Fig. 1 **Sleep-enhacing effect of 10 pmol uridine icv infused for a nocturnal 10-h period in otherwise continuously icv saline-infused freely moving rats. Changes in NREM sleep (top) and REM sleep (bottom) in the environmental 12-h dark period as compared to the baseline (100%). D: duration of sleep episodes, N: number of sleep episodes, T: total amount of sleep. *p < 0.05, **p < 0.01 Significantly different from the corresponding baseline values (Student's t-test).**

generated. Presumably the REM sleep-enhancing activity of uridine may be evoked by the lower region of the brainstem, since it is reported: (1) the action of cholinergic influence on the induction of REM sleep is mediated by γ-amino-butylic acid (GABA)-ergic input to the locus coeruleus (Alam and Mallick, 1992); (2) the cessation of firing of serotonergic dorsal raphe neurons caused by GABA is a key controlling event of REM sleep (Nitz and Siegel, 1997); and (3) uridine may facilitate GABAergic neurotransmission via the uridine receptor coupled with the GABA-benzodiazepine-adenosine receptor-chloride channel complex (Yamamoto et al., 1993). The above findings also indicate that the LPOA is not essentially required by MDP for its NREM sleep-enhancing and REM sleep-suppressing activities.

The somnogenic activity of DSIP, which was extensively studied during 1980's in different laboratories, is largely variable and depends on the dosage (as evidenced by a typical bell-shaped dose-response relation), route and

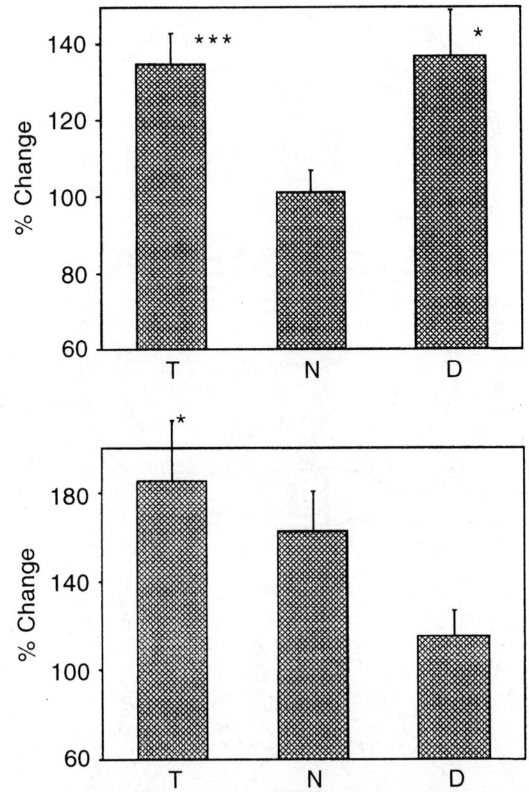

Fig. 2 Sleep-enhacing effect of 25 nmol oxidized glutathione (GSSG) icv infused for a nocturnal 10-h period in otherwise continuously icv saline-infused freely moving rats. Changes in NREM sleep (top) and REM sleep (bottom) in the environmental 12-h dark period as compared to the baseline (100%). For abbreviations, see the legend of Figure 1. *$p < 0.005$, ***$p < 0.001$ Significantly different from the corresponding baseline values (Student's t-test).

timing of administration, species and physiological or pathological situations of recipients, and even sources of synthetic products (Inoué, 1995b). In this respect, we found that synthetic analogs such as [D-Tyr^1]DSIP, [D-Ala^2]DSIP, [D-Ala^2, Try^6]DSIP are NREM/REM sleep-enhancing, whereas [D-Ala^2]DSIP$_{1-6}$ is REM sleep-enhancing (Inoué et al., 1988, 1997). The somnogenic activity of these analogs is differential and appears immediately or slowly after administration, indicating that there is structural dependency. It is suggested that the sleep-promoting action of DSIP analogs is mediated through interactions with other neuropeptides such as growth hormone-releasing hormone (GRH), luteinizing hormone-releasing hormone, α-melanophore-stimulating hormone, corticotropin-like intermediate lobe peptide (CLIP) and others (Kovalzon, 1994).

In addition to the established NREM sleep-suppressive activity of corticotropin-releasing hormone (CRH), Marrosu et al. (1990) reported that

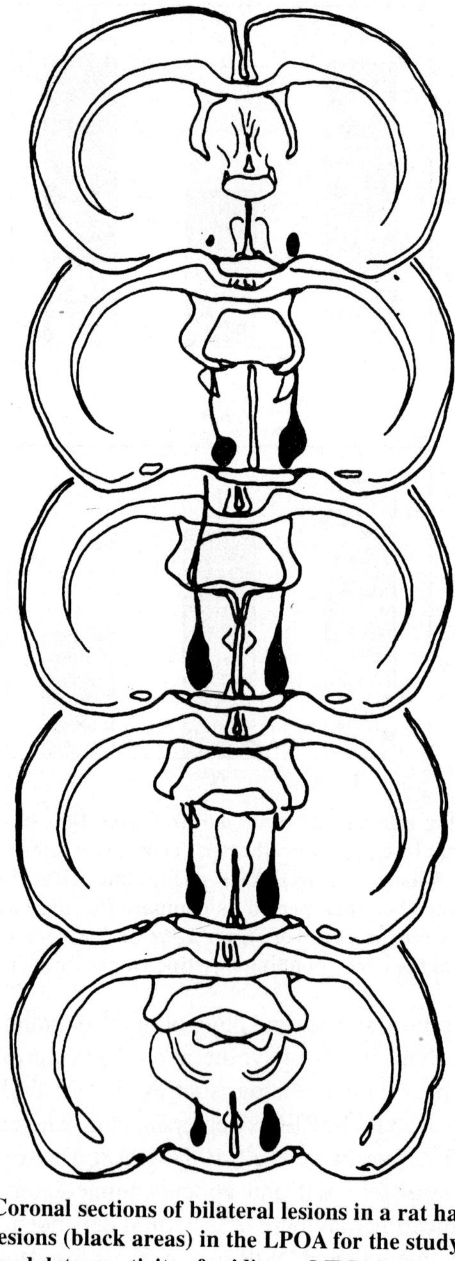

Fig. 3 **Coronal sections of bilateral lesions in a rat having bilateral lesions (black areas) in the LPOA for the study on the sleep-modulatory activity of uridine or MDP alone or in combination (see text).**

REM sleep rebound is enhanced by icv administration of CRH in REM sleep-deprived rats.

REM Sleep-suppressing Substances

It is well known that cytokines such as interferon α, interleukin 1 (IL-1) and tumor necrosis factor as well as the related substances such as bacterial and viral products may enhance selectively NREM sleep and selectively suppress REM sleep in rabbits, simultaneously bringing about a striking fever (Krueger, 1990). Exceptionally, however, Mullington et al. (1994) reported that granulocyte-macrophage colony stimulating factor, a neutrophil-stimulating cytokine, induces an enhancement of NREM sleep and also REM density in humans. In addition, we recently found that icv administration of granulocyte-macrophage colony stimulating factor, the novel somnogenic cytokine, induces an enhancement of not only NREM sleep but also REM sleep in rats (Kimura et al., 1997).

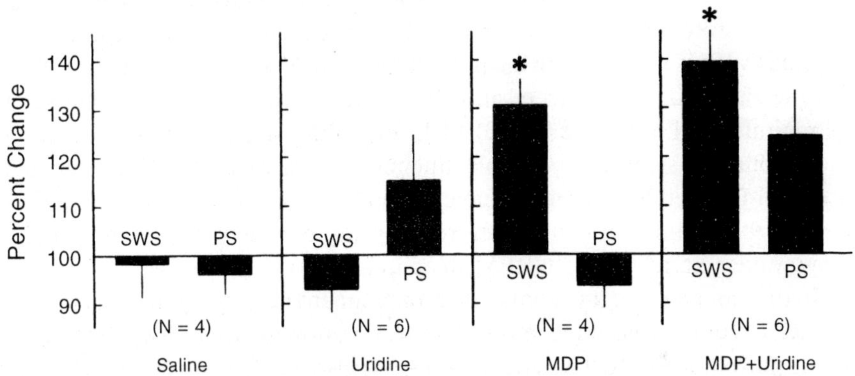

Fig. 4 Changes in the 12-h nocturnal amount of NREM sleep (SWS) and REM sleep (PS) in LPOA-lesioned rats, receiving a nocturnal 10-h icv infusion of saline, uridine (10 pmol/10 h), MDP (2 nmol/10 h) and uridine in combination with MDP. *$p < 0.05$. Significantly different from the corresponding baseline values (Student's t-test).

Substances Which Selectively Affect REM Sleep

REM Sleep-Enhancing Substances

As early as in 1973, Torii and his coworkers demonstrated that methylheptyl-γ-bromoacetoacetate (γ-Br) selectively induces REM sleep in encéphale isolé cats. Later, they observed the same effect in normal cats (Torii et al.,1990). They also found that plasma levels of this substance fluctuate in phase with REM sleep in cats and humans (Okudaira et al., 1985).

Up to 1988, the effect of growth hormone (GH), prolactin (PRL), GRH and somatostatin (SRIF) on sleep was not clearly elucidated (Inoué, 1989, Roky et al., 1995). However, it was noted that the enhancement of REM sleep by GH is sometimes attributed to the secondary effects of SRIF via a feedback control mechanism for GH secretion, and that GH secretion is stimulated with the co-existence of DSIP. Danguir (1986) reported that icv infusion of SRIF selectively increases REM sleep in rats. Hypothalamic SRIF levels as measured by *in vivo* voltammetry in unrestrained rats exhibit a decrease in sleep and an increase in wakefulness (Danguir and De Saint-Hilaire Kafi,

1989). A possible involvement of serotonin (Imeri et al., 1990) and GABA (Nagaki et al., 1990) is suggested with respect to the relation of SRIF with the neurotransmitter systems. A definite circadian rhythmicity is known for SRIF-like immunoreactivity in the suprachiasmatic nucleus (Fukuhara et al., 1993). Recently, Toppila et al. (1997) reported that total sleep deprivation increases SRIF- and GRH-messenger RNA in the rat hypothalamus, suggesting a close relationship between NREM/REM sleep rebound and an elevated secretory activity of SRIF as well as GRH.

Obál et al. (1989) reported that subcutaneous administration of PRL selectively increases REM sleep in rabbits. They further demonstrated that antiserum to PRL selectively suppresses REM sleep in rats (Obál et al., 1992), and that a close relationship between PRL and systemic vasoactive intestinal peptide (VIP) is suggested in the promotion of REM sleep (Obál et al., 1994). On the other hand, Valatx et al. (1990) suggested a correlation between extrapituitary (i.e. hypothalamic) PRL and REM sleep through estimating the amount of sleep in hypoprolactinemic mutant rats, hypohysectomized estradiol-induced hyperprolactinemic rats and normal rats. In contrast, Spiegel et al. (1995) demonstrated a close relationship between PRL secretion and slow wave activity during NREM sleep in humans.

In this connection, it is known that, in pregnant rats, dynamic daily surges of PRL occur to cause luteal progesterone secretion for continuing pregnancy until approximately 10 days after mating. It is also well known that, at the early phase of pregnancy in rats, a dopaminergic agonist bromocriptine specifically inhibits PRL release from the pituitary without affecting luteinizing hormone secretion. We observed a marked enhancement of nocturnal sleep in pregnant (Kimura et al., 1996) and pseudopregnant (Zhang et al., 1995) rats and suggested a possible involvement of PRL in the modulation of sleep. We further found that the pregnancy-associated sleep enhancement is blocked by repeated administration of bromocriptine (Zhang et al., 1996). Thus, it seems likely that PRL is responsible for the regulation of sleep during pregnancy.

Up to 1988, several authors reported a sleep-enhancing activity of VIP, for NREM sleep and REM sleep in rats and for only REM sleep in cats (see Inoué, 1989). Mirmiran et al. (1988) observed that continuous icv infusion of a competitive VIP receptor antagonist selectively suppresses diurnal REM sleep in rats. On the basis of this finding they concluded that VIP participates in the generation and maintenance of REM sleep. Obál et al. (1989) reported that VIP and its related peptide histidine methionine selectively enhances REM sleep (and wakefulness at a high dose) in rabbits. They suggested that these effects are mediated via release of PRL. Pacho-Cano et al. (1990) demonstrated that VIP restores both NREM and REM sleep in basal forebrain-lesioned cats. The investigators emphasized that VIP exerts a powerful modulatory influence on REM sleep. This notion was substantiated by later studies in cats (Jiménez-Anguiano et al., 1993) and rats (Bredow et al., 1994), but not in humans (Steiger et al., 1994).

For the sleep-modulatory activities of proopiomelanocortin components, Chastrette et al. (1988) demonstrated that adrenocorticotropic hormone (ACTH) or corticotropin, desacetyl-α-melanophore-stimulating hormone ($ACTH_{1-13}$) and CLIP ($ACTH_{18-39}$) differentially enhance wakefulness, NREM sleep and REM sleep, respecively, in rats. Later, Léger et al. (1994) reported that CLIP-containing neurons are synaptically connected with serotonergic neurons in the brainstem and may regulate REM sleep via modulating the activity of these neurons.

Goldstein (1984) speculated that arginine vasotocin (AVT) plays a role in inhibiting the maturation of the brain by enhancing REM sleep in developing cats.

REM Sleep-suppressing Substances

According to Fehm and Born (1990), ACTH reduces REM sleep in humans. Guldner et al. (1990) observed that an $ACTH_{4-9}$ fragment suppresses deep stages of NREM sleep and REM sleep, and increases wakefulness in humans. Born et al. (1992) demonstrated that vasopressin (AVP) regulates human sleep by reducing REM sleep.

In contrast to the assumption of Goldstein (see above), Pavel et al. (1977) reported that AVT induces an enhancement of NREM sleep and a suppression of REM sleep in cats. This report was followed by several investigations on the sleep-modulatory activity of AVT in humans and animals resulting in controversial conclusions (see Inoué, 1989). Later, Pavel and Adrien (1989) demonstrated that AVT dose-dependently increases quiet sleep (precursor of NREM sleep) and suppresses active sleep (precursor of REM sleep) in new born cats. Interestingly, the icv administration of AVT antiserum contrarily increases active sleep.

Up to date, the effect of nitric oxide (NO) on sleep is controversial. Inhibition of brain NO production by icv or intraperitonel administration of an NO synthesis inhibitor decreases both NREM and REM sleep in rats (Kápas et al., 1994a) and in rabbits (Kápas et al., 1994b), while stimulation of brain NO production by NO donors increases NREM sleep but sometimes suppresses REM sleep (Kápas and Krueger, 1996). On the contrary, inhibition of NO production by microdialytic administration of the same NO synthesis inhibitor in the ventromedial thalamus increases NREM sleep and, dramatically, REM sleep in rats (H. Matsumura, personal communication).

Conclusion

Evolutionally, REM sleep appears more ancient than NREM sleep, manifesting various primitive and physiologically less well organized characteristics. This may explain the fact that a limited number of endogenous substances are selectively involved in the physiological regulation of REM sleep. In other words, the majority of sleep substances which may later have become involved in the regulation of sleep during the course of the evolution of NREM sleep. Hence, they could not acquire the entity of modulating selectively REM sleep. Never the less, some substances exhibit a selective REM sleep-enhancing or -suppressive activity. It must be an interesting and important task to unveil the specific role played by these substances in the humoral regulation of sleep with special reference to the brain mechanisms.

Acknowledgment

This study was supported by a Special Coordination Fund for Promoting Science and Technology from the Science and Technology Agency, Prime Minister's Office to SI.

References

Alam, M.N. and Mallick, B.N. Involvement of GABA in pontine area acetylcholine induced rapid eye movement sleep, Abstract, Int. Conf. Sleep-Wakefulness, New Delhi, 1992, p. 39.

Born, J., Kellner, C., Uthgennannt, D., Kern, W. and Fehm, H.L. Vasopressin regulates human sleep by reducing rapid-eye-movement sleep, Am. J. Physiol., 262 (1992) R295–R300.

Bredow, S., Kacsóh, B., Obál, F., Jr. and Krueger, J.M. VIP and PACAP induce PRL mRNA expression in rat hypothalamus, J. Sleep Res., 3 (Suppl. 1) (1994) 32.

Cravatt, B.F., Prospero-Garcia, O., Siuzdak, G., Giang, D.K., Henriksen, S.J., Gilula, N.B., Boger, D.L. and Lerner, R.A. The chemistry and biology of 9(z)-octadecenamide (oleamide): a sleep-inducing lipid. In: O. Hayaishi and S. Inoué (Eds.), *Sleep and Sleep Disorders: From Molecule to Behavior*, Academic Press, Tokyo, 1997, pp. 29–45.

Chastrette, N., and Cespuglio, R. Effects hypnogénes de la des-acetyl-α-MSH et du CLIP (ACTH 18-39) chez le rat, C.R. Acad. Sci., 301 (1985) 527–530.

Chastrette, N., Clement, H.W., Prevautel, H. and Cespuglio, R. Proopiomelanocortin components: differential sleep-waking regulation? In: S. Inoué and D. Schneider-Helmert (Eds.), *Sleep Peptides: Basic and Clinical Approaches*, Japan Scientific Societies Press, Tokyo/Springer-Verlag, Berlin, 1988, pp. 27–52.

Danguir, J., Instracerebroventricular infusion of somatostatin selectively increases paradoxical sleep in rats, Brain Res., 367 (1986) 26–30.

Danguir, J. and De Saint-Hilaire Kafi, S. Reversal of desipramine-induced suppression of paradoxical sleep by a long-acting somatostatin analogue (octreotide) in rats, Neurosci. Lett., 98 (1989) 154–158.

Ehlers, C.l., Reed, T.K. and Henriksen, S.J. Effects of corticotropin-releasing factor and growth hormone-releasing factor on sleep and activity in rats, Neuroendocrinology, 42 (1986) 467–474.

Fukuhara, C., Shinohara, K., Tominaga, K., Otori, Y. and Inouye, S.T. Endogenous circadian rhythmicity of somatostatin like-immunoreactivity in the rat suprachiasmatic nucleus, Brain Res., 606 (1993) 28–35.

Fehm, H.L. and Born, J. Interactions between the hypothalamus-pituitary-adrenal (HPA) system and sleep in humans. In: J. Horne (Ed.), *Sleep '90*, Pontenagel Press, Bochum, 1990, pp. 379–383.

Goldstein, R. The involvement of arginine vasotocin in the maturation of the kitten brain, Peptides, 5 (1984) 25–28.

Guldner, J., Rothe, B., Knisatschek, H., Steiger, A. and Holsboer, F. Influence of the ACTH-(4-9)-fragment HOE 427 on sleep-EEG and sleep associated secretion of cortisol and growth hormone. In: J. Horne (Ed.), *Sleep '90*, Pontenagel Press, Bochum, 1990, pp. 48–50.

Inoué, S., Biology of Sleep Substances. CRC Press, Boca Raton, Florida, 1989.

Inoué, S., Challenges to the ancient concept of humoral sleep regulation. In: S.Y. Liu and S. Inoué (Eds.), *Sleep—Ancient and Modern*, Shanghai Scientific and Technological Literature Publishing House, Shanghai/Asian Sleep Research Society, Tokyo, 1995a, pp. 55–88.

Inoué, S., Pharmacology of the CNS Peptides. In: A. Kales (Ed.), *Handbook of Experimental Pharmacology: Pharmacology of Sleep*, Springer-Verlag, New York, 1995b, pp. 243–277.

Inoué, S., Kimura, M., Honda, K. and Komoda, Y. Sleep peptides: general and comparative aspects. In: S. Inoué and D. Schneider-Helmert (Eds.), *Sleep Peptides: Basic and Clinical Approaches*, Japan Scientific Societies Press, Tokyo/Springer-Verlag, Berlin, 1988, pp. 1–26.

Inoué, S., Kimura, M., Kovalzon, V.M., Mikhaleva, L.L. and Prudchenko, L.A. Structure-function interrelationships in the hypnogenic effects of the delta sleep-inducing peptide (DSIP) and its analogs, Neurochmistry, 14 (1997) 168–174.

Ishimori, K. True cause of sleep: a hypnogenic substance as evidence in the brain of sleep-deprived animals, Tokyo Igakkai Zasshi, 23 (1909) 429–457.

Jiménez-Anguiano, A., Báez-Saldana, A. and Drucker-Colín, R. Cerebrospinal fluid (CSF) extracted immediately after REM sleep deprivation prevents REM rebound and contains vasoactive intestinal peptide (VIP), Brain Res., 631 (1993) 345–348.

Jouvet, M., Buda, C., Cespuglio, R., Chastrette, N., Denoyer, M., Sallanon, M. and Sastre, J.P. Hypnogenic effects of some hypothalamo-pituitary peptides, Clin. Neuropharmacol., 9 (Suppl. 4) (1986) 465–467.

Kápas, L., Fang, J. and Krueger, J.M. Inhibition of nitric oxide synthesis inhibits rat sleep, Brain Res., 664 (1994a) 189–196.

Kápas, L. and Krueger, J.M. Nitric oxide donors SIN-1 and SNAP promote nonrapid-eye-movement sleep in rats, Brain Res. Bull., 41 (1996) 293–298.

Kápas, L. Shibata, M., Kimura, M. and Krueger, J.M. Inhibition of nitric oxide synthesis suppresses sleep in rabbits, Am. J. Physiol., 266 (1994b) R151–R157.

Kimura, M., MaCann, S.M., Marks, G.A., Shaffery, J.P., Kodama, T., Zhang, S.Q. and Inoué, S. A novel somnogenic cytokine: granulocyte-macrophage colony stimulating factor (GM-CSF), Soc. Neurosci., Abstr., 23 (1997) 792.

Kimura, M., Zhang, S.Q. and Inoué S. Pregnancy-associated sleep changes in the rat, Am. J. Physiol., 271 (1996) R1063–R1069.

Kimura-Takeuchi, M. and Inoué S. Lateral preoptic lesions void slow-wave sleep enhanced by uridine but not by muramyl dipeptide in rats, Neurosci. Lett., 157 (1993) 17–20.

Kovalzon, V.M. DSIP: sleep peptide or unknown hypothalamic hormone?, J. Evol. Biochem. Physiol., 30 (1994) 195–199.

Krueger, J.M. Somnogenic activity of immune response modifiers, Trends Pharmacol. Sci., 11 (1990) 122–126.

Legendre, R. and Piéron, H. Recherches sur le besoin de sommeil consecutif à une veille prolongée, Z. Allgem. Physiol., 14(1913) 235–262.

Léger, L., Bonnet, C., Cespuglio, R. and Jouvet, M. Immunocytochemical study of the CLIP/ACTH-immunoreactive nerve fibres in the dorsal raphe nucleus of the rat, Neurosci. Lett. 174 (1994) 137–140.

Marrosu, F., Gessa, G.L., Giagheddu, M. and Fratta, W. Corticotropin-releasing factor (CRF) increases paradoxical sleep (PS) rebound in PS-deprived rats, Brain Res., 515 (1990) 315–318.

Mirmiran, M., Kruisbrink, J., Bos, N.P.A., Van der Werf, D. and Boer, G.J. Decrease of rapid-eye-movement sleep in the light by intraventricular application of a VIP-antagonist in the rat, Brain Res., 458 (1988) 192–194.

Mullington, J., Korth, C., Holsboer, F. and Pollmächer, T. Granulocyte colony stimulating factor enhances nocturnal sleep in men, J. Sleep Res., 3 (Suppl. 1) (1994) 175.

Nagaki, S., Kato, N., Minatogawa, Y. and Higuchi, T. Effects of anticonvulsants and gamma-aminobutyric acid (GABA)-mimetic drugs on immunoreactive somatostatin and GABA contents in the rat brain, Life Sci., 46 (1990) 1587–1595.

Nitz, D. and Siegel, J. GABA release in the dorsal raphe nucleus: role in the control of REM sleep, Am. J. Physiol., 273 (1997) R451–R455.

Obál, F., Jr., Kacsóh, B., Alföldi, P., Payne, L., Markovic, O., Grovenor, C. and Krueger, J.M. Antiserum to prolactin decreases rapid eye movement sleep (REM sleep) in the male rat, Physiol. Behav., 52 (1992) 1063–1068.

Obál, F., Jr., Opp, M., Cady, A.B., Johannsen, L. and Krueger, J.M. Prolactin, vasoactive intestinal peptide, and peptide histidine methionine elicit selective increases in REM sleep in rabbits, Brain Res., 490 (1989) 292–300.

Obál, F., Jr., Payne, L., Kacsoh, B., Opp, M., Kapás, L., Grosvenor, C.E. and Krueger, J.M. Involvement of prolactin in the REM sleep-promoting activity of systemic vasoactive intestinal peptide (VIP), Brain Res., 645 (1994) 143–149.

Okudaira, N., Koike, H., Inubushi, S., Abe, A., Yamane, M., Yanagisawa, I. and Torii, S. Organic bromine substance: an endogenous REM sleep modulator? In: S. Inoué and A.A. Borbély (Eds.), *Endogenous Sleep Substances and Sleep Regulation*, Japan Scientific Societies Press, Tokyo/VNU Science Press, Utrecht, 1985, pp. 237–248.

Pacheco-Cano, M.T., García-Hernández, F., Prospéro-García, O. and Drucker-Colín, R. Vasoactive instestinal polypeptide induces REM recovery in insomniac forebrain lesioned cats, Sleep, 13 (1990) 297–303.

Pavel, S. and Adrien, J. Vasotocin increases quiet sleep and suppresses active sleep in newborn cats. Opposite effects after vasotocin immunoneutralization, Brain Res. Bull., 23 (1989) 463–466.

Pavel, S., Psatta, D. and Goldstein, R. Slow-wave sleep induced in cats by extremely small amounts of synthetic and pineal vasotocin injected into the third ventricle of the brain, Brain Res. Bull., 2 (1977) 251–254.

Polc, P., Schneeberger, J. and Haefely, W. Effect of the delta sleep-inducing peptide (DSIP) on the sleep-wakefulness cycle of cats, Neurosci. Lett., 9 (1978) 33–36.

Riou, F., Cespuglio, R. and Jouvet, M. Endogenous peptides and sleep in the rat: II Peptides without significant effect on the sleep-waking cycle, Neuropeptides, 2 (1982a) 255–264.

Riou, F., Cespuglio, R. and Jouvet, M. Endogenous peptides and sleep in the rat: III The hypnogenic properties of vasoactive intestinal polypeptide, Neuropeptides, 2 (1982b) 265–277.

Roky, R., Obál, F., Jr., Valatx, J.L., Bredow, S., Fang, J., Pagano, L.P. and Krueger, J.M. Prolactin and rapid eye movement sleep regulation, Sleep, 18 (1995) 536–542.

Schoenenberger, G.A., Maier, P.F., Tobler, H.J. and Monnier, M. A naturally occurring delta-EEG enhancing nonapeptide in rabbits. X. Final isolation, characterization and activity test, Pflügers Arch., 369 (1977) 99–109.

Spiegel, K., Luthringer, R., Follenius, M., Schaltenbrand, N., Macher, J.P., Muzet, A. and Brandenberger, G. Temporal relationship between prolactin secretion and slow-wave electroencephalic activity during sleep, Sleep, 18 (1995) 543–548.

Steiger, A., Schier, T., Colla-Müller, M., Frieboes, R., Guldner, J., Wiedermann, K. and Holsboer, F. Dose-dependent effects of vasoactive intestinal polypeptide (VIP) on sleep EEG and prolactin release in man, J. Sleep Res., 3 (Suppl. 1) (1994) 240.

Toppila, J., Alanko, L., Asikainen, M., Tobler, I., Stenberg, D. and Porkka-Heiskanen, T.

Sleep deprivation increases somatostatin and growth hormone-releasing hormone messenger RNA in the rat hypothalamus, J. Sleep Res., 6 (1997) 171–178.

Torii, S., Mitsumori, K., Inubushi, S. and Yanagisawa, I. The REM sleep-inducing action of a bromine substance: an endogenous REM sleep modulator?, Psychopharmacologia, 29 (1973) 65–75.

Torii, S., Okudaira, N., Inubushi, S., Koike, H., Hagino, K. and Kudo, K. Enhancement of REM sleep by gamma-Br microinjection into the pontine brainstem. In: S. Inoué and J.M. Krueger (Eds.), *Endogenous Sleep Factors,* SPB Academic Publishing bv, Hague, 1990, pp. 155–163.

Valatx, J.L., Roky, R. and Paut-Pagano, L. Prolactin and sleep regulation. In: J. Horne (Ed.), *Sleep '90*, Pontenagel Press, Bochum, 1990, pp. 346–348.

Yamamoto, I., Kimura, T., Watanabe, K., Kondo, S. and Ho, I.K. Receptors of uridine and its derivatives, Sleep Res., 22 (1993) 512.

Zhang, S.Q., Kimura, M. and Inoué, S. Sleep patterns in cyclic and pseudopregnant rats, Neurosci. Lett., 193 (1995) 125–128.

Zhang, S.Q., Kimura, M. and Inoué, S. Bromocriptine-induced blockade of pregnancy affects sleep patterns in rats, Neuroimmunomodulation, 3 (1996) 219–226.

Rapid Eye Movement Sleep
B.N. Mallick and S. Inoué (Eds)

17. Brain Energy Production and Sleep Occurrence

Raymond Cespuglio, Hélène Faradji-Prevautel and Larissa Netchiporouk

Claude Bernard University, INSERM Unit 480, 8 avenue Rockefeller
F-69373 Lyon Cedex 08

SUMMARY

Regarding the mechanisms of energy production, glucose is generally accepted as the main fuel for the synthesis of ATP in endotherms. In anaerobic conditions and throughout glycolysis, glucose is first metabolized into pyruvate. At the end of this step there is a positive balance of 2 moles of ATP. Still in anaerobic conditions, pyruvate can be subjected to lactic fermentation which lasts into lactate. In aerobic conditions and throughout the Kreb's cycle and the respiratory chain, however, pyruvate can contribute to the production of 34 additive moles of ATP. The degradation of one mole of glucose thus produces 36 moles of ATP. The general yield is 40% and the 60% of the potential energy remaining is dissipated in heat.

Cold-blooded ectothermic species (fishes, amphibians, reptiles) are submitted to ambient variations of the temperature and possess a low rate of energy production. Warm-blooded endothermic species (birds, mammals), however, are capable of maintaining their internal temperature constant. Their homeothermic status is first paid in terms of a permanent need of energy and this is illustrated: by muscles shivering, which can compensate the heat loss in allowing the heat productive hydrolysis of ATP; by hibernation, an extreme case of heterothermy, resulting from a brutal decrease in the possibilities of energy transaction.

In ectotherms, while a clear rest-activity cycle is reported, the existence of sleep remains controversial. In endotherms, however, wakefulness, slow wave sleep (SWS) and paradoxical sleep (PS) are clearly evidenced. SWS appears as a state saving energy while during PS the energy consumption is as high as or higher than during wakefulness.

It is noteworthy that in endotherms, already challenged by the energy production necessary to maintain their temperature constant, phylogeny allowed the emergence of PS, a state also requiring energy to occur.

Key Words: ectotherms, endotherms, energy, glucose, ATP, aerobic, anaerobic, sleep.

Introduction

With respect to sleep function, a restorative role of sleep has been one among the first hypotheses suggested (Adams and Oswald, 1977; Adams, 1980). Such a role appears, indeed, quite natural since intuitively it is easy to feel that a recovery occurs during sleep. On the basis of experimental evidence, it can now be further specified that sleep restoration concerns more the brain than the body since quiet waking is sufficient for resting the body and particularly the motor systems, but insufficient for satisfying the need of sleep (Horne, 1985; Benigton and Heller, 1995). The question of how sleep is restorative, immediately arises. According to the brain studies conducted in this respect, it appears clearly that restorative processes are concerned with energy storage and production in which glucose and glycogen play a determinant part (Benington and Heller, 1995).

In the present paper, we have first recalled the basic biochemical pathways through which an availability of the energy is rendered possible. The importance of the energy supply for homeothermic mammals has then been considered. Finally, the specificity of each sleep and waking states towards the energetic aspect has been analyzed on the basis of the data now available.

Biochemical Pathways for Energy Production

The energy necessary for the functioning of cellular elements comes mainly from ATP (adenosine triphosphate) which is a biochemical form of energy immediately available. The basic problem lies in the ATP renewal which can be solved only through the supply of energetic substrata represented, in the case of eukaryotic cells, by organic nutrients. The molecules constituting these nutrients represent a potential stock of energy whose degradation allows the production of the energy necessary for the synthesis of ATP. While the cellular elements are capable to metabolize compounds like sugars, proteins, amino acids and lipids, it must be noted that among sugars, glucose is the main energetic compound at the basis of ATP production (see in: Lehninger, 1977; Alberts et al., 1986; Siesjö, 1978).

Glycolysis

Regarding glucose ($C_6H_{12}O_6$), whatever the conditions of oxygen supply (aerobic or anaerobic), the first metabolic step defined as glycolysis lasts in the production of two moles of pyruvate (CH_3-CO-COOH) per mole of glucose. Throughout this biochemical process, the glucose dehydrogenation achieved in several successive steps (about 9) ensures a transfer of protons and electrons to a transporter (e.g. NAD^+/NADH, nicotinamide adenine dinucleotide, oxidised and reduced form). At this step, the energetic balance sheet is at two moles of ATP produced per mole of glucose metabolized. Through a minor extent, other simple sugars like mannose and fructose may also gain entry into glycolysis. Again through a minor extent, glucose can also be oxidized to pentose phosphate in the soluble cytoplasm by a sequence of enzymes and the electron transporter is then NADPH (NADH phosphorylated) (see in: Lehninger, 1977; Alberts et al., 1986; Siesjö, 1978). Finally, regarding brain

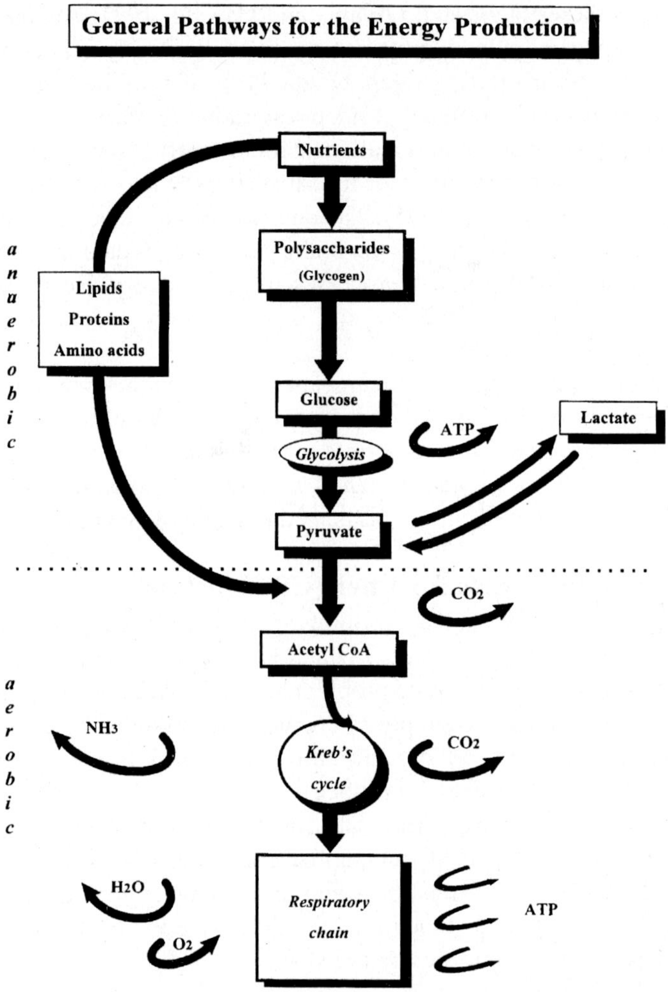

Fig. 1 Glucose, derived directly from nutrients or polysaccharides (glycogen), is the main source for the synthesis of ATP (adenosine triphosphate) in endotherms. In anaerobic conditions and throughout glycolysis, it is metabolized into pyruvate. At this step there is a positive balance of 2 moles of ATP. Still in anaerobic conditions, pyruvate can be subjected to lactic fermentation to produce lactate. In aerobic conditions, throughout the Kreb's cycle and the respiratory chain, pyruvate is totally oxidized into CO_2 and H_2O and contribute to the production of 34 additive moles of ATP. The total degradation of one mole of glucose produces 36 moles of ATP and the general energetic yield is close to 40%, the 60% of the potential energy remaining is dissipated in heat. Lipids, proteins and amino acids can also be involved in the production of ATP. They enter in the pathway of the phosphorylative oxidation in the form of acetyl-coenzyme A (Acetyl CoA). The general metabolism of the nutrients thus outlast into 2 main issues—the production of ATP and the production of heat. ATP—adenosine triphosphate; O_2—oxygen; NH_3—amine radical; H_2O—water; CO_2—carbon.

glycolytic processes it must be specified that they take place mainly in glial cells and particularly in astrocytes. In such elements, glucose is also stored in glycogen in order to respond quickly to the energy demand of the neuronal activity (Tsacopoulos and Magistretti, 1996).

Lactic fermentation

In anaerobic conditions, pyruvate produced by glycolysis is transformed into lactate (CH_3-CHOH-COOH). In these conditions pyruvate plays the role of a proton acceptor and the electrons produced by glycolysis are transferred to a transporter. While lactate is an end product that can be eliminated through blood and urine, it can, however, be again converted into pyruvate through the activity of lactate dehydrogenase. Pyruvate produced in this way is normally used through the oxidative phosphorylation to produce ATP. Thus, throughout glycolysis and lactic fermentation, both taking place in the cytosol, only two moles of ATP are produced per mole of glucose. In admitting that glucose is the main source for the energy production and according to the known energetic values of the moles of glucose and ATP, respectively 2840 k Joules and 30.5 k Joules, it is easy to calculate the yield in ATP for glycolysis and lactic fermentation, i.e. $30.5 \times 2 \times 100/2840 = 2\%$. In strict anaerobic conditions, the yield in ATP is thus very poor since 98% of the potential energy is lost and part of it is dissiapted in heat (see in: Lehninger; Alberts et al.; Siesjö).

Oxidative phosphorylation

It starts after glycolysis and takes place, under aerobic conditions, within the internal membrane of the mitochondria. Pyruvic acid, produced by glycolysis, is first transformed into acetyl-coenzyme A (CH_3-CO-coenzyme A: acetyl CoA). Together, with oxaloacetate ($HOOCCOCH_2COOH$) this radical contributes to the synthesis of citric acid which enters the Kreb's cycle where it is submitted to successive decarboxylations and dehydrogenations. Each decarboxylation allows the release of a carbon dioxide molecule while each dehydrogenation transfers a proton to a transporter which reduces it. Finally, through the redox processes of the proton motive respiratory chain, the reduced transporters are then again oxidized, thus regenerated, and the potential energy released converted by the ATPases into 34 moles of ATP. The resulting protons and electrons, together with oxygen, produce 6 moles of H_2O. It must also be further specified that lipids, proteins and amino acids can also contribute to the formation of acetyl CoA and thus participate as nutrients to the production of ATP through the oxidative phosphorylation (see in: Lehninger, 1977; Alberts et al., 1986; Siesjö, 1978). Finally, acetyl CoA, besides its major involvement in ATP production, might also cross the mitochondrial membrane to contribute together with choline to the synthesis of acetylcholine within the cytoplasm. This process, however, still remains controversial (Tucek, 1984; Lefresne et al, 1978).

Conclusion

The total degradation of one mole of glucose including glycolysis, Kreb's cycle and respiratory chain, produces 36 ATP. The general energetic yield is correct and close to 40%, the 60% remaining being dissipated in heat. The general metabolism thus lasts into two main issues: the production of ATP and the production of heat.

Importance of the Energy Supply for Homeothermy

As regards the dependence of the homeothermy towards energy, the phylogenetic evolution from pre-mammals (fishes, amphibians and reptiles) to birds and mammals point out the following important features:

—while pre-mammals are ectothermic cold-blooded species submitted to ambient variations of the temperature, mammals and birds become endothermic warm-blooded and are capable of maintaining their internal temperature constant in spite of environmental changes (Karmanova, 1982)

—the mammals homeothermy occurs together with the losing of the neurogenesis and several other aromorphoses (heart division into an arterial and a venous half, intensification of the respiration, etc.) necessary for the maintenance and the control of an intensified metabolism (Karmanova, 1982); in this respect, it is documented that for a constant body temperature, the resting metabolic rate of a reptile, as compared with mammals, is lower (Guppy et al., 1987).

—homeothermy is concomitant with the appearance of paradoxical sleep (PS) which, in addition to waking (W) and slow wave sleep (SWS), constitutes a new functional state for the brain.

Homeothermy-energy challenge

According to the above aspects and together with several other reports of the literature (see in: Karmanova, 1982; Heller et al., 1988; Berger and Phillips, 1993) it is first pointed out that the independence towards the environmental conditions allowed by homeothermy is paid in terms of a permanent need of energy. The metabolic rate of an organism, even at rest, is profoundly affected by whether or not it does work to maintain its body temperature, i.e. whether or not it is ectothermal or endothermal. Homeotherms adapt to environmental changes in temperature in modifying their metabolism through homeostatic and behavioral adaptation, i.e. changes in the blood flow, in the posture, induction of shivering or sudation and panting. In this respect, it can be noticed that in extremely low temperatures, shivering compensates for the heat loss in allowing in muscles the hydrolysis of ATP which is heat productive. On the contrary, when the temperature increases, sudation and/or panting facilitate the heat loss. All these adaptive aspects are under hypothalamic control and involve peripheral and central thermosensors (see in: Simon et al., 1986; Purves et al., 1992; Parmeggiani, 1985; Sawaya and Ingvar, 1989). It is also noteworthy that in homeothermic mammals, within brown adipose tissue, the mitochondrial respiration is naturally uncoupled from the synthesis of ATP and the energy produced is totally dissipated in heat. While the functional aspects attached to the brown adipose tissue are not yet fully understood, it is now, nethertheless, specified that brown adipose tissue is particularly useful for hibernating, mainly during their emerging process from the hibernation, as well as for animals or humans to adapt to cold conditions (Nicholls, 1976; Konings and Michels, 1980; Himms-Hagen, 1976). A clear challenge thus exists between the processes of energy production and homeothermy and this aspect is particularly well illustrated by hibernation which is not just a response to a lowering of temperature but an extreme case of temporal heterothermy resulting from a brutal decrease in the possibilities of energy transaction.

Paradoxical sleep

In addition to the above considerations, it is also remarkable that the evolution processes allowed the emergence of PS (Jouvet, (a), 1972; (b), 1976), a state of sleep rendering even more complex the situation of homeotherms as regards the energy production. The species related to this category of vertebrates must, indeed, supply the energy required for their homeothermic status but also that necessary for PS occurrence (Frank et al., 1987; Lydic et al., 1991), as will be discussed in the next subheading. It is thus clear that such a difficult challenge has not been chosen through the phylogenetic evolution only on the basis of inconveniences but certainly because PS is a particular state during which determinant functional processes do occur.

Conclusion

According to the above considerations it appears clearly that the pathways for energy and heat production are crucial for the exitence of homeotherms. A better understanding of the functions attached to PS would be easier in considering through phylogeny the precise conditions in which this state of sleep originates.

Specificity of the Sleep States Towards the Energetic Aspect

Ectotherms

Referring to the conventional sleep criteria established in homeothermic species, it appears still difficult to accept a clear existence of sleep in ectothermic species. While rest-activity periods are correctly distinguished in fishes and amphibians, the studies related to the electrical activity of their brain (EEG) do not report marked differences between rest or activity. The EEG in reptiles, however, is described with some features of synchronization, deceleration of the heart rate and reduction in the muscle tone. Correlations with the rest-activity cycle are, nevertheless, controversial in some reports. Whatever the difficulties in this respect, it is reported that in ectotherms forced activity is followed by longer resting periods. This may indicate the existence of

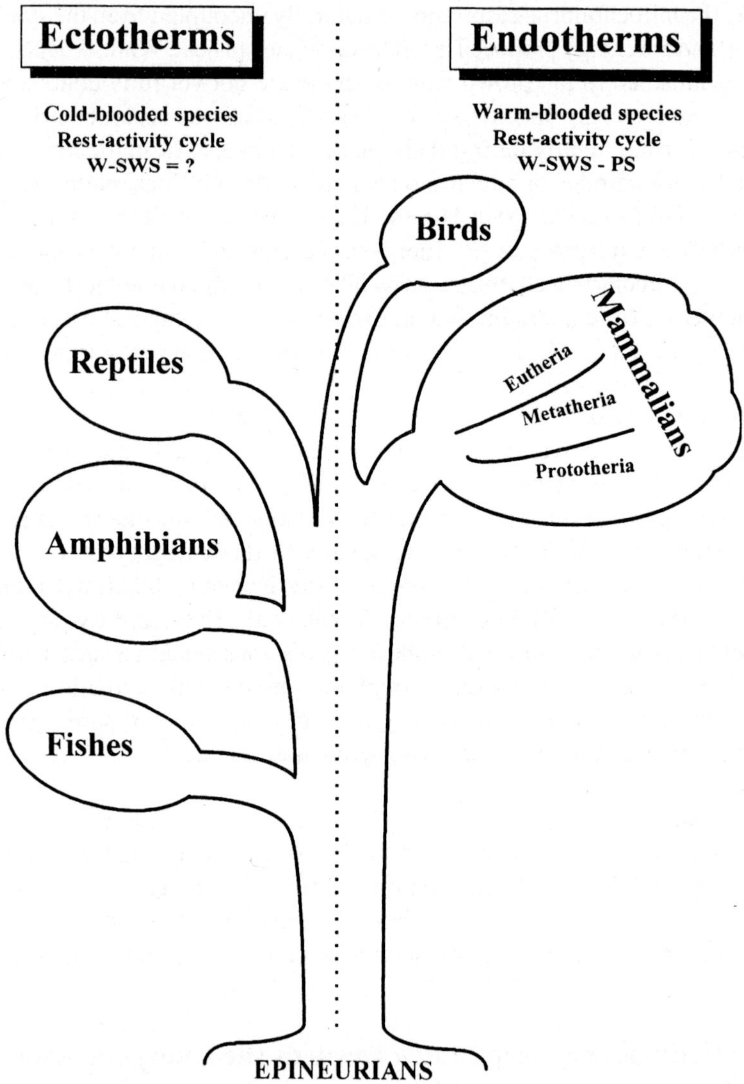

Fig. 2. Ectotherms (fishes, amphibians, reptiles) are cold-blooded species submitted to ambient variations of the temperature. A clear rest-activity cycle has been reported in the related species, but the existence of sleep-wake states still remains controversial. Endotherms are warm-blooded species (birds, mammals) capable of maintaining their internal temperature constant. Mammals include eutheria (placentals), metatheria (marsupials) and prototheria. Monotremes are the living species of the prototheria (platypus and echidna) and recent DNA hybridization analyses indicate that they diverge from the other lines of mammals. A PS-like state with a synchronized EEG has been reported by Siegel. The occurrence of PS is concomitant with the homeothermic status. W—wakefulness; SWS—slow wave sleep; PS—paradoxical sleep; EEG—electroencephalogram.

homeostatic processes regulating their rest-activity cycle. PS, however, has never been described convincingly in these species (see in: Karmanova, 1982; Campbell and Tabler, 1984; Jouvet, (c), 1994 a). Finally, concerning metabolism, the rate of production of the energy is clearly lower in ectotherms than in endotherms. In this respect, experimental evidence suggests that lactate production is lower in the resting lizard than in resting mammals (Guppy et al.,1987), while in fishes gluconeogenesis is reported to be upregulated during activity (Weber and Haman, 1996).

Endotherms-slow wave sleep

For the endothermic related species the question is generally better documented and regarding SWS a negative correlation has been evidenced between its amount and the body weight of the species. This aspect has been interpreted in terms of metabolism since its basal values are more elevated in mammals with a low body weight (Allison and Cicchetti). For this purpose, the model of hibernaning mammals is again interesting since, when these animals start hibernating in order to save energy, they exhibit increased amounts of SWS (Walker and Berger, 1980). Regarding the metabolites related to the energy production, during SWS as compared to wakefulness, there is a general decrease in the utilization of glucose accompanied by a reduced oxygen consumption and a hypothermia probably resulting from a heat loss (Reich et al., 1972; Kennedy et al., 1982; Giuditta et al., 1984; Roussel et al., 1980; Marsden and Wildschiodtz, 1994). This decrease seems to be reflected in the processes of the phosphorylative oxidation (Giuditta et al., 1984 and, while controversial, in the lactate pathway (Reich et al., 1972). At the same time, glycogen content increases, thus ensuring a glial storage of the unmetabolized glucose (Karnovsky et al., 1983; Reich et al., 1972). Studies related to the regional metabolic activity during the sleep-wake cycle, and performed with [^{14}C] 2-deoxyglucose, also confirm that during SWS mean cerebral metabolism reflects, indeed, a situation of energy conservation. During this state of sleep, however, it is further emphasized that the general decrease in the energetic metabolism suggested is not homogeneous throughout the entire brain and while a low metabolism occurs in sensory relays, the cerebellum, the thalamus and the cortex, increased metabolic rates are determined in the hypothalamus and the hippocampus (Ramm and Frost, (a), 1983; (b), 1986). Evaluation of glucose consumption, however, does not indicate whether the increases in glucose metabolism reported run through the phosphorylative oxidation or the lactate pathways. For this purpose, specific microsensors for glucose (Netchiporouk et al., 1996)., lactate (Shram et al., 1998) and ATP, possessing a high anatomical resolution are necessary. In this respect, in using optical sensors, an increased laser induced NADH fluorescence has been measured within the nucleus raphe dorsalis (nRD) during SWS (Mottin et al., 1997. This effect has been interpreted as a possible local activation of the oxidative energy production.

Paradoxical sleep

Its relationships with the energetic mechanisms are first characterized by the fact that during its occurrence the energy consumption is as high or even higher than during wakefulness. In this respect, it is reported that during PS, glucose consumption increases while glycogen contents decrease (Karadzic and Mrsulija, 1969; Giuditta et al., 1984; Franck et al., 1987; Maquet et al., 1990; Jouvet, (d), 1994b). Regional studies conducted with [^{14}C] 2-deoxyglucose further underline the existence of positive changes in glucose metabolism during PS in specific areas like the hippocampus, the substancia nigra, the extrapyramidal motor system and reticular formation. These sites are different from those described for SWS (Ramm and Frost, (a), 1983; Lydic et al., 1991). Regarding the metabolic pathways concerned, the involvement of an oxidative metabolism is supported by the decrease in PS observed during hypoxia (Huertas and McMillin, 1968; Baker and McGinty, 1979) or the increase obtained when an increased O_2 availability is allowed to pontine cats (Arnulf et al., 1991). This view is also strengthened by the data obtained after administration of chloramphenicol, a compound inhibiting the first site of the oxidative phosphorylation and reducing PS very efficiently either in the cat (Petitjean et al., 1979) or in the rat (Chastrette et al., 1990). Data obtained with optical sensors reporting an increased laser induced NADH fluorescence during PS within the nRD and the cortex (Mottin et al., 1997) are also in favor of this view. Again in the same sense, the acetylcholine synthesis dependence upon the oxidative metabolism can be evoked. Acetylcholine is, indeed, a neurotransmitter involved in PS executive mechanisms (Sakai, 1988; Mallick and Thakkar, 1992). According to this last aspect, it is noteworthy that most of the cholinergic PS executive neurons are rich in pyruvate dehydrogenase, an enzyme producing acetyl CoA, the basal substratum for both acetylcholine synthesis and phosphorylative oxidation pathway activity (Gibson and Shimada, 1980). Finally, regarding the lactate pathway, while only a few papers are available in this respect, it seems likely that the ATP produced in such anaerobic conditions might be, contrary to wakefulness, reduced during PS (Shimizu et al., 1966; Thurston and Hauhart, 1989).

Conclusion

According to the above statements and as previously suggested (Jouvet, (d), 1994 b) it appears likely that PS might be an energy-gated state of sleep and that the ultradian rhythm through which it occurs might be determined by the oxidative energy produced and stored during the SWS period preceding each PS episode.

Acknowledgments

This work was supported by INSERM unit U480, Grant Rhône-Alpes Region (97-99). We thank also C. Limoge for improving the English text and G. Debilly for computerizing the figures.

References

Adams, K. and Oswald, I. Sleep is for tissue restoration, J. Royal Coll. Physic. Lond., 11 (1977) 376–388.

Adams, K., Sleep as a restorative process and a theory to explain why, Prog. in Brain Res., 53 (1980) 289–305.

Alberts, B., Bray, D., Lewis, J., Raff, M., Roberts, K. and Watson, J.D. Biologie Moléculaire de la Cellule. Flammarion Médecine-Science Press, Paris, 1986.

Allison, T. and Cicchetti, D.V. Sleep in mammals: ecological and constitutional correlates, Science, 194 (1976) 732–734.

Arnulf, I., Sastre, J.P., Buda, C. and Jouvet, M. Is paradoxical sleep ultradian rhythm energy-gated in pontine cat?, Sleep Res., 20 (1991) 569.

Baker, T.L. and McGinty, D.J. Sleep-waking patterns in hypoxic kittens, Devl. PsychoBiol., 12 (1979) C377–C389.

Benington, J.H. and Heller, C. Restoration of brain energy metabolism as the function of sleep, Progress in Neurobiology, 45 (1995) 347–360.

Berger, R.J. and Phillips, N.H. Sleep and energy conservation, NIPS, 8 (1993) 276–281.

Campbell, S.S. and Tobler, I. Animal sleep: a review of sleep duration across phylogeny, Neuroscience and Biochem, Reviews, 8 (1984) 269–300.

Chastrette, N., Cespuglio, R. and Jouvet, M. POMC-derived peptides and sleep in the rat. Part 1: hypnogenic properties of ACTH derivatives, Neuropeptides, 15 (1990) 61–74.

Franck, G., Salmon, E., Poirier, R., Sadzot, B. and Franco, G. Etude du métabolisme glucidique cérébral régional chez l'homme au cours de l'éveil et du sommeil, par tomographie à émission de positons, Rev. Electroenceph. Neurophysiol. Clin., 17 (1987) 71–77.

Gibson, G.E. and Shimada, M. Studies on the metabolic pathways of the acetyl group for acetylcholine synthesis, Biochem. Pharmacol., 29 (1980) 167–174.

Giuditta, A., Capano, C.P. and Zucconi, G.G. The Neurochemical approach to the study of sleep. In: A Lajtha (Ed.), *Handbook of Neurochemistry,* Plenum Press, New York, 1984, pp. 443–476.

Guppy, M., Bradshaw, S.D., Fergusson, B., Hansen, I.A. and Atwood, C. Metabolism in lizards: low lactate turnover and advantages of heterothermy, Am. J. Physiol., 253 (1987) 77–82.

Heller, H.C., Musacchia, X.J. and Wang, L.C.H., Living in the Cold, Elsevier Press, New York, Amsterdam, London, 1988, pp. 1–573.

Himms-Haggen, J. Cellular thermogenesis, *Ann. Rev. Physiol,* 38 (1976) 315-351.

Horne, J.A., Mammalian sleep function with particular reference to man. In: A.M. Van Nostland Reinolds (Ed.), *Sleep Mechanisms and Functions in Humans and Animals—An Evolutionary Perspective,* Ltd: Wokingham, England, 1985, pp. 262–311.

Huertas, J. and McMillin, J.K. Paradoxical sleep: effect of low partial pressures of atmospheric oxygen, 159 (1968) 745–746.

Jouvet, M. (a) Le discours biologique, La revue de médecine, 16 (1972) 1003–1063.

Jouvet, M. (b) Histoire naturelle du rêve, C.R. Acad. Sci, Paris, 9 (1976) 1–20.

Jouvet, M. (c) Phylogeny of the states of sleep, Acta psychiat. belg., 94 (1994a) 256–267.

Jouvet, M. (d) Paradoxical sleep mechanisms, Sleep, 17 (1994b) 77–83.

Karadzic, Y. and Mrsulja, B. Deprivation of paradoxical sleep and brain glycogen, J. Neurochem., 16 (1969) 29–34.

Karmanova, I.G., Evolution of sleep. In: W.P. Koella (Ed.), *Evolution of Sleep,* Karger Press, New York, 1982, pp. 1–147.

Karnovsky, M.L., Reich, P., Anchors, J.M. and Burrow, B.L. Changes in brain glycogen during slow-wave sleep in the rat, J.Neurochem., 41 (1983) 1498–1501.

Kennedy, C., Gillin, J.C., Mendelson, W., Suda, S., Miyaoka, M., Ito, M., Nakamura, R.K., Storch, F.I., Pettigrew, K., Mishkin, M. and Sokolov, L. Local cerebral glucose utilization in non-piraid eye movement sleep, Nature, 297 (1982) 325–327.

Konings, W.N. and Michels, P.A.M. Electron-transfer-driven solute translocation across bacterial membranes. In: C.J. Knowles (Ed.), *Diversity of Bacterial Respiratory Systems*, CRC Press, 1980, pp. 33–86.

Lefresne, P., Beaujouan, J.C. and Glowinski, J. Origin of the acetyl moiety of acetylcholine in rat striatal synaptosomes: a specific pyruvate dehydrogenase involved in Ach synthesis?, Biochimie, 60 (1978) 479–487.

Lehninger, A.L., Biochimie. Flammarion Médecine Science Press, Paris, 1977.

Lydic, R., Baghdoyan, H.A., Hibbard, L., Bonyak, E.V., De Joseph, M.R. and Hawkins, R.A. Regional brain glucose metabolism is altered during rapid eye movements in cat: a preliminary study, J. Comp. Neurol., 304 (1991) 517–529.

Mallick, B.N. and Thakkar, M. Effect of REM sleep deprivation on molecular forms of acetylcholinesterase in rats, NeuroReport, 3 (1992) 676–678.

Maquet, P., Dive, D., Salmon, E., Sadzot, B., Franco, G., Poirier, R., Von Frenckell, R. and Franck, G. Cerebral glucose utilization during sleep-wake cycle in man determined by positron emission tomography and 2-fluoro-2-deoxy-D-glucose method, Brain Res., 513 (1990) 136–143.

Marsden, P.L. and Wildschiodtz, G. Cerebral oxygen metabolism during the sleep-wake cycle in man, J. Sleep Res., 3 (1994) 154.

Mottin, S., Laporte, P., Jouvet, M. and Cespuglio, R. Determination of NADH in the rat brain during sleep-wake states with an optic fibre sensor and time-resolved fluorescence procedures, Neuroscience, 79 (1997) 683–693.

Netchiporouk, L.I., Shram, N.F., Jaffrezic-Renault, N., Martelet, C. and Cespuglio, R. *In vivo* brain glucose measurements: differential normal pulse voltammetry with enzyme-modified carbon fiber microelectrodes, Anal. Chem., 68 (1996) 4358–4364.

Nicholls, D.G. Brown fat mitochondria, Trends Biochem. Sci., 1 (1976) 128–130.

Parmeggiani, P.L., Interactions between temperature and sleep regulations. In: H.C. Heller, X.J. Musacchia and L.C.U. Wang, (Eds.) *Living in the Cold,* Elsevier Press, New York, 1985, pp. 177–184.

Petitjean, F., Buda, C., Janin, M., David, M. and Jouvet, M. Effets du chloramphenicol sur le sommeil du chat, comparaison avec le thiamhénicol, l' erythromycine et l'oxytétracycline, Psychopharmacol., 66 (1979) 147–153.

Purves, W.K., Orians G.H., Heller, H.C. and London, J. Physiologie, homéostasie et regulation de la température corporelle. In: Flammarion (Ed). *Le Monde du Vivant,* Paris, 1992, pp. 741–764.

Ramm, P. and Frost, B.J. (a) Regional metabolic activity in the brain during sleep-wake activity, Sleep, 6 (1983) 196–216.

Ramm, P. and Frost, B.J. (b) Cerebral and local cerebral metabolism in the cat during slow wave and REM sleep, Brain Res., 365 (1986) 112–124.

Reich, P., Geyer, S.J. and Karnovsky, M.L. Metabolism of brain during sleep and wakefulness, J. Neurochem., 19 (1972) 487–497.

Roussel, B., Dittmar, A. and Chouvet, G. Internal temperature variations during the sleep-wake cycle in the rat, Waking and Sleeping, 4 (1980) 63–75.

Sawaya, R. and Ingvar, D.H. Cerebral blood flow and metabolism in sleep, Acta Neurol. Scand., 80 (1989) 481–491.

Sakai, K. Executive mechanisms of paradoxical sleep, Arch. Ital. Biol., 479 (1988) 225–240.

Shimizu, H., Tabushi, K., Hishikawa, Y., Kakimoto, Y. and Kaneko, Z. Concentration of lactic acid in rat during natural sleep, Nature, 26 (1966) 936–937.

Shram, N.F., Netchiporouk, L.I., Martelet, C., Jaffrezic-Renault, N. and Cespuglio, R. In vivo voltammetric detection of rat brain lactate with carbon fibre microelectrodes coated with lactate oxidase, Anal. Chem., (1998) in press.

Siegel, J.M., Sleep in monotremes: implications for the evolution of REM sleep. In: O. Hayaishi and S. Inoué (Eds.), *Sleep and Sleep Disorders: from Molecule to Behavior,* Academic Press, Tokyo, New York, 1997, pp. 113–128.

Simon, E., Pierau, F.K. and Taylor, D.C. Central and peripheral thermal control of effectors in homeothermic temperature regulation, *Physiological Reviews,* 66 (1986) 235–300.

Siesjö, B.K., Brain Energy Metabolism. John Wiley and sons (Eds.), Interscience Publication, New York, 1978, pp. 151–379.

Thurston, J.H. and Hauhart, R.E. Effect of momentary stress on brain energy metabolism in weaning mice: apparent use of lactate as cerebral metabolic fuel concomitant with a decrease in brain glucose utilisation, Metabolic Brain Disease, 4 (1989) 177–186.

Tsacopoulos, M. and Magistretti, P.J. Metabolic coupling between glia and neurons, J. Neuroscience, 16 (1996) 877–885.

Tucek, S. Problems in the organization and control of acetylcholine synthesis in brain neurons, Prog. Biophys. Molec. Biol., 44 (1984) 1–46.

Walker, J.M. and Berger, R.J. Sleep as an adaptation for energy conservation functionally related to hibernation and shallow torpor, Prog. Brain Res., 53 (1980) 255–278.

Weber, J.M. and Haman, M. Pathways for metabolic fuels and oxygen in high performance fish, Comp. Biochem. Physiol., 113 (1996) 33–38.

Rapid Eye Movement Sleep
B.N. Mallick and S. Inoué (Eds)

18. Rapid Eye Movement Sleep: From cerebral metabolism to functional brain mapping

Pierre Maquet[1] and Christophe Phillips[2]

[1] Cyclotron Research Centre, University of Liége, Belgium

[2] Department of Neurology, CHU Sart Tilman, Liége, Belgium

SUMMARY

This artical presents an overview of the brain metabolic and hemodynamic explorations of REM sleep. Early studies were essentially successful in describing the variations of the global level of cerebral blood flow and energy metabolism. In animals as in man, both were shown to be similar during REM sleep and wakefulness.

Recent technological advances allowed the observation of the regional cerebral blood flow distribution in REM sleep in humans. The results suggest that the executive processes of REM sleep are similar in humans and in animals. They also show significant activation of limbic structures: amygdala, hippocampus, anterior cingulate cortex. More sophisticated analyses of the data can shed some light on neuronal interactions during REM sleep. As an example, results suggesting a modulation of temporal cortex by the amygdala during REM sleep are presented.

Future research in functional neuroimaging will attempt to explore functional (and, hopefully, effective) connectivity between cerebral areas involved in sleep processes.

Key Words: Cerebral metabolism, cerebral blood flow, REM sleep, positron emission tomography, statistical parametric mapping, connectivity.

Introduction

This article focusses on the estimation of brain metabolism and blood flow by various techniques during rapid eye movement sleep (REM sleep) sleep. Cerebral blood flow can be estimated by methods with short time constant. In consequence, blood flow has very often been used as a marker of cerebral activity during REM sleep. In contrast, REM sleep does not easily lend itself to biochemical investigation. This is due to the short duration of REM sleep episodes in animals and the limited access to cerebral biochemistry in humans. In this respect, brain imaging has proved useful in providing an easy access to cerebral metabolism during sleep.

During the last years, neuroimaging techniques have been particularly successful in describing brain areas involved in perception, movement or cognitive functions (Frackowiak et al., 1997). These techniques have recently been applied to human sleep. The aim of this chapter is to offer an historical overview of the applications of brain imaging techniques applied to sleep research, whereby scientific interest has moved from brain biochemistry toward functional neuroanatomy and neuronal connectivity.

Cerebral Energy Metabolism During REM Sleep in Animals

Cerebral blood flow was the first parameter investigated during sleep as a measure of brain activity. Whatever the technique used, cerebral blood flow was shown to be similar or even higher in REM sleep than in wakefulness (Baust, 967; Goas et al., 1969; Seylaz et al., 1975; Shapiro and Rosendorff, 1975; Sugawara and Ootani, 1976; Dufour and Court, 1977; Santiago et al., 1984; Richardson et al., 1985; Lenzi et al., 1987; Parisi et al., 1988; Abrams et al., 1990; Cote and Haddad, 1990; for review, see Franzini, 1992).

Cerebral energy metabolism during REM sleep has been more difficult to estimate, using the deoxyglucose method. In cats, using a regression analysis between regional cerebral glucose consumption and the percent weighted time spent in REM sleep during the uptake period, Ramm and Frost failed to observe any significant correlation. They summarised their results as follows: "metabolism during REM sleep is unlike wakefulness and is not unlike slow wave sleep" (Ramm and Frost, 1986).

The first description of the distribution of regional cerebral blood flow is due to Baust who observed that in REM sleep, blood flow increases in rhombencephalon while it decreases in mesencephalon (Baust, 1967). Activation brainstem structures was also reported by Cote and Haddad (1990) and Richardson et al. (1985). Reivich et al. (1968), using the antipyrine method, were the first report a significant activation in multiple cerebral areas from brainstem to cortex.

To explore brain energy metabolism during REM sleep in animals, authors relied on methods requiring shorter uptake periods or on foetal preparations that exhibit large amount of REM sleep (or, more exactly, active sleep). Using [^{14}C] glucose, Lydic et al. (1991) were able to estimate glucose metabolism in periods as short as 5–7 min. The glucose brain metabolism, estimated in 3 cats in REM sleep, was not significantly different from metabolic rates observed in 3 control awake cats. In pre-term foetal lambs, cerebral glucose utilisation was first shown to be positively correlated with the percent weighted time spent in active sleep (Abrams et al., 1988). Then, it was reported that brain glucose metabolism is 37% higher in low voltage than in high voltage state (Chao et al., 1989). Similar results were observed concerning the cerebral oxygen consumption (Richardson et al., 1985).

The distribution of glucose metabolism was also reported in animals. In rats or in cats, REM sleep was associated with a important activity in the reticular core (Ramm and Frost, 983; Lydic et al., 99), in thalamic nuclei (Lydic et al., 1991) and in the limbic system. Within the limbic system, the structures activated varied: cingulate gyrus (Lydic et al., 1991), hippocampus proper (Ramm and Frost, 1983; Ramm and Frost, 1986), mammilary bodies (Lydic et al., 1991).

Cerebral Energy Metabolism During Human REM Sleep

The cerebral glucose metabolism can be measured in humans by positron emission tomography, using the [^{18}F] fluorodeoxyglucose (FDG) method. The very first observation of cerebral utilisation possibly in REM sleep was reported by Heiss et al. (1985). They observed a global 6% increase in brain glucose consumption, as compared to the metabolism at wake, in a subject who recalled a dream upon awakening. It was later shown that the levels of cerebral glucose metabolism are similar during REM sleep and wakefulness (Franck et al., 1987; Buchsbaum et al., 1989; Maquet et al., 1990), in good agreement with the data obtained in animals. The same results were reported with the cerebral oxygen consumption, as measured by the Kety-Schmidt technique (Madsen et al., 1991).

Cerebral glucose metabolism essentially reflects the energy requirements of synaptic activity (Jueptner and Weiller, 1995). The high level of brain glucose consumption during REM sleep was interpreted as an indicator of an important neuronal activity (Maquet et al., 1990), in keeping with neurophysiological data (Steriade and McCarley, 1990).

In contrast to global metabolic levels which were reliably estimated with FDG method, the regional distribution of cerebral glucose metabolism during REM sleep was not precisely described in humans. This may be due to several factors: poor spatial resolution of the first generation scanners, low anatomical precision in the neuroanatomical placement of regions of interest (ROI), statistical limitations of the ROI method (Ford et al., 1991). However, a feature frequently reported during REM sleep was the activation of temporal and occipital areas that was putatively attributed to the subjects' oneiric activity (Meyer et al., 1987; Maquet et al., 1990; Madsen et al., 1991).

Distribution of Regional Cerebral Blood Flow During REM Sleep in Humans

The estimation of cerebral glucose metabolism by the fluorodeoxyglucose method needs a cerebral steady state for 45 min. This technical requirement restricted the description to long lasting average changes in glucose metabolism during sleep. More importantly, the physical half-life of ^{18}F (108 min) hampers the repetition of the measurements in the same subject during a single session.

In contrast, ^{15}oxygen-labeled water method, due to the very short half-life of ^{15}oxygen (123 sec), allows multiple (up to 12) acquisitions in the same subject, under various conditions, thus allowing the comparison of different brain states. Furthermore, acquisition times are reduced to 1 or 2 min, giving access to the study of briefer physiological processes. The estimated physiological parameter is the regional cerebral blood flow, which is an indirect marker of neuronal activity (Jueptner and Weiller, 1995).

In the same time, powerful tridimensional statistical techniques were developed, the most popular of which is the Statistical Parametric Mapping (SPM, Wellcome Institute for Cognitive Neuroscience, London (Frackowiak et al., 1997)). They can test, at the voxel level, for significant regional signal modifications. With these new methods in hand, a more precise description of the functional neuroanatomy of human sleep could be obtained.

At present, two groups have reported on regional distribution of cerebral blood flow during REM sleep (Maquet et al., 1996; Braun et al., 1997). Their results are generally in agreement. In both studies, the functional neuroanatomy of REM sleep is characterised by the significant activation of pontine and mesencephalic tegmentum, limbic structures, anterior cingulate cortex and by the relative deactivation of prefrontal areas, parietal cortex and precuneus/ posterior cingulate cortex (Fig. 1).

The activation of meso-pontine tegmentum is in excellent agreement with the mechanisms of REM sleep generation and maintenance described in animals, that put a special emphasis on the role of cholinergic nuclei in this region (for review, see Jones, 1990; Steriade and McCarley, 1990). These results suggest that REM sleep is generated in normal humans as in other mammals.

Brainstem structures are known to monosynaptically activate thalamic nuclei (Steriade et al., 1990), which we found significantly activated, although only on the left side. The lateralisation of the thalamic activation has no obvious physiological basis. It could be due to a bias specific to the sample studied or, less likely, could represent a genuine asymmetric thalamic activation in strongly right-handed subjects.

Discrepancies appear in the exact activations within the limbic system: we observed a very localized and highly significant signal within both amygdaloid complexes, whereas Braun et al. (1997) report an activation of hippocampus and parahippocampal gyrus. It should be stressed that, for instance on the right side, these activation sites are distant from less than 10 mm (amygdala coordinates in the Talairach space: $x = 6$; $y = -4$; $z = -12$ in (Maquet et al., 1996); hippocampus coordinates: $x = 22$; $y = -8$; $z = -12$ in (Braun et al., 1997)). Besides technical differences (the scanner, the handling of the data, the types of comparison, the statistical threshold...were different), these discrepancies might indicate that the functional neuroanatomy of REM sleep is not unique and immutable but, on the contrary, that the cerebral areas active during REM sleep might be variable, essentially depending on individual factors, such as previous waking experience.

Following text will discuss the amygalar activation and show that: (i) in animals, close relationships have been observed between REM sleep and amygdalar activity, and (ii) the distribution of cortical activity suggest an modulation of the cortex by the amygdala during REM sleep.

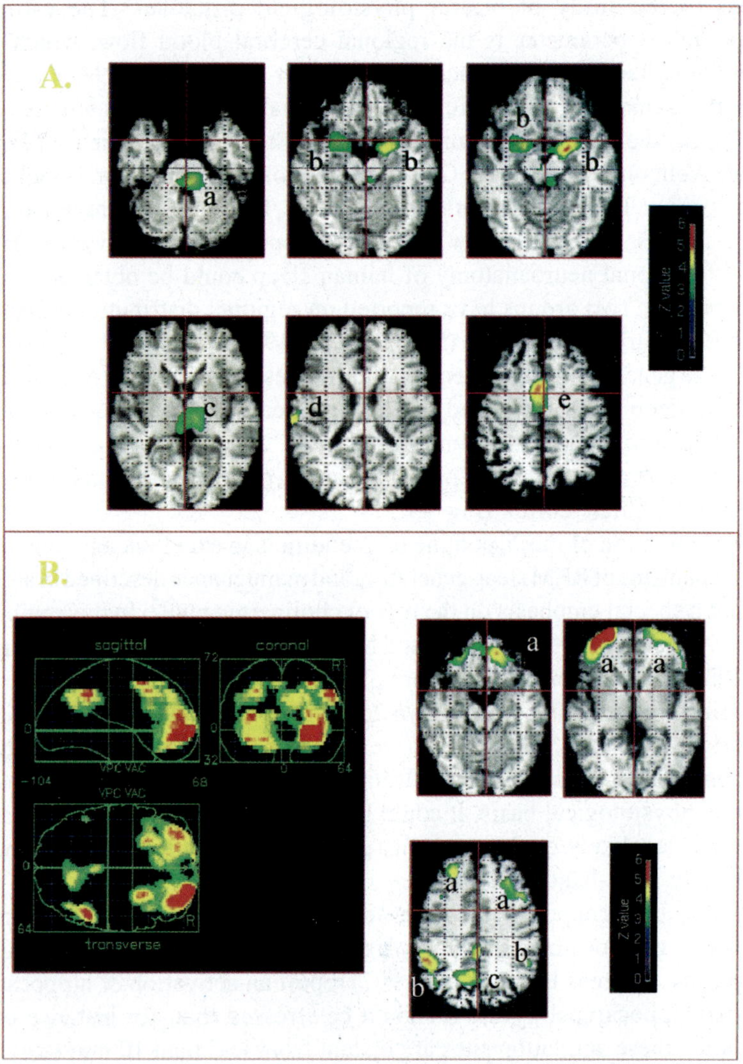

Fig. 1. Functional neuroanatomy of REM sleep. Adapted from (Maquet et al., 1996), with permission. A. Cerebral areas significantly activated during REM sleep, displayed in transverse sections on a MRI scan normalised to the stereotaxic space of Talairach and Tournoux (Talairach and Tournoux, 1988) at –28, –6, –12, 4, 16 and 40 mm from bicommissural plane. Patient's right is on the left of the sections. a, pons; b, amygdaloid complexes; c, thalamus; d, right parietal operculum; e, anterior cingulate cortex. B. Cerebral areas significantly deactivated during REM sleep Left panel, glass brain showing significant voxels in the stereotaxic space of Talairach and Tournoux (Talairach and Tournoux, 1988). Left panel, the same results displayed in transverse sections on a MRI scan normalised to the same stereotaxic space at –16, 0 and 32 mm from bicommissural plane. Patient's right is on the left of the sections. a, prefrontal cortex. b, parietal areas; c, precuneus/posterior cingulate cortex.

Several publications suggests that amygdaloid complexes are involved REM sleep modulation. An early observation reports that brief daily amygdalar stimulations increase REM sleep in REM sleep deprived rats (Smith and Miskiman, 1975). Furthermore, the electrical stimulation of central nucleus of the amygdala increases PGO frequency (Calvo et al., 1987). Likewise, carbachol injection within the same nucleus increases REM sleep and PGO waves (Calvo et al., 1996) and serotonin injection causes a rapid sleep change from slow wave sleep to REM sleep (Sanford et al., 1995).

Our results suggest that amygdaloid complexes may modulate the cortical activity during REM sleep. It is remarkable that significantly activated cortical areas [anterior cingulate cortex (Maquet et al., 1996; Braun et al., 1997), parietal lobule (Maquet et al., 1996), inferior temporal and fusiform gyri (Braun et al., 1997)] all receive an important number of amygdalar projections (Amaral and Price, 1984; Amaral et al., 1992; Aggleton, 1993), whereas he deactivated cortical regions [prefrontal and parietal cortices (Maquet et al., 1996; Braun et al., 1997)] receive few or no amygdalar efferents.

The absence of significant temporal and occipital activation in our experiment (Maquet et al., 1996) might be considered as an argument against our hypothesis of a cortical modulation by the amygdala in REM sleep. However, this negative result could be due to the group analysis that might have averaged out individual posterior activations. Furthermore, other authors have reported such activations (Meyer et al., 1987; Maquet et al., 1990; Madsen et al., 1991; Braun et al., 1997). Finally, next section shows that our hypothesis can be formally tested and post hoc analysis of our data lends further support to our hypothesis.

Neuronal Connectivity In Human REM Sleep

The final aim of functional brain imaging is to describe as precisely as possible the spatio-temporal course of neuronal interactions, at a macroscopic level. Recent advances in statistical analysis go in this direction and provide the logistic background necessary to explore functional relationships between cerebral areas (Friston et al., 1997). Analyses of psychophysiological interaction look for the modulation of one cerebral area by another, under a specific experimental context.

Such analysis on our already published data shows a significant positive interaction between amygdalar blood flow and occurrence of REM sleep in the temporal cortex (Fig. 2). These results suggest a modulation of temporal cortex by the amygdala in, and only in, REM sleep, in keeping with our hypothesis. Although similar interactions were observed on both sides (Maquet and Phillips, accepted for publication), a significant interaction was also shown in some more posterior regions on the right side (Fig. 2). Interestingly enough, our results do not show such an interaction within brainstem. They do not provide evidence that amygdaloid complexes would favour REM sleep by stimulating brainstem areas, as suggested by animal studies. However, these negative results could be due to artifacts (e.g., partial volume effects on

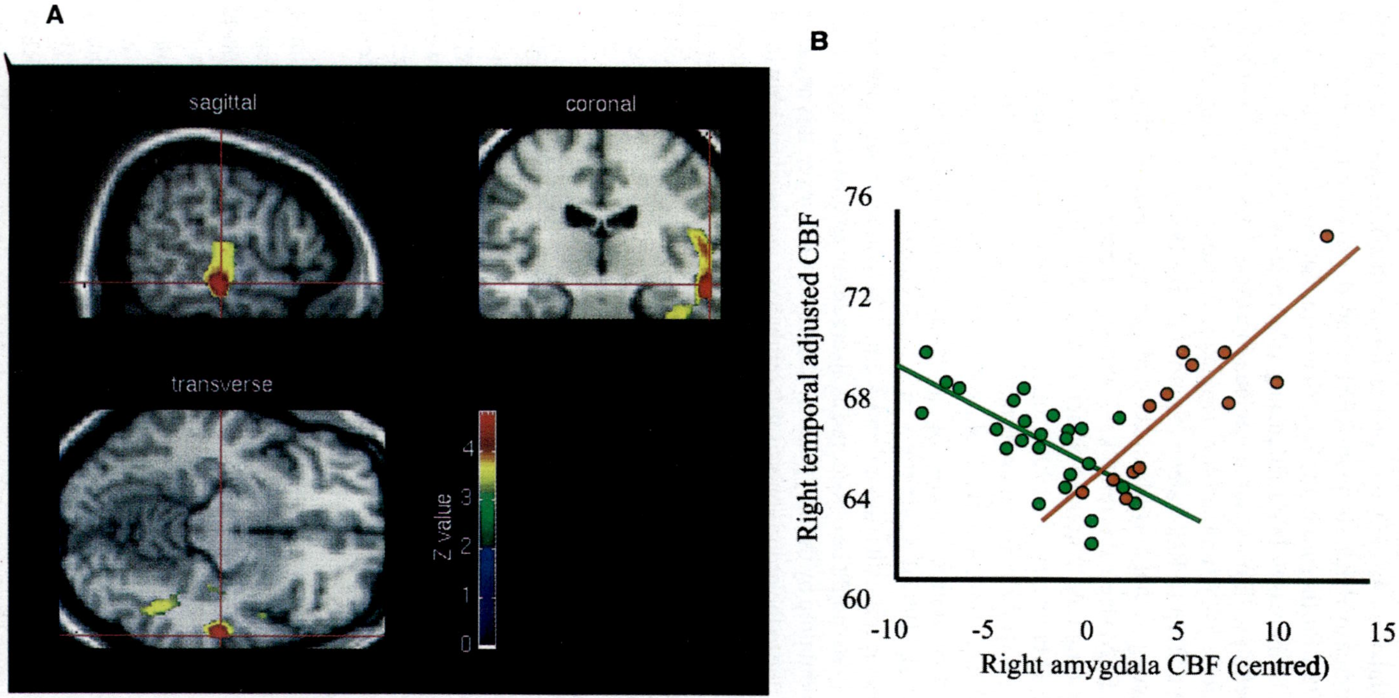

Fig. 2. **Psychophysiological interactions: (A) modulation of right temporal activity ($Z = 4.80$; $P_{corrected} = 0.007$; coordinates : $x = 58$; $y = -20$; $z = -12$ mm) by the right amygdala within the specific context of REM sleep. The significant area is displayed on an MRI image, within the stereotaxic space of Talairach and Tournoux. Patient's right is on the right of transverse and coronal sections. (B) The plot shows the relations, in terms of regional blood flow, between right amygdalar and the significant temporal voxel in REM sleep (red) and in the other states of vigilance (green).**

small structures, insufficient number of observations) rather than to the absence of the physiological processes.

Conclusions

Functional brain imaging has reached unprecedented sophistication that allows to investigate, at the macroscopic level, the dynamic interactions that occur between cerebral areas in humans, in particular during sleep. In view of their contribution to other fields of cognitive neuroscience, these techniques should provide an original contribution to the study of human sleep.

Acknowledgements

PM is Research Associate at the Fonds National de la Recherche Scientifique de Belgique (FNRS). CP is Research Engineer at the University of Liége. Personal research presented in this paper has been supported by the ESRS-Synthelabo grant 1994, by FNRS grants, by University of Liége Research grants and by the Fonds Médical Reine Elisabeth.

References

Abrams, R.M., Hutchison, A.A., Jay, M.T., Sokoloff, L. and Kennedy, C. Local cerebral glucose utilization non-selectively elevated in rapid eye movement sleep of the fetus, Developmental Brain Research, 40 (988) 65–70.

Abrams, R.M., Post, J.C., Burchfield, D.J., Gomez, K.J., Hutchison, A.A. and Conlon, M. Local cerebral blood flow is increased in rapid-eye-movement sleep in fetal sheep, American Journal of Obstetrics and Gynecology 162 (1990) 278–281.

Aggleton, J.P. The contribution of the amygdala to normal and abnormal emotional states, Trends in Neuroscience, 16 (1993) 328–333.

Amaral, D.G. and Price, J.L. Amygdalo-cortical projections in the monkey *(Macaca fascicularis),* Journal of Comparative Neurology, 230 (1984) 465–496.

Amaral, D.G., Price, J.L., Pitkänen, A. and carmichael, S.T. Anatomical organization of the primate amygdaloid complex. In J.P. Aggelton(Ed.), *The amygdala: neurobiological aspects of emotion, memory and mental dysfunction,* Wiley-Liss, New York, 992, pp. 1–66.

Baust, W. Local blood flow in different regions of the brain-sem during natural sleep and arousal, Electroencephalography and Clinical Neurophysiology, 22 (1997) 369–372.

Braun, A.R., Balkin, T.J., Wesensten, N.J., Carson, R.E., Varga, M., Baldwin, P., Selbie, S., Belenky, G. and Herscovitch, P. Regional cerebral blood flow throughout the sleep-wake cycle-An (H20)-0-5 PET study, Brain, 120 (1997) 1173–1197.

Buchsbaum, M.S., Gillin, J.C., Wu, J., Hazlett, E., Sicotte, N., Dupont, R.M. and Bunney, W.E. Regional cerebral glucose metabolic rate in human sleep assessed by positron emission tomography, Life Sciences, 45 (1989) 1349–1356.

Calvo, J.M., Badillo, S., Morales-Ramirez, M. and Palacios-Salas, P. The role of the temporal lobe amygdala in ponto-geniculo-occiupital activity and sleep organization in cats, Brain Research 403 (1987) 22–30.

Calvo, J.M. Simon-Arceo, K. and Fernandez-Mas, R. Prolonged enhancement of REM sleep produced by carbachol microinjection into the amygdala, Neuroreport, 7 (1996) 577–580.

Chao, C.R., Hohimer, A.R. and Bissonnett, J.M. The effects of electrocortical state on cerebral carbohydrate metabolism in fetal sheep, Developmental Brain Research, 49 (1989). 5–5.

Cote, A. and Haddad, G.G. Effects of sleep on regional blood flow distribution in piglets, Pediatric Research, 28 (1990) 218–222.

Dufour, R. and Court, L. Le débit cérébral sanguin au cours du sommeil paradoxal du lapin, Archives Italiennes de Biologie, 15 (1977) 57–76.

Ford, I., McColl, J.H., McCormack, A.G. and McCrory, S.J. Statistical issues in the analysis of neuroimages, Journal of Cerebral Blood Flow and Metabolism, 11 (1991) A89–A95.

Frackowiak, R.S.J., Friston, K.J., Frith, C.D., Dolan, R.J. and Mazziotta, J.C. Human brain function. Academic Press, San Diego, 1997.

Franck, G., Salmon, E., Poirrier, R., Sadzot, B. and Franco, G. Etude du metabolisme glucidique cérébral regional, chez l'homme, au cours de l'éveil et du sommeil, par tomographie á emission de positrons, Revue d' EEG et de Neurophysiologie Clinique, 17 (1987) 71–77.

Franzini, C., Brain metabolism and blood flow during sleep, Journal of Sleep Research, (1992) 3–16.

Friston, K.J., Buechel, C., Fink, G.R., Morris, J., Rolls, E. and Dolan, R.J. Psychophysiological and modulatory interactions in neuroimaging, Neuroimage, 6 (1997) 218–229.

Goas, J.Y., Seylaz, J., Mamo, H., McLeod, P., Caron, J.P. and Houdart, R. Variations du débit sanguin du cortex de l'homme au cours du sommeil, Revue Neurologique, Paris, 120 (1969) 159–176.

Heiss, W.D., Pawlik, G., Herholz, K., Wagner, R. and Wienhard, K. Regional cerebral glucose metabolism in man during wakefulness, sleep, and dreaming, Brain Research 327 (1985) 362–366.

Jones, B.E. Paradoxical sleep and its chemical/structural substrates in the brain, Neuroscience, 40 (1990) 637–656.

Jueptner, M. and Weiller, C. Review: does measurement of regional cerebral blood flow reflect synaptic activity? Implications for PET and fMRI, Neuroimage, 2 (995) 148–156.

Lenzi, P., Cianci, T., Guidalotti, P.L., Leonardi, G.S. and Franzini, C. Brain circulation during sleep and its relation to extracerebral hemodynamics, Brain Research, 415 (1987) 14–20.

Lydic, R., Baghdoyan, H.A., Hibbard, L., Bonyak, E.V., DeJoseph, M.R. and Hawkins, R.A. Regional brain glucose metabolism is altered during rapid-eye movement sleep in the cat: a preliminary study, Journal of Comparative Neurology, 304 (1991) 517–529.

Madsen, P.L., Holm, S., Vorstrup, S., Friberg, L., Lassen, N.L. and Wildschiodtz, G. Human regional cerebral blood flow during rapid-eye-movement sleep, Journal of Cerebral Blood Flow and Metabolism, 1 (1991) 502–507.

Madsen, P.L., Schmidt, J.F., Wildschiodtz, G., Friberg, L., Holm, S., Vorstrup, S. and Lassen, N. L. Cerebral O_2 metabolism and cerebral blood flow in humans during deep sleep and rapid-eye-movement sleep, Journal of Applied Physiology, 70 (1991) 2597–2601.

Maquet, P., Dive, D., Salmon, E., Sadzot, B., Franco, G., Poirrier, R., von Frenckell, R. and Franck, G. Cerebral glucose utilization during sleep-wake cycle in man determined by positron emission tomography and [18F] 2-fluoro-2 deoxy-D-glucose method, Brain Research 513 (1990) 136–143.

Maquet, P., Peters, J.M., Aerts, J., Delfiore, G., Degueldre, C., Luxen, A. and Franck, G. Functional neuroanatomy of human rapid eye movement sleep and dreaming, Nature, 383 (1996) 163–166.

Maquet, p. and Phillips, C. Functional brain imaging of human sleep, Journal of Sleep Research, (Accepted for publication).

Meyer, J.S., Ishikawa, Y., Hata, T. and Karacan, I. Cerebral blood flow in normal and abnormal sleep and dreaming, Brain and Cognition, 6 (987) 266–294.

Parisi, R.A., Neubauer, J.A., Frank, M.M., Santiago, T.V. and Edelman, N.H. Linkage between brain blood flow and respiratory drive during rapid-eye-movement sleep, Journal of Applied Physiology, 64 (1988) 1457–1465.

Ramm, P. and Frost, B.J. Regional metabolic activity in the rat brain during sleep-wake activity, Sleep, 6 (1983) 196–216.

Ramm, P. and Frost, B.J. Cerebral and local cerebral metabolism in the cat during slow wave and REM sleep, Brain Research, 365 (1986) 12–24.

Reivich, M., Isaacs, G., Evarts, E. and Kety, S. The effect of slow wave sleep and REM sleep on regional cerebral blood flow in cats, Journal of Neurochemistry, 15 (1968) 301-306.

Richardson, B.S., Patrick, J.E. and Abduljabbar, H. Cerebral oxidative metabolism in the fetal lamb: relationship to electrocortical state, American Journal of Obstetrics and Gynecology, 53 (1985) 426-431.

Sanford, L.D., Tejani-Butt, S.M., Ross, R.J. and Morrison, A.R. Amygdaloid control of alerting and behavioral arousal in rats: involvement of serotonergic mechanisms, Archives Italiennes de Biologie, 134 (1995) 81–99.

Santiago, T.V., Guerra, E., Neubauer, J.A. and Edelman, N.H. Correlation between ventilation and brain blood flow during sleep, Journal of Clinical Invetigations, 73 (1984) 497–506.

Seylaz, J., Pinard, E., Mamo, H., Goas, J.Y., Luft, A. and Correze, J.L. Human cerebral blood flow during sleep. In: D.H. Ingvar and N.A. Lassen (Eds.)."*Brain Work: the coupling of function, metabolism, and blood flow in the brain*" , Alfred Benzon Symposium VIII, Munksgaard, Copenhague, 1975, pp. 235–252.

Shapiro, C.M. and Rosendorff, C. Local hypothalamic blood flow during sleep, Electroencephalography and Clinical Neurophysiology, 39 (1975) 365–369.

Smith, C.T. and Miskiman, D.E. Increases in paradoxical sleep as a result of amygdaloid stimulation, Physiology and Behavior, 15 (1975) 17–19.

Steriade, M., Datta, S., Paré, D., Oakson, G. and CurroDossi, R. Neuronal activities in brain-stem cholinergic nuclei related to tonic activation processes in thalamocortical systems, Journal of Neuroscience, 10 (1990) 2541–2559.

Steriade, M. and McCarley, R.W. Brainstem control of wakefulness and sleep. Plenum press New York, 1990).

Sugawara, H. and Ootani, K. Studies on the electroencephalogram in ruminants. VII. Brain temperature, brain oxygen tension and blood flow in the common artery during paradoxical sleep in goats, Japanese Journal of Veterinary Science, 38 (976) 41–48.

Talairach, J. and Tournoux, P. Co-planar stereotaxic atlas of the human brain. George Thieme Verlag, Stuttgart, 1988.

Rapid Eye Movement Sleep
B.N. Mallick and S. Inoué (Eds)

19. REM Sleep and Apnea

David W. Carley[1,2] and Miodrag Radulovacki[2]

Department of Medicine[1] and Department of Pharmacology[2]
College of Medicine, University of Illinois at Chicago, Chicago, Illinois 60612

SUMMARY

Sleep apnea syndrome affects at least 3%–5% of the adult population in this country leading to significant morbidity and mortality. The prevalence of apnea is dramatically elevated in the elderly with recent estimates ranging to over 50%. Despite these facts, the mechanisms of apnea genesis and the impact of aging on these mechanisms remain poorly understood. This uncertainty stems in part from a paucity of appropriate animal models to study spontaneous apneas in all stages of sleep.

Rapid eye movement (REM) sleep is associated with apneas, reductions in respiratory and upper airway motor outputs, and increased variability of respiration, heart rate and blood pressure. These autonomic fluctuations are transient, usually ranging from 10 to 60 sec in duration, and are strongly associated with other "phasic" phenomena of REM sleep. Pontogeniculo-occipital (PGO) waves are very sensitive markers for the phasic bursts of activity which are widespread throughout the brain, including the brainstem, during REM sleep. The rapid eye movements which define REM sleep occur in association with bursts of PGO waves. The timing of PGO waves and eye movements also strongly correlates with changes in tidal volume, respiratory frequency, thoracoabdominal movements, diaphragm and upper airway muscle activation and apnea expression.

A variety of sensory stimuli can elicit PGO-like activity and rapid eye movements in man and in animals. These evoked events appear to arise from the same generators as spontaneous PGO waves. Moreover, acoustic stimuli can produce a host of cardio-respiratory changes, including apnea genesis and apnea termination, even in the absence of generalized cortical activation.

The mechanisms by which REM-related brainstem phasic events impact respiratory control are of both basic and clinical interest. Do PGO waves themselves reflect activity which is essential to the respiratory effects, or are they an epiphenomenon? Are pontine phasic event generator sites coupled to respiratory systems by ascending connections, descending connections or both? What neurotransmitter systems are involved? Answering these and related questions is an area with intrinsic basic appeal and great clinical relevance.

Key Words: Apnea, PGO-wave, respiration, arousal, acoustic-stimulation, drug-treatment.

Sleep Apnea Syndrome

Sleep apnea syndrome (SAS) is a significant public health concern, with a prevalence of at least 3%–5% in the United States (c.f. Dempsey, 1996). These patients may exhibit central apneas, operationally defined as cessation of respiratory effort for more than 10 sec; obstructive apneas, characterized by continued inspiratory efforts against an occluded upper airway; mixed apneas, which present with an initial central component followed immediately by an obstructive component; hypopneas, physiologically significant transient reductions of airflow; or even more subtle fluctuation of upper airway resistance which require measures of both airflow and pressure in the airways to detect. Apnea may be a normal phenomenon during transitions from wakefulness to sleep and during rapid eye movement (REM) sleep in man (Webb, 1974). However, the incidence of sleep-related apnea covers a continuous spectrum from 0–5 apneas per h in healthy young adults to over 100 apneas per h of sleep in some patients with severe sleep apnea syndrome.

The high prevalence of sleep-related breathing disorders in the elderly is now well established (Hoch et al., 1990; Ancoli-Israel and Kripke, 1991; Redline and Young, 1993). Although sleep apnea syndrome is most commonly diagnosed among patients in the fifth to seventh decades of life, studies of more general populations reveal a clear age-related increase in sleep apnea that extends to at least the eighth decade (Hoch et al., 1990; Redline and Young, 1993). Prevalence rates for apnea index greater than 5 per h are wide ranging: 28%–67% for elderly males and 20%–54% for elderly females (Redline and Young, 1993).

Although available data are not fully consistent, hypothesized consequences of sleep apnea are hypertension, coronary heart disease, myocardial infarction, stroke, dementia, depression, cognitive dysfunction, sexual dysfunction, and injury due to accidents. Compared to the US general population, Partinen et al. (1990) demonstrated sleep apnea patients to have twice the prevalence of hypertension, three times the prevalence of heart disease and four times the prevalence of cerebrovascular disease. Untreated sleep apnea has also been associated with mortality (Partinen, 1988; He et al., 1988).

Apneas and hypopneas are inherently short-lasting (i.e. 10–60 s), or "phasic" events (Hanly, 1992; Krieger and Kurtz, 1978). This leads to the concept that apnea and hypopnea may be driven by transient inhibition or dysfacilitation of the respiratory rhythm generator or motor systems (Hanly, 1992). This view may also help to explain why central and obstructive apneas are most frequent during REM sleep (Lugaresi et al., 1978), because REM sleep is associated with frequent phasic events which may be inhibitory to respiration (see sections 2 and 3 below; also for a recent review see Pack, 1994). The sleep state dependence of apnea has not been systematically studied in the elderly, but it is clear that REM sleep becomes increasingly fragmented with aging (Hoch et al., 1988; Reynolds et al., 1980). This disruption of REM sleep architecture may be both a cause and a consequence of apnea in elderly sleep apnea patients.

No general agreement exists regarding mechanisms underlying the generation or termination of apneas during sleep in young adults or in the elderly. Although a systematic review of this literature is beyond the scope of this chapter, it is now clear however that, in general, patients with primarily obstructive apnea have redundant pharyngeal tissue with relatively narrow and collapsible upper airways (for recent reviews see Shepard et al., 1991; Schwab, 1996). These anatomical defects can predispose to upper airway collapse during sleep, launching the vicious cycle of sleep-apnea-arousal-ventilation which is the hallmark of the syndrome. Still, the change in behavioral state from wakefulness to sleep is necessary for airway collapse even in the most severe presentations of sleep apnea. Numerous investigations of state dependent activation of upper airway muscles demonstrate the importance of active and coordinated control of motor outputs to the upper airway in avoiding airway collapse or flow limitation (for a review see Kuna, 1991).

These studies also highlight an important gap in our knowledge: little is known about the brainstem and forebrain mechanisms mediating state dependent changes in respiratory control. Virtually nothing is known about the effects of aging on these central processes and nothing at all is known regarding possible defects in these mechanisms associated with sleep apnea syndrome. This lack of knowledge stems in part from a paucity of animal models to study naturally occurring apnea across all behavioral states. During the past decade two animal models have been systematically characterized;.NREM and REM related central apnea in the rat (c.f. Mendelson et al., 1988; Carley et al., 1996a, 1996b, 1996c, 1996d, 1997a; Radulovacki et al., 1996, 1997) and REM related obstructive apnea in the English Bulldog (c.f. Hendricks et al., 1987, 1991; Veasey et al., 1996). Relevant findings from these animal models will be noted below.

Neural Systems Which Generate REM Sleep Phasic Events Including PGO Waves

An understanding of the phasic respiratory behaviors characteristic of REM sleep should naturally be built upon our current knowledge of the mechanisms governing REM sleep and its intrinsic phasic phenomena, such as ponto-geniculo-occipital (PGO) waves. These topics are the focus of other chapters in this book, so that a brief review will suffice here. PGO waves are large amplitude field potentials which can be recorded from the pons, lateral geniculate and occipital cortex of cats; an observation which gives the wave it name. PGO waves are very sensitive markers for the phasic events of REM sleep.

Basically, PGO wave generation has been linked to two groups of cholinergic cells in the rostral pons, the laterodorsal tegmental nucleus (LDT) and the pedunculo-pontine-tegmental (PPT) nucleus. Lesions out of both LDT and PPT caused a reduction of PGO waves and eye movements which was correlated to the number of cholinergic cells destroyed (Webster and Jones, 1988). More specifically, small lesions in the caudal PPT abolish PGO waves without affecting REM sleep or REM-related postural muscle atonia (Shouse and Siegel, 1992).

Conversely, cholinergic stimulation of the caudal PPT leads to profound and long-lasting increases in both REM sleep and PGO wave generation in all sleep/wake states (Datta et al., 1992). At least 2 types of bursting cells have been identified in the caudal PPT which fire just prior to PGO waves and eye movements (for a review see Datta, 1995). Virtually all of this work regarding the neural basis of PGO wave generation has been performed in cats.

Field potentials fulfilling all criteria for PGO waves can also be recorded from the region of the locus coeruleus in rats (c.f. Farber et al., 1980). In fact, the pontine distribution of sites at which PGO waves can be recorded is similar in cat and rat (Marks et al., 1980). Although geniculate PGO waves have not been demonstrated the rat, single unit recordings in the dorsal lateral geniculate nucleus of rats demonstrated that discharge rates in 70% of these cells exhibit a temporal relationship with hindbrain PGO waves (Marks et al., 1981). This finding supports the existence of a phasic event system in the rat.

Changes in Respiratory Control During REM Sleep

As first recognized by Aserinsky and Kleitman (1953) REM sleep is uniquely characterized by two major phenomena: (1) atonia of postural muscles resulting from active inhibition of motor neurons, and (2) transient, or "phasic", events which are widespread in the brain and are typified by the rapid eye movements, for which REM sleep was named. It is important to recognize that there exist many other pheripherally observable correlates of these central neural events. Phasic perturbations of respiratory pattern and motor outputs, sympathetic outflow, heart rate, blood pressure and coronary blood flow are all characteristic of REM sleep (c.f. Verrier et al., 1996).

REM sleep leads to substantial changes in respiratory control. With respect to wakefulness, ventilation is consistently decreased during REM sleep whereas changes in respiratory frequency are more variable (Dempsey, 1996; Stradling and Phillipson, 1986). Although respiration during NREM sleep is determined primarily by the homeostatic control mechanisms, during REM sleep it is clear that there are phasic influences arising outside the homeostatic controllers that dramatically affect ventilation. The intermittent and variable presentation of these phasic events within and between REM periods underlies the fact that the greatest change in ventilation during REM sleep is increased breath to breath variability (Pack, 1994).

Human studies clearly demonstrate that during REM sleep ventilation is driven primarily by abdominal motions with rib cage motions being significantly attenuated with respect to NREM sleep (Millman et al., 1988; Neilly et al., 1991). This is particularly true during phasic events (Millman et al., 1988; Neilly et al., 1991). These changes in rib cage and abdomen motion reflect changes in the motor outputs to the diaphragm and intercostal muscles. Although respiratory muscles do not exhibit the dramatic atonia of most postural muscles during REM sleep, there are motor units even within the diaphragm which are completely silent during REM sleep (Sieck et al., 1984).

In animals, high amplitude field potentials can be recorded from the pons, lateral geniculate and occipital cortex and have been termed ponto-geniculo-occipital (PGO) waves. These waves are very sensitive markers for the phasic events of REM sleep. Orem first described "fractionations" of the diaphragm EMG in association with PGO waves in the cat (1980). These 40–100 ms pauses in muscle activity occur in clusters and are strongly cross-correlated with the timing of PGO waves. Similar diaphragm EMG fractionation is observed following acoustically evoked PGO waves (c.f. Sanford et al., 1992); an effect which is most prominent during REM sleep. These findings imply an important connection between the mechanisms generating PGO waves and motor outputs to the diaphragm.

Kline et al. (1986) also showed a strong temporal association between PGO waves and the diaphragmatic electromyogram (EMG) in cats. During REM sleep the rate of rise in inspiratory activity was highly variable with respect to slow-wave sleep. Transient (several breath) decrements and transient increments in the rate of rise were associated with PGO activity (Kline et al., 1986). Similar patterns of decremented and incremented EMG and compartmental movement have been associated with rapid eye movements during REM sleep in human studies (Millman et al., 1988; Wiegand et al., 1991). Because brainstem field potentials cannot be recorded, eye movements are the most commonly employed marker for phasic REM events in human studies.

The operation of the respiratory rhythm generator is also altered in association with PGO waves. Transient PGO-related decrements and increments in diaphragm activity are associated with lengthening and shortening of inspiratory duration, respectively (Kline et al., 1986). Baust et al. (1972) demonstrated that when isolated PGO waves occur during inspiration, breath time increases by 25%–30% in cats. In a human study, Neilly et al. (1991) found a strong increase in respiratory rate in association with eye movements. In an early study, Aserinsky described 10%–20% changes in respiratory rate and tidal volume with flurries of eye movements; changes which could even mimic Cheyne-Stokes respiration (Aserinsky, 1965). The disparity of these findings emphasized the fact that brainstem phasic events typified by PGO waves may exert both excitatory and inhibitory effects on respiration. Hence, the most consistent relationship between phasic events and respiration is increased breath to breath variability.

Relationship Between REM Sleep Phasic Events and Apneas

Changes in respiratory control associated with REM phasic events are relevant to the occurrence of apnea, hypopnea and upper airway flow limitation. The English bulldog, with significant upper airway anatomical abnormalities, exhibits obstructive sleep apneas during REM sleep (Hendricks et al., 1987; Hendricks et al., 1991; Veasey et al., 1996). In this animal model, tonic reductions in motor output are insufficient to produce apnea; rather, apneas occur in association with further phasic decrements in muscle activities coincident with eye movements.

In the rat, eye movements are associated with irregular activation of the diaphragm and upper airway muscles, and with central apnea (Sherrey and Megirian, 1980). In addition, the rate of apneas is highest during REM sleep in the rat (Mendelson et al., 1988; Carley et al., 1996a, 1996b) as in man (Lugaresi et al., 1978). Similar eye movement-related changes have been noted in man including: (1) reductions in the motor outputs to upper airway muscles (Weigand et al., 1991); (2) increased upper airway resistance (c.f. Gaultier, 1994); (3) decreased minute ventilation (Gould et al., 1988); and (4) apnea (Hanly, 1992). Thus, understanding the neural basis for these phasic respiratory phenomena is both intrinsically interesting and of great clinical relevance.

Figures 1 to 4 provide the first direct evidence relating PGO waves to the occurrence of apnea and hypopnea. We instrumented a group of 3 month old Sprague-Dawley rats with electrodes for recording cortical EEG, nuchal muscle EMG and pontine PGO waves using bilateral electrodes stereotaxically targeted

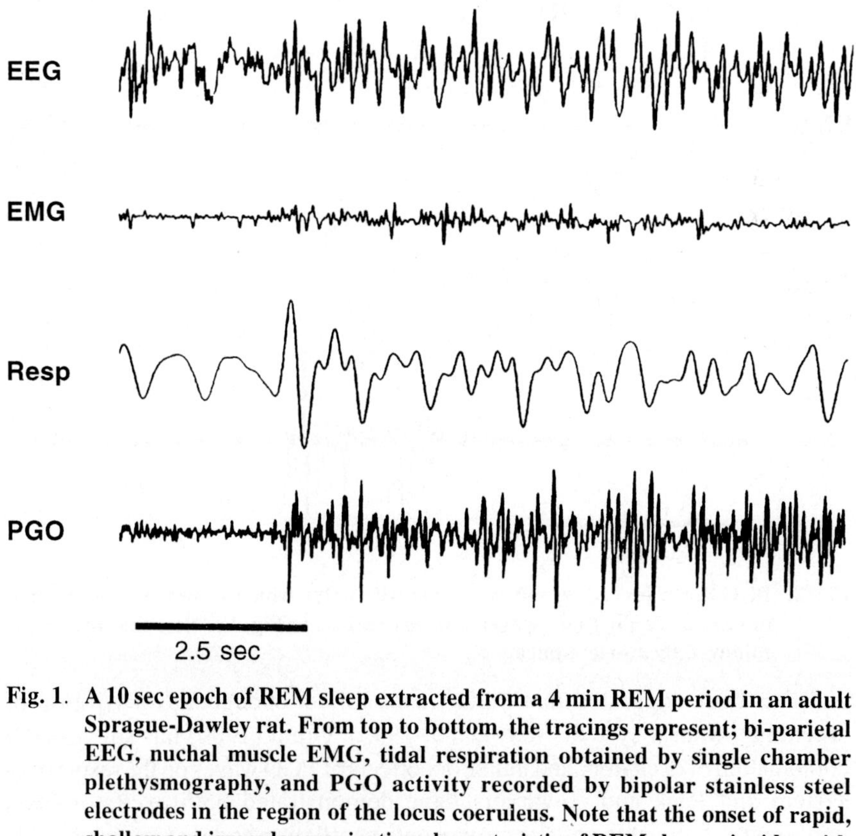

Fig. 1. **A 10 sec epoch of REM sleep extracted from a 4 min REM period in an adult Sprague-Dawley rat. From top to bottom, the tracings represent; bi-parietal EEG, nuchal muscle EMG, tidal respiration obtained by single chamber plethysmography, and PGO activity recorded by bipolar stainless steel electrodes in the region of the locus coeruleus. Note that the onset of rapid, shallow and irregular respiration characteristic of REM sleep coincides with the onset of a flurry of PGO waves.**

to the region of the locus coeruleus. Figure 1 represents a 10 sec epoch of REM sleep extracted from the middle of a 4 min REM period. The upper tracing in this and Figs. 2 to 4 is the bi-parietal cortical EEG obtained using skull screws. The next tracing is nuchal muscle EMG, followed by respiration obtained by single chamber plethysmography. The bottom tracing is PGO activity recorded by bilateral stainless steel electrodes in the region of the locus coeruleus. Figure 1 depicts the pattern of rapid, shallow and irregular respiration typically associated with human and rat REM sleep. Notice that the beginning of this record is characterized by regular tidal breathing and an absence of PGO waves. Coincident with the onset of PGO waves, respiration becomes highly erratic, suggesting that the PGO-related phasic events of REM sleep may be the primary source of respiratory instability. This observation is consistent with the relationships between eye movements and respiration observed during human sleep.

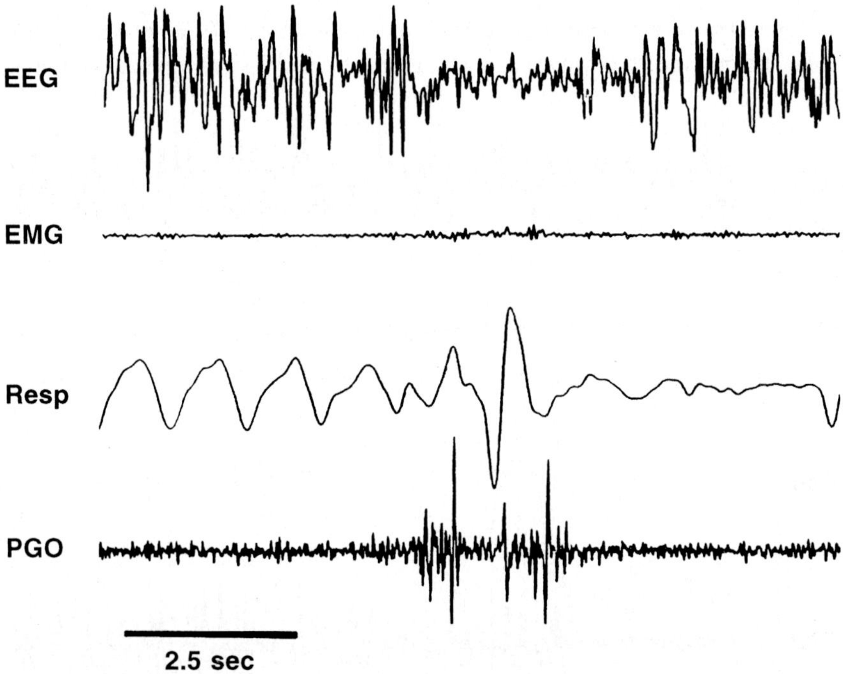

Fig. 2. PGO flurry preceding and associated with a sigh which occurs in a background of regular respiration (waveforms ordered as in Fig. 1.). Note that the sigh is followed by a brief apnea.

It has been suggested that PGO waves reflect endogenous activation of ascending arousal systems (c.f. Datta, 1995). Orem (1980) has suggested a combination of excitatory and inhibitory effects of PGO waves on the respiratory system and Pack and coworkers have demonstrated both excitatory and inhibitory relationships of eye movements ventilation and respiratory muscle activity (Millman et al., 1988; Kline et al., 1986). Figure 2 illustrates a PGO flurry preceding and associated with an augmented breath, or sigh. This event occurs in a background of regular respiration and the sigh is followed by a brief apnea. Thus, even when PGO-related respiratory changes are initially excitatory, they can precipitate apnea or hypopnea.

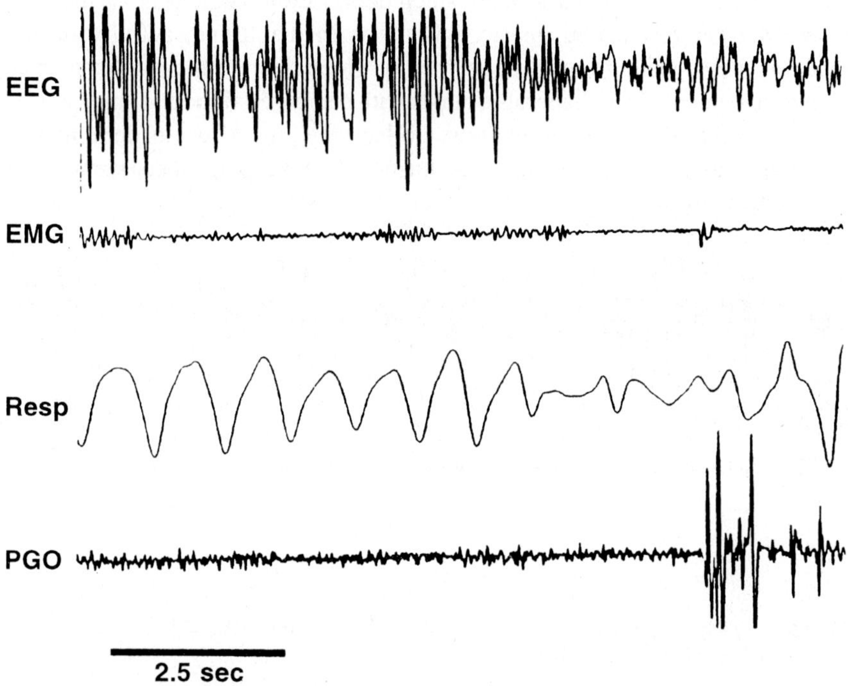

Fig. 3. A 10 sec record depicting REM onset immediately associated with apnea/hypopnea. This apnea is reversed at the moment the first volley of PGO waves occurs.

Another apparently excitatory correlation between PGO waves and apnea is illustrated in Fig. 3. In this 10 sec record REM sleep onset is associated with apnea/hypopnea which is reversed at the moment the first volley of PGO waves occurs.

Figure 4 demonstrates that PGO waves can also have inhibitory correlations with respiratory pattern. In Fig. 4 regular respiration is interrupted by an apnea of 7–8 breaths duration. This apnea is coincident with an intense flurry of PGO waves, but no arousal or state change is observable in the EEG or EMG.

Effects of Sleep Deprivation on Sleep Apnea

Acute sleep deprivation precipitates numerous changes in sleep homeostatic mechanisms and autonomic function. Twenty four hour sleep deprivation leads to acute diathesis of sleep apnea in man (Persson and Svanborg, 1996) and sleep disordered respiration in a canine model of experimental airway obstruction (O'Donnell et al., 1994). Forty eight hour REM sleep deprivation in normals induced an increased frequency of breaths in which ventilation was reduced below the range for tonic REM sleep. These events were associated with longer REM periods and an increase in total phasic activity during REM (Neilly et al., 1992). In cats, acute sleep deprivation resulted in an increase in REM sleep-associated phasic alterations in diaphragm activity during recovery REM sleep (Veasey et al., 1993). Sleep deprivation also caused an increased percentage of diaphragm bursts with decreased slope and an increased number of diaphragm fractionations. Collectively, these results demonstrate that respiratory control mechanisms during REM sleep are sensitive to prior sleep deprivation.

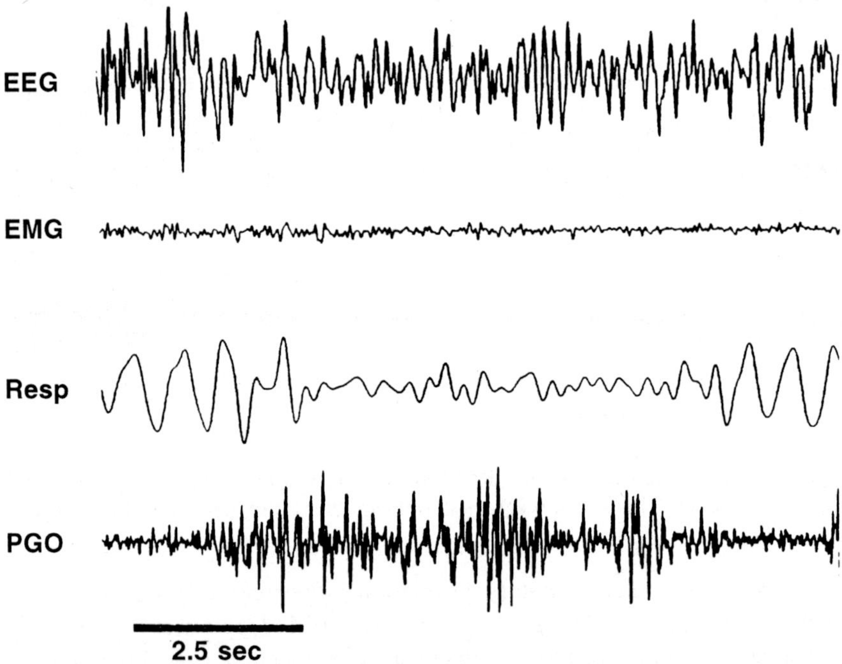

Fig. 4. Regular respiration interrupted by an apnea of 7–8 breaths duration. The apnea is coincident with an intense flurry of PGO waves, but no arousal or state change is visible in the EEG or EMG.

We examined the effects of 48 h REM sleep deprivation in Sprague-Dawley (Carley et al., 1996a), Wistar-Kyoto (Carley et al., 1996b) and SHR rats (Carley et al., 1996b). Recovery sleep exhibited equivalent increases in REM sleep in all strains. Proportional increases in REM-related apnea occurred in both normotensive strains, but SHR rats demonstrated a stronger, disproportionate, increase in REM apneas during recovery sleep. These observations support the hypothesis that states of low respiratory drive (e.g. hypertension) combined with conditions of intense phasic events (e.g. recovery from REM sleep deprivation) strongly promote respiratory instability and apnea.

Impact of Acoustic Stimulation on PGO Waves and Respiratory Control

Arousal from sleep elicited by brief acoustic stimulation exerts a substantial impact on autonomic control. Tone bursts leading to 3–15 sec arousals are associated with increased respiratory frequency and tidal volume (Carley et al., 1997; Dempsey, 1996; Horner, 1996), increased diaphragmatic and palatal muscle activation (Carlson et al., 1994) and increased heart rate and blood pressure (Davies et al., 1993; Horner, 1996). Moreover, the duration of obstructive apnea is reduced by acoustic stimulation applied during the airway occlusion (Basner et al., 1995). It is of interest that each of the above phenomena may be elicited by acoustic stimulation even when generalized electrocortical arousal is not apparent. This suggests that such acoustic orienting stimuli may elicit important autonomic responses by activating specific brainstem phasic response systems, including the PGO generation system.

PGO waves in baboons, guinea pigs, cats and rats can be elicited by sensory stimulation in all behavioral states. Evidence from Kaufman and Morrison (1981) suggests that spontaneous and elicited PGO waves recorded from the area of the locus ceruleus of the rat are produced by the same neurons. Elicited PGO waves were never recorded if spontaneous PGO waves were not recorded from the same electrode, or vice versa. The waveforms of spontaneous and elicited PGO waves were virtually identical. The occurrence and amplitude of elicited PGO waves was state dependent in a manner consistent with the conclusion that these waves are an electrophysiological sign of sensory responsiveness, and that their spontaneous appearance during and preceding REM sleep reflects endogenously generated reticular activation (Kaufman, 1983). More recently Sanford et al. (1992) demonstrated that acoustically evoked PGO waves do not habituate after more than 100 tones.

We used repetitive tone bursts similar to the paradigm of Kaufman (1983) which were sufficient to consistently elicit PGO waves without habituation in rats. By applying a single tone burst every 30 sec throughout a 6 h recording we caused a six-fold increase in apnea expression (Carley et al., 1996d). This effect was greatest during REM sleep, growing stronger throughout the recording such that during the final two hours almost every tone which occurred during REM sleep produced an apnea (Carley et al., 1996d). These findings clearly demonstrate the potent ability of PGO-eliciting acoustic stimulation to destabilize respiration, especially during REM sleep.

Pharmacological Suppression of REM Sleep Apneas in Rats

Intravenous injection of serotonin (5-HT), or 5-HT$_2$ or 5-HT$_3$ receptor agonists elicits transient apnea in anesthetized rats (Yoshioka et al., 1992). We recently determined that intraperitoneal bolus injection of 5-HT (.79 mg/kg) exacerbates the expression of spontaneous apnea in conscious Sprague-Dawley rats. As Fig. 5 shows, two to three-fold increase in apnea expression induced by 5-HT was restricted to REM sleep. Furthermore, this serotonin-induced anea expression was completely blocked by pretreatment with Ondansetron (GR38032F, Glaxo-Wellcome), a 5-HT_3 selective receptor antagonist, at a dose (.1 mg/kg) which had no effect when administered alone. Taken together, these findings argue that activation of 5-HT_3 receptors within the peripheral nervous system destabilizes respiration during REM sleep.

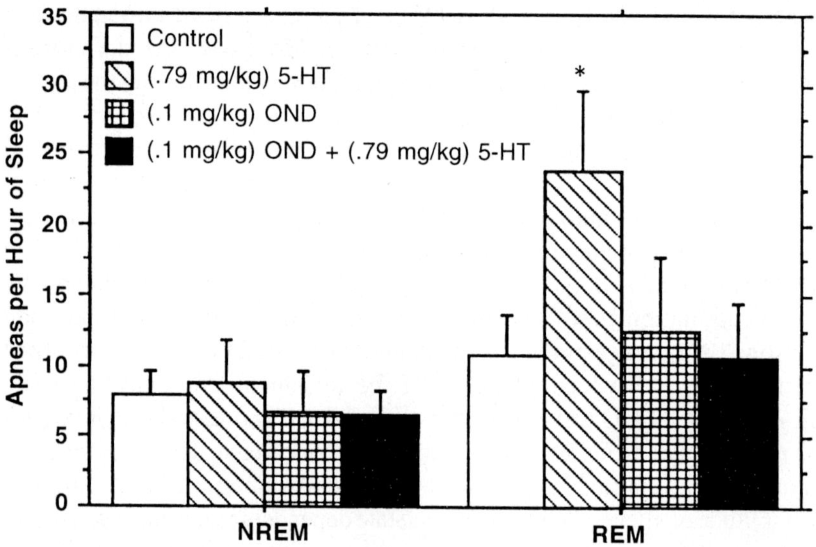

Fig. 5. Bolus injection of 5-HT (.79 mg/kg i.p.) exacerbates REM-related spontaneous apnea in rats (right-hand panel) without affecting apnea during NREM sleep (left panel). The impact of 5-HT on apnea expression is completely blocked by GR38032F, a 5-HT_3 receptor antagonist, at a dose which had no independent effect on apneas. $^*p < 01$.

To determine whether endogenous 5-HT may influence REM-related respiratory stability, we administered a higher (1 mg/kg) dose of Ondansetron. As depicted in Fig. 6, this treatment was associated with dramatic suppression of REM-related spontaneous apnea for at least 6 h. This supports the conclusion that endogenous serotonin, most probably acting at 5-HT_3 receptors in the peripheral nervous system, does indeed destabilize respiration during REM sleep in the rat. These observations also suggest that serotonin receptors may be an appropriate target for pharmacological management of sleep apnea in man.

Conclusions

REM sleep generates increased variability in virtually every autonomic overlaid on a background of skeletal muscle atonia. This variability may have adverse consequences for respiratory homeostasis including apnea and hypopnea, events associated with behavioral and vascular morbidity and mortality. Strong circumstantial evidence links the incidence of brainstem phasic events, typified

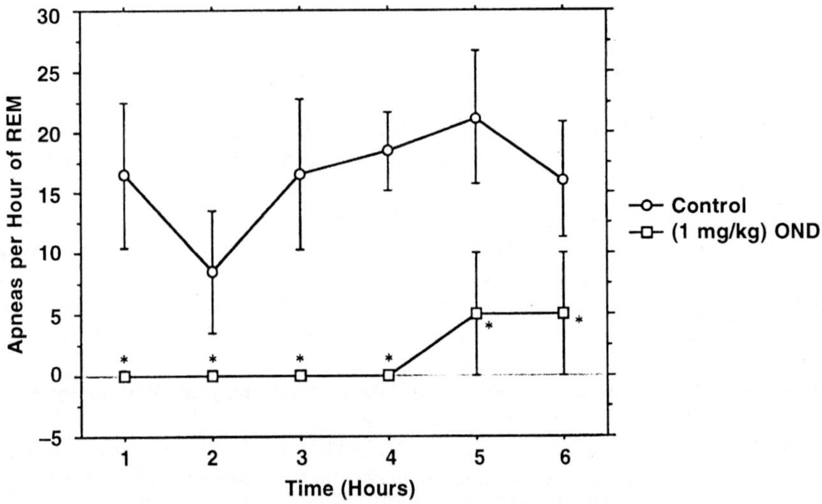

Fig. 6. Hourly expression of REM-related spontaneous apnea during 6 h recordings of rats pretreated with 5-HT$_3$ antagonist GR38032F. REM-related apnea was completely suppressed in all animals (N = 9) for at least 4 h $^*P <. 05$.

by PGO waves, to the occurrence of apnea and hypopnea but we know virtually nothing of the mechanisms which mediate this connection. Clearly, this is an important challenge to the field today; to understand the neural bases for sleep-related changes in control of respiration. Much has been learned about the fundamental mechanisms which regulate sleep states and REM related phasic events. We need to extend this approach to understand the bases of state dependent autonomic function. Animal model will be essential in this endeavor and recent advances have been made. Our group has illustrated one example of this strategy, by characterizing spontaneous respiratory instability during sleep in the rat and demonstrating association of spontaneous and evoked phasic events to respiratory instability in this model. Future work must capitalize on this and related animal models to understand the neural mechanisms of state dependent respiratory control, an endeavor which will yield basic and clinical benefits.

References

Ancoli-Israel, S. and Kripke, D.F. Prevalent sleep problems in the aged. Biofeedback and Self-Regulation, 16 (1991) 349–359.

Aserinsky, E. and Kleitman, N. Regularly occurring periods of eye motility, and concommitant phenomena during sleep, Science, 118 (1953). 273–274.

Aserinsky, E. Periodic respiratory pattern occurring in conjunction with eye movements during sleep, Science, 150 (1965) 763–766.

Basner, R.C., Önal, E., Carley, D.W. Stepanski, E.J. and Lopata, M. Effect of induced transient arousal on obstruvtive apnea duration, J. Appl. Physiol., 78 (1995) 1469–1476.

Baust, W., Holzback, E. and Zechlin, O. Phasic changes in heart rate and respiration correlated with PGO-spike activity during REM sleep, Pflugers Arch., 331 (1972) 113–123.

Carley, D.W., Trbovic, S. and Radulovacki, M. Effect of REM sleep deprivation on sleep apneas in rats, Exp. Neurol., 137 (1996a) 291–293.

Carley, D.W., Trbovic, S. and Radulovacki, M. Effects of REM sleep deprivation on apnea in spontaneously hypertensive (SHR) and normotensive Wistar rats, Physiol. Behav., 59 (1996b) 827–831.

Carley, D.W., Trbovic, S. and Radulovacki, M. Hydralazine reduces elevated sleep apnea index in spontaneously hypertensive (SHR) rats to equivalence with normotensive Wistar-Kyoto rats, Sleep, 19 (1996c) 363–366.

Carley, D.W., Trbovic, S., Monti, D. and Radulovacki, M. Effects of sleep fragmentation and clonidine administration on apnea in the rat, Res. Comm. Biol. Psychol. Psychiat., 20 (1996d) 95–111.

Carley, D.W., Applebaum, R., Basner, R.C., Önal, E. and Lopata, M. Respiratory and arousal responses to acoustic stimulation, Chest, 112 (1997) 1567–1571.

Carlson, D.M., Carley, D.W., Önal, E. Lopata, M. and Basner, R.C. Acoustically induced transient arousal increases pharyngeal and diaphragm muscle phasic EMG in normals, J. Appl. Physiol., 76 (1994) 1553–1559.

Datta, S., Calvo, J.M. Quatrochi, J.J. and Hobson, J.A. Cholinergic microstimulation of the peribrachial nucleus in the cat. I. Immediate and prolonged increases in ponto-geniculo-occipital waves, Arch. Ital. Biol., 130 (1992) 263–284.

Datta, S. Neuronal activity in the peribrachial area: relationship to behavioral state control, Neurosci. Biobehav. Rev., 19 (1995) 67–84.

Davies, R.J., Belt, P.J., Roberts, S.J., Ali, N.J. and Strandling, J.R. Arterial blood pressure responses to graded transient arousal from sleep in normal humans, J. Appl. Physiol., 74 (1993) 1123–1130.

Dempsey, R.J., Smith, C.A., Harms, C.A., Chow, C. and Saupe, K.W. Sleep-induced breathing instability. University of Wisconsin-Madison sleep and respiration research group, Sleep, 19 (1996) 236–247.

Farber, J., Marks, G.A. and Roffwarg, H.P. Rapid eye movement sleep PGO-type waves are present in the dorsal pons of the albino rat, Science, 209 (1980) 615–617.

Gaultier, C. Upper airway muscles and physiopathology of obstructive sleep apnea syndrome, Neurophysiol. Clin., 24 (1994) 195–206.

Gould, G.A., Gugger, M., Molloy, J., Tsara, V., Shapiro, C.M. and Douglas, N.J. Breathing pattern and eye movement density during REM sleep in humans, Am. Rev. Respir. Dis., 138 (1988) 874–877.

Hanly, P.J. Mechanisms and management of central sleep apnea, Lung, 170 (1992) 1–17.

He, J., Kryger, M., Zorick, F., Conway, W. and Roth, T. Mortality and apnea index in obstructive sleep apneas, Chest 89 (1988) 331–334.

Hendricks, J.C., Kline, L.R., Kovalski, R.J., O'Brien, J.A., Morrison, A.R. and Pack, A.I. The English bulldog: a natural model of sleep-disordered breathing, J. Appl. Physiol., 63 (1987) 1344–1350.

Hendricks, J.C. Kovalski, R.J. and Kline, L.R. Phasic respiratory muscle patterns and sleep-disordered breathing during rapid eye movement sleep in the English bulldog, Am. Rev. Respir. Dis., 144 (1991) 1112–1120.

Hoch, C.C., Reynolds, C.F., Kupfer, D.J. and Berman, S.R. Stab;ility of EEG sleep and sleep quality in healthy seniors, Sleep, 11 (1988) 521–527.

Hoch, C.C., Reynolds, C.F. Monk, T.H. Buysse, D.J. et al. Comparison of sleep-disordered breathing among healthy elderly in the seventh, eight and ninth decades of life, Sleep, 13 (1990) 502–511.

Horner, R.L. Autonomic consequences of arousal from sleep: mechanisms and implications, Sleep 19 (1996) S193–S195.

Kaufman, L.S. and Morrison, A.R. Spontaneous and elicited PGO spikes in rats, Brain Res., 214 (1981) 61–72.

Kaufman, L.S. PGO waves in rats in the non-paradoxical sleep states, Brain Res., 276 (1983) 73–80.

Kline, L.R., Hendricks, J.C., Davies, R.O. and Pack, A.I. Control of the activity of the diaphragm in rapid eye movement sleep, J. Appl. Physiol., 61 (1986) 1293–1300.

Krieger, J. and Kurtz, D. EEG changes before and after apnea, In: *Sleep Apnea Syndromes,* C. Guilleminault and W.C. Dement (Eds.) Alan R. Liss, NY, 1978, pp. 161–176.

Kuna, S.T. and Sant'Ambrogio, G. Pathophysiology of upper airway closure during sleep, JAMA 266 (1991) 1384–1389.

Lugaresi, E., Coccagna, G. and Mantovani, M. Hypersomnia with periodic apneas. In: *Advances in Sleep Research, Volume 4,* E. Weitzman (Ed.), Spectrum Publications, New York, 1978, pp. 68–70.

Marks, G.A., Farber, J. and Roffwarg, H.P. Metencephalic localization of ponto-geniculo-occipital waves in the albino rat, Exper. Neurol., 69 (1980) 667–677.

Marks, G.A., Farber, J. and Roffwarg, H.P. Phasic influences during REM sleep upon dorsal lateral geniculate nucleus unit activity in the rat, Brain Res., 222 (1981) 388–394.

Mendelson, W.B., Martin, J.V. Perlis, M. Giesen, H. Wagner, R. and Rapoport, S.I. Periodic cessation of respiratory effort during sleep in adult rats, Physiol. Behav., 43 (1988) 229–234.

Millman, R.P., Knight, H. Kline, L.R. Shore, E.T. Chung, D.C. and Pack, A.I. Changes in compartmental ventilation in association with eye movements during REM sleep, J. Appl. Physiol., 65 (1988) 1196–1202.

Neilly, J.B., Gaipa, E.A. Maislin, G. and Pack, A.I. Ventilation during early and late rapid eye movement sleep in normal humans, J. Appl. Physiol., 71 (1991) 1201–1215.

Neilly, J.B., Kribbs, N.B. Maislin, G. and Pack, A.I. Effects of selective sleep deprivation on ventilation during recovery sleep in normal humans, J. Appl. Physiol., 72 (1992) 100–109.

O'Donnell, C.P., King, E.D. Schwartz, A.R. Smith, P.L. and Robotham, J.L. Effect of sleep deprivation on responses to airway obstruction in the sleeping dog, J. Appl. Physiol., 77 (1994) 1811–1818.

Orem, J. Neuronal mechanisms of respiration in REM sleep, Sleep, 3 (1980) 251–267.

Pack, A.I., Changes in respiratory motor activity during rapid eye movement sleep. In: M., Partinen, A., Jamieson and C., Guilleminault. *Long-term outcome for obstructive sleep apnea syndrome patients,* Chest, 94 (1988) 1200–1204.

Partinen, M. and Guilleminault, C. Daytime sleepiness and vascular morbidity at seven-year follow up in obstructive sleep apnea patients, Chest, 97 (1990) 27–32.

Persson, H.E. and Svanborg, E. Sleep deprivation worsens obstructive sleep apnea. Comparison between diurnal and nocturnal polysomnography, Chest 109 (1996) 645–650.

Radulovacki, M., Trbovic, S. and Carley, D.W. Acute hypotension reduces sleep apneas in Zucker lean and Zucker obese rats, Sleep, 19 (1996) 767–773.

Radulovacki, M., Trbovic, S.M. and Carley, D.W. Cardiopulmonary interactions following REM sleep deprivation in Sprague-Dawley rats, Exp. Neurol., 145 (1997) 371–375.

Redline, S. and Young, T. Epidemiology and natural history of obstructive sleep apnea, ENT Journal, 72 (1993) 20–26.

Reynolds, C.F., Coble, P.A. Black, R.S. Holzer, B. Carroll, R. and Kupfer, D.J. Sleep disturbances in a series of elderly patients: polysomnographic findings, J. Am. Geriatr. Soc., 28 (1980) 164–70.

Sanford, L.D., Ball, W.A. Morrison, A.R. Ross, R.J. and Mann, G. Peripheral and central components of alerting: habituation of acoustic startle, orienting responses, and elicited waveforms, Behav. Neurosci., 106 (1992) 112–120.

Schwab, R.J. Properties of tissues surrounding the upper airway, Sleep, 19 (1996) S170–S174.

Seick, G.C., Trelease, R.B. and Harper, R.M. Sleep influences on diaphragmatic motor unit discharge, Exp. Neurol., 85 (1984) 316–335.

Shepard, J.W., Gefter, W.B. Guilleminault, C. Hoffman, E.A. Hoffstein, V. Hudgel, D.W. Suratt, P.M. and White, D.P. Evaluation of the upper airway in patients with obstructive sleep apnea, Sleep, 14 (1991) 361–371.

Sherrey, J.H. and Megirian, D. Respiratory EMG activity of the posterior cricoarytenoid, cricothyroid and diaphragm muscles during sleep, Respir. Physiol., 39 (1980) 355–365.

Shouse, M.N. and Siegel, J.M. Pontine regulation of REM sleep components in cats: integrity of the pedunculopontine tegmentum (PPT) is important for phasic events but unnecessary for atonia during REM sleep, Brain Res., 571 (1992) 50–63.

Stradling, J.R. and Phillipson, E.A. Breathing disorders during sleep, Quart. J. Med., 225 (1986) 3–18.

Veasey, S.C., Hendricks, J.C., Kline, L.R. and Pack, A.I. Effects of acute sleep deprivation on control of the diaphragm during REM sleep in cats, J. Appl. Physiol., 74 (1993) 2253–2260.

Veasey, S.C., Panckeri, K.A. Hoffman, E.A. Pack, A.I. and Hendricks, J.C. The effects of serotonin antagonists in an animal model of sleep-disordered breathing, Am. J. Respir. Crit. Care. Med., 153 (1996) 776–786.

Verrier, R.L., Muller, J.E. and Hobson, J.A. Sleep, dreams and sudden death: the case for sleep as an autonomic stress test for the heart, Cardiovasc. Res., 31 (1996) 181–211.

Webb, P. Periodic breathing during sleep, J. Appl. Physiol., 37 (1974) 899–903.

Webster, H.H. and Jones, B.E. Neurotoxic lesions of the dorsolateral pontomesencephalic tegmentum-cholinergic cell area in the cat. II. Effects upon sleep-waking states, Brain Res., 458 (1988) 285–302.

Weigand, L., Zwillich, C.W. Wiegand, D. and White, D.P. Changes in upper airway muscle activation and ventilation during phasic REM sleep in normal men, J. Appl. Physiol., 71 (1991) 488–497.

Yoshioka, M., Goda, Y. Togashi, H. Matsumoto, M. and Saito, H. Pharmacological characterization of 5-hydroxytryptamine-induced apnea in the rat, J. Pharmacol. Exp. Ther., 260 (1992) 917–924.

Rapid Eye Movement Sleep
B.N. Mallick and S. Inoué (Eds)

20. Thermoregulatory Control of the NonREM-REM Sleep Cycle

Ronald Szymusiak[1,3], Md. Noor Alam[1,2] and Dennis McGinty[1,2]

[1]Research Service (151 A3), Veterans Administration Medical Center, North Hills, CA

[2]Department of Psychology, [3]Department of Medicine, School of Medicine, University of California, Los Angeles, CA

SUMMARY

Manipulations of environmental temperature or of brain/body core temperatures exert powerful effects on the amount and timing of nonREM and REM sleep in mammals. Thermoregulatory variables appear to interact with sleep-regulatory mechanisms in several ways. First small changes in environmental temperature with respect to the thermoneutral zone can alter time spent in REM sleep, independent of changes in nonREM sleep parameters. These effects can be achieved by changing skin temperature in the absence of core body temperature changes, and may reflect a relatively direct thermosensitive modulation of brainstem REM sleep generating mechanisms. A second type of thermoregulatory modulation of REM sleep occurs primarily through effects on nonREM sleep, and is mediated by thermosensitive, sleep/wake-regulating cell types located in the preoptic/anterior hypothalamic area. In addition, hypothalamic warm-sensitive neurons can be hypothesized to regulate cycling between nonREM and REM sleep during sustained sleep episodes, via inhibitory modulation of monoaminergic afferents to brainstem cholinergic cell groups.

Key Words: nonREM sleep, body temperature, skin temperature, thermosensitivity, thermoregulation, preoptic area, anterior hypothalamus.

Introduction

A basic feature of mammalian sleep is the orderly alternation of periods of nonrapid-eye-movement sleep (quiet sleep, slow-wave sleep) with episodes of rapid-eye movement (REM) sleep. While total daily sleep amounts and the proportion of time spent in nonREM and REM sleep can vary substantially across species (Zepelin and Rechtschaffen, 1974), certain features of the nonREM-REM sleep cycle are more constant. First, in stable-state conditions (normal circadian entrainment, absence of sleep deprivation) and in the absence of pathology, major sleep episodes in adult animals begin with a period of nonREM sleep. This is true in species that exhibit highly consolidated sleep (e.g., humans and some nonhuman primates with one major sleep period/ day), and those with more distributed sleep (e.g., rats and cats). Second, within major sleep periods, an initial REM sleep episode emerges from nonREM sleep with a characteristic latency. While REM sleep latency may be somewhat variable in species with polyphasic sleep, in man this measure exhibits sufficient consistency between individuals to serve as a marker of pathology (Kupfer, 1976; Guilleminault, 1974). Third, following the initial REM sleep period, episodes of nonREM and REM sleep are alternately expressed with a species-specific periodicity (nonREM-REM cycle length).

As evidenced by the content of this volume, the brain mechanisms responsible for these features of mammalian sleep are complex. Considerable progress has been made in identifying neural circuits and the neuropharmacology of brainstem mechanisms responsible for the individual components of REM sleep (EEG activation, antigravity muscle atonia) and the generation of the REM sleep state. A critical role for brainstem cholinergic mechanisms in most major aspects of REM sleep control is firmly established, and interactions among brainstem cholinergic and monoaminergic cell groups are hypothesized to underlie alternations between nonREM and REM sleep (see Steriade and McCarley, 1990 for review).

While brainstem mechanisms are critically important in orchestrating the manifestations and timing of REM sleep, a variety of factors appear to modulate these brainstem mechanisms. Important among these are thermoregulatory variables. Manipulations of environmental temperature (T_{en}) or of central thermoregulatory/thermosensing control systems exert powerful effects on the amount and timing of nonREM and REM sleep. Indeed, Wehr (1992) has proposed a primary thermoregulatory function for REM sleep, i.e., to raise brain and core body temperature in compensation for falls in temperature occurring during nonREM sleep.

This chapter reviews the effects of thermoregulatory variables on the nonREM-REM sleep cycle. First, we will review the effects of peripheral and central temperature manipulations on REM sleep, the relationship of thermoregulatory stress to REM sleep expression, and the relative importance of peripheral versus core body temperature changes in REM sleep control. Second, we will review how thermosensitive neurons in the preoptic anterior/ hypothalamic area (POA) are involved in the regulation of nonREM sleep onset, sleep depth, and sleep maintenance. Third we will discuss how interactions among hypothalamic and brainstem thermoregulatory mechanisms may contribute to the expression of nonREM-REM sleep cycles.

Influences of Environmental Temperature and Thermoregulatory Stress on REM Sleep

Acute manipulations of (T_{en}) exert powerful effects on the amount and composition of sleep in mammals. Total sleep time is greatest at thermoneutral temperatures (T_{en} at which resting rates of heat production are approximately equal to passive rates of heart loss, and minimal thermoregulatory effort is required to maintain body temperature), and declines in the heat and the cold. This has been documented in cats (Parmeggiani and Rabini, 1970), rats (Schmidek et al., 1972), mice (Roussel et al., 1984), kangaroo rats (Sakaguchi et al., 1979), and man (Haskell et al., 1981). Rather than absolute T_{en} level, the critical variable appears to be the relationship of T_{en} to the thermoneutral zone. Thus, in animals that are cold acclimated and undergo a shift in the thermoneutral zone toward lower T_{en}, the cold-induced suppression of REM sleep in attenuated (Schmidek et al., 1972).

NonREM and REM sleep are not equally sensitive to changes in T_{en}. As originally demonstrated in cats (Parmeggiani, Rabini, 1970), REM sleep amounts are more severely reduced than nonREM sleep amounts in response to heat or cold exposure. In rats, changes in T_{en} of just a few °C can result in significant alterations in time spent in REM sleep, with little or no change in nonREM sleep amounts (Szymusiak and Satinoff, 1981). While, nonREM sleep is less sensitive than REM sleep to changes in T_{en} amounts of nonREM sleep, particularly deep nonREM sleep, do decline in response to significant heat or cold stress.

The sensitivity of REM sleep to changes in T_{en} can be partially understood in the context of changes in thermoregulatory control that occur during REM sleep. As for other homeostatic systems, thermoregulatory function is significantly altered during REM sleep, compared to waking or nonREM sleep (see Parmeggiani, 1980 for review). Physiological mechanisms of both cold defense and, to a lesser extent, heat defense are compromised during the REM sleep state. For instance, the increases in metabolic heat production observed in kangaroo rats in response to focal hypothalamic cooling during waking and nonREM sleep, is absent during REM sleep (Glotzbach and Heller, 1976). Cats exposed to a cold T_{en} will display shivering during waking and nonREM sleep, but shivering will stop at the onset of REM sleep (Parmeggiani and Rabini, 1967). Absence of shivering is not due simply to the suppression of muscle tone that accompanies REM sleep. Cats with pontine lesions that exhibit REM sleep without muscle atonia, shiver normally during waking and nonREM sleep when exposed to cold. However, this shivering activity stops with the onset of REM sleep (Hendricks, 1982). REM sleep appears to be associated with fundamental changes in the functioning of central thermoregulatory control systems, such that compensatory responses to cold stress are either absent or severely attenuated during this state.

Loss of heat defense ability in REM sleep is a less consistent finding. In appropriate changes in peripheral vasomotor tone (e.g., vasoconstriction in a warm T_{en}) are commonly observed during REM sleep (Parmeggiani et al., 1977). Thermoregulatory panting in response to heat exposure has been reported to be diminished during REM sleep, compared to waking and nonREM sleep in cats (Parmeggiani and Rabini, 1967). However, in man, sweating in response to heat exposure is preserved during REM sleep (Libert et al., 1982). Cats with pontine lesions display panting that is proportional to imposed heat loads during REM sleep without muscle atonia (Amini-Sereshki and Morrison, 1988).

In light of the diminished capacity to defend body temperature against thermal challenge, reductions in REM sleep under conditions of thermal stress would appear to be an adaptive response. How is this response achieved? One possibility is that increases or decreases in body (T_b) and/or brain (T_{br}) temperature occurring during sleep in animals exposed to extreme thermal environments, functions to prevent the expression of REM sleep. A second possibility is that the presence of ongoing thermoregulatory effector activity during nonREM sleep (e.g., shivering and huddled, heat conserving postures) are incompatible with the postural atonia of REM sleep, resulting in REM suppression. While these factors undoubtedly contribute, experiments in which core and peripheral temperatures are manipulated independently suggest that skin temperature (T_{sk}) changes can, in some circumstances, be of primary importance in mediating T_{en} effects on REM sleep.

One study examined the effects of brief T_{sk} changes on REM sleep in rats. When rats were exposed to a mild cold stress, transiently warming the skin with either convective or radiant heat during nonREM sleep significantly increased the frequency of occurrence of REM sleep, compared to control (no skin warming) nonREM sleep episodes (Szymusiak et al., 1980). Cooling the skin of mildly heat-stressed rats also promoted REM sleep. When rats were sleeping at a thermoneutral temperature, warming or cooling the skin suppressed REM sleep. Thus changes in T_{sk} toward thermoneutral levels can modulate REM sleep even when there are no changes in T_{br} or any obvious changes in thermoregulatory effector activity.

In a second study in rats, amounts of sleep were recorded for 3 h periods at different T_{en} within the zone of minimum resting metabolic rate (25–31°C). Within this T_{en} range, changes in thermoeffector activity were primarily changes in peripheral blood flow. However, REM sleep time varied more than two-fold, with a peak at 29°C (Szymusiak and Satinoff, 1981). Thus, changes in peripheral temperature, in the absence of significant alterations in T_b or T_{br}, and in the absence of changes in nonREM sleep time can profoundly alter REM sleep amounts.

Differential effects of core and peripheral temperatures on REM versus nonREM sleep were further demonstrated in a study of kangaroo rats. Animals were exposed to T_{en} of either 20°C or 30°C. At each T_{en}, hypothalamic temperature (T_{hy}) was clamped either above or below baseline levels via a chronically implanted water-perfused thermode. At T_{en} of 20°C, clamping T_{hy} 1–1.5°C above baseline increased time spent in nonREM sleep, but REM sleep amounts did not differ from those recorded at baseline levels of T_{hy} (Sakaguchi et al. 1979). In contrast, at T_{en} of 30°C, REM sleep amounts at baseline levels of T_{hy} were elevated compared to baseline values at 20°C, but increasing T_{hy} at 30°C yielded no further increases in REM sleep amounts.

The above findings demonstrate that changes in T_{sk} can modulate brainstem mechanisms that generate REM sleep, and can exert these effects on REM sleep independent of changes in nonREM sleep. Under some circumstances, changes in core body temperature or of critical hypothalamic thermosensitive regions can have effects on REM sleep (e.g., Parmeggiani et al., 1974; Szymusiak et al., 1991), but these variables appear to more readily influence nonREM sleep measures (see below).

The findings that changes in T_{sk} or T_b/T_{br} can alter REM sleep in the absence of changes in nonREM sleep, suggests a possible direct modulation of brainstem REM sleep-generating mechanisms by thermoresponsive neurons. Some experimental evidence can be found in support of such a hypothesis. Changes in T_{en} exerts a profound effect on the appearance of a "REM sleep-like state" in cats chronically maintained following a brainstem transection at the rostral pontine level. When a cycle of T_b and T_{br} change is imposed upon these animals by cycling T_{en} in an incubator, the appearance of a REM sleep-like state occurs most frequently around the T_{en} minimum (Jouvet et al., 1988). This REM sleep-like state is suppressed when the T_b of the pontine cats rose above 35°C. This finding demonstrates that pontine REM sleep-generating mechanisms can respond to T_{en} changes in the absence of input from forebrain thermoregulatory and/or thermosensing inputs. However the effects of T_{en} on the isolated pons are clearly different from those in the intact brain, as the profile of temperature changes that promote REM sleep in the pontine cat (low T_b/T_{br} and low T_{sk}), always evoke powerful suppression of REM sleep in the intact animals. Therefore, an intact forebrain is capable of modifying the response of pontine mechanisms to changes in $T_{en.}$

Cholinergic neurons of the laterodorsal pontine tegmentum and cholinoceptive neurons of the paramedian pontine reticular formation (PRF) play critical roles in REM sleep generation (see Steriade and McCarley, 1990 for review). Microinjection of cholinergic agonists in the PRF in cats evokes prolonged episodes of a REM sleep-like state (Baghdoyan et al., 1984). A substantial subpopulation of mPRF neurons (74%) studied in rats exhibit increases in discharge rate in response to warming of the skin (Farham and Douglas, 1985). The receptive fields for these responses are typically large, and bilateral in most cases. Also in rats, a majority of PRF neurons, including identified reticulospinal neurons, were found to exhibit local thermosensitivity (Asami et al., 1988). The potential for a hypothalamic modulation of the PRF is supported by the finding that a subpopulation of the locally thermosensitive neurons in the mPRF also responded to changes in T_{hy} (Asami et al., 1988).

A thermosensitive modulation of brainstem cholinergic and/or cholinoceptive neurons involved in REM sleep generation is supported by effects of T_{en} on the REM sleep-suppressing effects of atropine. Systemic administration of cholinergic antagonists acutely suppress REM sleep, and are presumed to do so via effects on brainstem cholinoceptive neurons (Shiromani et al., 1987). Adult cats given 0.5 mg/kg of the muscarinic receptor antagonist, atropine, displayed reductions in REM sleep that persisted for 8 h following drug injections, when adapted to and recorded at a T_{en} 23°C. However, in response to the same dose of atropine administered at 30°C, no significant decrease in REM sleep amounts from baseline were observed (Szymusiak et al., 1994). The mild hypothermic effect of atropine that was present at 23°C was attenuated at 30°C. One possible explanation for these results is that the activity of mPRF cholinoceptive neurons and/ or the activity of cholinergic neurons in the laterodorsal pontine tegmentum are increased at 30°C versus 23°C, thereby compensating for the muscarinic receptor blockade produce by atropine.

The hypothesis that brainstem REM sleep mechanisms are directly influenced by thermoregulatory processes is supported by additional evidence. Experimental damage to dorsolateral regions of the pontine tegmentum alters thermoeffector responses to both heat and to cold exposure. These animals display suppression of REM sleep when examined at 20–23°C, but no suppression of REM sleep at 30°C (Amini-Sereshki, 1991). Thus, the T_{en}-dependence of pontine lesion effects is much like the atropine-induced suppression of REM sleep described above. In both instances, recovery of REM sleep upon exposure to a thermoneutral T_{en} can be hypothesized to be the result of a thermosensitive modulation of pontine cholinergic and/or cholinoceptive systems involved in REM sleep control.

Thermoregulatory and Sleep Regulatory Mechanisms in the Preoptic/Anterior Hypothalamic Area

Abundant and varied evidence supports a hypothesized role for neurons within the preoptic/anterior hypothalamic area (POA) in the regulation of sleep onset and sleep maintenance. Electrolytic or neurotoxin-induced lesions of the POA yield persistent insomnia in both cats and rats (Szymusiak and Satinoff, 1984; Szymusiak et al., 1991; Sallanon et al., 1989). POA microinjection of benzodiazepines (Mendelson et al., 1989), adenosine (Ticho and Radulovacki, 1991), and prostaglandin D_2 (Ueno et al., 1982) acutely promote sleep in rats, Electrical stimulation in the lateral-ventral portion of the POA of cats can evoke sleep with EEG synchrony (Sterman and Clemente, 1962) and suppression of arousal-related neuronal discharge in the midbrain reticular formation (Szymusiak and McGinty, 1989). Neurons that display elevated discharge rates during sleep and EEG synchronization compared to waking have been described in the POA of cats, rabbits and rats (Alam et al., 1995a; Mallick et al., 1983; Alam et al., 1997; Findlay and Hayward, 1969). neurons that display elevated *c-fos* Protein immunoreactivity during sleep have been localized to the ventrolateral POA (Sherin et al., 1996).

The POA is also critically involved in the control of body temperature, It is among the most important thermosensing and thermointegrating sites in the mammalian brain. Focal POA warming or cooling can evoke compensatory thermoregulatory effector responses (see Satinoff, 1978 for review). Bilateral

POA ablation leads to persistent deficits in cold defense mechanisms and/or heat defense mechanisms (Satinoff et al., 1976; Szymusiak et al., 1991). The POA contains populations of thermosensitive neurons, i.e., those that srongly respond to increases and/or decreases in local temperature (see Boulant and Dean, 1986 for review). Warm-sensitive neurons have been found to outnumber cold sensitive neurons in most preparations. Studies of POA thermosensitive neurons in hypothalamic slice preparations indicate that a subpopulation of warm sensitive neurons are intrinsically thermosensitive, i.e., they maintain thermosensitivity under conditions of synaptic blockade (Kelso and Boulant, 1982). Cold sensitive neurons do not maintain thermosensitivity during synaptic blockade.

Thermoregulatory and sleep-regulatory functions of the POA are closely related. In cats, focal POA warming promotes sleep onset (Roberts and Robinson, 1969) and increases the depth of nonREM sleep, as evidenced by increased delta activity (0.3-4.0 Hz) in the neocortical EEG (McGinty et al., 1994). Activation of POA warm sensitive neurons by local warming or by chemical stimulation increases sleep in rats (Benedek et al., 1982; Sakaguchi et al., 1979). These findings demonstrate that acute activation of POA warm-sensitive neurons leads to acute elevations in sleep amounts.

Results of POA lesion studies demonstrate that POA warm-sensitvie neurons exert tonic control over sleep amounts. Cats with neurotoxin-induced lessions of the POA exhibit both insomnia and thermoregulatory abnormalities (Szymusial et al., 1991). The major thermoregulatory disturbance is impaired heat defense, i.e., they display an elevated core T_b threshold for panting when exposed to elevated ambient temperatures. POA-damaged cats will exhibit panting in response to increasing $T_{br,}$ but panting is initiated at T_{br} values 1.5–2.0°C above the baseline temperature threshold. When placed for several hours at an ambient temperature of 23°C, where T_b is maintained at baseline levels of 38.5–39.0°C, animals exhibit significant insomnia, with reductions in both nonREM and REM sleep amounts compared to baseline. However, when placed at an ambient temperature of 33°C, where T_b is maintained at 40.0–40.5°C, amounts of nonREM and REM sleep increase to control levels. These tonic sleep and thermoregulatory disturbances can both be interpreted to relfect neurotoxin-induced loss of POA warm-sensitive neurons.

These findings suggest that activation of POA warm sensitive cell types should accompany spontaneous transitions from wakefulness to sleep, i.e., a subpopulation of POA warm-sensitive neurons should also display sleep-related neuronal discharge. We have confirmed this in our studies of medial POA neuronal activity during wakefulness and sleep in cats and rats (Alam et al., 1995a; Alam et al., 1997). In cats, a majority of all POA warm-sensitive neurons studied (64%) also displayed spontaneously elevated discharge during nonREM sleep compared to waking. In contrast, 70% of medial POA cold-sensitive neurons had elevated discharge rates during waking compared to nonREM sleep. In both warm- and cold-sensitive cells types, changes in discharge rate were evident during state transition periods. Thus, discharge rate in warm-sensitive neurons beings to increase several seconds prior to the initial signs of EEG synchrony during wake to sleep transitions. This suggests that increasing discharge in medial POA warm-sensitive neurons, as well as diminishing activity of cold-sensitive cell types facilitate the transition from wake to sleep. Similar results were obtained during a study of warm- and cold-sensitive neurons in the diagonal band/medial POA region of rats (Alam et al., 1997).

It was furher shown that the thermosensitivity of warm-sensitive neurons (i.e., the magnitude of the increase in discharge rate/°C change in local temperature) is elevated during nonREM sleep compared to waking in cats (Alam et al., 1996). Thermosensitivity in cold-sensitive neurons declined during nonREM sleep with respect to waking. These changes in thermosensitivity help to amplify state-dependent changes in discharge rate during nonREM sleep onset as T_b and T_{br} fall. This could be a mechanism to help stabilize nonREM-waking transitions, and to help promote the development of consolidated nonREM sleep epidsodes.

Examination of medial POA neuronal thermosensitivity in REM sleep revealed that cold sensitive neurons did not retain thermosensitivity during this state, while thermosensitivity of warm-sensitive neurons in REM sleep was similar to waking values (Alam et al., 1995b). This finding explains why cold defense mechanisms are more impaired during REM sleep than are heat defense mechanisms, and suggests that some thermosensitive influence of hypothalamic origin can persist during REM sleep.

It should be emphasized that not all POA warm-sensitive and cold-sensitive neurons displayed sleep-wake state dependent discharge patterns. The remaining thermosensitive neurons can be hypothesized to mediate thermoregulatory responses independent of sleep-waking state. The existence of these two populations of POA thermosensitive neurons could explain the findings that, while most manipulations of POA have both sleep and thermoregulatory effects, some experimental manipulations can influence these functions independently (Mallick and Alam, 1992).

Following our series of studies on the sleep- and wake-related discharge of medial POA neurons, it was reported that within the ventrolateral POA of rats, of discrete population of neurons exhibited increased *c-fos* protein immunoreactivity during sleep (Sherin et al., 1996). As elevated expression of the *c-fos* gene has been associated with neuronal activation, this suggested that neurons with sleep-related discharge patterns should be highly localized to this portion of the POA. We examined the sleep-wake discharge profiles of neurons recorded throughout the dorsal to ventral extent of the lateral POA.

Consistent with the distribution of *c-fos* protein immunoreactive neurons, we found that neurons with sleep-related neuronal discharge were localized to the ventrolateral POA (Szymusiak et al., 1997). In fact >50% of the cells recorded in the ventral-most portions of the lateral POA displayed sleep-related discharge. Similar to medial POA neurons, sleep-related cells in this area also exhibited increases in discharge during sleep-wake transitions. Discharge of ventrolateral POA neurons was correlated with the depth of nonREM sleep, i.e., discharge rates increased significantly from light to deep nonREM sleep, coincident with increasing neocortical EEG delta power. Discharge rate of most ventrolateral POA neurons during REM sleep were similar to nonREM sleep values. Twenty eight percent of ventrolateral POA neurons with sleep-related discharge were also warm-sensitive, a proportion of warm-sensitivity similar to that observed among sleep-related cell types in the medial POA of rats.

The presence of warm-sensitivity in medial and ventrolateral POA neurons with sleep-related discharge and of cold sensitivity in medial POA neurons with waking-related discharge, provides a mechanistic explanation for the ability of focal POA warming to evoke EEG synchrony and sleep. It also explains the reduction in metabolic rate, activation of heat loss effectors and the decrease in T_b that normally accompany sleep onset.

Available anatomical and physiological evidence supports the hypothesis that POA sleep-related cell types function to promote sleep, in part, via descending inhibitory actions on arousal systems localized in the magnocellular basal forebrain, the posterior hypothalamus and the brainstem. A substantial projection from gamma-aminobutryic acid- (GABA) containing neurons in the POA to the posterior hypothalamus has been described in rats (Gritti et al., 1994). *C-fos* immunoreactive neurons in the ventrolateral POA project monosynaptically to histaminergic regions of the posterior hypothalamus (Sherin et al., 1996). Focal POA warming evokes suppression of neuronal activity in waking-related cells recorded in cholinergic regions of the magnocellular basal forebrain (Alam et al., 1995) and in the posterior hypothalamus of cats (Krilowicz et al., 1995). Electrical stimulation of medial and ventrolateral regions of the POA evoke suppression of waking-related discharge in the midbrain reticular formation (Szymusiak and McGinty, 1989). Electrical stimulation of the POA evokes GABA-mediated inhibitory postsynaptic potentials in posterior hypothalamic histaminergic neurons recorded in horizontal slices of the rat hypothalamus (Yang and Hatton, 1997; Steiniger et al., 1997a). In summary, activation of POA warm-sensitive neurons (and deactivation of cold sensitive cell types) evoked by local warming, or occurring spontaneously at wake-sleep transitions, leads to suppression of activity in multiple hypothalamic and brainstem arousals systems.

Interaction Among Hypothalamic and Brainstem Thermoregulatory Mechanisms and Control of the nonREM-REM Sleep Cycle

The preceding review demonsrates that manipulation of T_{en} and/or T_b can influence every major aspect of the nonREM-REM sleep cycle. Several types of temperature effects appear to exist. First, small changes in T_{en} and T_{sk} with respect to thermoneutrality can significantly alter time spent in REM sleep in the absence of changes in core T_b, of changes in thermoeffector activity, and of changes in nonREM sleep parameters. These changes of REM sleep amount are due to increased frequency of REM sleep occurrence. The duration of REM sleep episodes is not significantly altered. Thus, the intervals between REM sleep episodes is reduced at thermoneutral T_{en}. While changes in T_{sk} alone can be effective, T_b or T_{hy} manipulations can influence REM parameters in some circumstances. For example, focal hypothalamic warming during REM sleep in mildly cold-stressed cats significantly increases REM sleep episode duration (Parmeggiani et al., 974). Exposure of atropine-treated cats to a thermoneutral T_{en} blocks the transient hypothermic effect of the drug and restores REM sleep amounts to control levels (Szymusiak et al., 1994). Collectively, the above findings suggest that a relatively direct thermosensitive modulation of brainstem cholinergic and/or cholinoceptive mechanisms involved in REM sleep generation exists, with a prominent input from peripheral thermosensitive neurons.

A second type of thermoregulatory modulation of the nonREM-REM sleep cycle is primarily exerted through effects on nonREM sleep, and mediated by thermosensitive, sleep-wake regulating cell types of the medial and ventrolateral POA. As REM sleep normally emerges from consolidated nonREM sleep episodes, the sleep-promoting and sleep-maintenance functions of POA warm-sensitive neurons will necessarily impact on REM sleep expression. Thus, following POA damage in cats, amouns of both nonREM and REM sleep are reduced. Manipulations that acutely restore sleep in cats with POA damage, restore both nonREM and REM sleep (Szymusiak et al., 1991; Sallanon et al., 1989).

A third, more direct modulation of brainstem REM sleep-generating mechanisms by POA thermosensitive cell types may also exist. As reviewed above, thermosensitive neurons located in REM sleep-generating regions of the PRF are also responsive to changes in $T_{hy.}$ Focal POA warming can prolong the duration of REM sleep episodes. The enhanced thermosensitivity during nonREM sleep and the persistence of thermosensitivity during REM sleep in POA warm sensitive neurons (Alam et al., 1996; Alam et al., 1995b) suggests a possible function for these cell types in modulating nonREM-REM sleep transitions and REM sleep maintenance.

Such a modulation of POA warm-sensitive neurons on brainstem REM sleep mechanisms may involve inhibitory actions on histaminergic cell groups of the posterior hypothalamus and on serotonergic neurons of the dorsal raphe nucleus. Monoamines exert inhibitory effects on cholinergic neurons of the dorsolateral PRF (Luebke et al., 1992). Removal of monoaminergic inhibition of brainstem cholinergic neurons is thought to be a critical step in the induction of the REM sleep state (Steriade and McCarley, 1990). Both dorsal raphe neurons and putative histaminergic neurons of the posterior hypothalamus undergo cessation of neuronal discharge during transition periods form nonREM to REM sleep and during REM sleep (McGinty and Harper, 1976; Vanni-Mercier et al., 1984; Steininger et al., 1996). Monosynaptic projections from hypothalamic histaminergic neurons and from serotonergic dorsal raphe neurons to cholinergcic regions of the dorsolateral pons have been identified (Lin et al., 1996; Honda and Semba, 1994; Steininger et al., 1997b).

Anatomical and physiological evidence indicates that POA neurons can exert inhibitory effects on histaminergic neurons in the posterior hypothalamus. A substantial GABAergic projection from the POA to the posterior hypothalamus has been described in rats. Neurons in the ventrolateral POA that express *c-fos* protein immunoreactivity during sleep project to histaminergic neurons of the posterior hypothalamus. The majority of ventrolateral POA neurons with sleep-related discharge display similar discharge rates during nonREM and REM sleep, suggesting that they could contribute to suppression of histaminergic neuronal discharge during REM. Finally, focal POA warming can suppress waking related discharge in the posterior hypothalamus, and electrical stimulation of the POA evokes GABA-mediated inhibitory potentials in histaminergic neurons. While physiological evidence for an inhibitory modulation of dorsal raphe neurons by POA warm-sensitive cell types is lacking at this time, we have observed direct GABAergic projections from the medial and ventrolateral POA to the dorsal raphe nucleus in rats (unpublished observations).

In summary, thermoregulatory effects on the nonREM and REM sleep cycle are of several types, involving a possible direct peripheral thermosensitive modulation of brainstem REM sleep mechanisms, effects on nonREM sleep induction and maintenance mediated through hypothalamic thermosensitive neurons, and a hypothalamic modulation of monoaminergic afferents to cholinergic neurons in the dorsolateral pons. Such a multifaceted interaction of thermoregulatory and sleep-regulatory mechanisms explains the ability of manipulations of peripheral and/or core body temperature to strongly influence most critical features of the nonREM-REM sleep cycle.

References

Alam, M.N., McGinty, D. and Szymusiak, R. Neuronal discharge of preoptic/anterior hypothalamic thermosensitive neurons: relation to nonREM sleep, Am. J. Physiol. 38 (1995a) R1240–R1249.

Alam, M.N., McGinity, D. and Szymusiak, R. Preoptic/anterior hypothalamic neurons: thermosensitivity in rapid eye movement sleep, Am. J. Physiol., 38 (1995b) R1250–R1257.

Alam, M.N., McGinty, D. and Szymusiak, R. Preoptic/anterior hypothalamic neurons: thermosensitivity in wakefulness and non rapid eye movement sleep, Brain Res., 718 (1996) 76–82.

Alam, M.N., McGinty, D. and Szymusiak, R. Thermosensitive neurons of the diagonal band in rats: relation to wakefulness and nonrapid eye movement sleep, Brain Res., 752 (1997) 81–89.

Alam, M.N., Szymusiak, R. and McGinty, D. Preoptic/anterior hypothalamic warming alters spontaneous and evoked neuronal activity in the magnocellular basal forebrain. Brain Res 646: 221–230, 1995.

Amini-Sereshki, L., Effect of ambient temperature on sleep-waking cycle in cats with electrolytic dorsolateral pontine tegmental lesions, Neurosci., 45 (1991) 474–752.

Amini-Sereshki, L. and Morrison, A. Release of heat-loss responses in paradoxical sleep by thermal loads and by pontine tegemental lesions in cats, Brain Res., 40 (1988) 9–17.

Asami, T., Hori, T., Kiyohara, T. and Nakashima, T. Convergence of thermal signals on the reticulospinal neurons in the midbrain, pons and medulla ablongata, Brain Res. Bull., 20 (1988) 581–596.

Baghdoyan, H.A., Rodrigo-Angula, M.L., McCarley, R.W. and Hobson, J.A. Site-specific enhancement and suppression of desynchronized sleep signs following cholinergic stimulation of three brainstem regions, Brain Res., 306 (1984) 39–52.

Benedek, G., Obal, Jr., F., Lelkes, Z. and Obal F. Thermal and chemical stimulations of the hypothalamic heat detectors: the effects on the EEG, Acta Physiol. Acad. Sci. Hung., 60 (1982) 27–35.

Boulant, J.A. and Dean, J.B. Temperature receptors in the central nervous system, Ann.Rev.Physiol., 48 (1986) 639–654.

Farham, C.J. and Douglas, R.J. The response of neurons of the medial pontomedullary reticular formation of rats to peripheral thermal stimuli, Brain Res., 336 (1985) 107–115.

Findlay, A.R. and Hayward, J.N. Spontaneous activity of single neurones in the hypothalamus of rabbits during sleep and waking, J. Physiol., 201 (1969) 237–258.

Glotzhach, S.F. and Heller, H.C. Central nervous regulation of body temperature during sleep, Science, 194 (1979) 537–539.

Gritti, I., Mainville, L. and Jones, B.E. Projections of GABAergic and cholinergic basal forebrain and GABAergic peroptic-anterior hypothalamic neurons to the posterior lateral hypothalamus of the rat, J. Comp. Neurol., 339 (1994) 251–268.

Guilleminault, C., Narcolepsy Syndrome. In: M.H. Kryger, T. Roth and W.C. Dement, *Principles and practice of Sleep Medicine,* W.B. Saunders Co., Philadelphia, 1994, pp. 549–561.

Haskell, E.H., Palca, J.W., Walker, J.M., Berger, R.J. and Heller, H.C. The effects of high and low ambient temperatures on human sleep stages, Electroenceph. Clin. Neurophysiol., 51 (1981) 494–501.

Hendricks, J.C., Absence of shivering in the cat during paradoxical sleep without atonia, Exp. Neurol., 75 (1982) 700–710.

Honda, T. and Semba, K. Serotonergic synaptic input to cholinergic neurons in the rat mesopontine tegmentum, Brain. Res., 647 (1994) 299–306.

Jouvet, M., Buda, C., Debilly,G., Dittmar, A. and Sastre, J.P. La temperature centrale est le facteur principal de regulation du sommeil paradoxal chez le chat pontique, C.R. Acad. Sci., Paris, 306 (1988) 69–73.

Kelso, S.R. and Boulant, J.A. Effect of synaptic blockade on thermosensitive neurons in hypothalamic tissue slices, Am. J. Physiol., 243 (1982) R480–R490.

Krilowicz, B.L., Szymusiak, R. and McGinty, D. Regulation of posterior lateral hypothalamic arousal related neuronal discharge by preoptic anterior hypothalamic warming, Brain Res., 668 (1995) 30–38.

Kupfer, D.J., A psychobiologic marker for primary depressive disease, Biol. Psychiat., 11 (1976) 159–174.

Libert, J.P., Candas, V., Muzet, A. and Erhart, J. Thermoregulatory adjustments to thermal transients during slow wave sleep and REM sleep in man, J. Physiol., Paris, 78 (1982) 251–257.

Lin, J.S., Hou, Y., Sakai, K. and Jouvet, M. Histaminergic descending inputs to the mesopontine tegmentum and their role in the control of cortical activation and wakefulness in the cat, J. Neurosci., 16 (1996) 1523–1537.

Luebke, J.I., Greene, R.W., Semba, K., Kamondi, A., McCarley, R.W. and Reiner, P.B. Serotonin hyperpolarizes cholinergic low-threshold burst neurons in the rat laterodorsal tegmental nucleus in vitro, Proc. Natl. Acad. Sci., 89 (1992) 743–747.

Mallick, B.N. and Alam, M.N. Different types of norepinephrinergic receptors are involved in preoptic area mediated independent modulation of sleep-wakefulness and body temperature, Brain Res. 591 (1992) 8–19.

Mallick, B.N., Chhina, G.S., Sundaram, K.R., Singh, B. and Mohan Kumar, V. Activity of preoptic neurons during synchronization and desynchronization, Exp. Neurol., 81 (1983) 586–597.

McGinty, D., Szymusiak, R. and Thomson, D. Preoptic/anterior hypothalamic warming increases EEG delta frequency activity within non-rapid eye movement sleep, brain Res., 667 (1994) 273–277.

McGinty, D.J. and Harper, R.M. Dorsal raphe neurons: depression of firing during sleep in cats, Brain Res., 101 (1976) 569–575.

Mendelson, W.B., Martin, J.V., Perlis, M. and Wagner, R. Enhancement of sleep by microinjection of triazolam into the medial preoptic area, Neuropsychopharm., 2 (1989) 61–66.

Parmeggiani, P.L., Temperature regulation during sleep: a study in homeostasis. In: J. Orem and C.D. barnes (Eds.), *Physiology in Sleep.* Academic Press, New York, 1980, pp. 97-143.

Parmeggiani, P.L. and Rabini, C. Shivering and panting during sleep, Brain Res., 6 (1967) 789–791.

Parmeggiani, P.L. and Rabini, C. Sleep and environmental temperature, Arch. Ital. Biol., 108 (1970) 369–387.

Parmeggiani, P.L., Zamboni, G., Cianci, T., Agnati, L.F. and Ricci C. Influence of anterior hypo-thalamic heating on the duration of fast wave sleep episodes, Electroenceph. Clin. neurophysiol., 36 (1974) 465–470.

Parmeggiani, P.L., Zamboni, G., Cianci, T. and Calasso, M. Absence of thermoregulatory vasomotor responses during fast wave sleep in cats, Electroenceph. and Clin. Neurophysiol., 42 (1977) 372–380.

Roberts, W.W. and Robinson, T.C.L. Relaxation and sleep induced by warming of the preoptic region and anterior hypothalamus in cats, Exp. Neurol. 25 (1962) 282–294.

Roussel, B., Turrillo, P. and Kitahama K. Effect of ambient temperature on the sleep-waking cycle in two strains of mice, Brain Res., 294 (1984) 7–73.

Sakaguchi, S., Glotzbach, S.F. and Heller, H.C. Influence of hypothalamic and ambient temperatures on sleep in kangaroo rats, Am. J. Physiol., 237 (1979) R80–R88

Sallanon, M., Sakai, K., Denoyer, M. and Jouvet, M. Long lasting insomnia induced by preoptic neurons lesions and its transient reversal by muscimol injection into the posterior hypothalamus, Neurosci., 32 (1989) 669–683.

Satinoff, E., Neural organization and evolution of thermal regulation in mammals, Science, 201 (1978) 16–22.

Satinoff, E., Valentino, D. and Teitelbaum, P. Thermoregulatory cold-defense deficits in rats with medial preoptic lesions, Brain Res. Bull., 1 (1976) 553–565.

Schmidek, W.R., Hoshino, K., Schmidek, M. and Timo-Iouria, C. Influence of environmental temperature on the sleep/wakefulness cycle in the rat, Physiol. Behave., 8 (1972) 363–371.

Sherin, J.E., Shiromani, P.J., McCarley, R.W. and Saper C.B. Activation of ventrolateral preoptic neurons during sleep, Science, 271 (1996) 216–219.

Shiromani, P.J., Gillin, J.C. and Henriksen, S.J. Acetylcholine and the regulation of REM sleep: basic mechanisms and clinical implications for affective illness and narcolepsy, Ann. Rev. Pharmacol. Toxicol., 27 (1987) 137–156.

Steininger, T.L., Alam, M.N., Szymusiak, R. and McGinty D. State dependent discharge of tuberomammillary neurons in the rat hypothalamus, Sleep Res., 25 (1996) 28.

Steininger, T.L., Stevens, D.R., Hass, H.L., McGinty, D. and Szymusiak, R. Preoptic area inhibition of histaminergic tuberomammillary neurons in vitro, Sleep Res., 26 (1997a) 45.

Steininger, T.L., Wainer, B.H., Blakely, R. and Rye, D.B. Dorsal raphe nucleus projections to cholinerigic and noncholinergic neurons of the pedunculopontine tegmental nucleus: a light and electron microscopic anterograde tracing study, J. Comp. Neurol., 382 (1997b) 302–322.

Steriade, M. and McCarley, R.W. Brainstem Control of Wakefulness and Sleep. Plenum Press, New York, 1990.

Sterman, M.B. and Clemente, C.D. Forebrain inhibitory mechanisms: sleep patterns induced by basal forebrain stimulation in the behaving cat, Exp. Neurol., 6 (1962) 103–117.

Szymusiak, R., Alam, M.N., Steininger, T.L. and McGinty, D. Localization of neurons with sleep-related discharge within the ventral-lateral preoptic area of rats, Sleep Res., 26 (1997) 48.

Szymusiak, R., Danowski, J. and McGinty D. Exposure to heat restores sleep in cats with preoptic/anterior hypothalamic cell loss, Brain Res., 541 (1991) 134–138.

Szymusiak, R., Danowski, J. and McGinty, D. REM sleep-suppressing effects of atropine in cats vary with environmental temperature, Brain Res., 636 (1994) 115–118.

Szymusiak, R. and McGinty, D. Effects of basal forebrain stimulation on the waking discharge of neurons in the midbrain reticular formation of cats, Brain Res., 498 (1989) 355–359.

Szymusiak, R. and Satinoff, E. Maximal REM sleep time defines a narrower thermoneutral zone than does minimal metabolic rate, Physiol. and Behav., 26 (1981) 687–690.

Szymusiak, R. and Satinoff, E. Ambient temperature-dependence of sleep disturbances produced by basal forebrain damage in rats, Brain Res. Bull., 12 (1984) 295–305.

Szymusiak, R., Satinoff, E., Schallert, T. and Whishaw, I.Q. Brief skin temperature changes towards thermoneutrality trigger REM sleep in rats, Physiol. Behav., 25 (1980) 305–311.

Ticho, S.R. and Radulovacki, M. Role of adenosine in sleep and temperature regulation in the preoptic area of rats, Pharmacol. Biochem. Behav. 40 (1991) 33–40.

Ueno, R., Ishikawa, Y., Nakayama, T. and Hayaishi, O. Prostaglandin D2 induces sleep when microinjected into the preoptic area of conscious rats, Biochem Biophys. Res. Com., 109. (1982) 576–582.

Vanni-Mercier, M., Sakai, K. and Jouvet M. Neurones specifiques de l'eveil dans l'hypothalamus posterieur du Chat, C.R. Acad. Sci., Paris, 298 (1984) 195-200.

Wehr, T.A., Brain warming function for REM sleep, Neurosci. Biobehav Rev., 16 (1992) 372–397.

Yang, Q.Z. and Hatton, G.I. Electrophysiology of excitatory and inhibitory afferents to rat histaminergic tuberomammillary nucleus neurons from hypothalamic and forebrain sites, Brain Res., 773 (1997) 162–172.

Zepelin, H. and Rechtschaffen, A. Mammalian Sleep, Longevity and energy metabolism, Brain Behav. Evol., 10 (1974) 425–470.

Rapid Eye Movement Sleep
B.N. Mallick and S. Inoué (Eds)

21. REM Sleep Across Age and Sex

Kazuo Mishima, Tetsuo Shimizu and Yasuo Hishikawa

Department of Neuropsychiatry, Akita University School of Medicine,
1-1-1 Hondo, Akita-City, 010, Japan

SUMMARY

Over the past several decades, a large number of studies have examined age- and gender-related changes in human sleep, which has been discussed primarily from an ontogenetic viewpoint. However, less effort was focused on rapid eye movement (REM) sleep, and age- and gender-related changes in REM sleep properties are still often controversial. The most consistent age-related changes in human REM sleep is a diminished REM latency. Decreases in the amount of slow wave sleep (SWS) with advanced age may be related to the reduction in REM latency in elderly people. It is well known that human core body temperature (cBT) rhythms have a strong influence on REM sleep propensity. Phase advance of cBT rhythms, namely, the decrease in duration from sleep onset to the nadir time of the cBT rhythm may induce shortening of REM latency and duration of REM sleep episode in elderly people. Some investigators have reported that elderly women showed significantly longer REM latencies, a trend toward longer first REM sleep periods, and more REM sleep time compared with age-matched elderly men. The underlying mechanism is unknown; however, it may be caused partially by better preservation of SWS in elderly women compared with men.

REM sleep behavior disorder (RBD) is characterized by dissociation of REM sleep, namely, lack of muscle atonia during REM sleep. In RBD, patients apparently enact their dreams. Many cases of RBD have been reported in patients with brain stem tumors, and systemic degenerative diseases that mainly involve the brain stem. These patients had abnormal sleep states characterized by low voltage mixed frequency EEG patterns, rapid eye movements, and tonic activity in the chin muscle EMG, which state is called stage 1-REM with tonic EMG (stage 1-REM). Recent studies indicate that abnormal behavior during stage 1-REM is not uncommon in otherwise healthy elderly adults.

Key Words: REM sleep, aging, gender, body temperature, REM sleep behavior disorder, stage 1-REM.

Over the past several decades, a large number of studies have examined age-related changes in human sleep, which has been discussed primarily from an ontogenetic viewpoint. A substantial number of studies using the electroencephalographic (EEG) recordings of elderly people have found consistently age-related changes in the amount and distribution of sleep stages including decreases in total sleep time, slow wave sleep (SWS; deep sleep, delta sleep), and sleep efficacy. However, less effort was focused on rapid eye movement (REM) sleep, and age-related changes in REM sleep properties are still often controversial.

Similarly, there are even less studies that have examined gender-related differences in REM sleep even though gender has been shown to exert a strong influence on human sleep (Miles, 1980). In this section, we review previously reported normative data about age- and gender-related properties of human REM sleep mainly in adults (development of REM sleep is discussed in another section of this Book), and then present REM sleep behavior disorder in elderly people that is a type of parasomnia usually associated with REM sleep.

REM Sleep Duration

In 1966, Roffwarg and Dement (Roffwarg, 1966) summarized the available normative data concerning age-associated changes in the daily total sleep time and REM vs non-REM (NREM) sleep ratio (Fig. 1.) These data demonstrated a consistent decrease in total sleep time from 16 h per day in 1 to 15 day old neonates to 5.75 h per day in 90 yr old individuals. It is noteworthy that REM sleep time shows a steep decline from 8 h/day at birth to 1.85 h/day at 10 to 13 yr old in contrast to the relatively constant amount of NREM sleep time (8 h vs 7.15 h). After adolescence, NREM sleep consistently decreases from 6.8 h in adrescence to 4.9 h in old age. Many recent studies of NREM sleep physiology have confirmed a progressive reduction in the

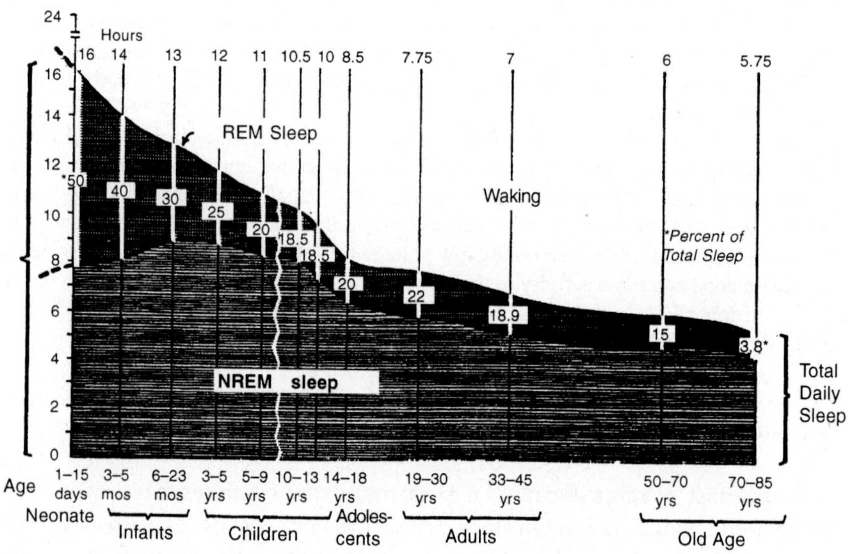

Fig. 1. Graph showing age-related changes in total amounts of daily sleep, daily REM sleep, and percentage of REM sleep (by Roffwarg et al., 1966).

percentage (Ehlers, 1989; Feinberg, 1974; Feinberg, 1980; Gillin, 1981; Kahn, 1969; Kales, 1967; Smith, 1977) and amplitude (Johnson, 1969; Kahn, 1969; Smith, 1977) of SWS with advanced age. These age-related decreases in SWS (stages 3 and 4), rather than changes in lighter sleep (stages 1 and 2), are thought to contribute to the deterioration of NREM sleep in elderly people. Both the absolute amount and percentage of REM sleep seem to show only small decreases after adlescence. In contrast to the drastic changes in REM sleep time from infancy to childhood, previous normative studies comparing polysomnography patterns between young and aged groups have yielded contradictory findings regarding significant age-related decreases in the amount and/or percentage of REM sleep (Feinberg, 1967; Gigli, 1996; Gillin, 1981; Hayashi, 1982; Prinz, 1977; Reynolds, 1991; Wauquier, 1992). Taken together, there is still some debate as to whether the absolute amount of sleep and REM/NREM proportion really change as a function of age after adlescence. The discrepancies among studies may be caused by a combination of factors including inter-individual variability, different recording techniques and interpretation, and influence of ambient conditions. In addition, we should acknowledge that many of the early studies failed to evaluate precisely sleep-related respiratory disturbances and/or periodic leg movements that are prevalent in elderly people, which are important factors involved in sleep deterioration with aging.

REM Latency/REM Sleep Distribution

In contrast to the controversial findings about absolute amount and percentage of REM sleep time, relatively consistent age-related changes in the time from sleep onset to the first appearance of REM sleep (REM latency) have been identified. Despite current debate (Gigli, 1996; Giles, 1990), many previous reports using the polysomnography method in young and elderly groups have revealed a diminished REM latency in elderly people compared with that in young healthy controls (Benca, 1992; Ehlers, 1989; Feinberg, 1974; Gillin, 1981; Hayashi, 1982; Reynolds, 1990; Wauquier, 1992). This age-related decrease in REM latency has also been reported to progress consistently in 60-, 70- and 80-yr olds (Reynolds, 1991).

Age-related decreases in the amount of first NREM sleep, especially decreases in SWS occurring before the first REM sleep, may be related to the reduction in REM latency in elderly people. It has been demonstrated that REM sleep starts with approximately 90 min cycles and gradually increases throughout the night according to the circadian time keeping mechanism (Czeisler, 1980; Czeisler, 1980). The amount of NREM sleep time is assumed to be dtermined primarily by the length of prior wakefulness according to a homeostatic balance. SWS tends to consolidate during the first 90–180 min of the night and exerts a kind of pressure that forces the first REM sleep backward. Generally, in many elderly people, REM sleep occurs at the beginning of the night simultaneously with a remarkable decrease in the amount of

SWS (Hayashi, 1982; Wauquier, 1992). Namely, elderly people show an intranight forward shift of REM sleep in contrast to younger people where the duration of REM sleep is longer at the end of the night. These same changes in sleep architecture have been reported in depressive patients, and the changes become more pronounced with advancing age (Giles, 1990; Gillin, 1981; Kupfer, 1982; Reynolds, 1990; Ulrich, 1980). In addition, it has been shown by power spectral analysis that age-related changes in slow wave activity are most prominent during the first 100 min of sleep (Ehlers, 1989). These findings support the assumption that decreases in SWS pressure cause changes in temporal distribution of REM sleep (Feinberg, 1974). Findings that an age-related reduction of REM latency is positively correlated with changes in the first NREM sleep rather than REM activity (Ehlers, 1989; Feinberg, 1974) have also supported this speculation.

One additional mechanism that contributes to the reduction of REM latency in the elderly is chronobiological effects with aging. Human core body temperature (cBT) rhythms have a strong influence on both the timing and duration of sleep. In general, in an isolated environment (free of time cues), subjects often show free-running sleep-wake cycles and cBT rhythms with different mutual phase positions from those observed under entrained conditions, and they sometimes exhibit free-running rhythms with different circadian periodicities that cause internal desynchronization of these rhythms (Wever, 1979). The prevalence of their voluntary bedtime and sleep onset was shown to be highest around the nadir time of the cBT rhythm (Czeisler, 1980; Zulley, 1981). Reportedly, REM sleep propensity is also coupled closely to cBT rhythm (Czeisler, 1980). Subjects were placed on a schedule of 2 h of sleep and 1 h of imposed bedrest in an isolated environment. The results showed clearly that REM sleep propensity (shortening of REM latency, duration of REM sleep episode) is maximized when the bedrest time coincided with the nadir time of the cBT rhythm (Figs. 2a and 2b). Zulley and his colleagues (Zulley, 1980) have reported the same effect of cBT rhythm on REM sleep propensity when studying the same subjects in entrained or isolated environments. The effects of cBT rhythm on human sleep have been observed in subjects entrained to a 24 h environmental cycle, in which total sleep time and REM sleep were influenced more by the phase of the cBT rhythm at bed time than the duration of wakefulness before sleep (Akerstedt, 1981). Some reports has demonstrated that REM sleep propensity have a bimodal distribution in relation to the cBT rhythm phase (Carman, 1984; Czeisler, 1980; Czeisler, 1980). Carman et al. (Carman, 1984) observed two peaks of REM sleep propensity appearing approximately 20 min and 120 min after the nadir time of cBT rhythm. From these data, it is possible that chronobiological aging of the cBT rhythm could affect sleep architecture in the elderly. Weitzman and his colleagues (Weitzman, 1982) studied young adults (ages 23–30) and elderly males (ages 53–70) living under isolated conditions for 3–8 weeks with comparative measurements of sleep-wake cycles, sleep stages, and cBT rhythms.

The results demonstrated a reduction in the period and amplitude of the cBT rhythm during free-running in the older group, although no significant change in the sleep-wake rhythm was found. This phenomenon induced changes in

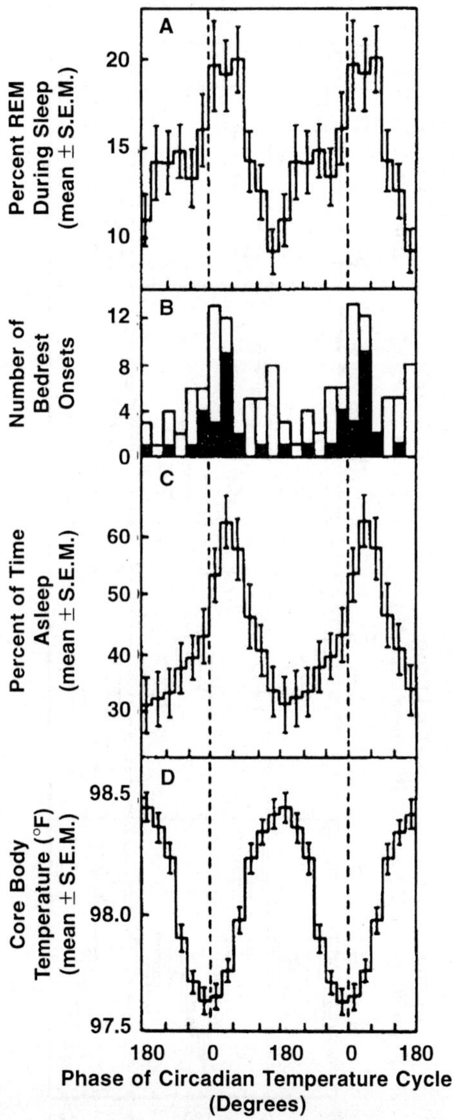

Fig. 2 **(a) Variations in the occurrence and internal organization of sleep with the circadian cBT cycle phase (Czeisler et al., 1980). In panel B, bedrest periods in which REM sleep episodes occurred within 10 min after bedrest onset are shaded black; those in which REM sleep episodes occurred within 30 min after bedrest onset are cross-hatched. Note that the peak of the REM sleep propensity rhythm (panel A) occurred on the rising slope of cBT cycle and was coincident with the phase intervals (0° to 90°) at which nearly all sleep-onset REM episodes occurred.**

the phase relationship of sleep-wake and cBT rhythms in several elderly subjects, namely, the duration from sleep onset to the nadir time of the cBT rhythm tends to be diminished in elderly people. These chronobiological changes could reduce the REM latency (Benca, 1992; Ehlers, 1989; Feinberg,

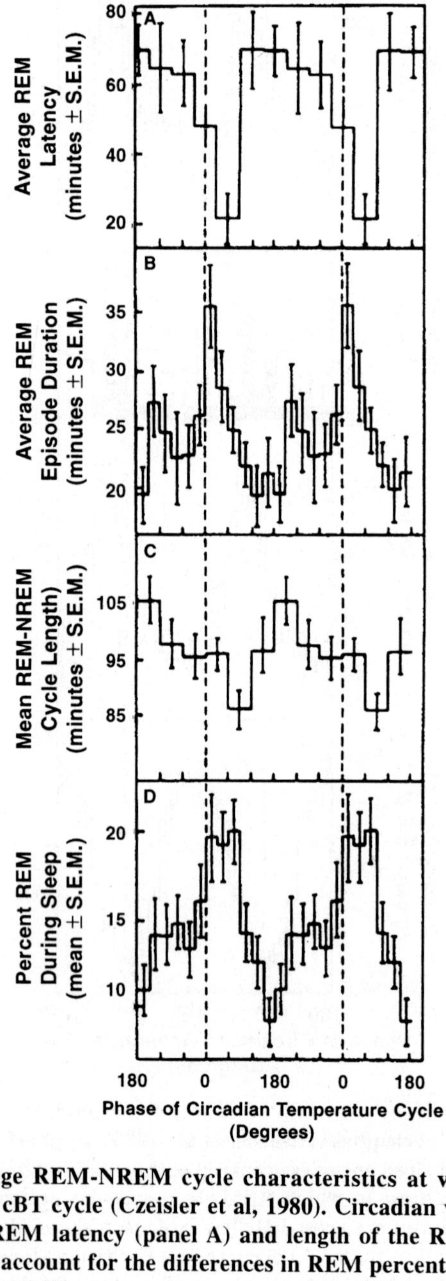

Fig. 2 **(b) Average REM-NREM cycle characteristics at various phases of the circadian cBT cycle (Czeisler et al, 1980). Circadian variations of both the average REM latency (panel A) and length of the REM episode duration (panel B) account for the differences in REM percent observed at different phases (panel D).**

1974; Gillin, 1981; Hayashi, 1982; Reynolds, 1990; Wauquier, 1992) and induce changes in the temporal distribution of REM sleep, such as a flattened distribution of REM sleep duration through the night (Hayashi, 1982) or an intranight forward shift of REM sleep duration that decreases during the course of the night in elderly people (Wauquier, 1992).

REM Sleep Across the Sex

Despite findings that gender has strong influences on sleep in healthy elderly persons (Miles 1980), there are very few studies that have identified a trend toward gender differences in REM sleep properties in humans. Previous reports failed to identify trends toward gender differences in either percentage of REM sleep, REM minutes, REM density, REM activity, or number of REM periods (Carskadon, 1982; Hoch, 1988; Hoch, 1987; Kupfer, 1978; Rediehs, 1990; Reynolds, 1985). However, some investigators have reported that elderly women showed significantly longer REM latencies (Kahn, 1969; Kahn, 1970; Rediehs, 1990), a trend toward longer first REM sleep periods (Rediehs, 1990), and more REM sleep time (Reynolds, 1991) compared with age-matched elderly men. The underlying mechanism is unknown; however, it may be caused partially by better preservation of SWS (particularly in the first NREM period) in elderly women compared with men (Rediehs, 1990; Reynolds, 1986; Reynolds, 1991; Wauquier, 1992., Webb, 1982). Wauquier et al. (Wauquier, 1992) reported a greater magnitude of gender differences with higher power densities in elderly women both during SWS and REM sleep studied using the EEG power spectral analysis method. They also found gender differences in reduction of REM latency simultaneous with higher amount of SWS in women compared with men. In addition, some studies have suggested that there are gender differences in the properties of circadian cBT rhythms (Campbell, 1989; Moe, 1991; Zepelin, 1987); elderly women exhibit advanced phases and higher amplitudes of the cBT rhythm compared with age-matched elderly men. It is possible that these gender-specific changes in the mutual phase relationship between cBT rhythm and sleep could also accelerate the altered REM latency among elderly men and women.

Abnormal REM Sleep and Aging

REM sleep is composed of two components: tonic (EEG desynchronization and atonia during REM sleep), and phasic (ponto-geniculate-occipital (PGO) waves, REMs muscle twitch during REM sleep) (Siegel, 1994). Animal experiments have demonstrated that EEG desynchronization, muscle atonia, and the phasic component of REM sleep are controlled by spatially close but distinct neural structures in the pontine tegmentum. Therefore, it is possible that dissociation of the components of REM sleep could occur if there were destruction, dysfunction, disinhibition, or hyperactivity in either of the neural structures controlling REM sleep. One example of symptoms with dissociation of REM sleep is REM sleep behavior disorder (RBD), which is characterized by lack of muscle atonia during dissociated REM sleep. In RBD, patients apparently enact their dreams. Many cases of RBD have been reported in patients with brain stem tumors (Barros-Ferreira, 1975), and systemic degenerative diseases, such as Shy-Drager syndrome, Parkinson's disease, olive-ponto-cerebellar degeneration (OPCA), and progressive supranuclear palsy, that mainly involve the brain stem (Mouret, 1975; Pareja, 1996; Plazzi, 1997; Shimizu, 1990; Tachibana, 1995; Wright, 1990). These patients had abnormal sleep states characterized by low voltage mixed frequency EEG patterns, rapid eye movements, and tonic activity in the chin muscle EMG.

Tachibana et al., (Tachibana, 1975) called this abnormal sleep state stage 1-REM with tonic EMG (stage 1-REM). The first episode of stage 1-REM apears after 60–90 min of NREM sleep, and then the episodes of NREM sleep and stage 1-REM alternate like a normal REM-NREM sleep cycle. During stage 1-REM, patients often show abnormal behaviors such as sleep talking, laughing, shouting, complex movements of the arms, or jumping out of bed. Some patients walk around or attack furniture or their sleeping partner. Injury to self or others is reported to occur during such abnormal behavior. This abnormal behavior is stopped by noise or somatic sensation caused by the patient's behavior. Patients are conscious immediately after such abnormal behavior, and they can usually describe the dream just prior to waking. The behavior in their dream is usually correlated to their actual behavior, thus the abnormal behavior is considered to be an enactment of the dream. An animal model of RBD has been reported (Jouvet, 1979). Cats with bilateral lesions of the pontine tegmentum, in which neural structures controlling muscle atonia during REM sleep are located, show REM sleep without muscle atonia. During REM sleep without muscle atonia, oneiric behaviors such as orientation and attack on hallucinated objects were observed repeatedly in these cats.

Recently, abnormal behavior during stage 1-REM was reported in otherwise normal elderly subjects, and this syndrome is called idiopathic RBD (Ishigooka, 1985; Schenck, 1987; Tachibana, 1991). Idiopathic RBD shares about 55% of RBD (Mahowald, 1994). Culebras et al. (Culebras, 1989) reported that lacunar infact in the white matter of the brain and the brain stem were observed in six patients with idiopathic RBD, however, this may be incidental because lacunar infarct is observed in many elderly subjects without RBD, and many patients with idiopathic RBD do not have lacunar infarcts. A more important point is that 38% of patients with idiopathic RBD developed Parkinson's disease within several years (Schenck, 1996). Uchiyama et al. (Uchiyama, 1995) reported a case of idiopathic RBD in which a postmortem examination revealed Lewy body disease. These findings indicate that RBD may herald various degenerative diseases involving the brain stem. The exact prevalence of RBD in the elderly population is not clear. Our preliminary field study in a community of elderly subjects over 65 years of age (N = 495) showed that 6.1% (N = 30) of subjects reported that somnambulistic behavior first appeared in their senescence. Home monitoring of these subjects using a Medilog machine revealed that 5 of 40 subjects showed stage 1–REM. These preliminary findings indicate that RBD is not uncommon in otherwise healthy elderly adults.

References

Akerstedt, T. and Gillberg, M. The circadian variation of experimentally displaced sleep, Sleep, 4 (1981) 159–169.

Barros-Ferreira, D.M., Chodkiewicz, J., Lairy, G.C. and Sazaulos, P. Disorganized relations of tonic and phasic events of REM sleep in a case of brain stem tumor, Electroenceph. Clin. Neurophysiol., 38 (1975) 203–207.

Benca, R.M., Obermeyer, W.H., Thisted, R.A. and Gillin, J.C. Sleep and psychiatric disorders: A meta-analysis, Arch. Gen. Psychiatry, 49 (8) (1992) 651–68.

Campbell, S.S., Gillin, J.C., Kripke, D.F., Erikson, P. and Clopton, P. Gender differences in the circadian temperature rhythms of helathy elderly subjects: Relationships to sleep quality, Sleep, 12 (1989) 529–536.

Carman, G.J., Mealey, L., Thompson, S.T. and Thompson, M.A. Patterns in the distribution of REM sleep in normal human sleep, Sleep, 7 (4) (1984) 347–55.

Carskadon, M.A., Brown, E.D. and Dement, W.C. Sleep fragmentation in the elderly: relationship to daytime sleep tendency, Neurobiol. Aging, 3 (4) (1982) 321–7.

Culebras, A. and Moore, J.T. Magnetic resonance findings in REM sleep behavior disorder, Neurology, 39 (11) (1989) 1519–23.

Czeisler, C.A., Weitzman, E.D., Moore, E.M., Zimmerman, J.C. and Knauer, R.S. Human sleep: its duration and organization depend on its circadian phase, Science, 210 (4475) (1980) 1264–7.

Czeisler, C.A., Zimmerman, J.C., Ronda, J.M., Moore, E.M. and Weitzman, E.D. Timing of REM sleep is coupled to the circadian rhythm of body temperature in man, Sleep, 2 (3) (1980) 329–46.

Ehlers, C.L. and Kupfer, D.J. Effects of age on delta and REM sleep parameters, Electroencephalogr. Clin. Neurophysiol., 72 (2) (1989) 118–25.

Feinberg, I., Changes in sleep cycle patterns with age, J. Psychiatr. Res., 10 (3–4) (1974) 283–306.

Feinberg, I., Fein, G. and Floyd, T.C. Period and amplitude analysis of NREM EEG in sleep: repeatability of results in young adults, Electroencephalogr. Clin. Neurophysiol., 48 (2) (1980) 212–21.

Feinberg, I., Koresko, R.L. and Heller, N. EEG sleep patterns as a function of normal and pathological aging in man, J. Psychiatr. Res., 5 (2) (1967) 107–44.

Gigli, G.L., Placidi, F., Diomedi, M., Maschio, M., Silvestri, G., Scalise, A. and Marciani, M.G. Sleep in healthy elderly subjects: a 24-hour ambulatory polysomnographic study, Int. J. Neurosci., 85 (3–4) (1996) 263–71.

Giles, D.E., Roffwarg, H.P., Rush, A.J. and Guzick, D.S. Age-adjusted threshold values for reduced REM latency in unipolar depression using ROC analysis, Biol. Psychiatry, 27 (8) (1990) 841–53.

Gillin, J.C., Duncan, W.C., Murphy, D.L., Post, R.M., Wehr, T.A., Goodwin, F.K., Wyatt, R.J. and Bunney, W.J. Age-related changes in sleep in depressed and normal subjects, Psychiatry Res., 4 (1) (1981) 73–8.

Hayashi, Y. and Endo, S. All-night sleep polygraphic recordings of healthy aged persons: REM and slow-wave sleep, Sleep, 5 (3) (1982) 277–83.

Hoch, C.C., Reynolds, C.F., Kupfer, D.J. and Berman, S.R. Stability of EEG sleep and sleep quality in healthy seniors, Sleep, 11 (6) (1988) 521–7.

Hoch, C.C., Reynolds, C.F., Kupfer, D.J., Berman, S.R., Houck, P.R. and Stack, J.A. Empirical note: self-report versus recorded sleep in healthy seniors, Psychophysiology, 24 (3) (1987) 293–9.

Ishigooka, J., Westendrop, F., Oguchi, T., Takahashi, A., Sumiyoshi, A. and Inami, M. Somnambulistic behavior associated with abnormal REM sleep in an elderly woman, Biol. Psychiatry, 20 (1985) 1003–1008.

Johnson, L., Lubin, A., Naitoh, P., Nute, C. and Austin, M. Spectral analysis of the EEG of dominant and non-dominant alpha subjects during waking and sleeping, Electroencephalogr. Clin. Neurophysiol., 26 (4) (1969) 361-70.

Jouvet, M. What does a cat dream about?, Trends Neurosci., 2 (1979) 280.

Kahn, E. and Fisher, C. The sleep characteristics of the normal aged male, J. Nerv. Ment. Dis., 148 (5) (1969) 477–94.

Kahn, E., Fisher, C. and Lieberman, L. Sleep characteristics of the human aged female, Compr. Psychiatry, 11 (3) (1970) 274–8.

Kales, A., Wilson, T., Kales, J.D., Jacobson, A., Paulson, M.J., Kollar, E. and Walter, R.D. Measurements of all-night sleep in normal elderly persons: effects of aging, J. Am. Geriatr. Soc., 15 (5) (1967) 405–14.

Kupfer, D.J., Reynolds, C.F., Ulrich, R.F., Shaw, D.H. and Cloble, P.A. EEG sleep, deression, and aging, Neurobiol., Aging, 3 (4) (1982) 351–60.

Kupfer, D.J., Spiker, D.G., Coble, P.A. and Shaw, D.H. Electroencephalographic sleep recordings and depression in the elderly. J. Am. Geriatr., Soc., 26 (2) (1978) 53–7.

Mahowald, M.W. and Schenck, C.H. REM Sleep Behavior Disorder. In: M.H. Kryger T. Roth and W.C. Dement (Eds.), *Principles and Practice of Sleep Medicine, 2nd ed.,* W.B. Saunders, Philadelphia, 1994.

Miles, L.E. and Dement, W.C. Sleep and aging, Sleep, 3 (2) (1980) 1–220.

Moe, K.E., Prinz, P.N., Vitiello, M.V., Marks, A.L. and Larsen, L.H. Healthy elderly women and men have different entrained circadian temperature rhythms, *J.* Am. Geriatr. Soc., 39 (4) (1991) 383–7.

Mouret, J., Differences in sleep in patients with Parkinson's disease, Electroenceph. Clin. Neurophysiol., 38 (1975) 653–657.

Pareja, J.A., Caminero, A.B., Masa, J.F. and Dobato, J.L. A. first case of progressive supranuclear palsy and pre-clinical REM sleep behavior disorder presenting as inhibition of speech during wakefulness and somniloquy with phasic muscle twitching during REM sleep, Neurologia 11 (8) (1996) 304–6.

Plazzi, G., Corsini, R., Provini, F., Pierangeli, G., Martinelli, P., Montagna, P., Lugaresi, E. and Cortelli, P. REM sleep behavior disorders in multiple system atrophy, Neurology, 48 (4) (1997) 1094–7.

Prinz, P.N. Sleep patterns in the healthy aged: Relationship with intellectual function, J. Gerontol., 32 (1977) 179–186.

Rediehs, M.H., Reis, J.S. and Creason, N.S. Sleep in old age: focus on gender differences, Sleep, 13 (5) (1990) 410–24.

Reynolds, C.F., Kupfer, D.J., Hoch, C.C., Stack, J.A., Houck, P.R. and Berman, S.R. Sleep deprivation in healthy elderly men and women: effects on mood and on sleep during recovery, Sleep, 9 (4) (1986) 492–501.

Reynolds, C.F., Kufer, D.J., Taska, L.S., Hoch, C.C., Sewitch, D.E. and Spiker, D.G. Sleep of healthy seniors: a revisit, Sleep, 8 (1) (1985) 20–9.

Reynolds, C.F., Kupfer, D.J., Thase, M.E., Frank, E., Jarrett, D.B., Coble, P.A., Hoch, C.C., Buysse, D.J., Simons, A.D. and Houck, P.R. Sleep, gender, and depression: an analysis of gender effects on the electroencephalographic sleep of 302 depressed outpatients, Biol. Psychiatry, 28 (8) (1990) 673–84.

Reynolds, C.F., Monk, T.H., Hoch, C.C., Jennings, J.R., Buysse, D.J., Houck, P.R., Jarrett,

D.B. and Kupfer, D.J. Electroencephalographic sleep in the healthy "old old": a comparison with the "young old" in visually scored and automated measures, J. Gerontol., 46 (2) (1991).

Roffwarg, H.P., Muzio, J.N. and Dement, W.C. Ontogenic development of the human sleep-dream cycle; The prime role of dreaming sleep in early life may be in the development of the central nervous system, Science, 152 (1966) 604–619.

Schenck, C.H., Bundlie, S.R. and Mahowald, M.W. Delayed emergence of a parkinsonian disorder in 38% of 29 older men initially diagnosed with idiopathic rapid eye movement sleep behavior disorder, Neurology, 46 (2) (1996) 388–93.

Schenck, C.H., Bundlie, S.R., Patterson, A.L. and Mahowald, M.W. Rapid eye movement sleep behavior disorder. A treatable parasomnia affecting older adults, JAMA, 257 (13) (1987) 1786–9.

Shimizu, T., Inami, Y., Sugita, Y., Iijima, S., Teshima, Y., Matsuo, R., Yasoshima, A., Egawa, I., Okawa, M., Tashiro, T. et al. REM sleep without muscle atonia (stage 1–REM) and its relation to delirious behavior during sleep in patients with degenerative diseases involving the brain stem, Jpn. J. Psychiatry Neurol., 44 (4) (1990) 681–92.

Siegel, J.M., Brain stem mechanisms generating REM sleep, In: M.H., Kryger T. Roth and W.C. Dement (Eds.), *Principles and Practice of Sleep Medicine, 2nd ed.,* W.B. Saunders, Philadelphia, 1994, 125–144.

Smith, J.R., Karacan, I. and Yang, M. Ontogeny of delta activity during human sleep, Electroencephalogr. Clin. Neurophysiol., 43 (2) (1977) 229–37.

Tachibana, M., Tanaka, K., Hishikawa, Y. and Kaneko, Z.A. Sleep study of acute psychotic state due to alcohol and meprobamate addiction. In: E.D. Weitzman (Ed.), *Advances in Sleep Research,* vol 2, Spectrum, New York, 1975, 175–205.

Tachibana, N., Kimura, K., Kitajima, K., Nagamine, T., Kimura, J. and Shibasaki, H. REM sleep without atonia at early stage of sporadic olivopontocerebellar atrophy, J. Neurol. Sci., 132 (1) (1995) 28–34.

Tachibana, N., Sugita, Y., Terashima, K., Teshima, Y., Shimizu, T. and Hishikawa, Y. Polysomnographic characteristics of healthy elderly subjects with somnambulism-like behaviors, Biol. Psychiatry, 30 (1) (1991) 4–14.

Uchiyama, M., Isse, K., Tanaka, K., Yokota, N., Hamamoto, M., Aida, S., Ito, Y., Yoshimura, M. and Okawa, M. Incidental Lewy body disease in a patient with REM sleep behavior disorder, Neurology, 45 (4) (1995) 709–12.

Ulrich, R.F., Shaw, D.H. and Kupfer, D.J. Effects of aging on EEG sleep in depression, Sleep, 3 (1) (1980) 31–40.

Wauquier, A. and Van, S.B. Aging of core and optional sleep, Biol. Psychiatry, 31 (9) (1992) 866–80.

Webb, W.B. Sleep in older persons: sleep structures of 50- to 60-year-old men and women, J. Gerontol., 37 (5) (1982) 581–6.

Weitzman, E.D., Moline, M.L., Czeisler, C.A. and Zimmerman, J.C. Chronobiology of aging: temperature, sleep-wake rhythms and entrainment, Neurobiol. Aging, 3 (4) (1982) 299–309.

Wever, R., The circadian system of man Springer-Verlag, Berlin; 1979.

Wright, B.A., Rosen, J.R., Buysse, D.J., Reynolds, C.F. and Zubenko, G.S. Shy-Drager syndrome presenting as a REM behavioral disorder, J. Geriatr. Psychiatry Neurol., 3 (2) (1990) 110–3.

Zepelin, H. and McDonald, C.S. Age differences in autonomic variables during sleep, J. Gerontol., 42 (2) (1987) 142–6.

Zulley, J. Distribution of REM sleep in entrained 24 hour and free-running sleep -wake cycles, Sleep, 2 (4) (1980) 377–89.

Zulley, J., Wever, R. and Aschoff, J. The dependence of onset and duration of sleep on the circadian rhythm of rectal temperature, Pflugers Arch., 391 (4) (1981) 314–8.

Rapid Eye Movement Sleep
B.N. Mallick and S. Inoué (Eds)

22. The Function of Fetal/Neonatal REM Sleep

Majid Mirmiran and Yolanda G.H. Maas

Netherlands Institute for Brain Research, Meibergdreef 33, 1105 AZ Amsterdam; Academic Medical Center, University of Amsterdam, Department of Neonatology, PO Box 22700, 1100 DE Amsterdam; Netherlands Ophthalmic Research Institute, Meibergdreef 9, 1105 AZ Amsterdam, The Netherlands

SUMMARY

The hypothesis is put forward that REM sleep in early life serves as: (1) an indicator for the degree of brain maturation and (2) the promoter of further brain development. This hypothesis, although not exclusive, differs from the theory of Roffwarg et al. that REM sleep substitutes for "wakefulness" during the period (early life) in which wakefulness is limited, from the theory of Crick and Mitchson, the "unlearning" hypothesis of REM sleep, from the theory of Jouvet, i.e. that REM sleep is a time for genetic read-out, and from the theory of Freud, i.e. dreams fulfil our wishes (in other words, activation of neuronal systems that were disproportionally activated during wakefulness).

Key Words: rapid eye movement sleep; brain development; fetus; neonate.

Following several decades of research on neuronal and neurochemical mechanisms underlying sleep and in particular rapid eye movement (REM) sleep with emphasis on the brainstem, it was interesting to read a review paper by Jouvet on the regulation of REM sleep by the hypothalamus (Jouvet, 1988). The role the hypothalamus plays in initiating the cascades of event necessary for the generation and maintenance of sleep is made very apparent by this review. However, although the review was devoted to the regulation of REM, it did not discuss the mechanisms underlying the circadian rhythm and functional significance of this state. In the present review we hope to complement Jouvet's paper by describing the hypothalamus regulation of the circadian rhythm of REM sleep and the functional significance of this state, particularly in early development.

REM sleep occupies a large proportion of time during early brain development (Jouvet-Monnier et al., 1969; Mirmiran et al., 1981; Roffwarg et al., 1966). In an altricial mammal, such as the rat, the high amount of REM sleep declines to a low level during the first four weeks of postnatal life, close to the adult value. Although the recent data of Franck et al. suggest the primitive form of sleep during the first two weeks of life in rat, this is indeed not REM sleep. In the guinea pig, a precocial mammal, prenatally high levels of REM sleep can be observed that decline to low (adult) values at birth, in proportion to brain maturation. These observations suggest that the amount of REM sleep can be used as an index for the level of brain development (Becker and Thoman, 1981, 1982; Denenberg and Thoman, 1981; Dinges et al., 1980; Mirmiran, 1986; Roffwarg et al., 1966). However, a recent study by us in which sleep states were longitudinally recorded in preterm infants at 30–40 weeks of postconceptional age did not reveal any change in the amount of time spent in REM sleep in this period of rapid human brain maturation (Fig. 1).

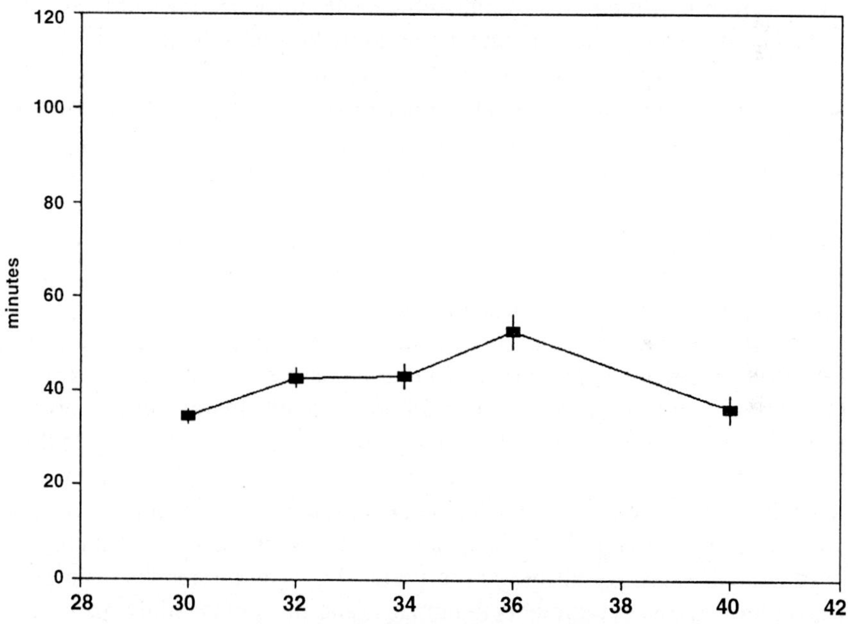

Fig. 1. The amount of REM in minutes (estimasted mean ± SE) in 309 2h-observations of 96 premature babies (gestation <30 weeks) between postmenstrual age week 30 and term corrected age.

The eye movements during REM sleep are one of the most distinctive characteristics of this state, and a good indicator of phasic central neuronal activity in the brain. Both isolated eye movements, as well as bursts, are phasically present during REM sleep. We have studied longitudinally the eye movement development in the REM of rats from postnatal day 10 (approximately the day at which behavioral states are differentiated in rats) to day 25 (approximately the day at which the large proportion of time occupied by REM sleep diminishes in rats). An increased frequency of eye movements was found around day 15 (eye opening), which was only due to changes in eye movement bursts. The burst density, burst duration, as well as the mean number of eye movements within a burst, increased around day 15–18, returning to baseline (day 10) level by day 25. Since pups reared in constant dark showed an increase in eye movement activity during REM sleep similar to that of pups exposed to diurnal light-dark alterations, the change seems to be part of an endogenous developmental timetable of REM sleep necessary for brain maturation, rather than to an increase of visual stimulation after eye opening and development of oculomotor system.

It is very difficult to make a comparison between human and animal studies, since the rapid development of the human brain that begins in the last trimester of gestation lasts at least until the first three months of postnatal life with more gradual maturation thereafter. Nevertheless, Birnholz (1981) reported the appearance of repetitive eye movements at 30 weeks of gestational age, and an increase in eye movement number until 40 weeks. Inoue et al., (1986) reported the appearance and increase of eye movement bursts in a similar developmental period (see also Birnholz, 1981; Bots et al., 1981; Horimoto et al., 1989; Okai et al., 1992). Data from premature babies, gathered by Petre-Quadens and De Lee (1974), show a peak for the frequency of eye movement intervals shorter than one second (i.e. bursts) at 40–41 weeks of conceptional age. All in all, the human data resemble those found in rats, with an increase in burst activity in the mid-developmental period and a subsequent decline at later stages of brain maturation.

Several criteria are required to assume a structure involvement in the generation of circadian rhythms. (a) The structure should be able endogenously to generate rhythms that are close to 24 hr. (b) Such an area must be able to influence other areas responsible for the generation of different overt rhythms (c) For entrainment to the light-dark cycle, it must receive light information from the environment. (d) Upon lesion or subsequent transplantation of such areas, circadian rhythm must disappear (Figs. 2 and 3) and reapear. At present, only the SCN satisfies all of these criteria. Based on the results of internal desynchronization experiments in humans (and the animal study mentioned above), it is generally believed that multiple oscillators are in fact involved in the generation of the mammalian circadian rhythm (contrary to a single oscillator found in invertebrates): it is not yet clear, however, whether these oscillators are situated within the SCN or outside the SCN boundaries (Rosenwasser and Adler, 1986).

In the original study of Roffwarg et al., (1966) two hypotheses were proposed to account for physiological significance of the high amount and intensity of REM sleep in early life. The first, so-called passive, hypothesis assumes that the immature central nervous system is unable to inhibit the REM sleep generator center; thus the decrease of REM sleep with age simply reflects the maturation of this central inhibition. The second active hypothesis favored by most investigators suggests that the function of REM sleep is to supply the

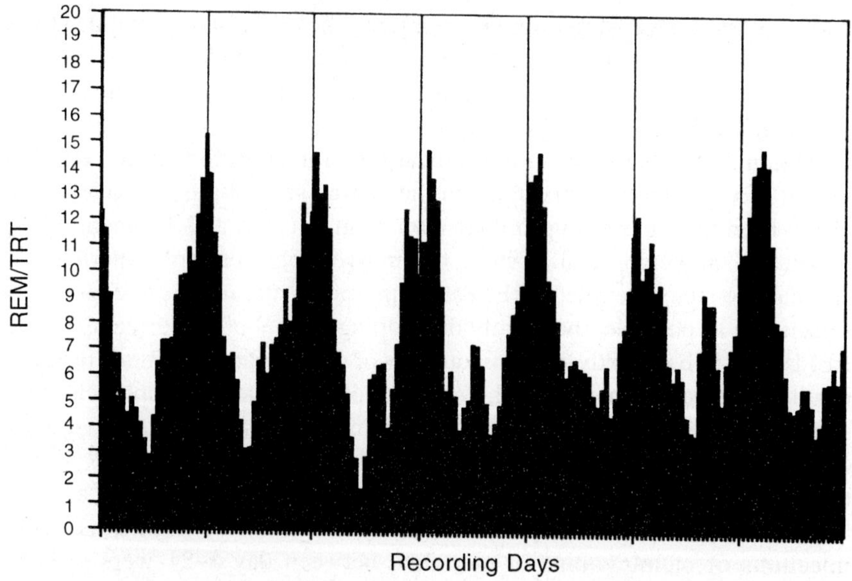

Fig. 2. The amount of REM (in min/hr of total recording time on seven successive days) is shown in an intact rat living in continuous darkness in an environment free of exogenous time cues.

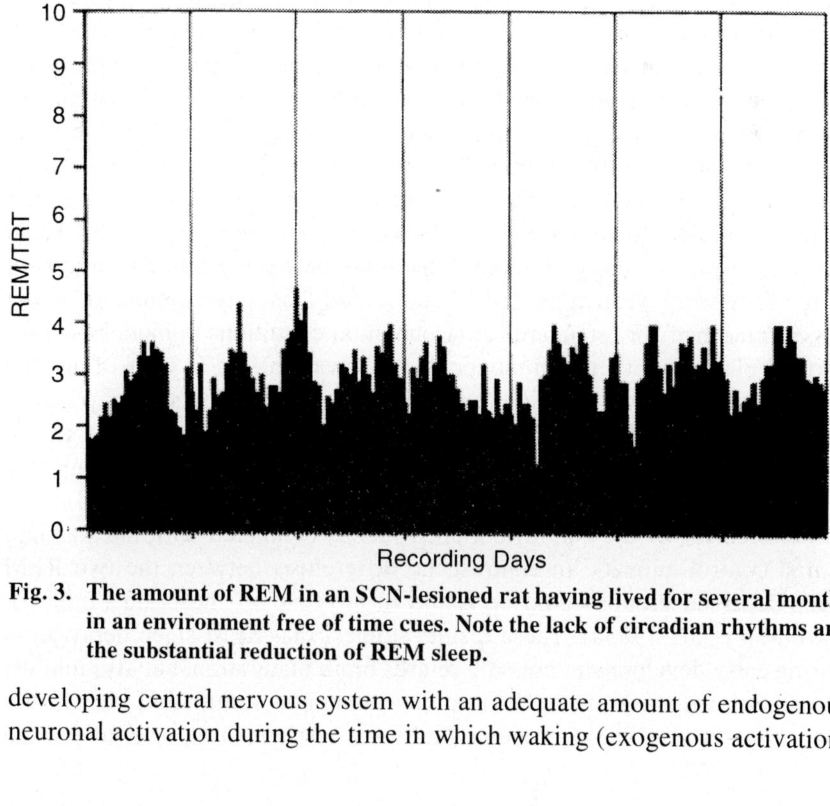

Fig. 3. The amount of REM in an SCN-lesioned rat having lived for several months in an environment free of time cues. Note the lack of circadian rhythms and the substantial reduction of REM sleep.

developing central nervous system with an adequate amount of endogenous neuronal activation during the time in which waking (exogenous activation) is very limited. Sleep-wake studies in infants showed a very strong negative correlation between the amount of REM sleep and quiet alertness in favor of the active hypothesis of REM sleep function in infancy (Denenberg and Thoman, 1981).

Although correlational studies indicating a relationship between e.g. the intensity of eye movement burst and the degree and subsequent maturation of the brain are interesting in their own right (Becker and Thoman, 1981, 1982; Van Someren et al., 1990), the most straightforward experimental approach to study the role of REM sleep in brain maturation is to deprive, as selectively as possible, the rapidly developing animal of its normal quota of REM sleep. Subsequently, the consequences of such treatment for brain function in adulthood can be investigated. We took a pharmacological approach using an anti-depressant drug, such as clomipramine, or an anti-hypertensive drug, clonidine, injected daily in rat pups during the second and third week of life, the period in which rapid rat brain maturation takes place and a high amount and intensity of REM sleep are found. In a series of experiments, daily injections of clomipramine or clonidine between day 8–21 were used to reduce the high level of REM sleep to less than the adult value throughout most of the experimental period (Mirmiran et al., 1981, 1983b). In adulthood, the neonatally REM sleep-deprived rats showed increased anxiety, reduced sexual activity and a disturbed sleep-wake pattern, the latter with higher amounts of time spent in REM sleep and a higher intensity of phasic activity, a pattern reminiscent of the REM intensity found in early neonatal life. Subsequent measurement of regional brain weights showed a significant reduction in the size of the cerebral cortex and brainstem, as compared with the respective control groups, in both the clomipramine- and the clonidine-treated rats. In addition, protein determination in the affected brain areas showed a proportional reduction (Table I; Mirmiran et al., 1983b). In another series of studies (Mirmiran et al., 1983a), rat pups were deprived of REM sleep by means of daily clonidine injections between 8 and 21 days after birth. They were weaned on day 22 and reared from day 28 onwards under either "enriched" or "standard" environmental conditions (a model used for cortical plasticity other than monocular deprivation). At 75 days of age the brains were dissected into eight parts and weighed. The mean group values were compared with those of two control groups which had been reared differentially, i.e. in the same fashion as the REM-deprived rats, but injected with the vehicle only. A selective weight increase was demonstrated in the cerebral cortex of the control-enriched rats as compared with the standard reared control animals. In contrast, no differences between the two REM sleep-deprived groups could be found in any part of the brain (Table II; Mirmiran et al., 1983a). These results suggest that REM sleep deprivation during early development not only retards brain maturation, but also inhibits the growth response of the brain to environmental stimulation later in life (Mirmiran et al., 1983a). Another study was performed in order to test the functional significance of REM sleep for the cerebral cortical or hippocampal maturation. Here, the rat pups were daily injected with clonidine between the ages of 8 and 21 days. To assess the long-term consequences of such treatment in adulthood, cortical or hippocampal extra-cellular discharges in response to micro-iontophoretically applied neurotransmitters were studied. No significant differences were found in the single cell response to glutamate, GABA or noradrenaline in the cerebral cortex of the clonidine-treated rats as compared with the age-matched controls. However, the magnitude of the GABAergic depression of the glutamate-induced neuronal responses was greater in the clonidine than in the control group (Mirmiran et al., 1990). This finding is in accordance with Vogel's hypothesis that REM sleep deprivation increases the inhibitory mechanism. In the hippocampus of neonatally clonidine-treated rats, a supersensitivity of the pyramidal cells to noradrenaline was found (Gorter et al., 1990). It is interesting in this regard that regulation of noradrenaline receptor sensitivity has been suggested to be a function of REM sleep in adulthood (Siegel and Rogawski, 1988; see also Mirmiran et al., 1988). In the homeostatic adaptation of receptor sensitivity to a persistent change in the intensity of incoming stimulation, the neuron makes use of "set-points", which themselves have been determined in early development by the amount of synaptic activation (also during REM sleep; Mirmiran et al., 1988). Furthermore, long-lasting hippocampal plasticity, as revealed in kindling epileptogenesis in adulthood, was significantly retarded in neonatally REM sleep-deprived rats in comparison with the saline-treated control group (Gorter et al., 1990). During early development REM sleep may serve to prevent precocious developmental responses to environmental stimulation, whereas later in development it promotes brain plasticity as a consequence of experience.

Within the framework of the hypothesis of hypothalamic regulation of the circadian sleep-wake rhythm, we suggest the following: the preoptic area, which is involved in regulating body temperature together with the SCN, influences the timing of sleep, i.e. time locked to the body temperature minima at least under natural (synchronized) conditions (Czeisler et al., 1980; Sewitch, 1987); Strogatz et al., 1986). Other areas of the hypothalamus such as the anterior lateral hypothalamus involved in the initiation of sleep and the posterior area involved in the initiation of wakefulness play important roles in these cascades of events (Jouvet, 1988). It could be said that the latter areas are either influenced by the SCN and/or themselves generate circadian rhythms and/or are under the influence of another still undiscovered clock. The studies described thus far show that REM sleep deprivation in a certain critical developmental period produces several long-lasting changes. Based on these findings, Vogel hypothesized that all these changes can be interpreted as symptoms of endogenous depression; in other words, in adulthood neonatally REM sleep-deprived rats develop depression.

Although there are several hypotheses on the function of REM sleep, most of them are based on, and limited to, adult data (for a review see Horne,

1989). A specific hypothesis on the function of REM-sleep in development has been proposed by Roffwarg et al., (1966). They suggested that the function of REM sleep during development might be stimulation of the brain in a period when waking life is limited in time and scope and offers little occasion for stimulation. Although this hypothesis is well supported by most experimental data (including ours), available data do not completely fit this hypothesis, which predicts a steady decline of REM phenomena, paralleling the decline in total amount of sleep during development: in fact, a short-lasting peak in REM phenomena is found (an inverted U-shape) in the amount of eye-movement bursts, which is not paralleled by a similar peak in total amount of sleep (Van Someren et al., 1990). Furthermore, sequelae of REM sleep reduction (which in most deprivation studies, pharmacological or instrumental, were compensated for by an increased wakefulness) are not counteracted. Thus, although we believe the endogenous CNS activation role of REM sleep proposed by Roffwarg et al. (1966), we do not consider REM sleep function to be equivalent to wakefulness in early life. A study carried out for several years by Marks et al., of the Roffwarg group, is of special interest in this regard (Marks et al., 1995). These investigators deprived kittens instrumentally of REM sleep. Mononuclear deprivation reduces the size of the lateral geniculate only during a critical period of brain development. REM deprivation amplified the environmental (monocular) deprivation induced brain deterioration.

It is interesting to note that although in this study all REM deprivation was compensated for by increased waking activity, developmental disturbances do occur. In other studies by Davene and Adrien (Davene and Adrien, 1984; also by Marks et al., 1995) only pontogeniculo occipital waves (PGO; characteristic phasic activity during REM sleep) were suppressed by lesioning brainstem PGO generator areas. A much smaller lateral geniculate nucleus size was found in mononuclearly deprived kittens after PGO lesion than in control monocularly deprived animals. Such lesions eliminate REM sleep state-specific neuronal activation in the lateral geniculate nucleus and not LGN activity induced during waking experiences (Davene and Adrien, 1987).

Any hypothesis on the function of fetal/neonatal REM sleep should cover this remarkable time-course in the development of REM phenomena. First, as the experiments of Vogel's and our own group pointed out, there appears to be a small time-window, in which REM sleep phenomena have a peak and deprivation is most detrimental for development. Secondly, if deprivation during this small time window take place, the detrimental effects: (1) are measurable only after a period of delay, (2) do not necessarily emerge concurrently and (3) might spontaneously disappear after a certain amount of time.

Many other functions suggested for REM sleep are briefly reviewed here since only few experimental studies have been done to support its significance in development. For instance, Wehr suggested that REM sleep is programmed to occur when CNS temperature is low and that it has a thermoregulatory function—that homeotherme uses REM sleep to produce heat in order to maintain or increase the temperature in the CNS during sleep (Wehr, 1992). A restorative function of sleep in general (Horne, 1989; see also Berger and Phillips, 1995) has been suggested for quiet sleep and might also be true for REM sleep during development. Neuronal activation might indeed be a mechanism by which REM sleep can promote growth and plasticity of the brain (Nelson et al., 1989; see also Marks et al, 1995). Another intriguing hypothesis is the one suggested by Vertes, the so-called life-sustaining function of REM sleep (1986). He suggests that the random endogenous activation of the brainstem during REM sleep prevents sustained brain inactivity, which might occur during sleep (switching from off to on or preparing to "wake up"). Cells in some regions of the brain may be damaged or their function disrupted if the levels of spontaneous activity fall below a critical point for prolonged periods of time. REM may serve to periodically "re-charge" certain vital systems (cardiorespiratory and thermoregulatory).

Many pioneer investigators (see Bloch, this volume) have suggested that REM sleep plays a role in learning and memory. The performance of REM sleep-deprived rats on intermittently rewarded learning tasks resembles that of animals, such as fish, which naturally lack REM sleep. There have recently been experimental studies which renewed the interest in this issue (Karni et al., 1994; Wilson and McNaughton, 1994). A different view was suggested by Kreuger and Obal (Krueger and Obal, 1993). These investigators proposed that on the basis of use-dependent synaptic stabilization the neuronal assembly not activated during wakefulness will be activated during sleep to prevent it from atrophy. This view resembles that of Freud who stated: "dreams are for fulfilling our wishes". However, there are recent experimental data of Wilson and McNaughton (1994), showing that the same type of hippocampal cells activated during training in a radial maze are indeed endogenously reactivated (probably for memory consolidation) during sleep. This is not in accordance with the point of view of Kreugel and Obal (1993). Both activation of neuronal groups that were disproportionally stimulated during wakefulness and sleep and consolidation of memory traces (that might be very depending on tasks or individuals; see below) are feasible, but the significance of these theories for fetal/neonatal REM sleep is very limited at present and required further study.

An intriguing hypothesis of REM sleep fuction has been put forward by Crick and Mitchison (1983). Based on the large amount of information (and association) that has to be stored in the brain, they proposed that in order to prevent the system from being overloaded (inducing fantasies, obsessions or hallucinations) an unlearning process is required to delete unwanted information. They suggested that during REM sleep, in which the brain is rather isolated from its normal input/output, a non-specific endogenous activation (e.g. PGO) in the brainstem might induce this so-called reverse learning or unlearning. To put it in a different way: we dream in order to forget. The result is that the dream (or some of its elements) is less likely to recur in the future. Still, the recall of the network after unlearning is less confused and more efficient. The investigators further indicated that the terms reverse learning and reverse unlearning are not ideal because they imply that one has to learn something first in order to unlearn it. But what does a fetus or preterm infants (with about 10 hours of daily REM sleep) learn that has to be unlearned? The answer is that during development, the semi-random process of making synaptic connections is likely to produce parasitic modes. It is these modes that must be unlearned (or deleted) in order to obtain a normally developed network.

Another hypothesis concerning REM sleep was proposed by Jouvet (1980). He suggested that an endogenous genetic programming of the brain occurs during REM sleep. In his view REM sleep is responsible for individual differences throughout the life span of an individual. There are indeed differences between genetically different strains of mice with regard to, e.g., the pattern of eye movements during REM sleep. Jouvet puts this hypothesis forward by stating that the genetic read-out, which is mostly related to "forward experience", is not necessarily adaptive to the latent historical situation of the individual. In this case, the periodic programming may or may not be in accordance with the processing of new information. In some cases, it may facilitate the acquisition of long-term memory, but in others it may have no effect or even inhibit or destroy traces of certain memories. In accordance with this prediction, suppression of REM sleep in two different strains of mice had positive vs. negative effects in an active avoidance task in a Y-maze (for references see Jouvet, 1980) and phenotype differences between the two strains were reduced by REM deprivation. The genetic programming function of REM sleep may not only take place during development. However, it would be interesting to examine whether during the fetal/neonatal period REM deprivation induces different results depending on the species strains used.

In many developmental studies in which hypotheses were put forward and experiments were designed to test the functional significance of biological phenomena, few findings emerge; this may be due to the fact that the central nervous system is selfish, conservative and redundant. Thus, many of our negative results may not simply mean that the biological phenomenon studied (in this case REM sleep) does not have an important function. On the contrary, it only suggests how a plastic brain can appropriately compensate for sequelae and hence functions normally, Furthermore, a sequela found by eliminating a biological phenomenon should not be regarded as its only function (see also Brown, 1995).

In summary, we believe that fetal/neonatal REM sleep has a very important function. It is not for nothing that a preterm/term baby spends about 10 hours in this state every day. This function might be fulfilled by the endogenous activation of the brain to influence: (1) neuronal growth, (2) synaptic plasticity, (3) learning and unlearning, (4) genetic read-out and individual differences and (5) cardiorespiratory and thermoregulatory centers. Furthermore, since the level of REM sleep at a given stage of development can be looked at in two contradictory ways, viz. as an index for both future brain growth potential and present brain maturity, attempts to use the amount of REM sleep as a predictive neurological outcome in neonatology/pediatrics may prove to be disappointing.

We would like to thank Miss Olga Pach for all her efforts in editing this manuscript. For the references regarding this review, the readers are referred to the references cited in two previous reviews by the first author on the same subject.

References

Becker, P.T. and Thoman, E.B. Rapid eye movement storms in infants: rate of occurrence at 6 months predicts mental development at 1 year, Science, 212 (1981) 1415–1416.

Becker, P.T. and Thoman E.B. Intense rapid eye movements during active sleep: an index of neurobehavioral instability, Dev. Psychobiol., 15 (1982) 203–210.

Birnholz, J.C. The development of human eye movement patterns, Science, **213** (1981) 679–681.

Bots, R.S.G.M., Nijhuis, J.G., Martin, Jr., C.B. and Prechtl, H.F.R. Human fetal eye movements: detection in utero by ultrasonography, Early Human Dev., 5 (1981) 87–94.

Brown, R. Muramyl peptides and the function of sleep (1994) (this volume).

Cornwell, A.C. and Laxminarayan, S. A sleep disturbance in high risk for SIDS infants, J. Sleep Res., 2 (1993) 110–114.

Crick, F. and Mitchson, G. The function of dream sleep, Nature, 304 (1983) 111–114.

Czeisler, C.A., Weitzman, E.D., Moore-Ede, M.C., Zimmerman, J.C. and Kanuer, R.S. Science 210 (1980) 1264–1267.

Daan, S., Barnes, B.M. and Strijkstra, A.M. Warming up for sleep?—ground squirrels sleep during arousals from hibernation, Neurosci. Lett., 128 (1991) 265–268.

Davene, D. and Adrien, J. Suppression of PGO waves in the kitten: anatomical effects of the lateral geniculate nucleus, Neuroscie. Lett., 45 (1984) 33–38.

Davene, D. and Adrien, J. Lesion of the pontogeniculate-occipital waves in Kittens. I. Effects on sleep and on unitary discharge of the lateral geniculate nucleus, Brain Res., 409 (1987) 1–9.

De Boer, S., Mirmiran, M., Van Haaren, F., Louwerse, A. and Van de Poll, N.E. Neurobehavioral teratogenic effects of clomi-pramine and alpha-methyldopa. Neuroteratol. Toxicol., 11 (1989) 7784.

Denenberg, V.H. and Thoman, E.B. Evidence for a functional role of active (REM) sleep in infancy, Sleep, 4 (1981) 185–191.

Dinges, D.F., Dadvis, M.M. and Glass, P. Fetal exposure to narcotics: neonatal sleep as a measure of nervous system disturbances, Science, 209 (1980) 619–621.

Gorter, J.A., Kamphuis, W., Huisman, E., Bos, N.P.A. and Mirmiran, M. Neonatal clonidine treatment results in long-lasting changes in noradrenaline sensitivity and kindling epileptogenesis, Brain Res., 535 (1990) 62–66.

Hoppenbrouwers, T., Hodgman, J., Arakawa, K., Giedel, S.A. and Sterman, M.B. Sleep and waking state in infancy: normative studies, Sleep 11 (1988) 387–401.

Horimoto, N., Koyanagai, T., Nagata, S., Nakahara, H. and Nakano, H. Concurrence of mouthing movement and rapid eye movement/non-rapid eye movement phases with advance in gestation of the human fetus, Am. J. Obstet. Gynecol., 161 (1989) 344–351.

Horne, J. Why We Sleep; The Functions of Sleep in Humans and Other Mammals, Oxford University Press, Oxford, 1989.

Inoue, M., Koyanagi, T., Nakahara, H., Hara, K., Hori, E. and Nakano, H. Functional development of human eye movement in utero assessed quantitatively with real time ultrasound, Am. J. Obstet. Gynecol., 155 (1986) 170–174.

Jouvet, M. Paradoxical sleep and the nature-nurture controversy, Prog. Brain Res., 53 (1980) 331–346.

Jouvet, M. Arch, Ital. Biol., 126 (1988) 259–274.

Jouvet-Monnier, D., Astic, L. and Lacote, D. Ontogenesis of the states of sleep in rat, cat and guinea pig during the first postnatal month, Dev. Psychobiol., 2 (1969) 216–239.

Karni, A., Tanne, D., Rubenstein, B.S., Askenasy, J.J.M. and Sagi, D. Dependence of REM sleep of overnight improvement of a perceptual skill, Science, 265 (1994) 679–682.

Korner, A.F., Brown, B.W., Reade, E.P., Stevenson, D.K., Fernbach, S.A. and Thom, V.A. State behavior of preterm infants as a function of development, individual and sex differences, Infant Behav. Dev., 11 (1988) 111–124.

Koyanagi, T., Horimoto, N., Maeda, H., Kukida, J., Minami, K., Ueda, K. and Nakano, H. Abnormal behavioral patterns in the human fetus at term: correlation with lesion sites in the central nervous system after birth, J. Child. Neurol., 8 (1993) 19–26.

Krueger, J.M. and Obal, F. A neural group theory of sleep function, J. Sleep Res., 2 (1993) 63–69.

Lydic, R., McCarley, R.W. and Hobson, J.A. The time-course of dorsal raphe discharge, PGO waves and muscle tone averaged across multiple sleep cycles, Brain Res., 274 (1983) 365–370.

Mancia, M. On the beginning of mental life in the foetus, Int. J. Psycho-Anal., 62 (1981) 351–357.

Marks, G.A., Shaffery, J.P., Oksenberg, A., Speciale, S.G. and Roffwarg, H.P. A functional role for REM sleep in brain maturation (1994) (this volume).

Mirmiran, M. The importance of fetal/neonatal REM sleep, Eur. J. Obstet. Gynecol. Reprod. Biol., 21 (1986) 283–291.

Mirmiran, M. and Corner, M. Neuronal discharge patterns in the occipital cortex of developing rats during active and quiet sleep, Dev. Brain Res., 3 (1982) 37–48.

Mirmiran, M., Van de Poll, N.E., Corner, M.A., Van Oyen, H.G. and Bour, H.L. Suppression of active sleep by chronic treatment with chlorimipramine during early postnatal development: effects upon sleep and behavior in the rat, Brain Res., 204 (1981) 129–146.

Mirmiran, M., Uylings, H.B.M. and Corner, M.A. Effects of post-weaning environmental enrichment on regional brain development in the pre-weaning REM (active) sleep deprived rats, Dev. Brain Res., 7 (1983a) 102–105.

Mirmiran, M., Scholtens, J., Van de Poll, N.E., Uylings, H.B.M., Van der Gugten, J. and Boer, G.J. Effects of experimental suppression of active (REM) sleep during early development upon adult brain and behavior in the rat, Dev. Brain Res. 7 (1983b) 277–286.

Mirmiran, M., Feenstra, M.G.P., Dijcks, F.A., Bos, N.P.A. and Van Haaren, F. Functional deprivation of noradrenaline neurotransmission: effects of clonidine on brain development, Prog. Brain Res., 73 (1988) 159–172.

Mirmiran, M., Dijcks, F.A., Bos, N.P.A., Gorter, J.A. and Van der Werf, D. Cortical neuron sensitivity to neurotransmitters following neonatal noradrenaline depletion, Int. J. Dev. Neurosci., 8 (1990) 217–221.

Nelson, P.G., Yu, C., Fields, R.D. and Neale, E.A. Synaptic connections in vitro: modulation of number and efficacy by electrical activity, Science, 244 (1989) 585–587.

Nijhuis, J.G., Prechtl, H.F.R., Martin, C.B. and Bots, R.S.G.M. Are there behavioral states present in the human fetus?, Early Human dev., 6 (1982) 177–195.

Okai, T., Kozuma, S., Shinozuka, N., Kuwabara, Y. and Mizuno, M. A study on the development of sleep-wakefulness cycle in the human fetus, Early Human Dev., 29 (1992) 391–396.

Petre-Quadens, O. and De Lee, C. Eye movement frequencies and related paradoxical sleep cycles: developmental changes, Chronobiologia, 1 (1974) 347–355.

Roffwarg, H.F., Muzio, J. and Dement, W.C. Ontogenetic development of the human sleep-wakefulness cycle, Science, 152 (1966) 604–619.

Rosenwasser, A.M. and Adler, N.T. Neurosci. Biobehav. Rev. 10 (1986) 431–448.

Sewitch, D.E. Psychophysiology, 24 (1987) 200–215.

Siegel, J.M. and Rogawski, M.A. A function for REM sleep: regulation of noradrenergic receptor sensitivity, Brain Res. Rev., 13 (1988) 213–233.

Strogatz, S.H., Kronauer, R.E. and Czeisler, C.A. Sleep 9 (1986) 353–364.

Van Someren, E.J.W., Mirmiran, M., Bos, N.P.A., Lamur, A., Kumar, A. and Molenaar, P.C.M. Quantitative analysis of eye movements during REM-sleep in developing rats, Dev. Psychobiol., 23 (1990) 55–61.

Vertes, R.P. A life-sustaining function for REM sleep: a theory, Neurosci. Biobehav. Rev., 10 (1986) 371–376.

Vertes, R.P. and Perry, G.W. Sudden infant death syndrome: a theory, Neurosci. Biobehav. Rev., 17 (1993) 305–312.

Wehr, T.A. A brain warming function for REM sleep, Neurosci. Biobehav. Rev., 16 (1992) 379–397.

Wilson, M.A. and McNaughton, B.L. Reactivation of hippocampal ensemble memories during sleep, Science, 265 (1994) 676–679.

Winson, J. The biology and function of rapid eye movement sleep, Current Opinion Neurobiol., 3 (1993) 243–248.

Yavari, P., Vogel, G.W. and Neill, D.B. Decreased raphe unit activity in a rat model of endogenous depression, Brain Res., 611 (1993) 31–36.

Rapid Eye Movement Sleep
B.N. Mallick and S. Inoué (Eds)

23. REM Sleep Deprivation Alters Factors Affecting Neuronal Excitability: Role of norepinephrine and its possible mechanism of action

Birendra N. Mallick, H.V. Anupama Adya and Stephen Thankachan

School of Life Sciences, Jawaharlal Nehru University, New Delhi 110 067, India

SUMMARY

The precise function(s) and mechanism(s) of action of rapid eye movement (REM) sleep is not known. Most of the studies have been done by observing effects of its deprivation on the parameter under investigation. Based on the results obtained from studies where effects of its deprivation were investigated on factors affecting neuronal excitability, a possible role of REM sleep stage in maintaining neuronal excitability has been discussed in this chapter. It has been suggested that REM sleep deprivation alters neuronal excitability by increasing Na-K ATPase activity. This increase in the enzyme activity is likely to be mediated by norepinephrine. Further, as possible mechanism(s) of action, it has been proposed that norepinephrine action is mediated through a_1 adrenoceptor and also by chelation of Ca^{++} ions in the medium.

Key Words: adrenoceptor; calcium ions; excitability; flower pot method; Na-K ATPase; norepinephrine; REM sleep deprivation.

Introduction

Sleep-wakefulness, a behavioral phenomenon, has been objectively classified and quantified primarily based on electrophysiological signals. The signals are electroencephalogram (EEG) recorded from the brain, the electromyogram (EMG) from muscles and the electroocculogram (EOG) recorded due to eye movements. As mentioned in other chapters also, based on these electrophysiological parameters a paradoxical state was identified within the sleep state. This state has objectively been identified by the characteristic electrophysiological signs viz. desynchronization of the EEG, rapid eye movement, atonia in the antigravity muscles and pontogeniculo-occipital waves. Since, some of the characteristic electrophysiological signals for identification of REM sleep resembled that of wakefulness while some others that of sleep state, this state has also been termed either as paradoxical sleep or rapid eye movement (REM) sleep or active sleep or desynchronized sleep. Dreams are associated with this stage of sleep and hence it is also known as dream sleep. It repeats a number of times during sleep, however, its frequency of occurrence and duration per bout increase with the progress of sleep (Kleitman, 1939). It occupies about 15–20% of the total sleep time in different species (Zepelin, 1989). Although, the REM sleep has been identified across species, all the identifying characteristic signs may not be uniformly expressed in different species in the evolutionary ladder. Until recently, REM sleep was considered to be relatively of later origin because it could not be identified in monotremes (Allison, 1972), however, with the recent findings that REM sleep like state is also present in the platypus, an egg laying mammal, it is likely that this state must have evolved more than 100 million years ago (Siegel et al., 1997). Thus, it may possibly be present in other living organism(s), although, the characteristic signs for its identification may vary. Therefore, this stage may be a basic and an instinct behavior.

Mention of a similar behavioral state can be found in ancient vedic literature, the *Mandukya Upanishad*. However, REM sleep state was experimentally identified by its present characteristic features about half a century ago (Aserinsky and Kleitman, 1953). Nevertheless, its mechanism of generation, physiological importance and functional significance are yet to be completely understood. Since this state has been identified in very early species of evolution, it is likely to serve an important basic function in maintaining normal physiology in living being. Although the basic mechanism and causal relationship are yet unknown, REM sleep has been reported to affect the regulation of several physiological functions (Alberts, 1975; Vogel, 1975). It has also been found that there is a compulsion within the physiological system to recover the lost REM sleep and its prolonged deprivation becomes fatal (Rechtschaffen et al., 1989). These suggest that the nature regards REM sleep as one of the very important and absolutely necessary physiological processes which cannot be ignored. Nevertheless, even after half a century of its discovery, the mechanism of generation and function(s) of REM sleep continue to be a challenge to the sleep-wakefulness researchers.

REM Sleep Selective Neurons in the Brain

Transection (Jouvet, 1962; Siegel et al., 1984) and lesion (Jones, 1979; Sakai, 1980) studies have shown that the dorsolateral pontine area is responsible for REM sleep. Subsequently, single unit recordings from the brain stem have shown neurons having unique pattern of firing in relation to sleep-wakefulness including REM sleep (Sakai, 1985). Some neurons alter firing rate during non-REM sleep, some alter in relation to REM sleep, while others remain unaltered during sleep and wakefulness. Among the REM sleep related neurons, some increase firing during REM sleep, the REM-ON neurons (McCarley and Hobson, 1971; Sakai, 1980); while others decrease their firing, the REM-OFF neurons (Chu and Bloom, 1974; McGinty and Harper, 1976; Aston-Jones and Bloom, 1981; Jacobs, 1986). However, REM sleep is identified by the presence of several discrete electrophysiological correlates viz. EEG desynchronization, PGO waves, muscle atonia and rapid eye movement (as mentioned above). Accordingly, neurons related to PGO waves (Datta, 1998) and muscle atonia (Lai and Siegel, 1998) have been identified. Although EEG remains desynchronized during wakefulness as well as REM sleep, only recently it has been shown that there are likely to be separate neurons (or groups of neurons) which may be responsible for EEG desynchronization during wakefulness and REM sleep (Mallick et al., 1998). In order to study neuronal specificity to EEG desynchronization during wakefulness and REM sleep, single neuronal activities were correlated to desynchronization of the EEG during spontaneous wakefulness and REM sleep and also during such behaviors induced experimentally in freely moving, normally behaving cats. It was found that on induction of wakefulness (Fig. 1) and REM sleep (Fig. 2), when the EEG was desynchronized, the neurons behaved in a manner similar to that of their behavior during spontaneous wakefulness and REM sleep, respectively. Thus, there are neurons whose firing may be selectively correlated to different correlates of REM sleep.

Various studies including stimulation, lesion, chemical injection, deprivation etc. have been conducted to investigate and understand the mechanism of generation and functions of REM sleep. However, unlike other treatments, the deprivation method is non-invasive, reasonably specific to REM sleep loss and involves the least, if at all, external treatment and is relatively near normal situation in the sense that often one experiences certain amount of REM sleep deprivation (i.e. reduction or loss of REM sleep) during normal day to day life. Although the method is not without any limitation, the positive aspect is that suitable control experiments may be carried out to rule out the non-specific effects reasonably. The significance of REM sleep on normal physiology has more often been studied by investigating the effects of its deprivation on physiological parameters. Such studies have revealed that REM sleep deprivation adversely affects several functions and the altered states return to normal level with recovery of REM sleep. Several functions involving brain, nervous system, mental faculty, behavior, etc. are adversely affected by the loss or deprivation of REM sleep (Alberts, 1975, Vogel, 1975). Since many brain and mental functions including brain excitability are altered during REM sleep and following its deprivation, it was hypothesized that one of the functions of REM sleep may be to maintain brain excitability, however, its mechanism of action is not known. The possible role of REM sleep in maintaining neuronal excitability and its possible mechanism of action will be discussed in this chapter. The knowledge has mostly been gathered from REM sleep deprivation studies. Hence, before proceeding further the method and validity of the latter will be discussed in brief.

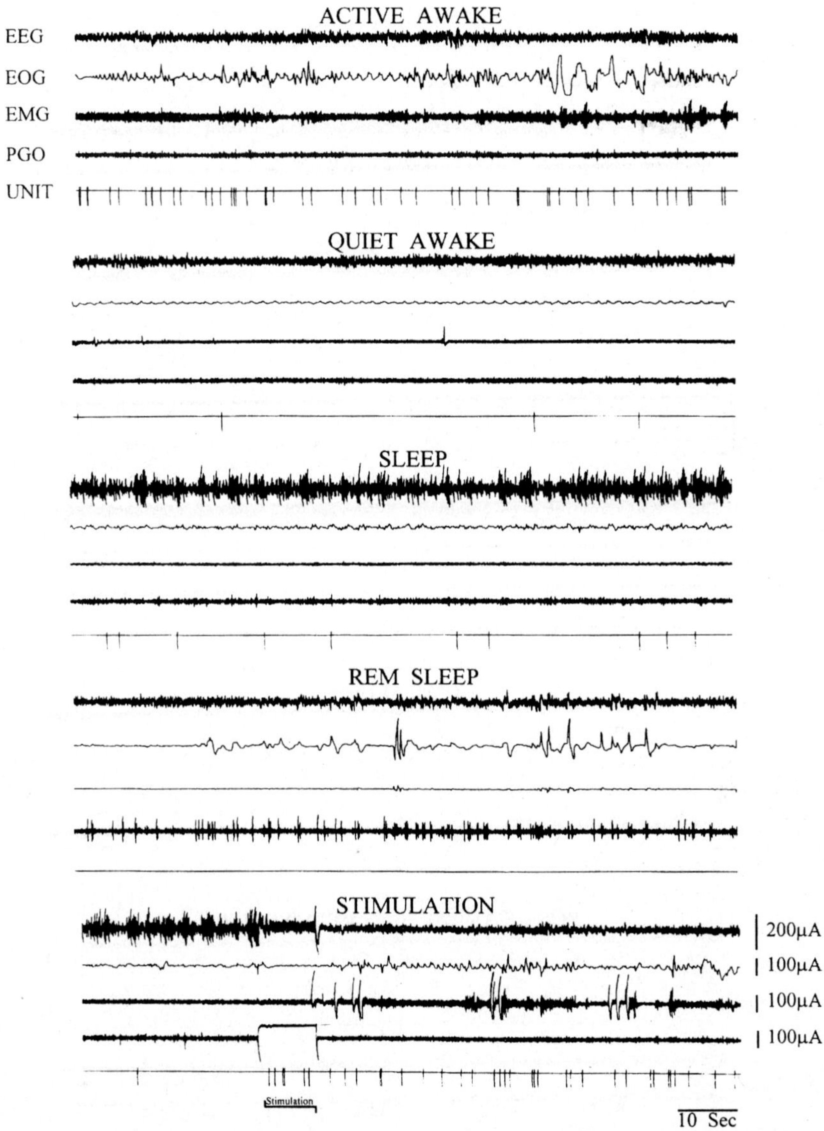

Fig. 1. A representative neuronal activity during spontaneous sleep-waking behavior and midbrain reticular formation (MRF) stimulation induced waking are shown in different sets of traces. The sleep-waking behavior has been classified based on characteristic EEG, EOG, EMG and PGO signals simultaneously recorded in a polygraph. The neuronal activity is the window discriminator output. The spontaneous activity of this neuron showed increased firing during waking, reduced during sleep and ceased during REM sleep, a typical REM-OFF neuron. The neuron increased its firing during midbrain reticular formation stimulation induced waking. Duration of MRF stimulation is marked by arrows. The stimulation parameters were 100 Hz, 200 μA, 300 μsec pulses. EEG—electroencephalogram; EOG—electrooculogram; EMG—electromyogram; PGO—pontogeniculo-occipital wave; UNIT—window discriminator output corresponding to the neuronal activity.

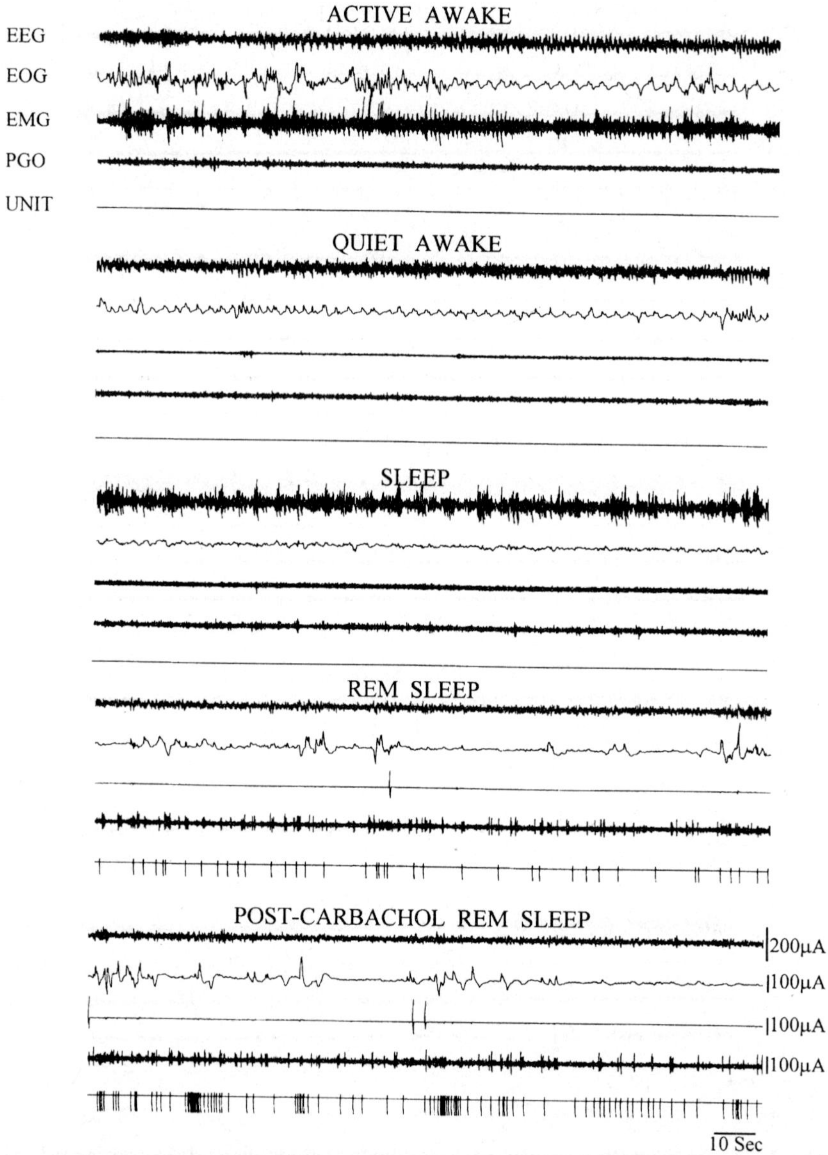

Fig. 2. Behavior of a REM-ON neuronal activity during spontaneous sleep-waking cycle and pontine reticular carbachol stimulation induced REM sleep like state when there was EEG desynchronization. The window discriminator output corresponding to the neuronal firing along with simultaneous changes in EEG, EOG, EMG and PGO waves were recorded in different channels of a polygraph. This neuron did not show any spontaneous firing during any of the sleep-waking states except during REM sleep. The neuron showed a similar behavior during carbachol stimulation induced REM sleep like state. Abbreviations same as in Fig. 1.

REM Sleep Deprivation Studies and Validity of Such Experiments

Several methods have been used for understanding the brain areas involved and their role in regulation of REM sleep. In the transection and lesion studies, role of the brain area under study in either the tonic or the phasic events could be investigated and the effects of recovery could not be studied. The REM sleep deprivation protocol, a powerful strategy to study its functions, on the other hand, takes care of most of those disadvantages to a reasonable extent. It is probably therefore, the deprivation studies have contributed significantly in elucidating and understanding the functional significance of REM sleep.

Normally, spontaneous REM sleep episodes are short lasting. Therefore, effect of single REM sleep episode or its deprivation might not be significant enough to be detected. However, the effects are likely to be cumulative resulting in amplification of the effects on the parameter(s) under study. Accordingly, by increasing or by decreasing the amount of REM sleep, an accumulation or a depletion of its effect is expected. As it is known that different correlates (signs) of REM sleep are generated by different brain areas or groups of neurons, it is very difficult to stimulate (chemically or electrically) simultaneously and precisely all those areas to generate all the signs expressed during spontaneous REM sleep. Often on chemical stimulation one or more of its major correlates (signs or symptoms) can be made to express for varying duration. Therefore, it has often been expressed as REM *sleep like state.* Nevertheless, a REM sleep like phenomenon, after carbachol microinjection into the pontine area has been reported (Baghdoyan et al., 1985) however, the effective sites for such induction are intermingled with non effective sites (Deurveilher et al., 1997). It is also true that normally REM sleep follows deep sleep. However, during induced REM sleep, although signs of REM sleep may be expressed, the timing and sequence of its occurrence (in relation to sleep-wakefulness paradigm) is lacking. Similarly, there are limitations with the electrical stimulation studies also. First, it is generally an invasive technique as the insertion of the electrode during implantation is likely to cause a mechanical damage to the neurons and fibres resulting in secondary effect; second, it is difficult to know the strength of the stimulating current which may be considered normal/natural; third, the extent of stimulating current cannot be localized precisely; fourth, accommodation of the stimulation, if any, cannot be taken care of. Added to these, it may be argued that in nature, the duration and frequency of any physiological phenomenon have evolved to an optimal level after a great deal of trial (and error) which would be ideal and best suited for that particular system. An increase in the phenomenon, REM sleep in this case, may not necessarily express a significant change in the parameter under study compared to that of withdrawing the same. This is because the intensity of an increase in the parameter under study may not be significant beyond the optimum level at which it already exists. Hence, experimental study by increasing the REM sleep may not be very effective approach.

The deprivation approach has the advantage of overcoming the draw backs mentioned above in understanding the functions of REM sleep. The major advantages of this method of study are that it is a non-invasive technique. The parameter(s) under study, if related to REM sleep, is likely to be amplified proportionate to the period of deprivation. As some degree of REM sleep loss is experienced by almost everybody, deprivation can probably be considered a near natural phenomenon in terms of applicability of the knowledge for the ultimate goal of human welfare. However, this method also has its own limitations. For instance, it is difficult to quantify the amount of stress (if at all), to compensate for the small loss of slow wave sleep during REM sleep deprivation and to determine the optimum period of REM sleep deprivation required to obtain the significant effects. These limitations are overcome to a great extent by the use of appropriate and effective control experiments. The effect of possible stress is most often the major criticism for the REM sleep deprivation studies and one wonders if induction of REM sleep could be an effective answer to it. A counter argument to this could be that, since in nature, any phenomenon including its duration has evolved to optimally benefit as per the requirement of the living being/organism, increasing the duration of that phenomenon by external means, REM sleep for instance in this case, may also be stressful. The quality and the intensity of stress in cases of deprivation and induction of REM sleep may, however, differ and can be argued. Since, there is hardly any study where functions of REM sleep have been investigated with an increase in REM sleep, the knowledge has mostly been gathered by observing studying the effects of REM sleep deprivation. Although there are several methods for REM sleep deprivation, the flower pot method has mostly been used for REM sleep deprivation studies. Hence, before proceeding further, the method of deprivation with its merits and demerits will be discussed for convenience and ease of understanding.

Flower-pot method for REM sleep deprivation

Among the various methods available, the flower pot method has most extensively been used for REM sleep deprivation studies. This is the most popular and rather effective technique for REM sleep deprivation. It is also known as water tank, platform, or pedestal technique. It was first used in cats (Jouvet-Mounier et al., 1964) and has since been extensively used in different species (Hicks et al., 1977; Oniani et al., 1988). In this method, the animals are maintained on small raised platforms, typically inverted flower pots, surrounded by water. The animals can sit, crouch, change position and move freely on this artificial island. However, the animals cannot have enough room for complete relaxation. As soon as the animals enter into REM sleep, due to atonia in the postural muscles, they fall into the surrounding water resulting in waking. Thus, although the animals can enjoy sleep, occurrence of REM sleep is prevented. In this method, the animals quickly learn to wake up at the onset of REM sleep and hence do not frequently fall into the water. Thus, in this method, the animals can have non-REM sleep but not REM sleep. This technique is very effective, inexpensive, procedurally simple and at the same time allows a number of animals to be simultaneously deprived of REM sleep.

It has been concluded that it is the ratio of platform size to the body weight of the animal which decides the amount of reduction in REM sleep (Mendelson, 1974). A large platform enabling the animals to curl up, move and have both slow wave sleep as well as REM sleep without falling into the water is most effectively used as control. Movement restriction controls are carried out to rule out the effects of restricting movement of the animals on small platform by maintaining the animals in a smaller cage where they can have both the slow wave sleep and REM sleep. Another control group which has frequently been used to eliminate the effect of increased muscular activity on the small platforms is the swimming control. Although it is difficult to equate the increased muscle activity on the small platforms and that of by swimming, attempts have been made to study the differential response by studying the effect of 2 and 5 h of swimming (Gulyani and Mallick, 1993). This platform technique has most widely and universally been used for REM sleep deprivation studies. Mendelson (1974) recorded the EEG for 96 hrs from four groups of rats: (a) normal or baseline, (b) on small platform (diameter 6.5 cm), (c) on large platform (diameter 12.5 cm) and (d) 1 h swimming. He reported that rats on small platform had 57% as much REM sleep as baseline during the first 24 h, without any change in non-REM sleep, while rats on large platform on the first day had 55% REM sleep (compared to baseline). The REM sleep increased to baseline value by the fourth day in the large platform control animals. The rats on the small platform continued to have significantly reduced REM sleep on the fourth day. The swimming control rats had no reduction in REM sleep or non-REM sleep at any stage.

REM Sleep Deprivation and Alteration in Neuronal Excitability

REM sleep deprivation has been found to cause various behavioral changes including fatigue (Kollar et al., 1969), irritability, inability to concentrate and anxiety (Dement, 1960). It also showed poorer retention of memory than normal rats in a passive avoidance test (Stern and Morgane, 1974). The REM sleep deprivation caused a decrease in threshold for induction of electroconvulsive shock in rats (Cohen and Dement, 1965). The threshold for electroconvulsive shock in control rats ranged from 21–40 ma (mean 26.2) and after 6 days of REM sleep deprivation by the platform method, the threshold ranged from 17.2–27 ma (mean 20.2). The threshold returned to the baseline after the recovery. An increase in shock induced fighting behavior was observed in rats after REM sleep deprivation (Morden et al., 1968). The mean number of fightings, in response to a range of shock intensities, delivered as a train of brief electric shock pulses at the footpads, were found to increase with an increase in the duration of deprivation. Loss of sleep, mainly the paradoxical phase, resulted in an increase in brain excitability as seen by a decrease in threshold to seizures (Owen and Bliss, 1970). Evoked potential, as an indicator of neural excitability, was studied during progressive REM sleep deprivation. An increase in amplitude of entorhinal potentials was observed from day 1 of deprivation, reaching an average of 374% of the predeprivation on the 7th day and returned to control amplitude after 2 nights of sleep in cats. Since there was a decrease in excitability of primary sensory afferent pathways, it was concluded that REM sleep deprivation may lead to paleocortical excitability and/or to increase in inhibition responsible for sensory filtering (Satinoff et al., 1971). Thus, REM sleep derivation is likely to have a generalized effect on neuronal responsiveness. Although the alteration in gross brain excitability following REM sleep deprivation was known, the effects at the cellular and molecular levels and their mechanism(s) of action have started unravelling only recently.

In addition to alteration in firing rate of neurons during REM sleep (described earlier in this chapter), its deprivation also affected the firing rate and responsiveness of brain stem neurons. It was observed that, following REM sleep deprivation in cats, the REM sleep selective neurons viz. the REM-ON and the REM-OFF neurons increased and decreased their firing rate, respectively—possibly a compensatory effect (Mallick et al., 1989). Although the firing rates of non-REM sleep related neurons were not significantly affected after REM sleep deprivation, it reduced the auditory stimulus bound responsiveness of neurons which were sensitive to auditory stimulus (Mallick et al., 1991). Thus, the REM sleep deprivation was experimentally shown to affect neuronal responsiveness when exposed to the stimulus to which it was sensitive to. Since REM sleep deprivation affected responsiveness of REM sleep related as well as unrelated neurons, it is likely that the deprivation might have a generalized effect on the brain. The responsiveness of neurons is an expression of neuronal excitability and it was affected by REM sleep deprivation. Hence, it was proposed that REM sleep deprivation possibly alters neuronal excitability, in other words, in normal situation one of the functions of REM sleep is possibly to maintain neuronal excitability.

Possible Cellular Changes for Alteration in Neuronal Excitability

The transmembrane potential is a reflection of neuronal excitability status. It is well established that the latter may be modulated by Na-K ATPase (Trachtenberg et al., 1982) and calcium level (Kuffler et al., 1984). The latter also alters secretion of neurotransmitters (Thomas and Almers, 1992). Therefore, Na-K ATPase, chloride ATPase activities and total, free as well as bound calcium concentrations were estimated in control and REM sleep deprived rat brain. Following REM sleep deprivation there was a generalized increase in the Na-K ATPase activity throughout the brain, however, the ponto-medullary area were first to be affected (Gulyani and Mallick, 1993). The chloride sensitive Mg-ATPase activity increased although the chloride insensitive Mg-ATPase activity remained unaltered (Mallick and Gulyani, 1993). Both the free and bound calcium concentrations were affected throughout the rat brain after REM sleep deprivation (Mallick and Gulyani, 1996). In general, there was a decrease in Ca^{++} concentration except in bound Ca^{++} in the brain stem only. Two days of REM sleep deprivation caused a decrease in free and an increase in bound Ca^{++} levels in the brain stem, which was first to be affected. The free Ca^{++} concentration decreased always and this decrease in free calcium levels after four days of REM sleep deprivation was significantly higher than that of following two days of deprivation. The bound calcium concentration increased in the brainstem but decreased in the cerebellum and the cerebrum after four days deprivation. The ratio between the bound to free calcium levels, though was maintained in cerebrum and cerebellum, was disturbed in the brain stem after the deprivation (M allick and Gulyani, 1996). Thus, REM sleep deprivation altered the factors responsible for maintenance of neuronal excitability. These supported the view that REM sleep is likely to maintain neuronal excitability. It is possibly therefore, from an evolutionary perspective, sleep (including REM sleep) is a polycyclic phenomenon. This view although may be supported by the polycyclic nature of sleep behavior in lower animals, does not quite hold good for humans, where, at least in adults it is not polycyclic. It is probably therefore, an afternoon nap is beneficial and helps improving efficiency and output (Bonnet and Arand, 1995).

As mentioned above, if REM sleep deprivation was to have a generalized effect on the brain, it should affect a common and basic property of the neuron. It has been seen (above) that Na-K ATPase activity is increased after REM sleep deprivation. Since Na-K ATPase is a transmembrane integral protein molecule, it was hypothesized that deprivation might affect neuronal membrane fluidity—a basic neural property. This was also likely because alteration in membrane fluidity has been reported to affect the Na-K ATPase activity and vice versa (Lebel and Schatz, 1990). Hence, membrane fluidity was estimated in control and REM sleep deprived rat brain microsome and synaptosome. It was found that the membrane fluidity altered in a manner so that the membrane became more rigid after the deprivation (Mallick et al., 1995). This view is supported by the fact that REM sleep is reduced with aging (Jouvet-Mounier et al., 1969; Bonnet and Arand, 1989) when the fluidity is also reduced (Viani et al., 1991; Sawada et al., 1992) and also that the drug centrophenoxine, which prevented aging process, also increased fluidity (Wood et al., 1986). Interestingly, activities of other membrane bound enzymes viz. acetylcholinesterase (Mallick and Thakkar, 1992), monoamine oxidase (Thakkar and Mallick, 1993), 5-nucleotidase (Thakkar and Mallick, 1996), also were affected after the deprivation. However, it is difficult to comment on the cause and effect relationship between the membrane fluidity and the enzyme activity. Also c-fos expression has been found to be altered after REM sleep (Shiromani et al., 1992) and its deprivation (Cirelli et al., 1995). This c-fos expression has been directly correlated with neuronal activation (Sagar et al., 1988; Yamuy et al., 1993). Therefore, from the studies mentioned above, it may be said that REM sleep deprivation alters neuronal excitability.

Increased Level of Norepinephrine in the Brain Following REM Sleep Deprivation

It has been mentioned above that MAO-A activity decreased following REM sleep deprivation (Thakkar and Mallick, 1993). This decrease is likely to increase the level of norepinephrine (NE) at the projected sites. An increase in NE turnover after REM sleep deprivation has been reported (Pujol, 1968; Mark et al., 1969). This is supported by the finding that, after REM sleep deprivation, there is an increase in tyrosine hydroxylase activity (Sinha et al., 1973), the enzyme responsible for NE synthesis. It is also known that the NE-ergic REM-OFF neurons cease firing during spontaneous REM sleep (Sakai, 1985) and continue to fire during REM sleep deprivation (Mallick et al., 1989). This continued firing of the NE-ergic neurons during REM sleep deprivation is likely to increase the level of NE, at least locally, at the projection sites in the brain. This view may also be supported by the finding that a mild continuous electrical stimulation of the locus coeruleus induced an effect similar to that of REM sleep deprivation. There was a significant reduction in REM sleep during the stimulatory period and a rebound increase during the post-stimulatory period (Singh and Mallick, 1996a). It may be noted that REM sleep deprivation also caused a rebound increase in REM sleep during the recovery period (Dement, 1960). Since locus coeruleus is rich in NE-ergic neurons, its stimulation is likely to increase NE in the brain which might be responsible for deprivation related effects. All these studies put together strongly suggest an increase in the level of NE in the brain after REM sleep deprivation although a direct study to show an increase NE after REM sleep deprivation is lacking.

REM Sleep Deprivation Induced Increase in Na-K ATPase Activity Might be Mediated by the Increased Norepinephrine

It has been seen above that Na-K ATPase activity is increased after REM sleep deprivation. Although direct study is lacking, all other studies (mentioned above) show that NE is likely to increase in the brain after REM sleep deprivation. On the other hand, independent study has reported that NE exerts a stimulatory effect on Na-K ATPase activity (Swann, 1983). Hence, it was proposed that the REM sleep deprivation induced increase in Na-K ATPase activity could be mediated by an increased level of NE. This is supported by *in vivo* and *in vitro* studies where NE-ergic α_1 blocker, prazosin, could prevent REM sleep deprivation induced increase in Na-K ATPase acitivity (Gulyani and Mallick, 1995). An interesting observation was that although the NE-ergic blocker could prevent REM sleep deprivation induced relative increase in Na-K ATPase activity, the enzyme activity was not completely blocked (Gulyani and Mallick, 1995). Similarly, the locus coeruleus stimulation induced reduction in REM sleep could also be prevented, however, by propranolol, a beta adrenoceptor blocker (Singh and Mallick, 1996b). These suggest that the Na-K ATPase activity is likely to be maintained by more than one factor and mechanism(s).

Possible Cellular Mechanism of REM Sleep Deprivation Induced Alteration in Neuronal Excitability—A synthesis

It has been mentioned above that following REM sleep deprivation there was an increase in Na-K ATPase activity which was likely to be due to increased NE. It was also found that REM sleep deprivation decreased calcium concentration. This is reasonable because calcium has an inhibitory action on Na-K ATPase activity (Davis and Vincenzi, 1971). The results also suggest that the REM sleep deprivation most likely increased NE in the brain and the

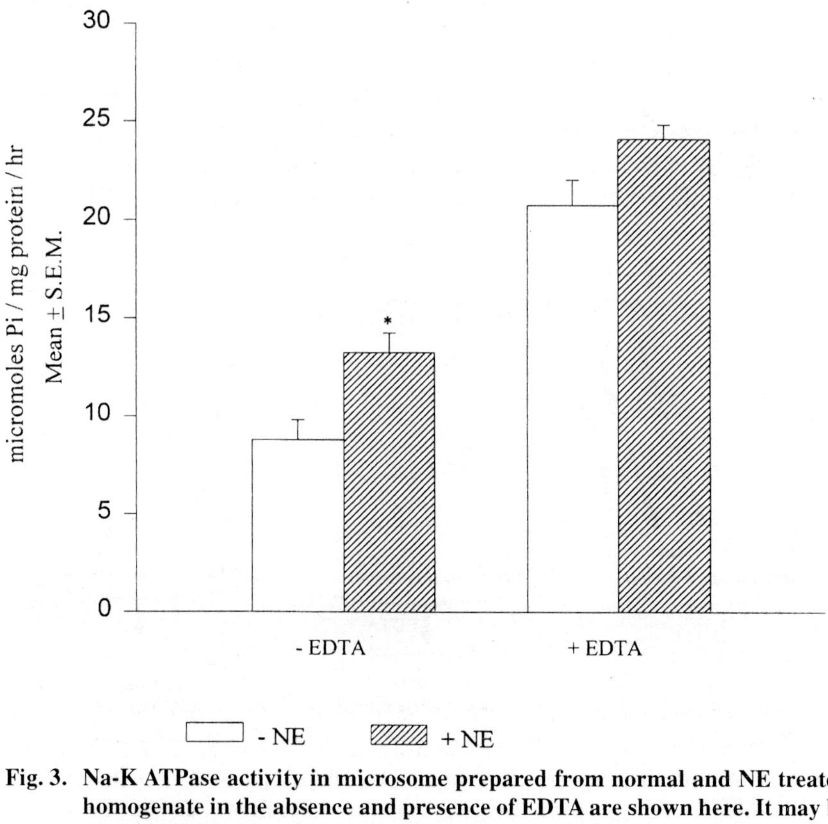

Fig. 3. Na-K ATPase activity in microsome prepared from normal and NE treated homogenate in the absence and presence of EDTA are shown here. It may be noted that NE treatment increased the enzyme activity in the medium without EDTA but it could not increase in the presence of EDTA. * – $p < 0.05$.

latter might be responsible, at least in part, for the deprivation induced increase in Na-K ATPase activity. Therefore, attempts were made to investigate if NE induced increase in Na-K ATPase activity may be mediated through removal of Ca^{++}. The NE induced increase in Na-K ATPase activity was estimated *in vitro* in presence and absence of calcium chelator, Ethylene diamine tetra acetic acid (EDTA). It was found that NE *in vitro* could increase Na-K ATPase activity in the absence of EDTA in the medium and it was ineffective in presence of EDTA (Fig. 3). Besides, NE could increase Na-K ATPase activity if added in the homogenate and then synaptosome prepared. On the other hand, if NE was added to the medium after preparation of the synaptosome, the increase in the enzyme activity could not be observed (Fig. 4). These observations suggest that NE possibly chelates the calcium ions present in the medium and thereby increases Na-K ATPase activity (Adya and Mallick, 1997). This view may be supported by the fact that NE is reported to have a Ca^{++}chelating action (Godfraind et al., 1972). Furthermore, as also mentioned earlier, part of the NE induced increase in Na-K ATPase activity could be blocked by α-1 adrenoceptor blocker. To explain, the microsomal Na-K ATPase activity was found to increase following NE treatment of the homogenate, however, the increase was not observed when the homogeante was pretreated with α-1 adrenoceptor blocker-Prazosin (Fig. 4). Therefore, in addition to the calcium chelating action of NE, it is also likely to act through α-1 adrenoceptor to increase Na-K ATPase activity. Whether these two mechanisms work independently or form part of the same pathway is under investigation. Finally, synthesis of all the information so far discussed suggest that REM sleep

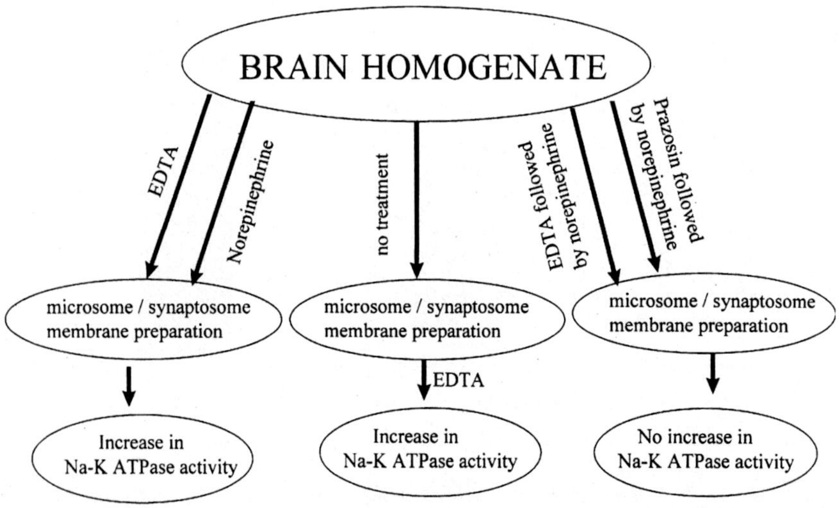

Fig. 4. Schematic diagram depicting the effect of NE, EDTA, and NE in the presence of EDTA or Prazosin on Na-K ATPase activity are shown. The enzyme activity was estimate in microsome and synaptosome prepared from brain homogenate with or without the treatment.

deprivation increases Na-K ATPase activity and reduces Ca^{++} concentration. These changes alter the neuronal excitability and ultimately affect physiological processes. This deprivation induced increase in Na-K ATPase activity is likely to be mediated by an increase in NE. The action of NE is likely to be mediated through α-1 adrenoceptor and by chelating Ca^{++} (Fig. 5).

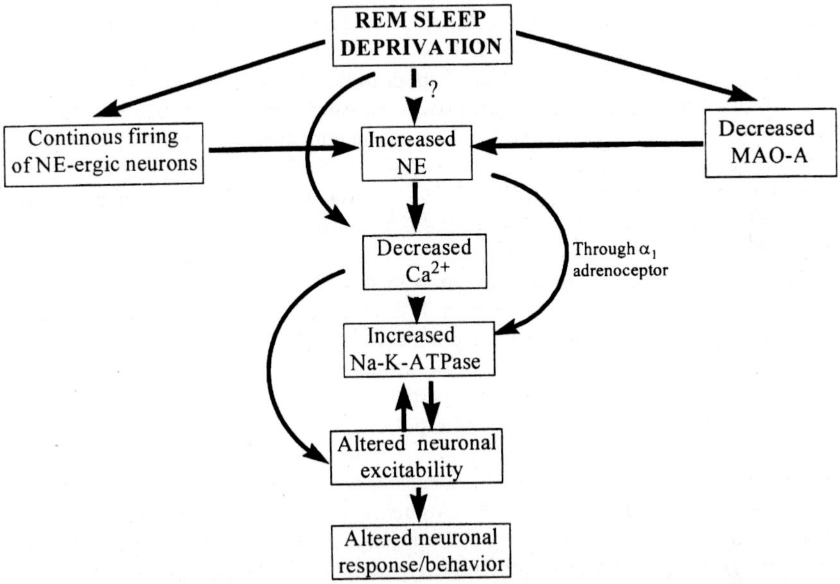

Fig. 5. Schematic diagram showing possible cellular mechanism(s) of rapid eye movement sleep deprivation induced changes in neuronal excitability and behavior.

Acknowledgements

Fundings from CSIR, DBT, DST and ICMR are acknowledged.

References

Adya, A. and Mallick, B.N. Possible molecular mechanism of norepinephrine induced increase in Na-K ATPase activity in rat brain microsome, Second Congress of the Asian Sleep Research Society, Jerusalem, (1997) 56.

Alberts, I.B. REM sleep deprivation, Biol. Psychiat., 10 (1975) 341–351.

Allison, T., van Twyver, G. and Goff, W.R. Electrophysiological studies of the echidna, Tachyglossus aculeatus, Arch. Ital. Biol., 110 (1972) 145–184.

Aserinsky, E. and Klietman, N. Regularly occurring periods of eye motility and concomitant phenomena during sleep, Science, 118 (1953) 273–274.

Aston-Jones, G. and Bloom, F.E. Activity of norepinephrine containing locus coeruleus neurons in behaving rats anticipates fluctuations in the sleep-waking cycle, J. Neurosci., 1 (1981) 876–886.

Baghdoyan, H.A., McCarley, R.W. and Hobson, J.A. Cholinergic manipulation of brain stem reticular systems: Effects on desynchronized sleep generation. In: A. Wauquier,

J.M. Monti, J.M. Gaillard and M. Radulovacki (Eds.), *Sleep: Neurotransmitters and Neuromodulators,* Raven Press, New York, 1985, pp. 57–68.

Bonnet, M.H. and Arand, D.L. Consolidated and distributed nap schedules and performance, J. Sleep Res., 4 (1995) 71–77.

Bonnet, M.H. and Arand, D.L. Sleep loss in aging, Clin. Geriatr. Med., 5 (1989) 405–420.

Chu, N.S. and Bloom, F.E. Activity patterns of catecholamine-containing pontine neurons in the dorsolateral tegmentum of unrestrained cats, J. Neurobiol., 5 (1974) 527–544.

Cirelli, C., Pompeiano, M. and Tononi, G. Sleep deprivation and *c-fos* expression in the rat brain, J. Sleep Res., 4 (1995) 92–106.

Cohen, H.B. and Dement, W.C. Sleep: Changes in threshold to electroconvulsive shock in rats after deprivation of the "paradoxical" phase, Science, 150 (1965) 501–504.

Datta, S. PGO wave generation: mechanism and functional significance, (1998) Chapter in this book.

Davis, P.W. and Vincenzi, F.F. Ca^{++} ATPase activation and Na-K ATPase inhibition as a function of Ca^{++} concentration in human red blood cell membranes, Life Sci., 10 (1971) 401-406.

Dement, W.C. Effect of dream deprivation, Science, 131 (1960) 1705.

Deurvilher, S., Bernard, H. and Hennevin, E. Pontine microinjection of carbachol does not reliably enhance paradoxical sleep in rats, Sleep, 20 (1997) 593–607.

Godfraind, T., Koch, M.C. and Verbeke, N. The action of EGTA on catecholamine stimulation of rat brain Na-K ATPase, Biochem. Pharmacol., 23 (1974) 3505–3511.

Gulyani, S. and Mallick, B.N. Effect of rapid eye movement sleep deprivation on rat brain Na–K ATPase activity, J. Sleep. Res., 2 (1993) 45–50.

Gulyani, S. and Mallick, B.N. Possible mechanism of REM sleep deprivation induced increase in Na-K ATPase activity, Neuroscience, 64 (1995) 255–260.

Hicks, R.A., Okuda, A. and Thomsen, D. Depriving rats of REM sleep: the identification of a methodological problem, Am. J. Physiol., 90 (1977) 95–102.

Jacobs, B.L., Single unit activity of locus coeruleus neurons in behaving animals, Prog. Neurobiol., 27 (1986) 183–194.

Jones, B.E. Elimination of paradoxical sleep by lesions of the pontine gigantocellular tegmental field in the cat, Neurosci. Lett., 13 (1979) 285–293.

Jouvet-Mounier, D., Vimont, P., Delorme, F. and Jouvet, M. Etude da la privation selective de la phase paradoxale de sommeil chez le chat, Copmt. Rend. Soc. Biol., 158 (1964) 754.

Jouvet, M., Recherches sur les structures nerveuses et les mecanismes responsables des differentes phases du sommeil physiologique, Arch. Ital. Biol., 100 (1962) 125–206.

Jouvet-Mounier, D., Asctic, L. and Lacote, D. Ontogenesis of states of sleep in rat, cat and guinea pig during first postnatal month, Dev. Psychobiol., 2 (1969) 216–239.

Kleitman, N., Sleep and wakefulness, University of Chicago Press, Chicago, (1939).

Kollar, E.J., Pasnau, R.O., Rubin, R.T., Naitoh, P., Slater, G.G., and Kales, A. Psychological, psychophysiological, and biochemical correlates of prolonged sleep deprivation, Am. J. Psychiatry 126 (4) (1969) 488–497.

Kuffer, S.W., Nicholls, J.G. and Martin, R.A. Control of membrane permeability. In: S.W. Kuffer, J.G. Nicholls, and R.A. Martin (Eds.), *From Neuron to Brain,* Sinauer Associates Inc., Massachusetts, 1984, pp. 133–166.

Lai, Y.Y. and Siegel, J.M. Muscle atonia in REM sleep, (1998). Chapter in this book

Lebel, C.P. and Schat, R.A. Altered synaptosomal phospholipid metabolism after toluene: possible relationship with membrane fluidity, Na-K adenosine triphosphatase and phospholipid methylation, J. Pharmacol. Exp. Ther., 253 (1990) 1189–1197.

Mallick, B.N., Fahringer, H. and Siegel, J.M. REM sleep deprivation reduces auditory evoked inhibition of dorsolateral pontine neurons, Brain Res., 552 (1991) 333–337.

Mallick, B.N. and Gulyani, S. Rapid eye movement sleep derivation increases chloride sensitive Mg-ATPase activity in rat brain, Pharmacol. Biochem. Behav., 45 (1993) 359–362.

Mallick, B.N. and Gulyani, S. Alterations in synaptosomal calcium concentrations after rapid eye movement sleep deprivation in rats, Neuroscience, 75 (1996) 729–736.

Mallick, B.N., Siegel, J.M. and Fahringer, H. Changes in pontine unit activity with REM sleep deprivation, Brain Res., 515 (1989) 94–98.

Mallick, B.N. and Thakkar, M. Effect of REM sleep deprivation on molecular forms of acetylcholinersterase in rats, NeuroReport, 3 (1992) 676–678.

Mallick, B.N., Thakkar, M. and Gangabhagirathi, R. Rapid eye movement sleep deprivation decreases membrane fluidity in the rat brain, Neurosci. Res., 22 (1995) 117–122.

Mallick, B.N., Thankachan, S. and Islam, F. Differential responses of brain stem neurons during spontaneous and stimulation induced desynchronization of the cortical EEG in freely moving cats, Sleep Res. Online, (1998) in press.

Mark, J., Heiner, L., Mandel, P. and Godin, Y. Norepinephrine turnover in brain and stress reactions in rats during paradoxical sleep deprivation, Life Sci., 8 (1969) 1085–1093.

McCarley, R.W. and Hobson, J.A. Single neuron activity in cat gigantocellular tegmental field: selectivity of discharge in desynchronized sleep, Science, 174 (1971) 1250–1252.

McGinty, D.J., and Harper, R.M Dorsal raphe neurons: Depression of firing during sleep in cats, Brain Res., 101 (1976) 569–575.

Mendelson, W.B., The flower pot technique of rapid eye movement (REM) sleep deprivation, Pharmacol, Biochem. Behav., 2 (1974) 553–556.

Morden, B., Conner, R., Mitchell, G., Dement, W. and Levine, S. Effects of rapid eye movement (REM) sleep deprivation on shock-induced fighting, Physiol. Behav., 3 (1968) 425–432.

Oniani, T.N., Lortkipanidze, N.D., Magaloblishvilli, M.M., Maisuradze, L.M., Oniani, L.T., Babilode, M.R. and Gvasalia, M.G. Neurophysiological analysis of paradoxical sleep deprivation. In: T. Oniani (Ed.), *Neurobiology of sleep wakefulness cycle,* Metsiereba, Tibilisi, 1988, pp. 19–42.

Owen, M. and Bliss, E.L. Sleep loss and cerebral excitability, Am. J. Physiol., 218 (1970) 171–173.

Pujol, J.F., Mouret, J., Jouvet, M. and Glowinski, J. Increased turnover of cerebral norepinephrine during rebound of paradoxical sleep in the rat, Science, 159 (1968) 112–114.

Rechtschaffen, A., Bergmann, B.N., Everson, C.A., Kushida, C.A. and Gilliland, M.A. Sleep deprivation in the rat: X. Integration and discussion of the findings, Sleep, 12 (1989) 68–87.

Sagar, S.M., Sharp, F.R. Curran, T. Expression of c-fos protein in brain: metabolic mapping at cellular level, Science, 240 (1988) 1328–1331.

Sakai, K., Some anatomical and physiological properties of ponto-mesencephalic tegmental neurons with special reference to the PGO waves and postural atonia during paradoxical sleep in cat. In: J.A. Hobson and M.A.B. Brazier (Eds.), *The Reticular Formation Revisited,* Raven Press, New York, 1980, pp. 427–447.

Sakai, K., Anatomical and physiological basis of paradoxical sleep. In: D.J. McGinty, A. Morrison, R. Drucker-Colin and P.L. Parmeggiani (Eds.), *Brain mechanisms of sleep,* Raven Press, New York, 1985, pp. 111–137.

Satinoff, E., Drucker-Collin, R.R. and Hernandez-Peon, R. Paleocortical excitability and sensory filtering during REM sleep deprivation, Physiol. Behav., 7 (1971) 103–106.

Sawada, M., Sester, U. and Carlson, J.C. Superoxide radical formation and associated biochemical alterations in the plasma membrane of brain, heart, and liver during the lifetime of the rat, J. Cell. Biochem., 48 (1992) 296–304.

Shiromani, P.J., Kilduff, T.S., Bloom, F.D. and McCarley, R.W. Cholinergically induced REM sleep triggers Fos-like immunoreactivity in dorsolateral pontine regions associated with REM sleep, Brain Res., 580 (1992) 351–357.

Siegel, J.M., Manger, P.R., Nienhuis, R., Fahringer, H.M. and Pettigrew, J.D. The platypus has REM sleep, *Sleep Res.,* 26 (1997) 177.

Siegel, J.M., Nienhuis, R. and Tomaszewski, K.S. REM sleep signs rostral to chronic transections at the ponto-medullary junction, Neurosci. Lett., 45 (1984) 241–246.

Singh, S. and Mallick, B.N. Mild electrical stimulation of pontine tegmentum around locus coeruleus reduces rapid eye movement sleep in rats, Neurosci. Res., 24 (1996a) 227–235.

Singh, S. and Mallick, B.N. Reduction in rapid eye movement sleep on stimulation of locus coeruleus is mediated by β adrenoceptor in rats, XIth National Symposium of the Indian Society of Chronobiology, New Delhi, (1996b) 30.

Sinha, A.K., Ciaranello, R.D., Dement, W.C. and Barchas, J.D. Tyrosine hydroxylase in the rat brain following "REM" sleep deprivation, J. Neurochem., 20 (1973) 1289–1290.

Stern, W.C. and Morgane, P.J. Theoretical view of REM sleep function: maintenance of catecholamine systems in the central nervous system, Behav. Biol., 11 (1974) 1–32.

Swann, A.C., Stimulation of brain Na^+, K^+ -ATPase by norepinephrine in vivo: prevention by receptor antagonists and enhancement by repeated stimulation, Brain Res., 260 (1983) 338–341.

Thakkar, M. and Mallick, B.N. Effect of REM sleep deprivation on rat brain monoamine oxidases, Neuroscience, 55 (1993) 677–683.

Thakkar, M. and Mallick, B.N. Effect of rapid eye movement sleep deprivation on 5' - nucleotidase activity in the rat brain, Neurosci. Lett., 206 (1996) 177–180.

Thomas, P. and Almers, W. Exocytosis and its control at the synapse, Curr. Opin. in Neurobiol., 2 (1992) 308–311.

Trachtenberg, M.C., Packey, D.J. and Sweeney, T. In vivo functioning of Na^+, K^+ activated ATPase, Curr. Topics Cell. Reg., 19 (1982) 159–217.

Viani, P., Cervato, G., Fiorilli, A. and Cestaro, B. Age-related difference in synaptosomal peroxidative damage and membrane properties, J. Neurochem., 56 (1991) 253–258.

Vogel, G.W., A review of REM sleep deprivation, Arch. Gen. Psychiat., 32 (1975) 749–761.

Wood, W.G., Gorka, C., Armbrecht, H.J., Williamson, L.S. and Strong, R. Fluidizing effects of centrophenoxine in vitro on brain and liver membranes from different age groups of mice, Life Sci., 39 (1986) 2089–2095.

Yamauy, J., Mancillas, J.R., Monales, F.R. and Chase, M.H. c-fos expression in pons and medulla of cat during carbachol induced active sleep, J. Neurol., 136 (1993) 2703–2713.

Zepelin, H. Mammalian sleep. In: M.H. Kryger, T.H. Roth, and W.C. Dement (Eds.), *Principles and Practice of Sleep Medicine,* Saunders, Philadelphia, 1989, pp. 30–49.

Rapid Eye Movement Sleep
B.N. Mallick and S. Inoué (Eds)

24. REM Sleep Deprivation and Behavioral Changes

Gerald W. Vogel

Department of Psychiatry, Emory University School of Medicine, Atlanta, GA, USA

SUMMARY

This article reviews findings about the behavioral deficiencies and REM sleep (RS) abnormalities in endogenous depression (ED) to explore the general relation between RS and behavior.

In humans, two lines of evidence indicate a relationship between RS and ED: (1) Human ED has a number of RS abnormalities which can be interpreted as RS disinhibition (e.g., increased RS duration and decreased RS delay or latency), and (2) REM sleep deprivation (RSD) improves ED. In rats, evidence indicates that RSD of neonates produces adult depression. Like human endogenous depressives, the adult "depressed" rats have decreased sexual, aggressive, and pleasure seeking (intracranial self stimulation) activity; and increased RS with decreased REM latency. These findings indicate that RSD has opposite effects which depend upon age: RSD of neonates produces (adult) depression; while RSD of adults improves depression. A theory is proposed to account for this paradox. The elements of the theory are inferences based on evidence that is reviewed. The reviewed evidence suggests that a RS inhibitory process develops during the neonatal period; that the inhibitory process delays REM rebound and by this delay contributes, in an unknown way, to the activation of waking behaviors; that this mechanism is involved in the improvement of ED by RSD; and that the neuronal substrate of the mechanism lies in the ascending recticular activating system (ARAS).

Key Words: REM sleep deprivation, endogenous depression, ARAS.

This chapter reviews studies in humans and in animals about the relationship between endogenous depression (ED) and REM sleep (RS). The aim is to use findings about the behavioral deficiencies and RS abnormalities of ED as nature's "experiments" which may shed light on the underlying relationship between RS and behavior. After a very brief review of older findings in humans, I will focus on newer animal studies.

Human Studies on REM Sleep and Endogenous Depression

Two independent lines of evidence indicate a relationship between ED and

RS. First, ED has a reliable set of RS abnormalities. These include increased duration of tonic RS, increased frequency of phasic events in RS and an early occurrence of RS. The latter is indicated by an increased number of sleep onset REM periods, decreased REM latency, and increased RS early in the sleep period (Benca et al., 1992;Vogel et al., 1988; Kupfer, 1976; Vogel et al., 1980). Together, the RS abnormalities suggest that RS is disinhibited in ED. The second line of evidence about RS and ED concerns REM sleep deprivation (RSD). RSD by arousals improved ED (Vogel, 1975). Drugs that reduce RS produce different kinds of RSD. One kind of drug induced RSD has the 3 salient characteristics of RSD by arousals, viz, a RSD which is large (> 50% reduction), persistent (over weeks with only small tolerance to the REM suppressant effects), and followed by a REM rebound. We call drug induced RSD with these 3 characteristics arousal type RSD. Along with the finding that RSD by arousals improved ED, several other observations support the hypothesis that antidepressant drugs improve ED by arousal type RSD (Vogel, 1983): (1) About 25 of 30 antidepressant drugs produce arousal type RSD. (Vogel et al., 1990a); (2) No nonantidepressant drugs produce arousal type RSD (Vogel et al., 1990a); (3) Antidepressant drugs and RSD by arousals produce similar magnitude and time course of improvement of ED (Vogel et al., 1975); (4) Patients with ED unimproved by RSD via arousals were unimproved by imipramine (Vogel et al., 1975); and (5) Improvement of ED by RSD via arousals and by antidepressant drugs correlated with the same effect of RSD, viz, REM rebound (Vogel et al., 1977; Gillin et al., 1978). Thus, although a few antidepressant drugs may work by mechanisms other than RSD, the evidence suggests that most antidepressant drugs improve ED by arousal type RSD.

Animal Studies on REM Sleep and Endogenous Depression

In many mammalian species neonates have a large amount of RS which decreases as the brain matures (Jouvet-Monnier et al., 1969; Roffwarg et al., 1966). Cats, rats, and humans are examples. In the late 1970s, Mirmiran and Corner became interested in the functional significance of the large amount of neonatal RS (Mirmiran et al., 1980, 1981, 1983). Their strategy to expose the function of neonatal RS in rats was to determine the effects of neonatal RSD on adult rats. They produced neonatal RSD pharmacologically because at that time mechanical RSD was not technically possible. In their studies they administered the antidepressant drug clomipramine, a powerful REM suppressant, to neonatal rats (Mirmiran et al., 1980, 1981, 1983). The drug was administered daily to male neonates from rat age 8–21 days. Saline was administered in the same schedule to control rats. To abbreviate I will call clomipramine treated rats CLI rats and saline treated rats SAL rats. Gross maturation was similar in CLI and SAL rats. However, when mature, CLI and SAL rats had behavioral and RS differences. Behaviorally, compared with SAL rats, CLI rats had indications of less sexual activity (fewer ejaculations) and had changes in locomotor speed. In terms of RS, compared with SAL rats, CLI rats had more RS, more sleep onset REM periods, and impressionistically more phasic muscle twitches during RS. Mirmiran and Corner interpreted these findings to indicate that a high level of neonatal RS "plays an important role in the normal maturation of the brain" (Mirmiran et al., 1983).

We suggested a different interpretation of their findings. We suggested that adult CLI rats were an animal model of human ED (Vogel and Vogel, 1982) and that Mirmiran and Corner had observed only a small part of the depressive syndrome. Our hypothesis predicted that with a larger range of tests adult CLI rats would show the full syndrome of salient behavioral, RS, and physiological abnormalities found in human ED as well as the treatment responses and the course of the human disorder. Controlled tests of these predictions were carried out. Salient abnormal symptoms of human ED are decreased sexual and aggressive activities (Hamilton and White, 1959; Mendels and Cochrane, 1968; Nelson and Charney, 1981). Compared with adult SAL rats, adult CLI rats had a wide, pervasive range of sexual deficiencies (decreased mounts, intromissions, and ejaculations; increased mount latency and postejaculatory pause) (Neill et al., 1990). In shock induced fighting CLI rats were less aggressive than SAL rats (Vogel et al., 1988). Also a cardinal symptom of ED is diminished capacity for pleasure (Mendels and Cochrane., 1968). Hamlet says "How weary, stale, flat, and unprofitable seem to me all the used of this world." He is depressed and Shakespeare tells us that the essence of depression is a diminished capacity for pleasure and a diminished interest in the outside world. We measured intracranial self-stimulation (ICSS) rates in rats to determine pleasure-seeking activity. CLI rats had lower rates of ICSS than SAL rats (Vogel et al., 1990c). In terms of RS, human endogenous depressives have increased RS amounts (Benca et al., 1992), increased phasic activity during RS (Vogel et al., 1988), decreased REM latency (Kupfer), increased number of sleep onset REM periods (Vogel et al., 1988), and an abnormal temporal course of REM rebound (Vogel et al., 1980). CLI rats have all of the RS abnormalities (Vogel et al., 1990d). In terms of physiology, clinical evidence indicates that human endogenous depressives have decreased 5HT neurotransmission (Delgado et al., 1994). We measured firing rates of 5HT neurons in the DRN, the brain's major 5HT nucleus (Kinney et al., 1997). Firing rates were significantly lower in CLI rats than SAL rats—about 1/4 to 1/2 the control rates. Human endogenous depressives often have a peculiar physiological abnormality of core body temperature. Their body temperature is increased during sleep but normal during wake (Avery et al., 1982; Lund et al., 1983). Compared with SAL rats, CLI rats have the same abnormality of body temperature (Rosenthal and Vogel, 1994). The responses of the CLI rat sexual deficiency to imipramine treatment were measured. Imipramine improved and normalized their sexual activity (Vogel et al., 1995). Finally, the course of the CLI rat disorder was similar to the course of human

ED. Like ED (Mendels and Cochrane, 1968; Nelson and Charney, 1981), the CLI rat disorder began without any external precipitant like a stressful event (Vogel et al., 1990a). And like human ED, the course of the CLI rat disorder persisted for months and was autonomous. Experiential events that influenced SAL rat behaviors, like housing conditions and handling, did not influence CLI rat symptoms. In short, the adult CLI rats had salient behavioral, RS, and physiological abnormalities of human ED, as well as treatment responses and course of human ED. These similarities suggest that adult CLI rats are a reasonably valid animal model of human ED.

How did neonatal treatment with CLI produce these adult abnormalities? Mirmiran and Corner suggested that neonatal RSD by CLI produced the adult symptoms, (Mirmiran et al., 1980, 1981, 1983). However, effects of pharmacological RSD are confounded by other effects of the drug. For example, CLI affects aminergic neurotransmission by blocking both 5HT and norepinephrine reuptake, 5HT more than norepinephrine (Richelson, 1994). Thus, as also suggested by Mirmiran and Corner, it is possible that neonatal CLI produced the adult depression by its blockade of 5HT reuptake. Nonetheless, there is other evidence related to this issue of whether 5HT effects of CLI or RSD effects of CLI produce the adult depression.

Other drugs, principally antidepressant drugs, administered to neonates, produce symptoms of adult depression (Vogel et al., 1990a). The drugs include desipramine, zimeldine, and clonidine. Desipramine blocks norepinephrine reuptake (Richelson, 1994); Zimeldine blocks 5HT reuptake (Richelson, 1994); And clonidine is an alpha agonist which stimulates presynaptic inhibitory noradrenergic autoreceptors (Uhde et al., 1984). These diverse aminergic effects of the neonatally administered depressogenic drugs suggest that a specific aminergic effect of antidepressant drugs does not mediate the adult depression. In contrast, all four drugs, whose neonatal administration produces indicators of adult depression, produce large RSD (Vogel et al., 1990a). This common effects makes neonatal RSD a more parsimonious explanation of the adult depression than a presynaptic effect on amines. Consistent with this explanation, neonatally administered iprindole, an antidepressant drug with no RSD effects, does not produce adult depression (Vogel and Hagler, 1996). Of course, an unambiguous test of the hypothesis that neonatal RSD produces adult depression requires instrumental RSD of neonates. This will eliminate the unavoidable confounds of drug RSD. We recently developed a method for continuous (24 h/day), long-term (over weeks), instrumental RSD of neonatal rats (Feng et al., 1997a). Tests of the neonatal RSD hypothesis with this method will be done in the near future. A finding that neonatal RSD produces adult depression may have clinical significance. Maternal separation of infant monkeys produces infant RSD (Reite and Short, 1978). Thus, in humans, a disturbance of the maternal infant bond may, via infantile RSD, produce the depressive disposition.

In any case, for the sake of discussion, let us conclude that neonatal RSD produces adult depression. This conclusion presents us with a troubling paradox. Neonatal RSD produces adult depression but adult RSD improves depression. How can we reconcile these opposite effects of RSD? For clues, let us review the evidence about empirical correlates of the two opposite effects of RSD.

Paradoxical Effects of REM Sleep Deprivation

We begin with the improvement of ED by RSD. In that study patients with ED were treated for weeks with RSD by arousals (Vogel et al., 1975). But 4 days after the onset of the treatment—before improvement occurred—the patients were allowed a single night of uninterrupted sleep without RSD (Vogel et al., 1980). Nondepressed control subjects were treated in the same schedule. On the single night of uninterrupted sleep, total nocturnal REM rebound was the same in endogenous depressives and in nondepressed controls. However, the temporal distribution of REM rebound in the two groups was very different. REM rebound in the first REM period was significantly greater in ED than in controls; and REM rebound after the first REM period—so-called late REM rebound—was significantly less in ED than in controls (Vogel et al., 1980). Furthermore, late REM rebound in the endogenous depressives predicted improvement 3 weeks later. The correlation between late REM rebound and later improvement was about +0.70 (Vogel et al., 1980). Thus, ED patients with a large REM rebound in the first REM period and small REM rebound later did not improve while patients with a small REM rebound in the first REM period and a large REM rebound later in the night did improve.

These findings suggest that improvement of ED by RSD was related to the sustained storage of the excitation produced by RSD. My terminology here may not be right but I mean to convey the speculation that depression improvement depended on processes that delayed the discharge and sustained the excitation produced by RSD. The gist of this notion is the relation between stored RS excitation and waking behavior. The finding that behavioral improvement correlated with late REM rebound suggests that only the stored or undischarged RS excitation increased excitation of waking motivational behaviors.

Let us now consider findings about neonatal RSD. Here we shall see related phenomena about the delay of REM rebound. As previously mentioned, we have developed a method for long-term, instrument RSD of neonatal rats (Feng et al., 1997a). With this method, two groups of neonatal rats were REM sleep deprived for 48 h. In one group the RSD began at age 13 days; in the second group the RSD began at age 15–16 days. In the younger group, there was no REM rebound after 48 h of RSD. In the second group a small REM rebound occurred but all of it occurred in the first 12 h after RSD. In mature rats REM rebound was larger and occurred during both the first and second 12 h after RSD (Feng and Vogel, 1998). Thus, in neonates the capacity for

REM rebound after RSD showed developmental changes, one of which was the progressive increase in delayed REM rebound. Moreover, the decreased capacity for delayed REM rebound in neonates resembles the decreased capacity for delayed REM rebound in ED. These findings suggest the possibility that RS abnormalities in ED represent an incomplete maturation of the neonatal RS system (Vogel et al., 1997). Other findings support this hypothesis. We measured several RS variables with our new method of continuous (24 h/day), long-term (weeks) polysomnographic recordings of neonates (Feng et al., 1997b). Compared with normal adult rats, two week old neonatal rats have more RS, more sleep onset REM periods, shorter REM latency, and more phasic muscle twitches during RS. Compared with normal adults (Vogel and Feng, 1998), endogenous depressives have the same RS abnormalities, though RS differences from normal adults were quantitatively smaller in ED than in neonates. Thus, a whole set of findings are consistent with the hypothesis that the RS abnormalities in ED represent an incomplete maturation or an arrested development from the neonatal state. To the best of my knowledge, no other hypothesis parsimoniously explains the set of RS abnormalities found in ED. It is noteworthy that the incomplete maturation hypothesis is empirically testable in CLI rats. The hypothesis predicts that the brainstem neuronal substrates of RS generation, such as discharge patterns and responses in the pedunculo pontine nucleus and laterodorsal tegmental nucleus (see below review), will be similar in neonates and in adult CLI rats. We plan such tests.

These findings raise the question of the cause of the incomplete maturation of the RS system. Again studies of RS in neonates provide clues. Before examining them, it is important to establish that the changes involved in RS maturation are not trivial. They are quantitatively very large. You can get some idea of magnitude from the following data. RS is about 40% of total sleep time (TST) in two week old rats (Vogel et al., 1997a) but about 10% of TST in adult rats (Vogel et al., 1990d). Or RS occupies about 1/4 of each 24 h day in 2 week old rats but about 1/16 of each 24 h day in adult rats. Thus, neonatal rats have about 4 times as much RS as adult rats and RS maturation must produce a four-fold decreases in RS. With this background, let us consider studies that may shed light on processes involved in the maturation of the RS system. In preliminary studies with our method of continuous PSG recordings of neonatal rats, we measured changes in five RS variables from age 12–28 days (Vogel and Feng, 1998). The variables were 24 h RS duration, frequency of phasic muscle twitching, number of sleep onset REM periods, REM latency, and mean REM period duration. Over the period from age 12–28 days all measures of RS magnitude decreased together and all measures of RS delay increased together. Over the same age period late REM rebound (delayed REM rebound) also increased in parallel with the other indicators of RS delay (Feng and Vogel, 1998). The finding that all these changes occurred in parallel suggests a single process as the cause of the intercorrelated changes. The direction of the parallel changes—decreased magnitude and increased delay—

suggests an inhibitory process. Thus, these data suggest that development of a single RS inhibitory process is responsible for the observed maturational changes in RS.

For purposes of clarification, let me restate the properties of the proposed inhibitory process and the evidence supporting these properties: (1) The inhibitory process decreases RS—as evidenced in neonates by the parallel decreases of RS magnitude and increases of RS delay (Vogel and Feng, 1998), including REM rebound delay (Feng and Vogel, 1998); (2) The inhibitory process "stores" undischarged RS excitation—as evidenced in neonates by the correlation between RS decreases and late REM rebound (Feng and Vogel, 1998; Vogel and Feng, 1998); (3) Undischarged RS excitation can be available for immediate discharge—as evidenced by immediate or early REM rebound in ED (Vogel et al., 1980); or its discharge can be delayed and stored—as evidenced by late REM rebound in ED (Vogel et al., 1980); and (4) Undischarged RS excitation "stored" by the inhibitory process increases sustained activated behaviors—as evidenced by the positive correlation in ED between late REM rebound and improvement of depression (Vogel et al., 1980).

A Conjecture About the Opposite Effects of REM Sleep Deprivation

ARS inhibitory process with these properties can explain the paradox that neonatal RSD leads to depression and adult RSD improves depression. Suppose that neonatal RS promotes the development of the RS inhibitory process. Then, neonatal RSD blocks the inhibitory development and thereby diminishes the adult strength of the inhibitory process. This prediction is consistent with the finding that neonatal RSD causes adult RS disinhibition (Vogel et al., 1990d) and causes diminutions in sustained excitation of adult waking behaviors (Vogel et al., 1990b). On the other hand, the relatively constant level of RS after the neonatal period indicates that in adults the development of the RS inhibitory process is finished. RSD can no longer affect the magnitude of the inhibitory process. That magnitude is fixed by the end of the neonatal period. In adults RSD leads only to increased RS excitation, without affecting the magnitude of the RS inhibitory process. To the extent that the inhibitory process can delay discharge and store the increased excitation, the accumulated excitation increases sustained motivated behavior. Accordingly, the opposite effects of RSD at different maturational ages depend on the maturational state of the RS inhibitory process.

Of course much of this theory is conjecture. There is no direct evidence of the inhibitory process. The nature of the proposed inhibitory process is unknown. The nature of RS excitation is unknown. The nature of stored excitation is unknown. The process by which delayed discharge of RS excitation leads to (or stores) excitation of the neuronal substrate of motivated behavior is unknown. Is there any value to the theory? Maybe. Maybe because recent evidence suggests a common neuronal substrate of RS and of waking behavioral activation. Such a substrate offers an opportunity for empirical tests of predictions made by the theory. Consider the following:

ED is characterized by two very different kinds of symptoms: RS abnormalities and behavioral deficiencies. We would like a parsimonious explanation for these disparate sets of symptoms. The ascending reticular activating system (ARAS) is responsible for the underlying brain activation during RS and during waking behavior (Steriade, 1996). ARAS "originates in the upper brainstem reticular core and projects through synaptic relays in the thalamus to the cerebral cortex where it increases excitability" (Steriade, 1996). Transection (Jouvet, 1962; Siegel, 1989), lesion (Webster and Jones, 1988; Shouse and Siegel, 1992; Sakai et al., 1976), stimulation (Thakkar et al., 1996), unit recordings (Steriade et al., 1990a; Saito et al., 1997; Steriade et al., 1990b) and pharmacologic (Baghdoyan et al., 1987, 1984; Gnadt and Pegram, 1986) evidence indicates that in the operation of ARAS major sources of tonic and phasic activation in wake and RS are located in two cholinergic cell groups at the mesopontine border: the pedunculopontine tegmental nucleus (PPT) and laterodorsal tegmental nucleus (LDT). Thus, several lines of evidence indicate that during RS and wake, the PPT/LDT neurons *tonically* activate thalamocortical neurons and their cortical targets: (1) For instance, during wake and RS most (~ 75%) PPT/LDT neurons spontaneously fire tonically with virtually no high frequency bursts, and during wake and RS their discharge rate is much higher than during nonREM sleep (NRS). During the transitions from NRS to wake and from NRS to RS, spontaneous discharge rates of tonically firing neurons began to increase about one minute in advance of the transitions (Steriade et al., 1990b); (2) The enhanced level of PPT/LDT spontaneous discharge during wake and RS was accompanied by increased PPT/LDT excitability measured by responses to orthodromic and antidromic stimulation (Steriade et al., 1990b); (3) Stimulation of PPT/LDT produced, at the neuronal level, tonic depolarizing (excitatory) effects in thalamocortical relay neurons similar to the depolarization that occurred during wake and RS (Steriade and McCarley, 1990; Curro' Dossi et al., 1991); and (4) Stimulation of PPT/LDT produced at the network level—(a) prolonged enhancement in the responses of thalamocortical relay neurons to orthodromic (synaptic) stimulation (Steriade et al., 1990c); (b) potentiation of 40 Hz oscillation in the cortical EEG, which may underlie the coherent activation of spatially separate but functionally similar neurons (Steriade, 1996; Singer, 1993); and (c) cortical EEG desynchronization with a time course similar to the time course depolarization produced by PPT/LDT stimulation (Curro Dossi et al., 1991) in thalamic neurons. These findings about ARAS indicate that, as Steriade (1996) put it, "electrophysiologically REM sleep is virtually identical to waking." With specific regard to waking behavior, ARAS produces the cellular excitability that is a "prerequisite" for active behavioral states" (Steriade, 1981). Indeed, current evidence supports the view that the ARAS creates "the conditions for the organism to be *actively interested* in the outside world...

the readiness to receive only some kinds of stimuli to the exclusion of others and to perform the motor patterns of one definite instinct" (Dell, 1958). Thus the behavioral deficiencies in ED—among which *loss of interest* in the environment is salient—may be a result of diminished ARAS activity during wake, while the increased RS in ED may be result of augmented ARAS activity during RS. Consequently, to parsimoniously account for the disparate behavioral and RS symptoms of ED, we have proposed the hypothesis that alterations of ARAS are a substrate of both the RS and the behavioral abnormalities found in ED (Vogel et al., 1997b). The findings about brainstem sources of ARAS activity suggest that alterations in the excitability of PPT/LDT during RS and during wake may be a neuronal substrate of the RS and wake alterations in ED. The hypothesis can be tested in adult CLI rats. The hypothesis predicts that RS decreases the waking excitability of PPT/LDT, that RSD increases waking excitability of PPT/LDT, and that inhibitory input to the PPT/LDT during RS increases waking excitability. More specifically, the ARAS hypothesis predicts that in a subset of PPT/LDT neurons, excitability at the end of RS is directly related to sustained excitability during wake. Also, in this neuronal subset, the strength of the positive relationship (correlation) between RS excitability and wake excitability is directly related to the strength of an inhibitory process operating on the neuronal excitability of the PPT/LDT subset only during RS. Confirmation of these predictions will suggest that one function of adult RS is regulation of waking excitability of ARAS.

Acknowledgement

Work supported by NIMH grant MH40880. An earlier version of the paper was presented to the "International Conference in Clinical Neuroscience", Milan, Italy, Feb. 23–24, 1998.

References

Avery, D.H., Wildschiotz, G. and Refaelson, O.J. Nocturnal temperature in affective disorder, Journal of Affective Disorders, 4 (1982) 61–71.

Baghdoyan, H.A., Rodrigo-Argulo, M.L., McCarley, R.W. and Hobson, J.A. A neuroanatomical gradient in the pontine tegmentum for the cholinoceptive induction of desynchronized sleep signs, Brain Research, 414 (1987) 245–261.

Baghdoyan, H.A., Rodrigo-Argulo, M.L., McCarley, R.W. and Hobson, J.A. Site specific enhancement and suppression of desynchronized sleep following cholinergic stimulation of the brainstem region, Brain Research, 306 (1984) 39–51.

Benca, R.M., Obermeyer, W.H., Thisted, R.A. and Gillin, J.D. Sleep in psychiatric disorders: a meta-analysis, Archives of General Psychiatry, 49 (1992) 651–658.

Curro' Dossi, R., Pare, D. and Steriade, M. Short-lasting nicotine and long-lasting muscarinic depolarizing responses of thalamocortical neurons to stimulation of mesopontine cholinergic nuclei, Journal of Neurophysiology, 65 (1991) 393–406.

Delgado, P.L., Price, L.H., Miller, H.L., Salomon, R.M., Aghajanian, G.K., Neninger, G.R. and Charney, D.S. Serotonin and the neurobiology of depression. Effects of tryptophan depletion in drug free depressed patients, Archives of General Psychiatry, 51 (1994) 865–874.

Dell, P.D., Some basic mechanisms of the translation of bodily needs into behavior. In: G.E.W. Wolstenholme and M.O' Conner (Eds.) *Neurobiological Basis of Behavior,* Churchill, London, 1958, pp.187–203.

Feng, P., Obermeyer, W., Vogel, G.W. and Kinney, G.G. Nine days of computer controlled instrumental PS deprivation in maternal separated infant rats, Sleep Research, 26 (1997a) 655.

Feng, P. and Vogel, G. Ontogeny of REM rebound in neonatal rats, Sleep Research, (1998) in press.

Feng, P., Vogel, G.W., Kinney, G.G. and Obermeyer, W. Two weeks of continuous polysomnographic sleep recordings of maternally separated infant rats, Sleep Research, 26 (1997b) 656.

Gnadt, J.W. and Pegram, G.V. Cholinergic brainstem mechanisms of REM sleep in the rat, Brain Research, 384 (1986) 29–41.

Hamilton, M. and White, J.M. Clinical syndromes in depressive states, Journal of Mental Science, 105 (1959) 985–988.

Jouvet, M., Researches ser les structures nerveuses et lest mechanismes responsables des differentas phases du sommeil physiologique, Archives of Italian Biology, 100 (1962) 125–206.

Jouvet-Monnier, P., Astic, L. and Lacote, D. Ontogenesis of the states of sleep in rat, cat, and guinea pig during the first postnatal month, Developmental Psychobiology, 2 (1969) 216–239.

Kinney, G.G., Vogel, G.W. and Feng, P. Decreased dorsal raphe nucleus neuronal activity in adult chloral hydrate anesthetized rats following neonatal clomipramine treatment: Implications for endogenous depression, Brain Research, 756 (1997) 68–75.

Kupfer, D.J, REM latency: A psychobiologic marker for primary depressive disease, Biological Psychiatry, 11 (1976) 159–174.

Lund, R., Kammerloher, A. and Dirlich, G. Body temperature in endogenous depressed patients during depression and remission. In: F.K. Goodwin (Ed.), *Circadian Rhythms in Psychiatry,* Pacific Grove, Los Angeles, 1983, pp. 77–88.

Mendels, J. and Cochrane, C. The nosology of depression: the endogenous-reactive concept, American Journal of Psychiatry, 124 supplement (1968) 1–11.

Mirmiran, M., Scholtens, J., van de Poll, N.D., Uhlings, H.G.M., van der Gugten, J. and Boer, G.J. Effects of experimental suppression of active (REM) sleep during early development upon adult brain and behavior in the rat, Developmental Brain Research, 7 (1983) 277–286.

Mirmiran, M., van de Poll, N., Corner, M., Boer, G. and van Oyer, H. Lasting sequelae of chronic treatment with chlorimipramine during early postnatal development in the rat, *IRCS* Journal of Medical Science, 8 (1980) 200–202.

Mirmiran, M., van de Poll, N.E., Corner, M.A., van Oyer, H.G. and Boer, H.L. Suppression of active sleep by chronic treatment with chlorimipramine during early postnatal development: effects upon sleep and behavior in the rat, Brain Research, 204 (1981) 129-146.

Neill, D., Vogel, G., Hagler, M., Kors, D. and Hennessey, A. Diminished sexual activity in a new animal model of endogenous depression, Neuroscience and Biobehavioral Review, 14 (1990) 73–76.

Nelson, J.C. and Charney, D.S. The symptoms of major depressive illness, American Journal of Psychiatry, 138 (1981) 73–76.

Reite, M. and Short, R.A. Nocturnal sleep in separated monkey infants, Archives of General Psychiatry, 35 (1978) 1247–1253.

Richelson, E., Pharmacology of antidepressants—characteristics of the ideal drug, Mayo Clinic Proceedings, 69 (1994) 1069–1081.

Roffwarg, H.F., Muzio, J. and Dement, W.C. Ontogenetic development of the human sleep-wakefulness cycle, Science, 152 (1966) 604–609.

Rosenthal, M.S. and Vogel, G. W. The effects of a 3 day increase of ambient temperature on body temperature and REM sleep in an animal model of depression, Sleep, 17 (1994) 291–297.

Saito, H., Sakai, K. and Jouvet, M. Discharge patterns of the nucleus parabranchialis lateralis neurons of the cat during sleep and waking, Brain Research, 134 (1997) 59–72.

Sakai, K., Petijean, F. and Jouvet, M. Effects of pontomesencephalic lesions and stimulation upon PGO waves and EMPs in unanesthetized cats, Electroencephalography and Clinical Neurophysiology, 41 (1976) 49–63.

Shouse, M.N. and Siegel, J.M. Pontine regulation of REM sleep components in cats: integrity of pedunculopontine tegmentum (PPT) is important for phasic events but unnecessary for atonia during REM sleep, Brain Research, 571 (1992) 50–63.

Siegel, J.M., Brain stem mechanisms generating REM sleep. In: M.H. Kryger, T. Roth, and W.C. Dement (EDs.) *Principles and practice of sleep medicine,*_Saunders, Philadelphia, 1989, pp. 86–120.

Singer, N. Synchronization of cortical activity and its putative role in information processing and learning Annual Review of Physiology, 55 (1993) 349–374.

Steriade, M. Arousal: Revisiting the reticular activating system, Science, 272 (1996) 225–226.

Steriade, M. EEG desynchronization is associated with cellular events that are prerequisites for active behavioral states, Behavioral Brain Sciences, 4 (1981) 489–492.

Steriade, M., Datta, S., Pare'D., Oakson, S. and Curro Dossi, R. Neuronal activities in brainstem cholinergic nuclei related to tonic activation processes in thalamocortical systems, Journal of Neuroscience, 10 (1990b) 2541–2559.

Steriade, M., Jones, E.G. and Llinas, R.R. Thalamic Oscillations and Signaling, John Wiley and Sons, New York, 1990c, p.5.

Steriade, M. and McCarley, R. Brainstem control of wakefulness and sleep. Plenum, New York, 1990.

Steriade, M., Pare'D, Datta, S., Oakson, G. and Curro' Dossi, R. Different cellular types in mesopontine cholinergic nuclei related to ponto-genicula-occipital waves, Journal of Neuroscience, 10 (1990a) 2560–2579.

Thakkar, M., Portas, C. and McCarley, R.W. Chronic low amplitude electrical stimulation of the laterodorsal tegmentum nucleus of freely moving cats increases REM sleep, Brain Research, 723 (1996) 223–227.

Uhde, T.W., Siever, L.J. and Post, R.M. Clonidine: Acute challenge and clinical trial paradigms for the investigation and treatment of anxiety disorders, affective illness, and pain syndromes. In: R.M. Post and J.C. Ballenger (Eds.), *Neurobiology of Mood Disorders*, Williams and Wilkins, Baltimore, 1984, pp. 554–571.

Vogel, G.W., Evidence for REM sleep deprivation as the mechanism of action of antidepressant drugs, Progress in Neuropsychopharmacology of Biological Psychiatry, 7 (1983) 343–349.

Vogel, G.W., Buffenstein, A., Minter, K. and Hennessey, A. Drug effects on REM sleep and on endogenous depression, Neuroscience and Biobehavioral Review, 14 (1990a) 49–63.

Vogel, G.W. and Feng, P. Ontogeny of REM sleep in neonatal rats, Sleep Research, (1998) in press.

Vogel, G.W., Feng, P. and Kinney, G.G. Similar REM sleep characteristics in neonates and in endogenous depressives, Sleep Research, 26 (1997a) 170.

Vogel, G. and Hagler, M. Effects of neonatally administered iprindole on adult behaviors of rats, Pharmacology, Biochemistry and Behavior, 55 (1996)157–161.

Vogel, G., Hagler, M., Hennessey, A. and Neill, D. Imipramine improves sexual deficiencies in the chlorimipramine rat model of depression, Sleep Research, 24 (1995) 180.

Vogel, G., Hartley, P., Neill, D., Hagler, M. and Kors, D. Animal depression model by neonatal chlorimipramine: Reduction of shock induced aggression, Pharmacology and Biochemistry Behavior, 31 (1988) 103–106.

Vogel, G.W., Kinney, G.G. and Feng, P. Endogenous depression and the ascending reticular activating system (ARAS), Sleep Research, 26 (1997b) 603.

Vogel, G.W., McAbee, R., Barber, K. and Thurmond, A. Endogenous depression improvement and REM pressure, Archives of General Psychiatry, 34 (1977) 96–97.

Vogel, G., Neill, D., Hagler, M. and Kors, D. A new animal model of endogenous depression: a summary of present findings, Neuroscience and Biobehavioral Review, 14 (1990b) 85–92.

Vogel, G., Neill, D., Hagler, M., Kors, D. and Hartley, P. Decreased intracranial self-stimulation in a new animal model of endogenous depression, Neuroscience and Biobehavioral Review, 14 (1990c) 77–83.

Vogel, G., Neill, D., Kor, D. and Hagler, M. REM sleep abnormalities in a new animal model of endogenous depression, Neuroscience and Biobehavioral Review, 14 (1990d) 77–83.

Vogel, G.W., Roth, T., Gillin, J.C., Mendelson, W.C. and Buffenstein, A. REM sleep and depression. In: T. Oniani (Ed.), *Neurobiology of sleep-wakefulness cycle,* Metsniereba, Tbilisi, USSR, 1988, pp. 187–214.

Vogel, G.W., Thurmond, A., Gibbons, P., Sloan, K., Boyd, M. and Walker, M. REM sleep reduction effects on depression syndromes, Archives of General Psychiatry, 32 (1975) 765–777.

Vogel, G.W. and Vogel, F.A. A new animal model of human endogenous depression, Sleep Research, 11 (1982) 222a.

Vogel, G.W., Vogel, F., McAbee, R.S. and Thurmond, A.J. Improvement of depression by REM sleep deprivation, Archives of General Psychiatry, 37 (1980) 37.

Webster, H.H. and Jones, B.E. Neurotoxic lesions of the dorsolateral pontomesocephalic tegmentum-cholinergic cell area in the cat. II Effects upon sleep-wake stages, Brain Research, 458 (1988) 288–302.

Rapid Eye Movement Sleep
B.N. Mallick and S. Inoué (Eds)

25. Cellular and Molecular Changes Occurring During REM Sleep

O. Prospéro-García[1], L. Navarro[1], E. Murillo-Rodríguez[1], M. Sánchez-Alavez[1], R. Guzmán-Marín[1], M. Méndez-Díaz[1], M. Gómez-Chavarín[1], A. Jiménez-Anguiano[3] and R. Drucker-Colín[1,2]

[1]Depto. de Fisiología, Fac. de Medicina.

[2]Depto. de Neurociencias, Instituto de Fisiología Celular, Universidad Nacional Autónoma de México.

[3]Depto. de Biología de la Reproducción, Universidad Autónoma Metropolitana, Iztapalapa. México, D.F., México.

Since Aserinsky and Kleitman (1953) described rapid eye movement (REM) sleep in 1953, it is known that several events occur in the body during this stage, such as electroencephalographic desynchronization, postural atonia and eye movements. These variables have been the landmark for determining the occurrence of this stage; therefore, the description of their brain generators has been the goal of several studies. In general, basic research has been very successful, since at present time, we know in great detail the mechanisms that regulate them.

Electrophysiology, after behavior, has been the cornerstone to obtain most of our knowledge regarding sleep. However, REM sleep is by far more complex than that. In fact, it has been observed that several other variables exhibit changes when the subject is in this sleep stage. For example, secretion of hormones, hemodynamic as well as respiratory variations and changes in the electrical activity of cells belonging to specific nuclei, i.e. serotonin (5-HT)-, noradrenaline (NA)-, and acetylcholine (ACh)-containing neurons (McGinty and Szymusiak, 1988). Moreover, the release of several neurotransmitters associated to the sleep-waking cycle has been determined in several brain areas, implicating their action in the control of this cycle. However, despite all this progress, we still do not know how sleep changes take place and what the brain needs these changes for.

Seeking to further understand sleep mechanisms, we have investigated the cellular changes associated with any of the stages of the sleep-wake cycle (Borbely and Tobler, 1989; Prospero-Garcia, 1995). We now hope to discover the intimacy of the cell, disclosing the cellular changes causing sleep and, complementary, the changes that sleep induces in the cell. Will the understanding of all these cellular variations give us a clue to comprehend how the brain controls, produces and utilizes REM sleep? Will this information give us a chance to admire the execution of such a unique event as is REM sleep, and eventually to understand why we need it? Why does the brain need REM sleep to maintain the mind the way it is and to somehow set up everyday strategies to help us function efficiently all following wakefulness periods? The entire answer may not be within the cells. However, it will complement the picture in great detail. Above all, we are seeking to describe the cascade of events happening during REM sleep and eventually understand why these occur. This cascade of events may be initiated by the occurrence of a brain stimulus which triggers the short-lived and somehow fragile event of releasing neurotransmitters and activating second messengers, to ultimately cause the consolidation of an ephemeral event into a permanent change, such as could be a memory trace. For now, we have detected changes in the expression of proteins such as c-Fos. We have observed that these proteins are expressed in certain groups of cells that are more active during REM sleep. These findings have helped to better define the cerebral regions involved in REM sleep generation.

Due to the high variety of changes that occur at cellular level, it is pertinent to discuss the evidence describing how these events may be happening during REM sleep and eventually to understand why they occur in this stage.

Electrical Unit Activity During REM Sleep

In the 60's Jouvet (1969) and his group involved several brainstem nuclei in REM sleep modulation. Although, at present time, we know that none of them seem to really play an executive role in the generation or maintenance of this sleep stage, the importance of Jouvet's contribution resides in his description of the brainstem as the generator of REM sleep. Since then, research about anatomical substrates of REM sleep has been focused on the brainstem (Drucker-Colin and Prospero-Garcia, 1990).

The monoaminergic nuclei proposed by Jouvet as generators of REM sleep, Locus Coeruleus (LC), and Raphe Dorsalis (RD), change their activity during REM sleep (McGinty and Szymusiak, 1988). For example, most of the cells conforming these nuclei reduce 50% their activity when the subject is in slow-waves sleep and are completely silent during REM sleep. These facts suggest that monoamines may play a permissive role in REM sleep generation.

Cells of laterodorsal and pedunculopontine tegmental (LDT and PPT) nuclei behave quite differently. These ACh-containing cells increase their activity during REM sleep, and maintain it low during SWS as well as during wakefulness (Kayana et al., 1992).

These simultaneous changes occurring in monoaminergic (LC and RD) and cholinergic (LDT and PPT) nuclei, have been utilized to explain how the brainstem may be the one controlling the occurrence of SWS and REM sleep (Hobson et al., 1975). Several studies *in vitro,* have supported this hypothesis (Luebke et al., 1992; Williams and Reiaer, 1993).

Since changes in the electrical activity of cells associated to REM sleep are going to be extensively discussed elsewhere in this book, we will not describe them any further. Instead, we will address our attention to changes occurring within the cell.

Genes and Behavior

External stimuli may regulate gene expression utilizing basically two ways, i.e. releasing membrane-permeant molecules such as steroids and thyroid hormones, which bind to ligand-specific receptors in the nucleus, and therefore bring about alterations in target gene expression, or the second possibility is that the stimuli release a molecule that, although it does not gain access to the cell's interior, binds to a metabotropic receptor and therefore brings about changes in the second messenger level. As a result, the second messenger may either, alter directly the activity of constitutive transcription factors or it may rapidly activate immediate-early genes encoding transcription factors. These inducible transcription factors, in turn, regulate target genes which express their activity via protein synthesis.

Due to the fact that several behaviors are modulated by neuropeptides (Crawley and McLean, 1996), we can track down the potential genes encoding for these molecules. Therefore, we can, at least in theory, link behavior and genes. For example, the processes that has been best studied is memory. A model partially corroborated in fruit flies, mollusks and rodents has described a response of nuclear proteins in the processes of memory consolidation (Bach et al., 1995; Bailey et al., 1996; Bernabeu et al., 1997). In short, this model suggests that several neurotransmitters, including serotoninergic and glutamatergic agonists, facilitate Ca^{++} /Na^{+} conductance. This event triggers a cascade of other events which includes protein kinase A (PKA) phosphorylation. PKA, once phosphorylated, translocates into the nucleus where it phosphorylates one or more transcription factors that bind to the cyclic AMP regulatory element (CRE) located in the upstream region of cAMP inducible genes (Tamura et al., 1996). These transcriptional activators belong to the protein family of cyclic AMP response element binding (CREB) proteins. The phosphorylation of CREB results in the activation of effector genes which encode for different proteins.

This is an example of what could be happening in neurons that control a specific process by generating protein synthesis, such as memory consolidation. Of course, several proteins participating in the regulation of adapting strategies are encoded in the DNA. These adapting strategies modulated by proteins seek to satisfy basic needs, such as feeding, drinking, pain regulation, stress control, sex behavior, among several others (Crawley and McLean, 1996). Likewise, we can hypothesize that REM sleep occurrence is encoded in the DNA, since it is modulated by different proteins. For example, it has been well established that vasoactive intestinal polypeptide (VIP), corticotropin-like intermediate pituitary lobe peptide (CLIP), prolactine hormone (PrLH) and some unidentified macroproteins regulate REM sleep (Borbely and Tabler, 1989; Henriksen and Prospero-Garcia, 1993; Prospero-Garcia and Drucker-Colin, 1995).

The initial studies about genes and sleep using radioactive precursors reported changes in DNA and RNA synthesis (Giuditta et al., 1984). Other techniques, like staining with gallocyanin and detecting with spectrophotometry, seeking to evaluate amounts of these nucleic acids in different conditions have also been successfully used. For example, utilizing the latter approach it was observed that after 24 h of REM sleep deprivation (REMSD), rats show a decrease in the neuronal cytoplasmic RNA in several nuclei, i.e. RD, raphe pontis nucleus, LC and, the supraoptic nucleus (Giuditta et al., 1980). Normal levels were regained by the second day of REMSD. In contrast, total sleep deprivation (TSD) increases the content of RNA in cells of RD and LC. An increase in the synthesis of RNA associated to spontaneous REM sleep was also suggested with these initial studies (Demin, 1974). DNA changes have not been reported (Giuditta et al., 1984).

Actually, variations in DNA synthesis is not necessarily a finding we should expect as a consequence of REM sleep occurrence. In contrast, an increase of the messenger RNA (mRNA) levels of specific proteins would not be a surprise. In fact, genes encoding for proteins that are somehow involved in REM sleep modulation, must be active whenever the induction of REM sleep is needed. For example during selective REMSD. As we have aforementioned, there are many proteins that change their concentration or activity during this stage or as a consequence of not having it. Among them, neurotransmitter receptors, VIP and, macroproteins (Drucker-Colin et al., 1975; Spanis et al., 1977). It is also interesting that the activity of the Krebs cycle succinate-oxydating system is higher during sleep than during wakefulness (Hyden and Lange, 1965). No evidence has been reported regarding changes in the genes encoding these proteins, i.e. increased levels of RNAs.

However, at present time, scientists have been interested in the description of genes that are particularly sensitive to environmental stimuli. These are the immediate early genes (IEGs) (Morgan and Curran, 1991). Due to their high reactivity to stimulation, they have been used as markers to determine the cluster(s) of cells active while a given process is occurring. This is an interesting approach that does not exclude the searching for changes in the expression of mRNA of specific proteins, known to affect REM sleep.

Immediate Early Genes (IEGs)

Although Jouvet and his group (Jouvet, 1969) suggested that LC was a nucleus that regulated REM sleep, it is now known that LC regulates wakefulness (McGinty and Szymusiak, 1988). In fact, not only wakefulness as a general state, but also events requiring focused attention. One approach to determine the involvement of LC in the regulation of the sleep-waking cycle has been recording its cells across the cycle, lesioning it, and seeing how this lesion modifies neuronal gene expression in cerebral regions receiving afferences from this nucleus. Regarding the latter point, it has been indicated that several transcription factors that may be mediating long-term changes, i.e. IEGs encoding c-Fos and nerve growth factor-induced A (NGFI-A), are expressed in the cerebral cortex and hippocampus at higher level during forced wakefulness than during sleep. However, this increased expression is prevented by lesioning LC. This lesion also decreases CREB phosphorylation in the same cluster of cells (Cirelli et al., 1994; Cirelli et al., 1996).

All these data fit well to further complete the picture with facts previously described. For example, single cell activity of LC neurons is high during wakefulness (McGinty and Szymusiak, 1988). As a result of this activity a higher noradrenaline release in the cerebral cortex, among other cerebral regions, is expected. Since β adrenergic receptors are mainly distributed in the cerebral cortex, we can speculate that noradrenaline will bind this type of receptors. As a result, both protein Gs and adenylyl cyclase are activated, generating cAMP, which in turn, activates cAMP-dependent protein kinases. These proteins translocate into the nucleus and phosphorylate CREB, which activates the expression of c-Fos and NIFI-A. Clearly, this cascade of events may ultimately generate proteins that may constitute membrane receptors, neurotransmitter synthesizing enzymes or neuropeptides acting as neuromodulators.

An increase in c-Fos expression has been similarly detected in the hypothalamic ventrolateral preoptic nucleus, in this case, associated to spontaneously occurring sleep or after sleep deprivation in rats (Sherin et al., 1996). Similar results were found in cats by Jouvet and his group (Ledoux et al., 1996). Quite interestingly, the administration of a c-Fos antisense oligonucleotide into the proximity of this region, the medial preoptic area, increases wakefulness (Cirelli et al., 1995a,b). Moreover, c-Fos knockout mice have reduced Non-REM sleep and increased wakefulness but exhibit no changes in REM sleep (Shiromani et al., 1997).

As for c-Fos expression associated to REM sleep, it has been shown that the REM sleep increase caused by sleep deprivation or acoustic stimulation, facilitates the expression of c-Fos in several brainstem nuclei, i.e. PPT, parabrachial nucleus, subcoeruleus nucleus and, the solitary tract nucleus and hypothalamus (Fig. 1) (Merchant-Nancy et al., 1992; 1995). In addition, the microinjection of carbachol, a cholinergic drug, into the pontine reticular formation (PRF), triggers the appearance of REM sleep behavioral and electrophysiological signs and ignites the expression of c-Fos in cells belonging to PPT, LDT, LC and the medial PRF (Shiromani et al., 1995; 1996; Yamuy et al., 1993; 1995).

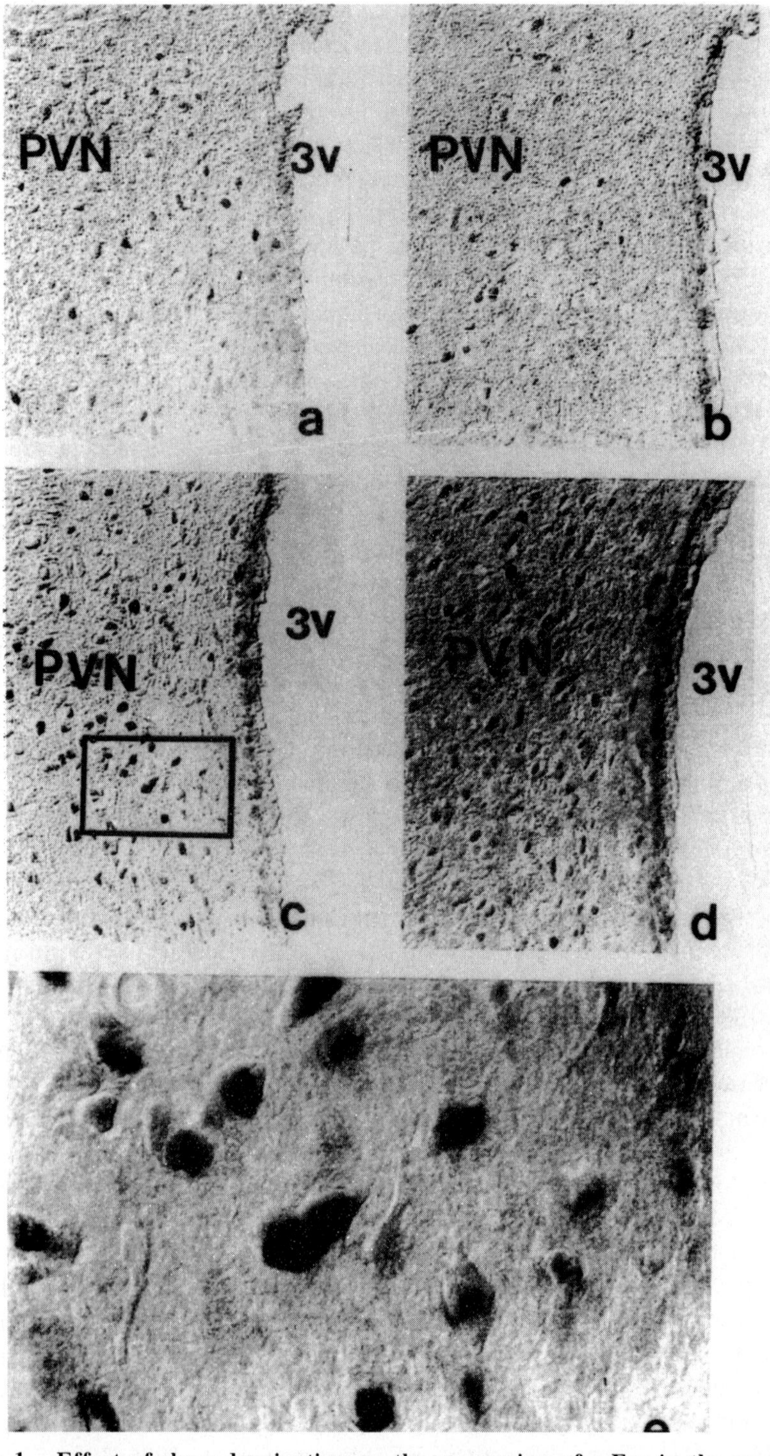

Fig. 1. Effect of sleep deprivation on the expression of c-Fos in the nucleus paraventricularis (PVN). Panel (a) and (b) show the expression of c-Fos in control animals, while (c) and (d) illustrate the effect of REM sleep deprivation and stress respectively. Panel (e) is a magnification of (c).

Although we have all these data that link acetylcholine, REM sleep occurrence and c-Fos expression, we can not easily integrate the picture we have described for noradrenaline and wakefulness. We are all aware of the impressive amount of data supporting acetylcholine's role in REM sleep generation (Gillin et al., 1993). Therefore, it is tempting to suggest that the stimulus that ignites cellular c-Fos expression during REM sleep is acetylcholine. According to this picture, acetylcholine stimulates any cholinoceptive neuron, including cholinergic cells, but does not restrict its activity to them. As a result, any kind of neurotransmitter-synthesizing neuron may be expressing c-Fos.

These data allow us to speculate about the intracellular processes triggered by acetylcholine, that may explain the expression of c-Fos. However, we still have some contradictory results. For example, all the original studies describing a relationship between ACh and REM sleep, have indicated that the receptor activated by ACh to trigger REM sleep is the muscarinic M2 (Gillin et al., 1993). In fact, we believe that carbachol binds to this receptor to trigger REM sleep and c-Fos expression. Hence, the main problem to put together this information is that the activation of the muscarinic receptors M2/M4 activates a Gi protein, which is inhibitory for adenylyl cyclase activity (Shuman et al., 1995). The resulting effect is a reduction of cAMP and therefore the restrain of all the following events. These data contradict evidences indicating that cAMP is high during sleep in general and even higher during REM sleep (Mottin et al., 1997). In fact, drugs that somehow facilitates adenylyl cyclase activity prevent carbachol REM sleep-inducing effect, but cause negligible effects on spontaneously occurring REM sleep (Shuman et al., 1995).

These contradictory data make us question the similarity between spontaneous and carbachol-induced REM sleep. Unfortunately, we do not have enough information to clarify this subject, now.

Changes in Neurotransmitters mRNAs

A few studies have been performed in relation to this subject. The basic hypothesis is that the mRNA of specific neurotransmitters synthesizing proteins changes across the sleep-waking cycle (Basheer et al., 1997). Choline acetyltransferase (ChAT), tyrosine hydroxylase (TH) and, glutamate decarboxylase (GAD) as well as norepinephrine transporter (NET) mRNA variations have been studied in rats. Results basically indicated that ChAT m RNA levels are higher in the basal forebrain and LDT nucleus during spontaneously occurring sleep than during wakefulness. As for the variations in TH and NET mRNA, an increase in both TH and NET mRNAs was detected after three days of REMSD. This augmentation returned to basal level after two days of sleep recovery. No detectable changes in GAD67 mRNA were observed in REM sleep deprived rats.

Other Gene Changes Associated to Sleep Deprivation

TSD induces an increase of NGFI-B and NGFI-A mRNA as well as an increase of NGFI-A protein levels. Moreover, TSD also induces a decrease in the levels of neurogranin mRNA, a protein that may be playing a role in postsynaptic signal transduction (Neuner-Jewe et al., 1995). In addition, TSD increases the level of IL-1β mRNA in the hypothalamus and in the brainstem (Mackiewiwcz et al., 1996). As for REMSD, it induces c-Fos and TH and NET mRNA changes, as we have aforementioned. In addition, REMSD increases galanin (Toppila et al., 1995) and somatostatin while reducing growth hormone releasing factor mRNA levels in the hypothalamus (Stenberg et al., 1997).

On the other hand, one function attributed to REM sleep is to propitiate memory consolidation (Prospero-Garcia et al., 1993; Skagaggs and McNaughton 1996). At the molecular level, it has been suggested that the calcium/calmodulin dependent protein kinase type II (CaMKII), plays an important role in memory consolidation (Bach et al., 1995; Bailey et al., 1996; Bernabeu et al., 1997). Seeking to determine if changes in this protein kinase may cause abnormalities in the expression of REM sleep, CaMKII knockout mice were studied. The main change observed in these mice was a shift of the theta band from fast to slow frequencies, during REM sleep (Thakkar et al., 1997).

Although sleep researchers have already initiated the description of gene activity associated to sleep, we still have to describe several other variations in genes encoding for proteins that affect sleep. For example, our studies have indicated that VIP is a neuropeptide which is highly sensitive to sleep manipulations (Drucker-Colin et al., 1988). We have observed that the intraventricular administration of VIP induces REM sleep in normal cats (Drucker-Colin et al., 1984) and restores it in parachlorophenylalanine (PCPA) treated (insomniac) cats (Prospero-Garcia et al., 1986). Moreover, we have described an increase in concentration of this molecule in the CSF of REM sleep deprived cats (CSFSD) (Jimenez-Anguiano et al., 1993). We actually believe that VIP is one of the molecules that once accumulated into the CSFSD, can be transferred into the brain of a recipient animal whereby inducing REM sleep. Some of our data lend support to this hypothesis, since we can block the CSFSD REM sleep inducing-effect by incubating it with VIP antibodies prior to its administration to recipient animals (Drucker-Colin et al., 1988). In addition, we have determined that VIP receptors seem to increase in number as a reaction to selective REMSD (Fig. 2) (Jimenez-Anguiano et al., 1996).

On the other hand, de Lecea et al. (1996) and others (Fukusumi et al., 1997) have recently described the existence of a new neuropeptide which seems to decrease REM sleep by antagonizing acetylcholine effects. This neuropeptide has been named cortistatin basically due to its homology with somatostatin and its particular distribution in the cerebral cortex and hippocampus of rats (Fig. 3). It has been observed in on-going studies, that cortistatin mRNA responds to sleep manipulations (Jimenez-Anguiano et al., 1997).

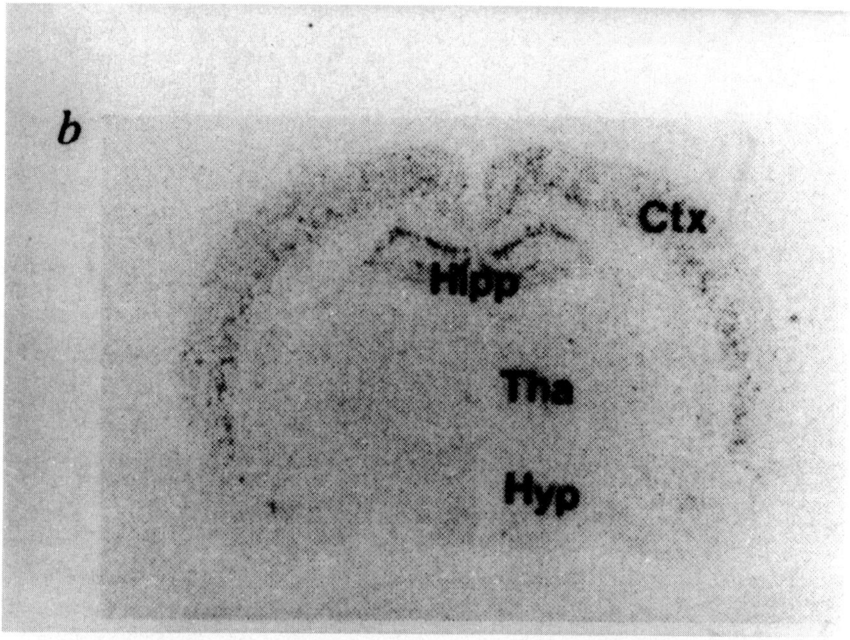

Fig. 2. This brain section illustrates the expression of the cortistatin mRNA in rats' brain.

These are two proteins, VIP and cortistatin, that posses modulating effects on REM sleep. We are in the process of determining variations in their mRNAs expression depending on the sleep-waking cycle.

On the other hand, Cravatt et al. (Cravatt et al., 1996; Lerner et al., 1994) have described a new molecule which seems to posses sleep-modulating actions. It has been named oleamide, basically because it is a lipid with an amide residue. Actually, it does not seem to be a crucial player in sleep control; yet, oleamide seems to be very important since it exerts cannabinoid effects by binding a (to be described) oleamide receptor or by facilitating the effect of anandamide. Anandamide is another lipid which we know binds cannabinoid receptors. A very interesting feature of oleamide is that it was isolated from the CSF of total sleep deprived cats. Cravatt et al. (1995) have already described the metabolizing enzyme which, in this context, is a good target protein to be evaluated during REMSD.

Genes and Sleep in Humans

Although it is important to describe all those changes in gene expression associated to sleep, we have to ask another crucial question: is there a gene(s) that controls sleep?

Since the pioneer work of Valatx (1972) describing changes in the sleep patterns of different mice strains, it was suggested that the occurrence and daily amount of sleep, particularly REM sleep, is under genetic control.

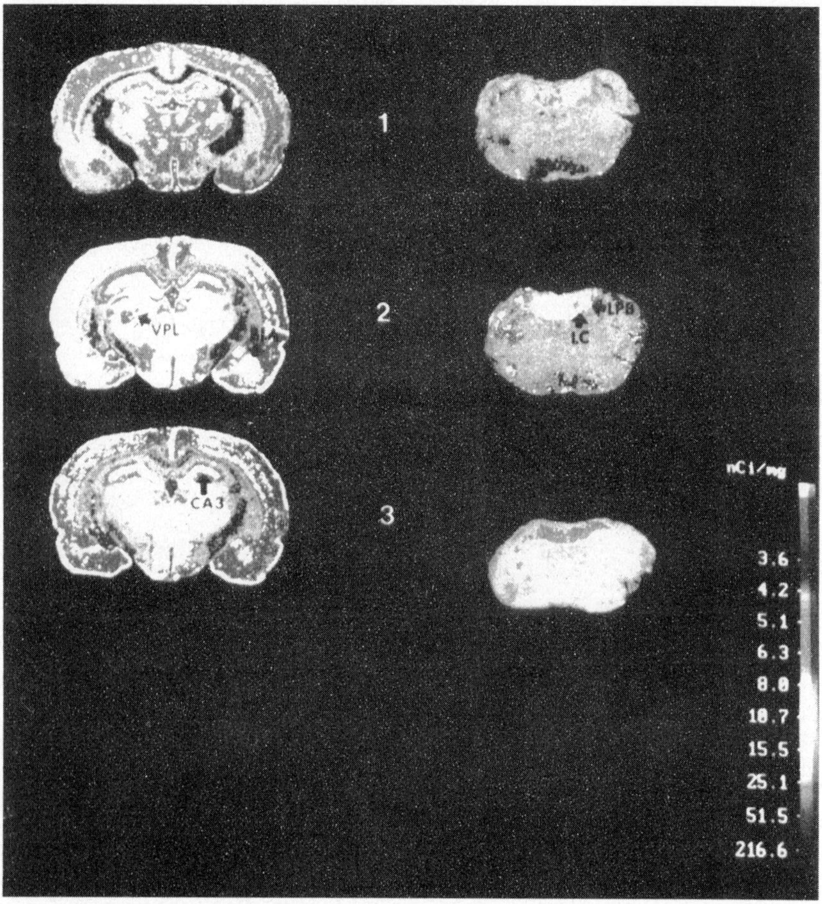

Fig. 3. Brain section 1 illustrates the expression of VIP receptors in rostral and caudal structures in normal animals. Sections 2 and 3 show the expression of VIP receptors after selective REM sleep deprivation.

Actually, recent studies have indicated that the spectrum of potency of delta and theta bands are different between mice strains (Franker et al., 1997) lending further support to Valatx's hypothesis.

There are just a few studies in humans corroborating this hypothesis. For example, we have observed that sleep patterns in identical twins are extremely alike compared to non-related people (Brailowsky et al., 1992; 1980). More recent studies in humans have corroborated that in addition to sleep patterns EEG features are inheritable (Beijsterveldt, 1996).

Of course, one way to know the genetic influence on sleep is by studying certain diseases. Among these are narcolepsy, REM sleep behavior disorder, and the fatal familial insomnia.

Although narcolepsy has never universally been considered an exclusively REM sleep disease, some authors require the presence of two or more sleep-onset REM periods in the Multiple Sleep Latency Test to make the diagnosis

(Guilleminault, 1994). Narcolepsy is a sleep disorder that basically is characterized by a tetrad of symptoms: excessive daytime sleepiness, cataplexy, sleep paralysis and, vivid hypnagogic hallucinations. Since the report published in 1986, indicating that 100% of the Japanese narcoleptics were human lymphocyte antigen (HLA) DR2 positive, there is an extensive literature indicating that narcolepsy is a genetic abnormality. Several studies have indicated that the DQβ0602 is the HLA marker for narcolepsy. In addition, an everyday increasing evidence has suggested that HLA genes may also be involved in the generation of the REM sleep behavior disorder (Schenck et al., 1996). This sleep abnormality is interesting basically because patients lack the ability to inhibit their postural tone during REM sleep, whereby acting out their dreams.

In addition the recently described sleep disease: fatal familial insomnia (Fiorino, 1996), a sleep disease that maintains the patient awake until his death, seems to be a result of the presence of an abnormal isoform of the prion protein (PrP) resistant to proteases (Telling et al., 1996). PrP is encoded by PRNP, a gene located on the short arm of the human chromosome 20. Actually recent experiments performed using mice lacking PrP have shown that this type of mice have abnormalities in both, the circadian motor activity rhythms and in the patterns of sleep, such as an abnormal sleep rebound after sleep deprivation and excessive sleep fragmentation compared to the wild type mice (Tobler et al., 1996). Hence, these facts suggest that sleep occurrence and possibly daily amounts are under gene control.

Discussion

In this chapter we have reviewed most of the experiments and results regarding sleep, particularly REM sleep, and gene expression, available in the literature. It is clear that the amount of studies is still limited, therefore we have a poor understanding of these processes. However, despite these limitations, we are beginning to reveal the molecular processes subserving REM sleep mechanisms and the molecular consequences of REM sleep occurrence.

In the context of determining the genetic basis of behavior, it is of great benefit for the biomedical research to further our knowledge on the genetics of REM sleep. This knowledge, in turn, will help us to understand the biochemical events happening in normal conditions and how they are altered by sleep disorders.

Learning about how cells behave electrically and biochemically during REM sleep, in addition to teach us new approaches to manipulate this behavior, it teaches us about how the brain functions in its innermost mechanisms. Once we gain a better comprehension of these, the challenge will be to utilize them to explain REM sleep as a behavior and possibly explain its functions, i.e. tissue restoration, memory consolidation, programming strategies, among others (McGinty and Szymusiak, 1988; Prospero-Garcia et al., 1993; Skagaggs and McNaugton, 1996).

Obviously, to determine which genes are involved in sleep regulation should be the main goal to go after. However, to understand why genes encoding for specific proteins have to be active during this stage is not a trivial goal, either. We all seek to discover which genes are active during REM sleep any why.

Acknowledgement

This work was partially supported by a grant from CONACyT (25488-M and 25128-M) and from FUNSALUD given to Dr. Oscar Prospéro-García and Dr. Luz Navarro. Authors wish to thank Miss Aldebarán Prospéro for her help in the translation of this manuscript.

References

Aserinsky, E. and Kleitman, N. Regularly occurring periods of eye motility and concomitant phenomena, during sleep, Science, 118 (1953) 273–274.

Bach, M.E., Hawkins, R.D., Osman, M., Kandel, E.R. and Mayford, M. Impairment of spatial but not contextual memory in CaMKII mutant mice with a slective loss of hippocampal LTP in the range of the *q* frequency, Cell, 81 (1995) 905–915.

Bailey, C.H., Bartsch, D. and Kandel, E.R. Toward a definition of long-term memory storage, Proc. Natl. Acad. Sci. USA, 93 (1996) 13445–13452.

Basheer, R., Ramanathan, L., Greco, M.A., McCarley, R.W. and Shiromani, P.J. Behavioral state dependent alterations in the steady state messenger RNAs of different neurotransmitter systems, APSS 11th Annual Meeting, (1997) 259.

Beijsterveldt. Heritability of human brain functioning as assessed by electroencephalography, Am. J. Hum. Gen. 58 (1996) 562–573.

Bernabeu, R., Bevilaqua, L., Ardenghi, P., Bromberg, E., Schnitz, P., Bianchin, M., Izquierdo, I. and Medina, J.H. Involvement of hippocampal cAMP/cAMP-dependent protein kinase signaling pathways in a late memory consolidation phase of aversively motivated learning in rats, Proc. Natl. Acad. Sci. USA, 94 (1997) 7041–7046.

Borbély, A.A. and Tobler, I. Endogenous sleep promoting substances and sleep modulation, Physiol. Rev., 69 (1989) 605–670.

Brailowsky, S., Prospero-Garcia, O., Salas, J. and Alvarez-Rueda, M. Identical twin studies indicate genetic influence on human sleep patterns, Neurosciences, 1 (1992) 83–88.

Brailowsky, S., Prospéro-García, O., Salas, J. and Shkurovich, M. Sleep in identical twins, Proc. Int. Union Physiol. Sci., 14 (1980) 336.

Cirelli, C., Pompeiano, M. and Tononi, J. In vivo antisense approaches to the role of immediate early gene expression in the brain, Reg. Peptides, 59 (1995a) 151–162.

Cirelli, C., Pompeiano, M. and Tononi, J. Neuronal gene expression in the waking state: a role for the locus coeruleus, Science, 274 (1996) 1211–1215.

Cirelli, C., Pompeiano, M., Arrighi, P. and Tononi, J. Sleep-waking changes after c-Fos antisense injections in the medial preoptic area, NeuroReport, 6 (1995b) 801–805.

Cirelli, C., Pompeiano, M., Arrighi, P. and Tononi, J. The locus coeruleus and immediate-early genes in spontaneous and forced wakefulness, Brain Res. Bull., 35 (1994) 589–596.

Cravatt, B.F., Giang, D.K., Mayfield, S.P., Boger, D.L., Lerner, R.A. and Gilula, N.B. Molecular characterization of an enzyme that degrades neuromodulatory fatty-acid amides, Nature, 384 (1996) 83–87.

Cravatt, B.F., Prospéro-García,O., Siuzdak, G., Henriksen, S.J., Boger, D.L. and Lerner, R.A. Chemical characterization of a family of brain lipids that induce sleep, Science, 268 (1995) 1506–1509.

Crawley, J.N. and McLean, S. (Eds.) *Neuropeptides: basic and clinical advances,* Ann. New York Acad. Sci. 780, New York, 1996.

de Lecea, L., Criado, J.R., Prospéro-García, O., Gautvik, K.M., Schweitzer, P., Danielson, P.E., Dunlop, C.L.M., Siggins, G.R., Henriksen, S.J. and Sutcliffe, J.G. A cortical neuropeptide with neuronal depressant and sleep-modulating properties, Nature, 381 (1996) 242–245.

Demin, N.N. Some features of the metabolism of brain-proteins and RNA during normal sleep and its artificial alterations. In: *Advances in Neurochemistry.* E.M. Kreps, (Ed.), Nauka, Leningrad, 1974, pp. 29–35.

Drucker-Colín, R., Bernal-Pedraza, J., Fernández-Cancino, F. and Oksenberg, A. Is vasoactive intestinal polypeptide (VIP) a sleep factor?, Brain Res., 278 (1984) 308–312.

Drucker-Colín, R., Spanis, C.W., Cotman, C.W. and McGaugh, J.L. Changes in protein perfusates of freely moving cats: relation to behavioral state, *Science,* 187 (1975) 963–965.

Drucker-Colin, R. and Prospéro-García, O. Neurophysiology of sleep. In: *Handbook of Sleep Disorders.* M.J. Thorphy (Ed.), Marcel Dekker, Inc., NY 1990, pp. 33–53.

Drucker-Colin, R., Prospéro-García, O., Arankowsky-Sandoval, G. and Pérez-Montfort, R. Gastropancreatic peptides and sensory stimuli as REM sleep factors. In: *Sleep Peptides: basic and clinical approaches.* S. Inoué, and D. Schneider-Helmert, (Eds.), Japan Soc. Press/Springer-Verlag, Japan-Berlin, 1988, pp. 74–94.

Fiorino, A.S. Sleep, genes and death: fatal familial insomnia, Brain Res. Rev., 22 (1996) 258–264.

Franken, P., Malafosse, A. and Tafti, M. Genetics of sleep EEG in mice: strain distribution pattern, APSS 11th Annual Meeting, 262, 1997.

Fukusumi, S., Kitada, C., Takekawa, S., Kisawa, H., Sakamoto, J., Miyamoto, M., Hinuma, S., Kitano, K. and Fujino, M. Identification and characterization of a novel human cortistatin-like peptide, Biochem. Biophys. Comm., 232 (1997) 157–163.

Gillin, J.C., Salin-Pascual, R., Velazquez-Moctezuma, J., Shiromani, P. and Zoltoski, R. Cholinergic receptor subtypes and REM sleep in animals and normal controls, Progr. Brain Res., 98 (1993) 379–387.

Giuditta, A., Neugebauer-Vitale, A., Grassi-Zucconi, G., Ambrosini, M.V. and Scaroni, R. Synthesis of brain RNA and DNA during sleep, Exp. Brain Res., Suppl. 8 (1984) 146–154.

Giuditta, A., Rutigliano, B. and Vitale-Neugebauer, A. Influence of synchronized sleep on the biosynthesis of RNA in two nuclear classes isolated from rabbit cerebral cortex, J. Neurochem., 35 (1980) 1259–1266.

Guilleminault, C. (Ed.), Narcolepsy, Sleep, Suppl. 8 (1994).

Henriksen, S.J. and Prospéro-García, O. Chemistry of sleep: petides. In: *Encyclopedia of sleep and dreaming.* M.A. Karskadon, (Ed.), Macmillan Publishing Co., New York, 1993, pp. 107–108.

Hobson, J.A., McCarley, R.W. and Wyzinski, P.W. Sleep cycle oscillation: reciprocal discharge by two brainstem neuronal groups, Science, 189 (1975) 55–58.

Hyden, H. and Lange, P.W. Rhythmic enzyme changes in neurons and glia during sleep and wakefulness, Prog. Brain Res. 18 (1965) 92–95.

Jiménez-Anguiano, A., Báez-Saldana, A. and Drucker-Colín, R. Cerebrospinal fluid (CSF) extracted immediatly after REM sleep deprivation prevents REM rebound and contains vasoactive intestinal peptide (VIP), Brain Res., 631 (1993) 345–348.

Jiménez-Anguiano, A., García-García, F., Mendoza-Ramírez, J.L., Durán-Vazquez, A. and Drucker-Colín, R. Brain distribution of vasoactive intestinal peptide receptors following REM sleep deprivation, Brain Res., 728 (1996) 37–46.

Jiménez-Anguiano, A., de Lecea, L., Sánchez-alavez, M., Criado, J., Palomero, M., Prado-Alcalá, R., Henriksen, S.J., Sutcliffe, J.G. and Prospéro-Garcia, O. Cortistatin, sleep and memory, APSS 11th Annual Meeting, 100, 1997.

Jouvet, M. Biogenic amines and the states of sleep, Science, 163 (1969) 32–41.

Kayama, Y., Ohta, M. and Jodo, E. Firing of possibly cholinergic neurons in the rat laterodorsal tegmental nucleus during sleep and wakefulness, Brain Res., 569 (1992) 210–220.

Ledoux, L., Sastre, J.P., Buda, C., Luppi, P.H. and Jouvet, M. Alterations in c-fos expression after different experimental procedures of sleep deprivation in the cat, Brain Res., 735 (1996) 108–118.

Lerner, R.A., Siuzdak, G., Prospéro-García, O., Henriksen, S.J., Boger, D.L. and Henriksen, S.J. Cerebrodiene: a brain lipid isolated from sleep-deprived cats, Proc. Natl. Acad. Sci. USA. 91 (1994) 9505–9508.

Luebke, J.I., Greene, R.W., Semba, K., Kamondi, A., McCarley, R.W. and Reiner, P.B. Serotonin hyperpolarizes cholinergic low-thresold burst neurons in the rat laterodorsal tegmental nucleus in vitro, Proc. Natl. Acad. Sci. USA, 89, (1992) 743–747.

Mackiewiwcz, M., Sollares, P.J., Ogilvie, M.D. and Pack, A.I. Modulation of IL-1*b* gene expression in the rat CNS during sleep deprivation, NeuroReport, 7 (1996) 529–533.

McGinty, D. and Szymusiak, R. Neuronal unit activity patterns in the behaving animals: brainstem and limbic system, Ann. Rev. Psychol., 39 (1988) 135–168.

Merchant-Nancy, H., Vazquez, J., Aguilar-Roblero, R. and Drucker-Colin, R. c-fos proto-oncogene changes in relation to REM sleep duration, Brain Res., 579 (1992) 342–346.

Merchant-Nancy, H., Vazquez, J., García, F. and Drucker-Colin, R. Brain distribution of c-fos expression as a result of prolonged rapid eye movement (REM) sleep period duration, Brain Res., 681 (1995) 15–22.

Morgan, J.J. and Curran, T. Stimulus transcription coupling in the nervous system: involvement of the inducible proto-oncogenes fos and jun, Ann. Rev. Neurosci., 14 (1991) 421–451.

Mottin, S., Laporte, P. and Cespuglio, R. Approach of the sleep energetic mechanisms by measuring the laser induced fluorescence of NADH, Neuroscience, 79 (1997) 683–693.

Neuner-Jehle, M., Rhyner, T.A. and Borbély, A.A. Sleep deprivation differentially alters the mRNA and protein levels of neurogranin in rat brain, Brain Res., 685 (1995) 143–153.

Oswald, I., The function of sleep in restoring the tissues. In: Sleep'86. W.P. Koella, F. Obál, H. Schulz, and P. Visser, (Eds.), Gustav Fischer Verlag, Stuttgart, New York, 1988, pp. 23–28.

Prospéro-García, O. and Drucker Colín, R. Control neural del ciclo sueno-vigilia. In: *Sueno and procesos cognitivos.* M.J. Ramos Platón, (Ed.), Editorial Síntesis, SA, 1995, pp. 131–160.

Prospéro-García, O., Miller, R.D. and Henriksen, S.J. Hippocampal interneuron activity in unanesthetized rats: relationship to the sleep-wake cycle, Neurosci. Lett., 156 (1993) 158–162.

Prospéro-García, O., Morales, M., Arankowsky-Sandoval, G. and Drucker-Colín, R. Vasoactive intestinal polypeptide (VIP) and cerebrospinal fluid (CSF) of sleep deprived cats restores REM sleep in insomniac recipients, Brain Res., 385 (1986) 169–173.

Schenck, C.H., Garcia-Rill, E., Segall, M., Noreen, H. and Mahowald, M.W. HLA class II genes associated with REM sleep behavior disorder, Ann. Neurol., 39 (1996) 261–263.

Sherin, J.E., Shiromani, P.J., McCarley, R.W. and Saper, C.B. Activation of ventrolateral preoptic neurons during sleep, Science, 271 (1996) 216–219.

Shiromani, P.J., Greco, M.A., Thakkar, M. and McCarley, R.W. c-fos knockout mice have reduce non-REM sleep, APSS 11th Annual Meeting, 260, 1997.

Shiromani, P.J., Malik, M., Winston, S. and McCarley, R.W. Time course of Fos-Like immunoreactivity associated with cholinergically induced REM sleep, J. Neurosci., 15 (1995) 3500–3508.

Shiromani, P.J., Winston, S. and McCarley, R.W. Pontine cholinergic neurons show FOS-like immunoreactivity associated with cholinergically induced REM sleep, Mol. Brain Res., 38 (1996) 77–84.

Shuman, S.L., Capece, M.L., Baghdoyan, H.A. and Lydic, R. Pertussis toxin-sensitive G proteins mediate carbachol-induced REM and respiratory depression, Am. J. Physiol., 269 (1995) R308–R317.

Skagaggs, W.E. and McNaughton, B.L. Replay of neuronal firing sequences in rat hippocampus during sleep following spatial experience, Science, 271 (1996) 1870–1873.

Spanis, C.W., Gutierrez, M.C. and Drucker-Colín, R. Neurohumoral correlates of sleep: further biochemical and physiological characterization of sleep perfusates, Pharmacol. Biochem. Behav., 7 (1977) 345–352.

Stenberg, D., Toppila, J., Asikainen, M., Alanko, L. and Porkka-Heiskanen, T. Differential effects on hypothalamic neuropeptide gene expression of total sleep deprivation in rats, APSS 11th Annual Meeting, 261, 1997.

Thakker, M.M., Rainnie, D.G., Hearn, E.F., Greene, R.W., McCarley, R.W. and Shiromani, P.J. Abnormal theta activity during REM sleep in *a*-calcium-calmodulin kinasa II knockout mice. APSS 11th Annual Meeting, 260, 1997.

Tamura, T., Konishi, Y., Makino, Y. and Mikoshiba, K. Mechanisms of transcriptional regulation and neural gene expression, Neurochem. Int., 29 (1996) 573–581.

Telling, G.C., Parchi, P., DeArmond, S.J., Cortelli, P., Montagna, P., Gabizon, R., Mastrianni, J., Lugaresi, E., Gambetti, P. and Prusiner, S.B. Evidence for the conformation of the pathologic isoform of the prion protein enciphering and propagating prion diversity, Science, 274 (1996) 2079–2082.

Tobler, I., Gaus, S.E., Deboer, T., Achermann, P., Fischer, M., Rulicke, T., Moser, M., Oesch, B., McBrides, P.A. and Manson, J.C. Altered circadin activity rhythms and sleep in mice devoid of prion protein, Nature, 380 (1996) 639–642.

Toppila, J., Stenberg, D., Alanko, L., Asikainen, M., Urban, J.H., Turek, F.W. and Porkka-Heiskanen, T. REM sleep deprivation induces galanin gene expression in the rat brain, Neurosci. Lett., 183 (1995) 171–174.

Valatx, Genetic studies of sleep in mice, Nature, 238 (1972) 226–227.

Williams, J.A. and Reiner, P.B. Noradrenaline hyperpolarizes identified rat mesopontine cholinergic neurons in vitro, J. Neurosci., 13 (1993) 3878–3883.

Yamuy, J., Mancillas, J.R., Morales, F.R. and Chase, M.H. c-Fos expression in the pons and medulla of the cat during carbachol-induced active sleep, J. Neurosci., 13 (1993) 2703–2718.

Yamuy, J., Sampogna, S., López-Rodríguez, F., Luppi, P.H., Morales, F.R. and Chase, M.H. Fos and serotonin immunoreactivity in the raphe nuclei of the cat during carbachol-induced active sleep: a doble-labeling study, Neuroscience, 67 (1995) 211–223.

Rapid Eye Movement Sleep
B.N. Mallick and S. Inoué (Eds)

26. Intensity of REM Sleep

Kazumi Takahashi

Department of Neuropsychiatry, Tokyo Medical and Dental University,
1-5-45 Yushima Bunkyo-Ku Tokyo, 113 Japan

SUMMARY

The concept of 'intensity of NREM and REM sleep' are the basis of reciprocal interaction model of sleep regulation. Not only REM sleep is inhibited by increased NREM sleep intensity but also NREM sleep intensity inhibited by increased REM sleep intensity.

Intensity of NREM sleep was measured by the amount of delta (slow) waves and intensity of REM sleep was measured by the duration of REM sleep episode and REM density. Progression of NREM sleep from stage-1 to stage-4 represent the intensity of NREM sleep and it parallels the increase of the amount of delta waves. Maximum delta wave activity occurs in the first NREM sleep episodes, typically followed by a gradual decline across subsequent NREM sleep periods in a descent-to-right pattern. Different from the intensity of NREM sleep, the intensity of REM sleep, represented by duration of REM episode and REM density, usually increase progressively across the REM periods in an ascent-to-right pattern. These pattern implies that the intensity of REM sleep increases across the night, and is highest in the early morning.

Besides the conventional parameters of REM intensity, there are other parameters on eye movements (EMs), which represent the magnitude of single EM. We measured a duration and an amplitude of single EM, from these measurements three parameters were calculated as follows; rotation, velocity and power of EM; By contrast with the an ascent-to-right pattern observed in conventional REM intensity, these new parameters increased significantly from the first to its peak value in the second REM episode, it then decreased in the third REM episode, making an inverted-v shaped pattern. Discrepancy suggests the magnitude of single EM reflects physiological mechanisms which are not described by the parameters conventionally used to describe REM sleep intensity.

Intensity of NREM and REM Sleep

Intensity of NREM Sleep

Physiological method for recording and analyzing sleep depends mostly on polysomnographs. Large amount of data have been analyzed according to Rechtschaffen and Kales criteria (Rechtschaffen and Kales, 1968) to draw sleep histograms. Sleep histograms shows the overview of sleep architecture; sleep onset, progression of NREM sleep and distribution of REM sleep.

Progression of NREM sleep from stage-1 to stage-4 represent the intensity of NREM sleep and it parallels the increase of the amount of delta waves. Since the introduction of computer into the analysis of polysomnograph the integrated amplitude of delta waves measured by period-amplitude analysis (advocated by Feinberg, 1974) and the power density of the delta band measured by FFT analysis (Borbéy, 1982) have mainly been used to reflect the "intensity of NREM sleep". It is well established that maximum delta wave activity occurs in the first NREM sleep episodes, typically followed by a gradual decline across subsequent NREM sleep periods in a descent-to-right pattern. Figure 1 shows an example of the progression of sleep stages and the integrated delta band (FFT analysis calculated every 20 sec) across a single night of a 25 yr old healthy young male. Dotted line shows the amount of delta waves in each NREM cycle. About 37% of delta power discharged in the first sleep cycle, 35% in the second, 13%, 10% in third and fourth, respectively.

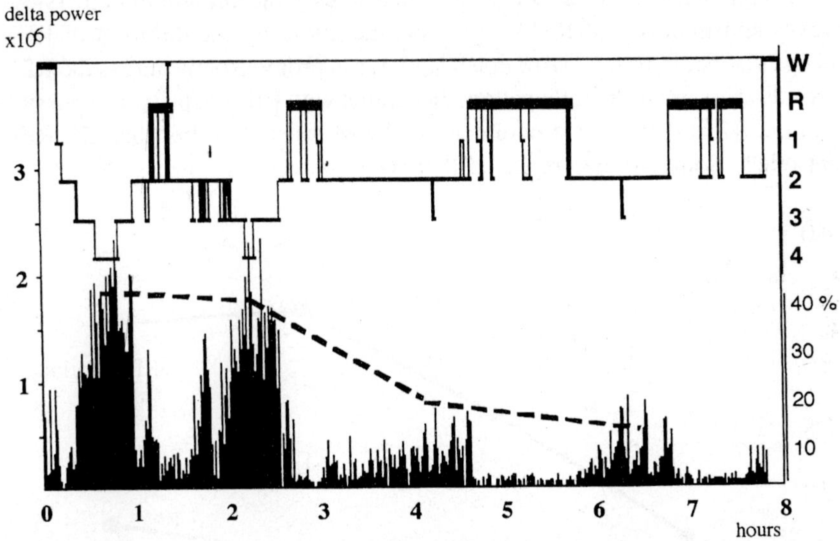

Fig. 1. Intensity of NREM sleep: An example of the progression of sleep stages and the power density of delta band (FFT analysis calculated every 20 sec) across a single night of a 25 year old healthy young male. Dotted line shows the amount of delta waves in each NREM cycle. Intensity of NREM sleep decreased in an descent-to-right pattern.

Intensity of REM sleep

Different from the intensity of NREM sleep, the intensity of REM sleep has not been discussed well, becauase of two reasons. First, there was no established definition of REM sleep intensity. Secondary, there was not as available parameters for measuring the quantitative changes of REM sleep as the delta waves in NREM sleep. Figure 2 illustrated averages of REM episode (duration of each REM episode) and REM density of each REM cycle from the average of 20 young male adults. As shown in the figure, these two parameters usually increase progressively across the REM periods in an ascent-to-right pattern (Benoit et al., 1974; Coble et al., 1987; Reynolds et al., 1985; Vitiello et al., 1984; Geisler et al., 1987; Aserinsky, 1969; Zimmermann et al., 1980). These pattern imply that the intensity of REM sleep increases across the night, and is highest in the early morning. The term'REM pressure' is also used in the same meaning of intensity which is an unknown mechanism that promote the emergence of REM sleep in opposition to the emergence of NREM sleep.

The concept of 'intensity of NREM and REM sleep' are the basis of reciprocal interaction model of sleep regulation. Not only REM sleep is inhibited by increased NREM sleep intensity after total sleep deprivation as shown by the REM episode reduction (Borbely, 1982). But also NREM sleep intensity inhibited by increased REM sleep intensity after selective REM sleep deprivation. That was shown by the significant reduction in the low frequency EEG activity in NREM sleep (Beerson et al., 1990). In these studies intensity of NREM sleep was measured by the amount of delta (slow) waves and intensity of REM sleep was measured by the duration of REM sleep episode. In Fig. 2, REM episode increase progressively across the REM periods in an ascent-to-right pattern, in parallel with REM density. Is it possible that the REM density is a parameter of REM intensity other than the REM episode? Some studies reported that REM density is not associated with

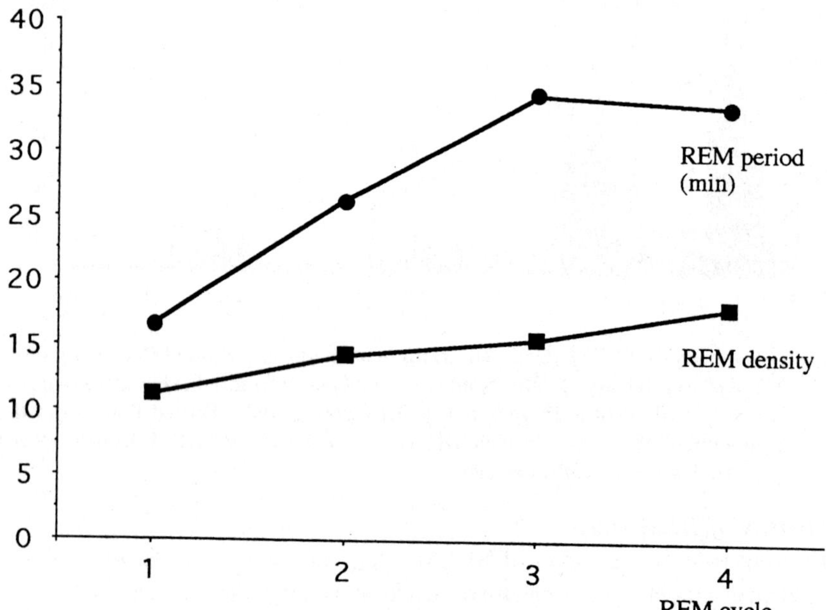

Fig. 2. Intensity of REM sleep: Average duration of REM episode and REM density of each REM cycle from 20 young male adults. These two parameters usually increase progressively across the REM periods in an ascent-to-right pattern. It implies that the intensity of REM sleep increases across the night in ascent-to-right pattern.

REM sleep intensity but is inversely related to delta wave (NREM) sleep propensity (Antonioli et al., 1981; Feinberg et al., 1987). In any case it interests us in relation to the intensity of NREM and REM sleep.

Intensity of REM Sleep (Incidence of EMs)

Parameters of REM sleep

It is not established which parameters of REM sleep effectively represent REM sleep intensity. The existing parameters of REM sleep are summarized as follows; they can be grouped into two categories. Tonic parameters include the timing and duration of REM sleep, such as the percentage of REM in total sleep, REM latency, and the duration of each REM sleep episode. Phasic REM sleep parameters include the number (REM activity) and incidence (REM density) of rapid eye movements (EMs; hereafter EM represents a single rapid eye movement in a REM period). In this chapter I focused on phasic REM sleep parameters as a parameter of REM sleep intensity and examined the possibility of extracting additional physiological information from polysomnograph of REM sleep.

REM density

The time course of changes in REM density across the night has been reported as a parameter of mental states. In healthy adults, as shown above, REM density usually increases progressively in successive REM episodes, changing in an ascent-to-right pattern. Coble et al. (1987) indicate that in childhood REM density is highest in the middle of the sleep period, producing an inverted-V-shape pattern, but changes to the ascent-to-right pattern in adolescence. These pattern has been reported to differ in different mental diseases. Depressive illness, which characteristically shows high REM density in the first REM episode, presents a V-shaped pattern (Kupfer, 1978; Reynolds et al., 1990). This V-shaped pattern changes following the administration of anti-depressant medications, but does not completely normalize (McPartland et al., 1979). In Alzheimer's disease, the REM density pattern presents an inverted-V dependent on the severity of the disease (Vitiello et al., 1984). In narcoleptic patients, REM density shows a flat pattern across the night (Geisler et al., 1987).

These reports suggest the utility of REM density and the necessity of more detailed study of the relationship between REM density and REM intensity.

Conventional REM density and new EM-frequency

REM activity and REM density are conventional parameters of phasic REM sleep. However, both represent only relative values; REM activity is not the exact number of EMs in REM sleep but the number of epochs which include EMs in. In the same manner REM density is a percentage of numbers of epochs which include EMs to the number of all epochs included in REM sleep. Duration of epoch usually is 20 sec, 30 sec or longer than that. The values of REM activity and REM density have been substituted for the true number and incidence of EMs (Aserinsky, 1971), because it was time-consuming and practically impossible to count the exact number of EMs by visual inspection of polysomnograph. As true values for number and incidence of EMs in REM sleep, we proposed EM-count and EM-frequency (Takahashi and Atsumi, 1997), EM-count is the true number of EMs; the total number of EMs in a period of time (e.g. each epoch, each REM episode, an entire night) and EM-frequency, the true incidence of EMs; the number of EMs per minute. To assess the sensitivity of refined parameter, EM-frequency, to conventional REM density, four values of REM density and EM-frequency were calculated from the same polysomnographs' data. Comparing these five values, relative values were calculated and potted in the figure, by dividing each value by the corresponding value from REM period 1. Figure 3 presents an example of the

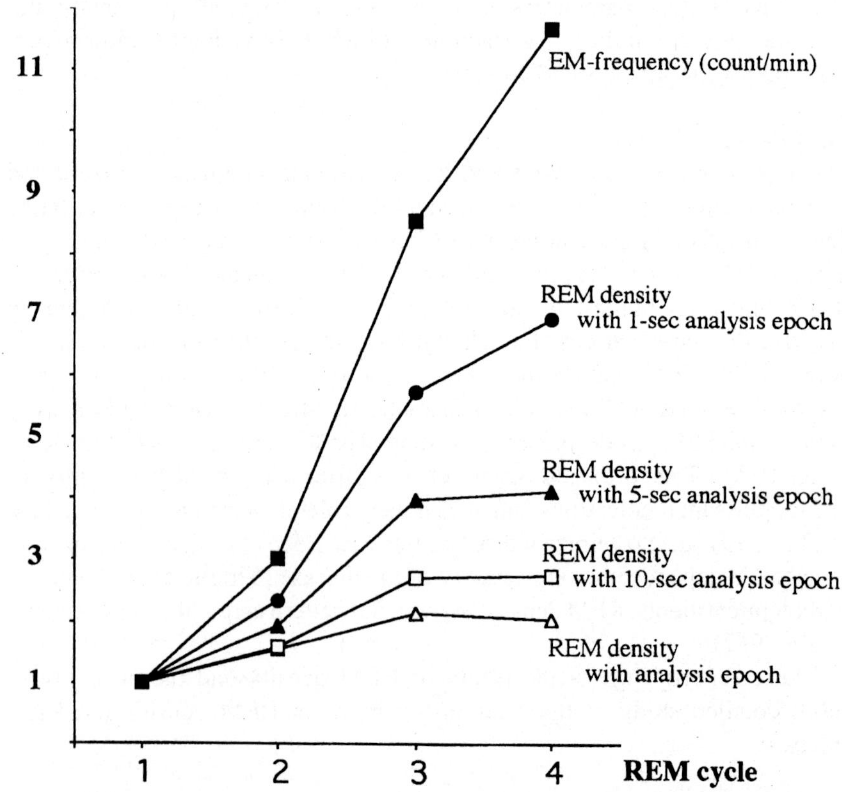

Fig. 3. REM density and EM-frequency: An example of the relative changes in REM density and EM-frequency across REM periods from a representative 25 year old male. All values of REM density increase across the night but differ from EM-frequency. New measure EM-frequency can detect fine changes between REM episodes.

relative changes in REM density and EM-frequency across REM periods from a representative 25 yr old male. For the calculation of REM density REM episodes were divided into 20, 10, 5 and 1 sec epochs and each was examined for the presence of at least one EM. All calculations of REM density increased significantly between the second and third REM episodes (ANOVA $p < 0.05$). REM density calculated for 20 sec intervals did not increase significantly between the second and fourth REM episodes. However, when computed for epochs shorter than 10 sec, REM density showed significant increments between REM periods 2 and 4 (ANOVA $p < 0.05$). Only EM-frequency increased significantly from the second to third, from second to fourth, and from third to fourth REM episode (ANOVA $p < 0.05$). The results indicate the utility of studying exact incidence of EMs.

Conventional REM activity and new EM-count

Measurements of the exact number of EM (EM-count) from 40 nights polysomnographs (second and third nights' data from three consecutive nights recordings of twenty healthy normal males) revealed interesting intra-individual variation. Although the EM-count for each night varied widely, with a mean of 1342.9 (range 258 to 3727, SD = 767.1), the number of EMs in the same subject were almost the same between second and third night of recording for 14 of 20 subjects. Figure 4 shows inter-individual and intra-individual variations in EM-count, with each subject's second and third night values connected.

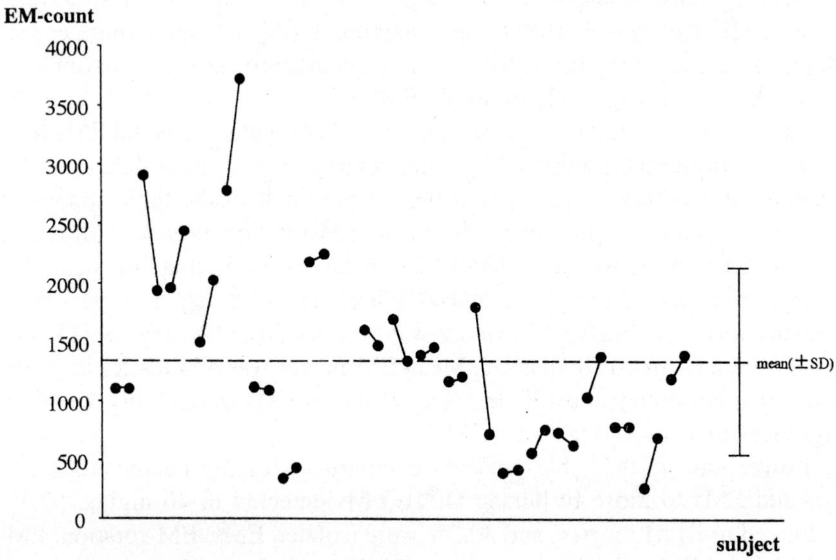

Fig. 4. EM-count (exact number of EMs): Inter-individual and intra-individual variations in EM-count, with each subject's second and third night values connected. In 14 of 20 subjects the number of EMs in second night is almost same as the EMs number in third night. Variability of EM-count was much less within individuals than between individuals.

Variability of EM-count was much less within individuals than between individuals.

The reason why almost the same value of EM-count was recorded in the same subject is not clear. On the other hand, in six of twenty subjects EM-count of second night differs from third night. In these case it is possible that intensity of REM sleep of whole night differs between second and third night. More precise comparison of intensity of NREM (delta wave), percentage of REM sleep and other parameters of polysomnographs may reveal differences of the quality of these six REM sleep.

New Parameters from the Single EM Measurement

Magnitude of single EM

Besides the number and incidence of EMs, there are other parameters on EMs, which represent the magnitude of single EM. We measured a duration and an amplitude of single EM (Takahashi and Atsumi, 1997), from these measurements three parameters were calculated as follows; rotation, velocity and power of EM;

1: EM-rotation, degrees of eyeball rotation,
2: EM-velocity, the angular velocity (in deg/sec) of eyeball rotation,
3: EM-power, defined as $(\text{EM-rotation})^2/\text{EM-duration}$.

EM-power is an estimate of the total energy expended during an EM (Kobayashi et al., 1979; Kobayashi, 1981). Large and quick EMs consume more energy than small and slow EMs. These three parameters are new parameters representing the magnitude of single EM.

Table 1 presents the mean characteristics of computer analyzed EMs from 40 night polysomnographs of 20 young healthy males. These data gives the readers an overview of new parameters. Upper half of the table shows the average of whole night sleep. The mean EM-rotation was 6.27 deg. The modal EM-rotation was 2–3 deg (18.9% of EMs were in this bin) with 80% of the values less than 9.0 deg and 91% less than 13.0 deg. The mean EM-velocity was 58.73 deg/sec. The mode was between 30 and 40 deg/sec (17.7%), 91% were less than 120 deg/sec. Mean EM-power was 525.85 deg^2/sec, the mode was between 0 and 100 deg^2/sec (26.4% of EMs were in this bin), 90% were less than 1400 deg^2/sec.

Lower half of the table presents comparisons of the characteristics of isolated EMs to those in bursts. Of all EMs detected in 40 nights, 69.8% belonged to REM 'bursts' and 30.2% were isolated EMs. EM-rotation, EM-velocity and EM-power in bursts were significantly larger than were isolated EMs (ANOVA $p < 0.001$). That is, EMs in bursts were larger, rotated more quickly and had greater power than isolated EMs.

Table 1 Measurements of Single EM

n = 53562			Mean ± S.D.	Min	Max
EM-rotation	(deg)	:	6.27 ± 4.86	1.23 (*)	61.51
EM-velocity	(deg/sec)	:	58.73 ± 40.9	8.85	491.98
EM-power	(deg^2/sec)	:	525.85 ± 884.0	9.08	19014.97

		EMs in Burst	Isolated EMs	All
Count (n)		37383 (69.8%)	16179 (30.2%)	53562 (100%)
EM-rotation	(deg)	6.77 ± 5.23**	5.13 ± 3.62	6.27 + 4.86
EM-velocity	(deg/sec)	63.21 + 44.13**	48.38 ± 29.76	58.73 ± 40.90
EM-power	(deg^2/sec)	611.57 ± 986.30**	327.81 ± 532.30	525.85 ± 884.0

*Threshold; $^{**}p < 0.001$

Upper half average magnitude of single EM measurements from total 53562 EMs in 40 nights polysomnograph records; EM-rotation, EM-velocity, EM-power. Lower half comparisons of the magnitude of EMs of isolated EMs to those in bursts. EMs in bursts were larger, rotated more quickly and had greater power than isolated EMs.

REM intensity and magnitude of single EM

It is possible that the magnitude of single EM reflect the intensity of REM sleep. Examination of REM bursts may help clarify the physiological meaning of the magnitude of EM in the concept of REM sleep intensity. Within a REM episode, EM frequency and magnitude always change in the same direction; during a REM burst both have high values, while EMs outside of bursts have low values. During REM bursts EOG activity may reflect particular patterns of brain neuron activity. For example, within REM bursts recorded from animals, PGO frequency closely corresponds to EM-frequency and is highly correlated with the intensity of neuronal activity. Comparisons of the relationships between precise EM measurements and findings such as PGO wave activity might strengthen comparisons of animal studies and human polysomnography (McCarley et al., 1983; Atsumi et al., 1987).

Figure 5 shows the changes in parameters of single EM and EM-frequency across REM episodes for all 40 recordings. By contrast with the an ascent-to-right pattern observed in EM-frequency, EM-rotation increased significantly from the first to its peak value in the second REM episode, it then decreased in the third REM episode, making an inverted-V-shaped pattern. EM-velocity and EM-power had the same pattern as EM-rotation. Thus, the most vigorous rapid eye movements occurred in the middle of the sleep period. In relation to REM sleep intensity, what do these new measures represent? Do they represent the intensity of REM sleep in ways comparable to conventional REM density? If so, these new measures should parallel conventional measures with an ascent-to-right increase across REM episodes. Alternatively, the magnitude of EMs might reflect mechanisms or processes different from conventional measures of REM sleep intensity. This seems to be the case, given the discrepancies between our new measures and conventional parameters.

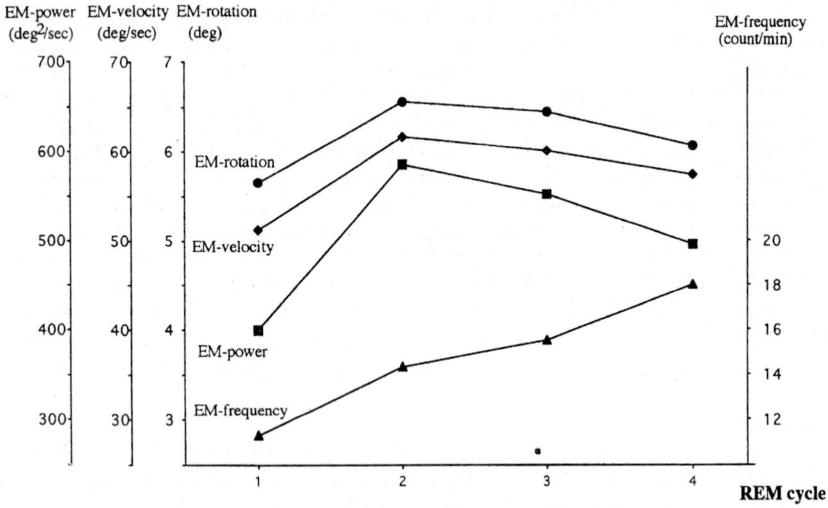

Fig. 5. Magnitude of EM-rotation: Changes in the duration of EM-frequency, EM-rotation, EM-velocity and EM-rotation across REM episodes for 40 night recordings. EM-frequency increased in an ascent-to-right pattern. By contrast, while EM-rotation, EM-velocity and EM-power had its peak value in the second REM episode, making an inverted-v shaped pattern. These three parameters represent magnitude of single EMs.

Discrepancy of Two Groups of Parameters

Discrepancy suggests the magnitude of single EM reflects physiological mechanisms which are not described by the parameters conventionally used to describe REM sleep intensity. One of the possible speculation is that; in the framework of reciprocal theory of NREM and REM sleep, both NREM intensity and REM intensity is highest at the beginning of sleep, however, because of priority of NREM sleep exhibition of REM sleep delayed. REM episode and REM density are the expression of the result of these antagonism. On the other hand, magnitude of single EM (EM-rotation) is the immediate expression of intensity of neuronal activity related to REM sleep.

Combinations of these measures could be used as parameters for investigations of mental activity during REM sleep. It is known that REM intensity can reflect different mental disorders such as schizophrenia (Reich et al., 1985) and depressive illness (Kupfer, 1978; Reynolds et al., 1990; Foster et al., 1976; Hudson et al., 1987). Automated measurement of the magnitude of single EMs in these mental disorders could provide much additional information in future such studies of REM sleep. New brain imaging technologies such as PET, functional MRI and near-infrared spectroscopy will soon reveal images of neuronal activity during REM sleep (Hoshi et al., 1994; Maquet et al., 1996). Precise EM measures could be used as concurrent measures of REM intensity.

References

Antonioli, M., Solano, L. and Torre, A., et al. Independence of REM density from other REM sleep parameters before and after REM deprivation, Sleep, 4 (1981) 221–5.

Aserinsky, E. The maximal capacity for sleep: rapid eye movement density as an index of sleep satiety, Biol. Psychiatry, 1 (1969) 147–59.

Aserinsky, E. Rapid eye movement density and pattern in the sleep of normal young adults, Psychophysiology, 8 (1971) 361–75.

Atsumi, Y., Tanaka, K., Shiraishi, H., Kojima, T. and Takahashi, R. Human REMs related potentials in REM sleep and lambda complex, Clin. Electroencephalogr., 29 (1987) 370–75.

Beerson, D.G.M., Dijk, D.J., Block, C.G.H. and Everhardus, I. REM sleep deprivation during 5 hours lead to an immediate REM sleep rebound and to suppression of non-REM sleep intensity, Electroencephalogr. Clin. Neurophysiol., 76 (1990) 114–22.

Benoit, O., Parot, S. and Garma, L. Evolution during the night of REM sleep in man, Electroencephalogr. Clin. Neurophysiol., 36 (1974) 245–251.

Borbéy, A.A. A two process model of sleep regulation, Human Neurobiology, 1 (1982) 195–204.

Coble, P.A., Reynolds, C.F., Kupfer, D.J. and Houck, P. Electroencephalographic sleep of healthy children. Part II: Findings using automated delta and REM sleep measurement methods, Sleep, 10 (1987) 551–562.

Feinberg, I. Changes in sleep cycle patterns with age. J. Psychiatr. Res., 10 (1974) 283–306.

Feinberg, I., Floyed, T.C. and March, J.D. Effect of sleep loss on delta EEG and eye movement density, Electroencephalogr. Clin. Neurophysiol., 67 (1987) 217–21.

Foster, F.G., Kupfer, D.J., Coble, P. and McPartland, R.J. Rapid eye movement sleep density, Arch. Gen. Psychiatry, 3 (1976) 1119–23.

Geisler, P., Meier-Ewert, K. and Matsubashi, K. Rapid eye movement, muscle twitches and sawtooth waves in the sleep of narcoleptic patients and controls, Electroencephalogr. Clin. Neurophysiol., 67 (1987) 499–507.

Hoshi, Y., Mizukami, S. and Tamura, M. Dynamic features of hemodynamic and metabolic changes in the human brain during all-night sleep as revealed by near-infrared spectroscopy, Brain Res., (1994) 257–62.

Hudson, J.I., Pope, H.G., Jonas, J.M., Stakes, J.W., Grochocinski, V., Lipinski, J.F. Kupfer, D.J. Sleep EEG in Bulimia, Biol. Psychiatry, 22 (1987) 820–8.

McCarley, R.W., Winkelman, J.W. and Duffy, F.H. Human cerebral potentials associated with REM sleep rapid eye movements: links to PGO waves and waking potentials, Brain Res., 274 (1983) 359–64.

Kobayashi, T., Fluctuation of REM activity as analyzed by REM moment. Sleep 1980, 5th Eur. Congr. Sleep Res., (1981) 311–13.

Kobayashi, T., Okuno, H., Endo, S. and Tsuji, Y. Computer analysis of rapid eye movements in REM sleep period, Iyodenshi-to-Seitaikogaku, 17 (1979) 222–29.

Kupfer, D.J. Application of EEG sleep for the differential diagnosis and treatment of affective disorders, Pharmakopsychiatr. Neuropsychopharmakol; 11 (1978) 17–26.

Maquet, P., Peters, J., Aerts, J., Delfiore, G., Dagueldre, C., Luxen, A. and Franck, G. Functional neuroanatomy of human rapid-eye movement sleep and dreaming, Nature, 383 (1996) 163–6.

McPartland, R.J., Kupfer, D.J., Coble, P., Shaw, D. and Spiker, D.G. An automated analysis of REM sleep in primary depression, Biol. Psychiatry, 14. (1979) 767–76.

Rechtschaffen, A., Kales, A. (Eds.), A manual of standardized terminology, techniques and scoring system for sleep stages of human subjects. Los Angeles: Brain Information Service/Brain Research Institute, University of California, 1968.

Reich, L., Weiss, B.L., Coble, P., McPartland, R. and Kupfer, D.J. Sleep disturbance in schizophrenia, Arch. Gen. Psychiatry, 32 (1985) 51–5.

Reynolds, C.F., Buysse, D.J., Kupfer, D.J., Hoch, C.C., Houck, R.R., Matzzie, J. and George, C.J. Rapid eye movement sleep deprivation as a probe in elderly subjects, Arch. Gen. Psychiatry, 47 (1990) 1128–36.

Reynolds, C.F., Kupfer, D.J., Taska, L.S., Hoch, C.C., Sewitch DE. and Spiker, D.G. Sleep of Healthy Seniors: A Revisit, Sleep, 8 (1985) 20–92.

Takahashi, K. and Atsumi, Y. Precise Measurement of Individual Rapid Eye Movements in REM Sleep of Humans, Sleep, 20 (9) (1997) 743–752.

Vitiello, M.V., Bokan, J.A., Kukull, W.A., Muniz, R.L., Smallwood, R.G. and Prinz, P.N. Rapid eye movement sleep measures of Alzheimer's-type dementia patients and optimally healthy aged individuals, Biol. Psychiatry, 19 (1984) 721–34.

Zimmermann, J.C., Czeisler, C.A. Weitzman, E.D. REM density is dissociated from REM sleep timing during free-running sleep episodes, Sleep, 2 (1980) 409–15.

Rapid Eye Movement Sleep
B.N. Mallick and S. Inoué (Eds)

27. Why We Believe What We Believe About REM-Sleep Regulation

Joel H. Benington

Department of Biology, Saint Bonaventure University, Saint Bonaventure, NY 14778, USA

Introduction

As described elsewhere in this volume, REM sleep is a highly activated state of the brain, comparable to the most active waking. Yet in this state, the subject lies quiescent and inattentive to the outside world. Its activated brain instead attends to a bizarre progression of dream imagery. To initiate REM sleep, the brain must become intensely activated without the subject's waking up. Specific physiological mechanisms appear to have evolved to make this possible. And even though these mechanisms appear to work well enough, the transition to REM sleep is often interrupted at the very beginning of a REM-sleep episode, making the brain return to nonREM sleep and try again.

The alternation between nonREM sleep and REM sleep is clearly an involved process, and what we already know of the physiological mechanisms underlying it indicates that they are complex and carefully regulated. The existence of such a process in a living thing implies adaptation for some purpose. It is highly unlikely that the vast majority of mammals would express REM sleep in such a characteristic and well regulated manner without some very good reason. And yet it is fair to say that we do not now have any clear idea of what the function of REM sleep is. This is not for want of trying: a number of hypotheses have been advanced over the years, but no hypothesis has been experimentally confirmed with enough authority to cause it to be preferred over all others.

There are two ways one could go about attempting to overcome our current ignorance of the function of REM sleep:

(1) one could propose a specific, physiologically detailed new hypothesis concerning the function of REM sleep and thereby hope to solve the problem at one stroke, or

(2) one could proceed more cautiously, considering first what we *do* confidently know about REM sleep and thereby narrow down the range of possible functions that REM sleep *could* serve.

The former approach has its uses. Even though most such hypotheses do not turn out to be right, they stimulate research and seldom fail to advance understanding, often in unexpected ways. If I had such a hypothesis to offer, I would certainly present it for consideration. The latter approach is not as flashy, but in lieu of plausible, specific hypotheses, it is probably our best approach to shedding some light on this difficult subject.

This chapter, re-examines the current thinking about REM sleep. and would argue the following:

(1) A consensus model of REM-sleep regulation has emerged over the past 40 years.
(2) This model was not the necessary conclusion of a series of experimental tests, but was instead agreed upon in a surprisingly casual way, presumably because it appeared intuitively obvious.
(3) The phenomena associated with REM sleep have come to be interpreted in light of this model, causing it to appear to be better substantiated than it is.
(4) There are in fact a number of robust findings that argue strongly against this model, but the consensus in favor of the currently dominant model of REM-sleep regulation is so strong that they have not had much impact on people's thinking.

Current View of REM-Sleep Regulation

The following ideas about REM-sleep regulation would be accepted readily by the vast majority of sleep researchers:

(A) REM sleep is homeostatically regulated

This implies that the amount of REM sleep expressed per day is regulated by mechanisms that increase REM-sleep expression after a period of deprivation. This is often modeled in terms of an increase in the 'need' or 'propensity' for REM sleep over time and the discharge of this need or propensity in REM sleep.

(B) REM sleep is physiologically necessary

As a rule, inconsequential events and processes are not homeostatically regulated. Temperature, water balance, and ion concentrations are regulated in mammals because physiological performance is reduced if they are not maintained within a narrow range of values. The existence of homeostatic regulation of REM-sleep expression implies that REM sleep is serving some essential function.

(C) The function of REM sleep is somehow related to waking (perhaps to the cognitive activities of waking)

The majority of hypothesized functions for REM sleep, and in particular the ones that receive the most attention have been waking-related. The idea that REM sleep is necessary for optimal cognitive function in waking is especially favored in the popular literature on sleep and has been the basis of a number of related hypotheses in the scientific literature (e.g., Crick and Mitchison, 1983; Jouvet, 1975). These hypotheses have spawned a great deal of biopsychological research, some of which has been published in prestigious journals quite recently (e.g., Karni et al., 1994).

(D) Homeostatic regulation of REM sleep must occur in relation to waking

This follows from (A) and (D). If the restorative function of REM relates to the activities of waking, then the need for REM sleep must accumulate in relation to the duration and/or quality of prior waking.

(E) Within sleep, the timing of REM sleep is controlled by an oscillator

The idea that there is a nonREM/REM sleep cycle is firmly entrenched in the literature. Generally, the implication is that the cycle is controlled by some sort of biological oscillator with an ultradian period, analogous perhaps to the circadian clock in the suprachiasmatic nucleus (SCN) in mammals.

Evidence for the Current View of REM-Sleep Regulation

These ideas are so well established, and so much research has been based on one or more of them, that many sleep researchers may not have given much thought to the nature of the evidence for them. As these ideas have all been around for more than 30 years, the initial findings that inspired them are buried in the early literature on sleep, in articles that are seldom consulted any longer. Since my research has suggested hypotheses that are at variance with some components of the current view of REM-sleep regulation, I have surveyed the earlier literature to determine how well founded those component ideas are. I have reviewed that literature extensively elsewhere (Benington, 1992; Benington and Heller, 1994a). In this article, I will only highlight the most important points in that development.

Let us take each of the above five ideas in turn and consider their antecedents.

(A) REM sleep is homeostatically regulated

Evidence for this idea came originally in 1960 with Dement's seminal study of selective REM-sleep deprivation in humans. That study demonstrated two things that have continued to be the basis for this idea ever since: (1) REM-sleep deprivation is followed by a compensatory 'rebound' increase in REM-sleep expression, (2) there is an increase in the frequency of attempts to enter REM sleep as the amount of REM-sleep deprivation increases. These findings have since been replicated repeatedly in other studies in both animals and humans. No other evidence of homeostatic regulation of REM sleep even approaches the robustness of these findings.

(B) REM sleep is physiologically necessary

On one hand, this idea could be seen as following logically from (A), as in the initial statement of this idea above. There have in addition been a number of attempts to demonstrate adverse consequences of REM-sleep deprivation. In Dement's original REM-sleep deprivation study (1960), there are allusions to mild 'behavioral changes', including anxiety, irritability, and difficulty in concentrating. There have since been untold studies purporting to demonstrate definite behavioral consequences of REM-sleep deprivation (for review, see Lewin and Singer, 1991). Most of these studies have been inspired by the general idea that REM sleep is somehow necessary for normal cognitive function. Another sort of evidence that REM sleep is physiologically necessary comes from long-term REM-sleep deprivation studies performed at the University of Chicago, in which rats die after several weeks of 'selective' REM-sleep deprivation (Kushida et al., 1989).

(C) Function of REM-sleep regulation is waking-related

This idea is in many ways the most interesting and pivotal element of the widely accepted of REM-sleep regulation, but there is little or no compelling evidence for this idea. At the very least, we can say that this idea was dominant very early, before any of the evidence that may now be advanced in its defense had been collected. It is already implicit in Dement's original REM-sleep deprivation study (1960), at a time when the only evidence at all for it was Dement's anecdotal reports of mild 'behavioral changes' associated with REM-sleep deprivation. It is also striking that no review I have seen of any aspect of REM sleep has bothered to defend this idea against any alternative. It apparently has just been accepted as a given by the sleep research community. This is not so surprising, since REM sleep is after all a sleep state, and one naturally supposes that sleep serves some function in relation to waking. But one does not expect the scientific community to arrive at conclusions in so casual a manner.

(D) Homeostatic regulation of REM sleep is in relation to waking

This idea is only reasonable if we accept (A) and (C). Given that homeostatic regulation can be modeled in terms of the accumulation and discharge of a 'need', if the function of REM sleep relates to waking, then the 'need' for REM sleep should accumulate in waking. If, however, the need for REM sleep can be shown to accumulate in nonREM sleep rather than waking, then we should logically conclude that the function of REM sleep concerns some aspect of nonREM sleep rather than waking. I was led to reconsider the current view of REM-sleep regulation by evidence that REM-sleep expression is homeostatically regulated in relation to nonREM sleep rather than waking.

(E) Within sleep, the timing of REM sleep is controlled by an oscillator Simply put, the evidence for this idea is that REM-sleep episodes occur at fairly regular intervals. The 'rhythmic' occurrence of REM sleep naturally suggests the idea of a 'sleep cycle', which in turn suggested the existence of a biological oscillator, along the lines perhaps of the circadian pacemaker in the SCN. However, an oscillator is only one way to produce a 'cycle', and not necessarily the most likely one. The following considerations suggest that the idea of an oscillator-controlled sleep cycle is worth rethinking.

(i) Seemingly rhythmic phenomena can be caused by processes that would not usually be described as oscillators. For example, common behaviors such as eating, drinking, and sexual activity will occur at fairly regular intervals in a constant enough environment. Like most biological phenomena that exhibit apparent rhythmicity, these are controlled homeostatically, by an accumulation of a need for gratification. A familiar non-biological example of such a process would be a refrigerator, which turns on at very regular intervals without being controlled by what one would usually call an oscillator. Rather, as long as the need in question accumulates and is discharged at constant rates, the system will 'cycle' between states at fairly regular intervals. A true oscillator such as the circadian clock in the SCN is seldom the mechanism underlying apparently rhythmic behaviors in biological systems.

(ii) At best, the timing of REM sleep could be said to be 'fairly regular'. It is not rigidly pacemaker-like by any means. Dement and Kleitman's original report on the electrographic manifestation of nonREM sleep vs. REM sleep (1957) illustrates the timing of REM-sleep episodes in a series of nights from a number of human subjects, and there is considerable variation in 'cycle' lengths. Other studies have shown that the standard deviation of sleep cycle length is typically about 1/4 of the mean, the second and third cycles during the night in humans are consistently longer than the first and fourth, and sleep cycle length is even more variable in non-human mammals (for review, see Benington, 1992). As one person has observed, "It seems unreasonable to imply rhythmicity when to account for 95% of the distribution, the range has to be of the order of 50–130 min" (Lewis, 1974)

(iii) The idea of an oscillator-controlled sleep cycle was personally important to Kleitman as a result of his highly speculative hypothesis of a basic rest-activity cycle (BRAC). According to this hypothesis, which was championed by Kleitman in spite of little or no evidence, the sleep cycle is simply one manifestation of an ultradian rhythm that persists during waking as well (for review, see Benington, 1992). It appears as though Kleitman's enthusiasm for the idea of an ultradian rhythm as a unifying explanation for the timing of REM sleep as well as other periodic phenomena in both humans and infants may have induced him to stress the idea of an oscillator in the absence of any concrete evidence for one.

(iv) The idea of an oscillator-controlled sleep cycle was further reinforced when Hobson and McCarley (1975) proposed a neurochemical mechanism for the cycle, involving reciprocal interactions between brainstem cholinergic and monoaminergic neurons. This hypothesis continued to garner attention in the sleep research community for many years, long after the original findings that inspired it had been discredited. The very fact that the most prominent physiological model of REM-sleep regulation was oscillatory in nature was bound to perpetuate the idea of an oscillator-controlled sleep cycle.

(v) As long as it is thought that REM sleep is homeostatically regulated in relation to waking, then either the sleep cycle is indeed controlled by an oscillator, or there must be two types of REM-sleep need, one that accumulates during waking and manifests itself in a REM-sleep rebound following deprivation, and another that accumulates during nonREM sleep and is responsible for the timing of REM sleep during a period of sleep. If, on the other hand, REM sleep is homeostatically regulated in relation to nonREM sleep, then the situation is vastly simplified, and both the amount and timing of REM sleep are controlled homeostatically by the same accumulation of REM-sleep need.

In brief, the evidence for (A) is quite good, (B) is less well established but seems perfectly reasonable given (A), and (C), (D), and (E) are poorly substantiated, having been established within the sleep research community virtually by fiat. This is not to say that they are not true, but it should encourage us at least to consider alternatives. Logically, the most straightforward alternative is that the function of REM sleep relates to nonREM sleep rather than to waking, that homeostatic regulation of REM-sleep expression is likewise in relation to nonREM sleep, and that the timing of REM sleep is controlled by accumulation of REM-sleep need during nonREM sleep.

The Hypothesis that REM Sleep is Functionally Related to NonREM Sleep

I have presented this hypothesis in detail elsewhere (Benington and Heller, 1994a), and have in that publication cited the other authors that have presented similar hypotheses over the past 30 years. In this chapter, I will give only a summary accounting of the evidence for this view.

First, the states of REM sleep and waking are nearly identical neurophysiologically, whereas REM sleep and nonREM sleep are quite different. It is difficult to imagine how REM sleep could be physiologically necessary to the optimal expression of another state that is as similar to it as waking is.

Not surprisingly, the functions that have been hypothesized for REM sleep generally involve either: (1) higher-order cognitive functions such as learning and memory which could be manifested quite subtly relative to the basic neurophysiological characteristics of a behavioral state (Crick and Mitchison, 1983; Jouvet, 1975), or (2) one of the few neurophysiological aspects of brain functioning that actually differ between waking and REM sleep, such as the activity of noradrenergic neurons (Siegel and Rogawski, 1988).

Second, the preponderance of evidence suggests that the regularity of the interval between REM-sleep episodes is measured in terms of total nonREM sleep time elapsed rather than total time elapsed (for review, see Benington and Heller, 1994a). Moreover, the 'clock' is not reset by waking. So either there is an oscillator that only runs during nonREM sleep and holds its place during intervening waking, or the timing of REM sleep is determined by accumulation of a need for REM sleep during nonREM sleep. The latter seems the more obvious explanation Yet so dominant was the idea of an oscillator-controlled cycle that the authors of these studies all referred to their hypothesis in terms of a 'sleep-dependent cycle'

In my own work, the interval between REM-sleep episodes in the rat has been shown to be punctuated often by one or more abortive attempts to enter REM sleep (Benington and Heller, 1994b). The apparent rhythmicity in REM-sleep timing (when measured according to the occurrence of REM-sleep episodes lasting longer than 30 sec) is underlain by a much less rhythmic occurrence of transitions to REM sleep. Strikingly, there is a very consistent, linear increase in the incidence of these transitions as a function of the time elapsed since the last REM-sleep episode of more than 30 sec. This strongly suggests that a propensity for REM sleep is accumulating progressively between REM-sleep episodes.

In another study that I worked on, drive to enter REM sleep in rats (as measured by the frequency of attempts to enter REM sleep), was found to increase rapidly during just two hours of highly selective REM-sleep deprivation in the rest period (Benington et al., 1994). At the end of the two-hour deprivation period, transitions to REM sleep occurred 44–60 times per hour of nonREM sleep. This suggests that just two hours of rest-period REM-sleep deprivation produces markedly elevated levels of REM-sleep drive, considerably above anything seen under normal sleep/wake conditions. This is consistent with the idea that the need for REM sleep accumulates during nonREM sleep and is normally discharged (in the rat) at circa 11 min intervals. It is difficult to reconcile with the idea that REM-sleep need accumulates over many hours of waking and is discharged over a series of REM-sleep episodes in each sleep period.

Finally, recent analyses of the ontogeny and phylogeny of sleep in mammals have suggested that nonREM sleep and REM sleep emerge in concert from an undifferentiated sleep state that is identical with neither the one nor the other. Marcos Frank and co-workers (1997) have suggested that 'active sleep' in pre-weanling rats is not REM sleep but that the electrographic characteristics of both slow-wave (nonREM) sleep and REM sleep emerge from a background of relatively undifferentiated EEG waveforms, beginning just before weaning. These findings suggest that nonREM sleep and REM sleep may be functionally interrelated, as opposed to the widespread view that each serves its own independent function in relation to waking.

Jerome Siegel and co-workers have recently published analyses of EEG recordings from two monotremous mammals, the echidna (Siegel et al., 1996) and platypus (Siegel et al., 1997), as well as single-unit recordings from the brainstem of the echidna (Siegel et al., 1996). Although there are differences in the manifestations of sleep states in these two species, Siegel et al. describe echidna sleep as a single state having characteristics of both placental nonREM and REM sleep, while they report that platypus sleep consists predominantly of a strikingly REM sleep-like state, characterized by muscle atonia, phasic EMG activity, and rapid eye movements. The one noteworthy difference between REM sleep in the platypus and placental REM sleep is that the EEG is of moderate to high amplitude in the platypus, "with cnsistently more power in all of the frequency bands assessed than that during waking states." In this respect, the REM sleep of the platypus is similar to the indeterminate sleep state of the echidna, only with more REM sleep-like motor and oculomotor manifestations. Siegel et al. scored 60% of platypus sleep time as REM sleep, based on EMG, EOG and ECG activity. The state making up the remainder of platypus sleep is not, however, markedly different from platypus REM sleep: EMG, EOG and ECG recordings exhibit some of the same REM sleep-like manifestations as in platypus REM sleep but to a lesser degree, and the EEG remains at moderate to high amplitude (J. Siegel, personal communication).

Thus, both echidna and platypus sleep lack the marked dichotomy between states that is seen in placental nonREM and REM sleep. And the more homogeneous sleep of both echidna and platypus exhibits characteristics typical of both nonREM sleep and REM sleep in placental mammals. It is not unreasonable to suppose that the more homogeneous sleep of monotremous mammals, like neonatal sleep, may represent a primordial condition out of which both nonREM sleep and REM sleep have emerged together. In the more primordial condition, there may be no need for a separate, highly activated state such as placental REM sleep, because the sleep state seen in monotremous mammals and neonates may not be characterized by as marked hyperpolarization and synchronization of neuronal circuits as placental nonREM sleep.

Conclusion

The hypothesis that the function of REM sleep concerns nonREM sleep rather than waking is a far cry from a concrete physiological hypothesis as to what that function is. However, the sort of function that would cause a need for REM sleep to accumulate during nonREM sleep is very different from one that would cause a need for REM sleep to accumulate during waking. If we as a community neglect one entire class of potential functions of REM sleep, we may make no progress towards the sort of concrete physiological hypothesis that will tell us in detail what REM sleep is for. I earnestly hope that this chapter will encourage more detailed consideration of what physiological aspects of nonREM sleep could result in a need for expression of the very different state of REM sleep.

References

Benington, J.H., REM-sleep homeostasis in the rat. In: *Department of Biological Sciences,* Stanford University, Stanford, California, 1992, pp.160.

Benington, J.H. and Heller, H.C. Does the function of REM sleep concern non-REM sleep or waking? Prog. Neurobiol., 44 (1994a) 433–449.

Benington, J.H. and Heller, H.C. REM-sleep timing is controlled homeostatically by accumulation of REM-sleep propensity during non-REM sleep, Am.J. Physiol., 266 (1994b) R1992–R2000.

Benington, J.H., Woudenberg, M.C. and Heller, H.C. REM-sleep propensity accumulates during two-hour REM-sleep deprivation in the rest period in rats, Neurosci. Lett., 180 (1994) 76–80.

Crick, F. and Mitchison, G. The function of dream sleep, Nature, 304 (1983) 111–113.

Dement, W., The effect of dream deprivation, Science, 131 (1960) 1705–1707.

Dement, W.C. and Kleitman, N. Cyclic variations in EEG during sleep and their relation to eye movements, body motility, and dreaming, Electroenceph. Clin. Neurophysiol., 9 (1957) 673–690.

Frank, M.G. and Heller, H.C. Development of REM and slow wave sleep in the rat, Am. J. Physiol., 272 (1997). R1792–9.

Hobson, J. A. McCarley, R.W. and Wyzinski, P.W. Sleep cycle oscillation: reciprocal discharge by two brainstem neuronal groups, Science, 189 (1975) 55–58.

Jouvet, M. The function of dreaming: a neurophysiologist's point of view. In: *Handbook of psychobiology* (Eds.), M.S. Gazzaniga and C. Blakemore, Academic Press, New York, 1975, pp. 499–527.

Karni, A., Tanne, D., Rubenstein, B.S., Askenasy, J.J. and Sagi, D. Dependence on REM sleep of overnight improvement of a perceptual skill, Science, 265 (1994) 679–682.

Kushida, C.A., Bergmann, B.M. and Rechtschaffen, A. (1989) Sleep deprivation in the rat: IV. Paradoxical sleep deprivation, Sleep, 12 (1989) 22–30.

Lewin, I. and Singer, J.L. Psychological effects of REM ("dream") deprivation upon waking mentation. In: *The mind in sleep: psychology and psychophysiology,* (Eds.), S.J. Ellman and J.S. Antrobus, John Wiley and Sons, Inc., New York, 1991, pp. 396–412.

Lewis, S., (1974) The paradoxical sleep cycle revisited. In: *Chronobiology* (Eds.), L.E. Scheving, F. Halberg and J.E. Pauly, Igaku Shoin, Tokyo, 1974, pp. 487–490.

Siegel, J.M., Manger, P.R., Nienhuis, R., Fahringer, H.M. and Pettigrew, J.D. The echidna Tachyglossus aculeatus combines REM and non-REM aspects in a single sleep state: implications for the evolution of sleep, J. Neurosci., 16 (1996) 3500–6.

Siegel, J.M., Manager, P.R., Nienhuis, R., Fahringer, H.M. and Pettigrew, J.D. The platypus has REM sleep, Sleep Res., 26 (1997) 177.

Siegel, J.M. and Rogawski, M.A. A function for REM sleep: regulation of noradrenergic receptor, sensitivity, Brain Res. Rev., 13 (1988) 213–233.

Rapid Eye Movement Sleep
B.N. Mallick and S. Inoué (Eds)

28. Hypothesis on REM Sleep from the Viewpoint of Neuronal Dynamics

Mitsuyuki Nakao and Mitsuaki Yamamoto

Laboratory of Neurophysiology and Bioinformatics, Graduate School of Information Sciences, Tohoku University, Sendai 980-8579, Japan

SUMMARY

The dynamics-transition of neuronal activities from flat (in slow wave sleep) to the $1/f$ (in REM sleep) power spectral density profile in a low frequency band was found during sleep cycle in the extended areas of the cat's brain. The simulation study was successfully performed using an interconnected neural network model in which each neuron was assumed to have an inhibitory postsynaptic input common to all neurons mimicking the input of an overall biasing from the aminergic/ cholinergic modulation system. Especially for the $1/f$ fluctuations during REM sleep, a statistical analysis showed that occurrence of long-lasting activations was essential for neurons to exhibit this dynamics, which, as for the network model, was achieved by reducing the postsynaptic inhibition. Under the weak inhibition, the network attractor was shown to have the metastability, in which there existed the several quasi-stable states (called metastable states) in the network. Under this condition, the network state was shown to be wandering among these metastable states, which underlies the 1/f fluctuations in the neuronal activities simulated. Because our modeling framework was also supported by pharmacological experiments under p-chlorophenylalanine treatment, thus obtained mechanism could be expected to apply to the actual neural network. This suggested metastability of actual neural networks during REM sleep might serve as a physiological basis responsible for "dreaming" which might contribute to the process of the consolidation and maintenance of memories.

Key Words: neuronal dynamics, $1/f$ fluctuations, REM sleep, slow wave sleep, metastability, network attractor, dynamics-transition, dreaming.

Introduction

There have been so many hypotheses on the physiological significance of rapid eye movement sleep (REM sleep, REMS), which are based on the physiological and pharmacological results. They include contributions of REM sleep to development and maintenance of neural circuits in the brain, survival in the severe environment, and higher-order brain function such as learning and memory (Hartmann, 1973; Horne, 1988). However, none of them has been established extensively. An appropriate model mediating between the results and the hypotheses may be necessary to show the reality of the hypotheses.

We have investigated dynamics of spontaneous neuronal activities recorded from various areas of the cat's brain during sleep. As a result, we discovered that slow neuronal dynamics in the widespread areas of the brain show the transition between the flat (or white) power spectral density (PSD) during slow wave sleep (SWS) and the $1/f$ PSD during REM sleep (Yamamoto et al., 1986a; Mushiake et al., 1988; Kodama et al., 1989a), which is called "dynamics-transition" in this paper. Here, the phenomenology of the dynamics-transition of single neuronal activities during sleep is first reviewed. Secondly, a possible generation mechanism of the neuronal dynamics is explored based on the neural network model in which each neuron has an inhibitory postsynaptic input common to all neurons. In addition, from the statistical point of view, an essential property of a time series is studied for the $1/f$ fluctuations of both the actual and modeled neuronal activities. Finally, the generation mechanism underlying the $1/f$ fluctuations is explored in the modeled neural network, where neurons are nonlinearly interacting. Through these studies, what is happening in actual neural networks is conjectured from the simulation results, and its possible contribution to the brain functions is discussed.

Dynamics-Transition of Neuronal Activities During Sleep

The phenomenon of the dynamics-transition was first demonstrated in the long-term spontaneous single neuronal activity of the mesencephalic reticular formation (MRF) of the cat during sleep (Yamamoto et al., 1986a). During slow wave sleep, a random spike density pattern was very often observed and almost a flat PSD was obtained for the low frequency range from approximately 0.01-1 Hz. On the other hand, for almost the same frequency range during REMS, a slowly fluctuating spike density pattern was observed and the PSD was inversely proportional to frequency. This kind of PSD is simply called the "$1/f$ PSD". Thus, an important characteristic of the cat's MRF neuronal activity is the "dynamics-transition" between the flat PSD during SWS and the 1/f PSD during REMS. The robustness of the characteristic of dynamic-transition between the 2 sleep states was verified for the 18 MRF neurons (Yamamoto et al., 1986a). Furthermore, the consistent characteristic of the transition has also been verified on a time axis during 24 h for a representative MRF neuron (Yamamoto et al., 1997) (Fig. 1). Each PSD has been calculated for a sample series of the neuronal activity extracted from all sustained episodes of both SWS and REMS. In the all episodes of the respective states, the MRF neurons display the flat and the $1/f$ PSD during SWS and REMS, respectively.

The $1/f$ PSD is more generally evaluated by a form of $1/f^b$ profile where exponent b is the slope of the PSD in a double logarithmic plot. When $b = 1$ the PSD corresponds to the exact 1/f profile, whereas $b = 0$ corresponds to the flat PSD. In Figure 1, the values of b are displayed in the upper part of

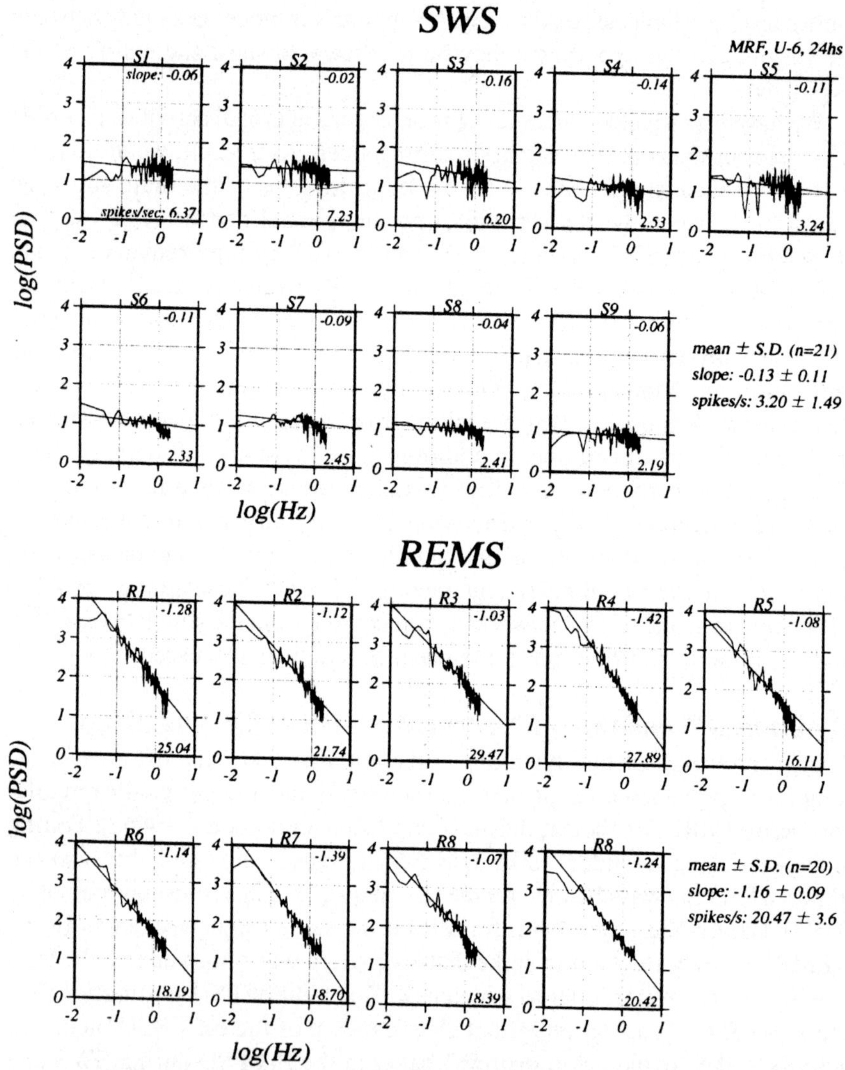

Fig. 1. Demonstration of the robustness of the neuronal dynamics-transition between episodes of slow wave sleep (SWS) and REM sleep (REMS) during 24 h for a single mesencephalic reticular formation (MRF) neuron. The MRF neuronal dynamics displays the white and the 1/*f* spectrum during slow wave sleep and REM sleep, respectively. 9 of 21 episodes for SWS, and 9 of 20 for REMS, were sequentially selected and displayed from the beginning, respectively. This neuron was healthy for more than a week. The ordinate indicates logarithm of power spectral density (PSD) in (spikes/s)^{2}Hz. The abscissa shows logarithm of frequency (Adapted from Yamamoto et al., 1997).

each PSD, with a mean (– 0.13) for the 21 stationary episodes of SWS, and with a mean (– 1.16) for the 20, REMS spisodes, being a small variability between episodes for the respective states.

As for the variability of the value of *b* during REMS among neurons, the mean values calculated from 19 MRF neuronal activities were distributed from – 0.56 to – 1.37 with a pooled mean, – 0.96. Each neuron tended to have its own value, which was hypothesized to indicate the structural specificity of the reticular network including the neuron under observation (Yamamoto, 1994a).

Figure 2 shows a summary of the dynamics-transition of single neuronal activities between the 2 sleep states in the 5 neuronal groups: the mesencephalic reticular formation, the primary somatosensory cortex, the ventrobasal complex of the thalamus, and the hippocampus (the theta and the pyramidal neurons) (Yamamoto et al., 1986b; Yamamoto et al., 1994b). In addition, the neurons recorded from the neo-cortex, areas 6 and 18, from the dorsal lateral geniculate nucleus region, have recently showed the similar characteristic (Yamamoto et al., 1995; Takahashi et al., 1997). It should be emphasized that all of these neuronal groups show a similar phenomenon of the dynamics-transition, despite of the structural and functional differences of the neuronal groups. Some common modulation mechanisms must operate under the dynamics-transition.

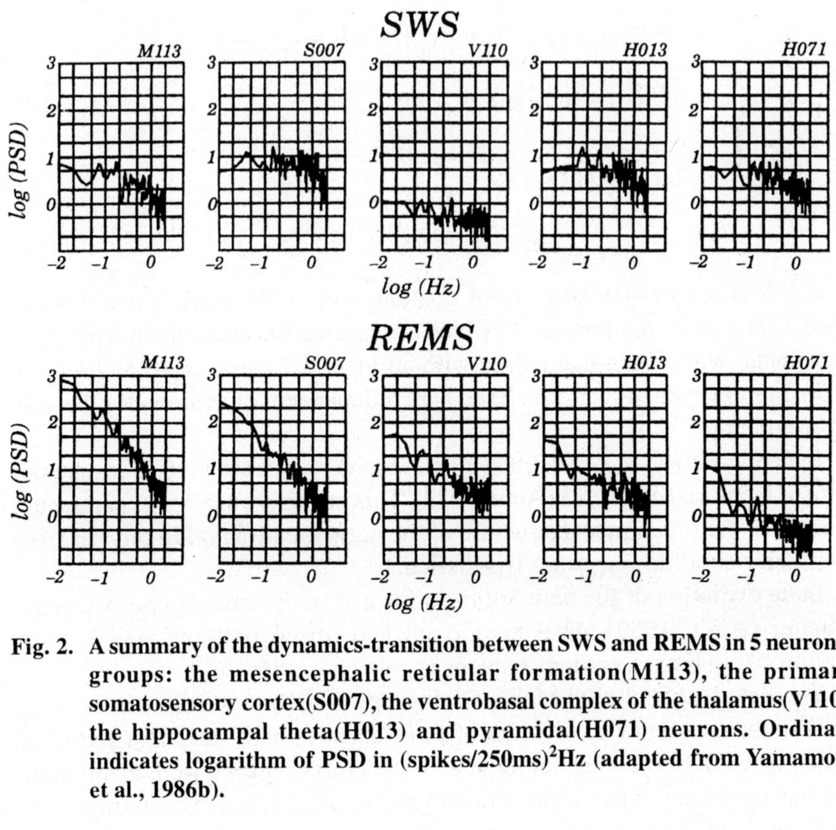

Fig. 2. A summary of the dynamics-transition between SWS and REMS in 5 neuronal groups: the mesencephalic reticular formation(M113), the primary somatosensory cortex(S007), the ventrobasal complex of the thalamus(V110), the hippocampal theta(H013) and pyramidal(H071) neurons. Ordinate indicates logarithm of PSD in $(\text{spikes}/250\text{ms})^2$Hz (adapted from Yamamoto et al., 1986b).

Neural Network Model of the Dynamics-Transition

It is interesting to investigate whether or not the dynamics-transition could originate from a neural network. We performed a simulation study using an asynchronous interconnected neural network model with a modulation input (Nakao et al., 1990). Each neuron was hypothesized to have an inhibitory postsynaptic input common to all neurons of the network to simulate the overall biasing from the global neuromodulation system such as the serotonergic, noradrenergic, and cholinergic system. The effect of overall biasing from the global neuromodulation system was assumed to be condensed into this common value of the postsynaptic inhibition. Furthermore, mutually independent random perturbations were applied to activate each neuron, mimicking various noise sources rapidly fluctuating.

Neural network model

Here, the model structure is reviewed briefly (Nakao et al., 1990). The neural network model consists of fully interconnected neuron-like elements (abbreviated as "neuron"). For the i-th neuron, the state evolution rule is defined as follows.

$$u_i(t+1) = \sum_{j=1}^{N} w_{ij} x_j(t) - h + \varepsilon_i(t+1) \tag{1}$$

$$x_i(t+1) = g(u_i(t+1)) \tag{2}$$

$$g(x) = \begin{cases} 1 & x \geq 0 \\ 0 & x < 0 \end{cases} \tag{3}$$

$$i = 1, 2, \ldots, N$$

where N is the number of neurons contained in the network; t is a discrete time, $\varepsilon_i(t)$ denotes random noises which are assumed as mutually independent, zero-mean white Gaussian noises with an identical variance σ^2 and h (> 0) is the inhibitory input which is fixed here independent of neurons for the sake of simplicity.

Synaptic weights w_{ij} are defined following the case of an associative memory with M memorized patterns (Amit, 1989). This possibly enables the parametric control of the fundamental structure of the network attractor. In this case, the symmetry condition $w_{ij} = w_{ji}$ is satisfied.

State evolution of the network is performed in the asynchronous (cyclic) manner (Amit, 1989). The memorized and initial patterns are given as equiprobable binary random sequences.

The data length, 10000 Monte Carlo steps (MCS), is selected to estimate PSD in the frequency band width of three decades with sufficient statistical reliability. PSDs of actual neuronal activities referred here were given in the similar frequency band width (Yamamoto et al., 1986a). Furthermore, the data length of the neuronal spike train analyzed was at most several hundred seconds. Comparing this actual data length with 10000 MCS, 1 MCS could correspond to several ten milliseconds. This could be regarded as a time unit during which a neuron keeps its state active (1) or inactive (0). The neuronal state may be determined responsible for the number of spikes during this time unit. This time resolution is presumably sufficient, considering that the firing rates of actual neurons under study were at most less than 100 spikes/sec, and the frequency range concerned is lower than 1 Hz (Yamamoto et al., 1986a).

Spectral analysis of neuronal activities in neural network model

Typical PSD profiles of single neuronal activities in the network model are shown in Fig. 3. Unless otherwise stated, the number of neurons: $N = 100$ and the number of memorized patterns: $M = 30$. In Fig. 3 activity of a single neuron is picked up from 100 neurons included in the network. The raster plot of $x_i(t)$ is shown together with the corresponding PSD, where a dot indicates $x_i(t) = 1$. As one can see, PSD changes its profile from the $1/f$ to the flat as the inhibitory input increase. Here, the parameter values, h and σ, are selected so that most of neurons in the network show the $1/f$ PSD profiles under the weak inhibition, and the flat PSD profiles under the strong. The time series: $x_i(t)$ responsible for the $1/f$ PSD shows larger and slower variations than that for the flat PSD. Naturally, the activity is reduced as the inhibitory input

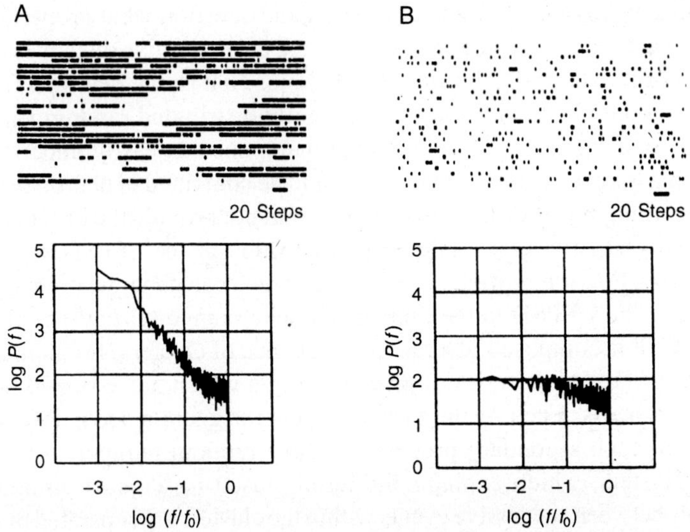

Fig. 3. Simulation results on the dynamics transition of neuronal activities in the neural network model under weak and strong inhibitions, *h*, where M = 30 and σ = 0.26. Raster plots of single activities and the corresponding PSDs for the picked up neuron are shown together. A shows the results for *h* = 0.40 and B for *h* = 0.50. In PSD, the frequency axis is normalized by f_0 which denotes an arbitrary standard frequency (Adapted from Nakao et al., 1997).

increases. As described previously (Nakao et al., 1990), the strong and the weak inhibitory inputs are responsible for SWS and REM in our framework, respectively. Qualitatively, the PSD profiles and the temporal characteristics of activities are well reproduced in our simulations. Through the simulations presented here, an inhibitory input is manipulated so that an activity under the strong inhibition is reduced about 10% of that under the weak inhibition. This proportion is selected as the limit of inhibition, because, under the stronger inhibition, the number of excitatory states could not be sufficient to statistically validate the estimated PSD.

Figure 4 shows σ-h dependency of the neuronal dynamics in terms of PSD. Within the appropriate range of σ and h, a larger h is shown to need a smaller σ for neuronal activities to exhibit the $1/f$ PSD, and vice versa. In this sense, these parameters exert compensatory effects on generating the $1/f$ PSD. Outside of this range, an excessive noise tends to randomize neuronal activities, which results in the flat PSD in the lower frequency range. Similarly, an excessive inhibition results in the flat PSD. These phenomena could be understood from the metastable structure of the network attractor (see section 3.4).

Cluster activities underlie the 1/f PSD

The single neuronal activities in the neural network model are shown to reproduce the dynamics-transition. Then, the question what properties of the neuronal activities are essential for exhibiting the respective dynamics naturally arises. For the flat PSD, the answer is simple. That is, the neuronal firing rate only has to be identically and independently distributed random variable on the time scale of interest. On the other hand, one can easily understand that somewhat longer-lasting excitations should be contained in neuronal activities for exhibiting the $1/f$ PSD. However, it would not provide the exact condition. In order to clarify the essential property for exhibiting the $1/f$ PSD, we introduce a clustering Poisson process (CPP) model which can generate point processes exhibiting the $1/f$ PSD in the limited frequency range (Grüneis et al., 1989).

The CPP is composed of a random sequence of clusters; for an illustration see Fig. 5. There is a series of primary events which are assumed to form a Poisson process. Each of the primary events triggers a secondary series of events. In each secondary process there are random number (m) of events called clusters, which explains the term' clustering Poisson process'. The intervals between successive events within the clusters are separated by random variables λ. Here, the secondary processes are also assumed to be Poisson processes (for more general cluster processes, see Takahashi et al. (1998)). The complete process is then the superposition of events originating from different clusters. It consists of an indistinguishable sequence of events, where the cluster formation is no longer obvious. The CPP model is an established framework for characterizing stochastic point processes which contain clusters of events (Takahashi et al., 1998).

Here, the spectral features of the CPP as concerns $1/f$ noise are of interest (Grüneis and Baiter, 1986). Let $\langle n_c \rangle$ be the parameter of the primary process; $\langle n_c \rangle$ can be regarded as the mean rate of cluster occurrence. The cluster-size, m, denotes the number of events in a cluster; then P_m is called the cluster-size distribuion and the mean number of events within a cluster is $\langle m \rangle$. The mean rate, $\langle n_{tot} \rangle$, of overall events is

$$\langle n_{tot} \rangle = \langle n_c \rangle \langle m \rangle. \tag{4}$$

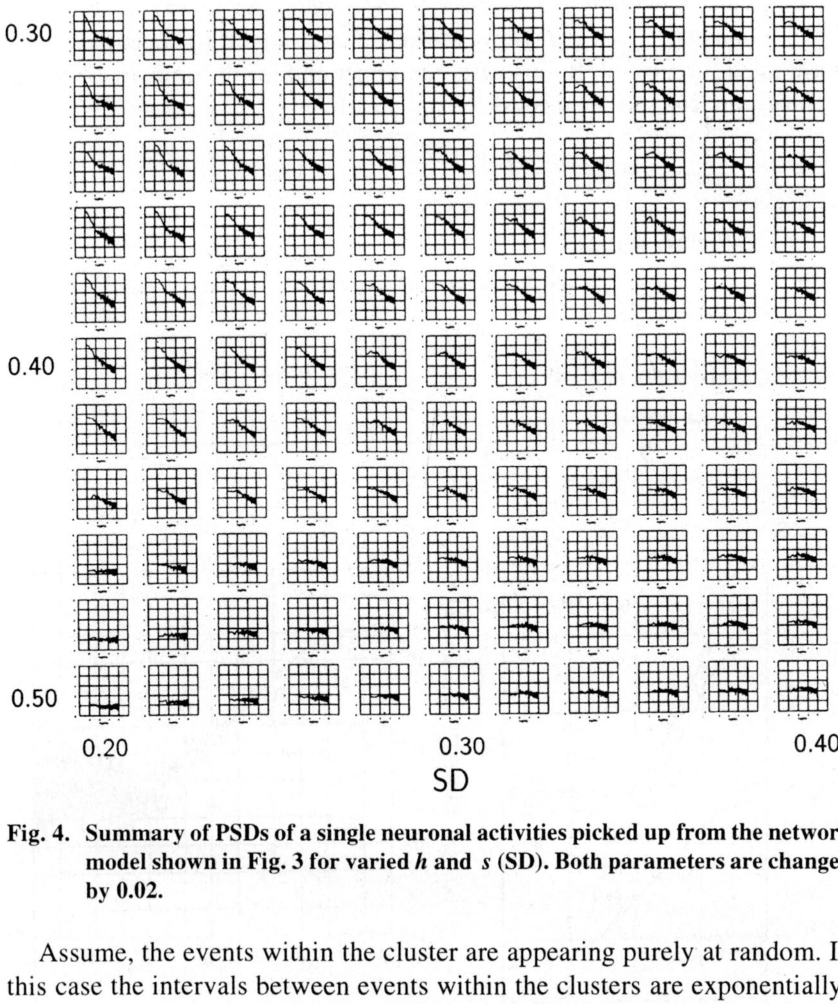

Fig. 4. Summary of PSDs of a single neuronal activities picked up from the network model shown in Fig. 3 for varied *h* and *s* (SD). Both parameters are changed by 0.02.

Assume, the events within the cluster are appearing purely at random. In this case the intervals between events within the clusters are exponentially-distributed. In terms of the characteristic function $u(f)$, the Fourier transformation of the probability density function of λ, and the cluster-size distribution P_m, the PSD (one-sided spectrum) of the CPP is given by

$$P(f) = 2\langle n_{tot} \rangle + 2\langle n_c \rangle Q(f), \tag{5}$$

whereby $$Q(f) = 2Re\left[\frac{u}{1-u}\right] + 2Re\left[\frac{u}{(u-1)^2}\left(\sum_{m=1}^{N_0} P_m u^m - 1\right)\right]. \tag{6}$$

$Q(f)$ describes mutual time relations among events within the cluster. The first term in (5) represents white noise; the second term in (5) is called excess noise due to cluster formation.

In the following a cluster-size distribution as

$$P_m \propto m^z \ (m = 1, 2, ..., N_0) \tag{7}$$

is of interest; here N_0, the maximum cluster-size has been introduced which limits the cluster to a finite size.

Numerical calculation of spectrum $P(f)$ according to (5) is shown in Fig. 5. For a discussion of $P(f)$, we are confined to a pure $1/f$ spectrum which is obtained for $z = -2$. In this case, the upper and lower limits to $1/f$ noise, f_1 and f_0, are roughly expressed for $N_0 >> 1$ by

$$f_1 \propto (\log N_0)^{-1} \tag{8}$$

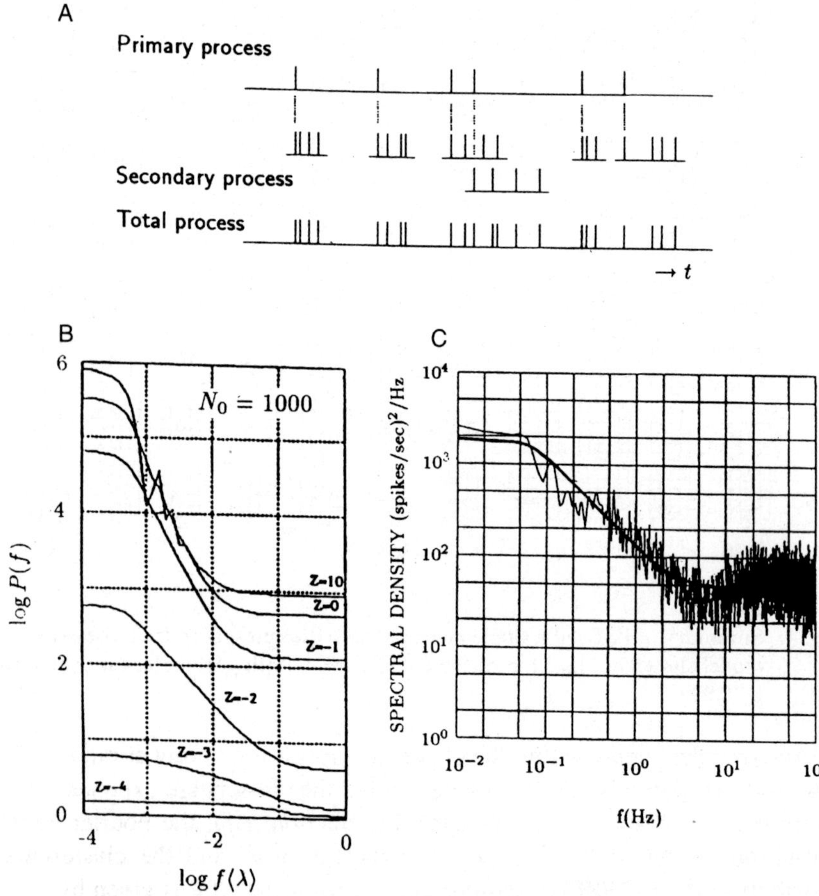

Fig. 5. Spectral structure of clustering Poisson process (CPP). A: Illustration of the structure of CPP. B: PSDs for varied *z*, the exponent of the power law distribution of cluster size (see eq. (7)). C: Similarity between PSDs of the CPP and the actual single neuronal activity in the mesencephalic reticular formation of the cat.

$$f_0 \propto N_0^{-1} \tag{9}$$

Similarly, the ratio between $P(0)$ and $P(\infty)$ are given by

$$\frac{P(0)}{P(\infty)} \sim \frac{N_0}{\log N_0} \tag{10}$$

This relation shows that the range of $1/f$ noise is only dependent on the maximum cluster-size N_0. The above results show that the parameters characterizing the statistics of the secondary process primarily shape the PSD of the CPP. The spectrum of the cluster process can be fitted to the spectrum of the neuronal spike train with reasonable success (Grüneis et al., 1989; Grüneis et al., 1993).

Shortly, it is indicated that the power-law distribution of the cluster size is essential for the CPP to exhibit the $1/f$ PSD. The cluster of events in CPP can be characterized by highly active states and inactive ones: superpositions of bursts of various durations. In fact, the neuronal activities during REMS in the cat's brain were observed to include highly active periods (Yamamoto et al., 1986a). On the other hand, in network model the cluster corresponding to a long-lasting excitatory period was clearly shown (Nakao et al., 1990). The distribution of the excitatory period of the single neuronal activity exhibiting the $1/f$PSD showed the long-tailed profile rather than a geometrical distribution (Nakao et al., 1990). These results also confirm the essential property with which the neuronal activities exhibit the $1/f$ fluctuations.

Dynamics of Neural Network in State Space

So far, we have studies the dynamics of neural network model which receives the global inhibition and the random noise. Here, the structural change of the network attractor associated with the inhibition is investigated through the state space analysis.

For the same symmetry network in Fig. 3, activities of all neurons (network activity) are briefly presented in Fig. 6A under the weak inhibition; B under the strong inhibition.

Under the weak inhibition (Fig. 6A), the network activity explicitly indicates that the regular and irregular patterns appear alternatively with varied durations. In the regular states, several different stripe patterns can be clearly seen. In contrast, only the irregular state becomes dominant under the strong inhibition. It can be shown that these stripe patterns correspond to the vicinities of equilibrium states under this condition, while the irregular pattern corresponds to the vicinity of the "0" state where all neurons are silent. Naturally, these equilibrium states except for the "0" state are different from the initially memorized ones due to the global inhibition. The closest reference equilibrium state to the current network state is determined every MCS in terms of a direction cosine (DC). DC here represents the "agreement" between a current network state $x(t)$ and a certain reference state. For the networks appeared in Fig. 3, reference equilibrium states including the "0" state are reached from 4000 statistically independent initial states under no noise, i.e., $\sigma = 0$. Here, the "0" state is denoted by $x_0 = [0, 0, ..., 0]$. Sixtythree and 2 equilibrium states are found for the networks in Figs. 3A and 3B, respectively. For all references, the closest reference to the current state is determined step by step by comparing the magnitude of the corresponding DCs.

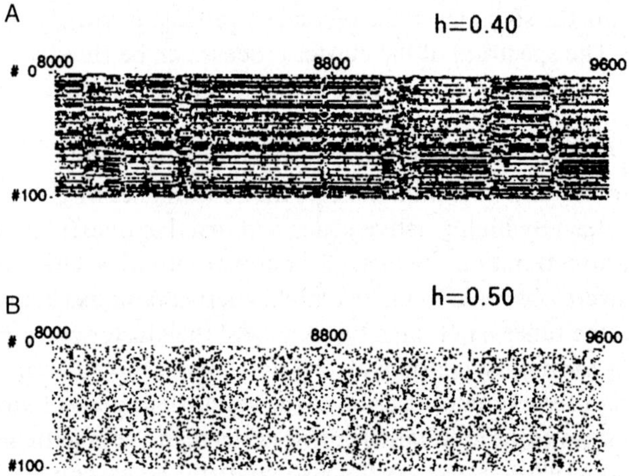

Fig. 6. Dynamics of network state evolution for the network appeared in Fig. 3. This shows the brief sequences of all neuronal activities, where the numbers on the left end denote neuron numbers, and the number on the top indicates the number of steps from the beginning of evolution (Adapted from Nakao et al., 1997).

Under the weak inhibition, the network state is wandering among the vicinities of the equilibrium states. Here, the equilibrium states are not absolutely stable, because intermittent transitions among them driven by the noise are observed. In this sense, they are denoted here as "metastable equilibrium states" or simply metastable states following the terminology of statistical physics (Amit, 1989). In spite of a constant drive by the noise, the network state is trapped in metastable states for a certain period. In the irregular states, the network is possibly wandering around the vicinity of the "0" state. While in the vicinity of the "0" state, each neuronal state is expected to be determined by an instantaneous value of the noise rather than inputs from the other neurons. This is presumably the reason why the spatio-temporal activity pattern looks random. Here, the "metastability" represents the structural properties of the network attractor in which such metastable equilibrium states dominantly exist. Therefore, the following description is possible concerning the dynamics transition of single neuronal activities. The globally applied inhibitory input modifies the structure of the network attractor. In the weakly inhibited case, the metastability of the network attractor becomes dominant so as to realize the $1/f$ fluctuations of single neuronal activities, and in the strongly inhibited case, the "0" state becomes the global attractor which underlies low and random activities. In other words, we suggest that these behavior reflect the geometrical structure of the network attractor: a "bumpy" structure becomes dominant under the strong inhibition, and rather monotonous structure dominates under the strong.

Discussion

The dynamics-transition of the neuronal activities during sleep was extensively observed in the cat's brain, which was subject to modeling by the artificial neural network with the global inhibition and the random noise. The simulation of the phenomenon based on this model was successful. Especially for the neuronal activities exhibiting the $1/f$ PSD during REMS, exploring the essential property for the neuronal activities to exhibit the $1/f$ PSD by the CPP model found that the long-lasting active states should be included. This property was shown to be shared by the actual neuronal activities and those in the neural network model. By the analysis of the state space activities of the neural network model showed that the structural change in the network attractor associated with the global inhibition underlied the reproduced neuronal dynamics transition: the metastable structure as pronounced under the weaker inhibition (REMS) and the '0' state was only attractive under the stronger inhibition (SWS).

1/f fluctuation of neuronal activity—A novel indicator of REM sleep

The $1/f$ fluctuations of the neuronal activities are found in the various regions of the cat's brain during REMS. From the statistical analysis on the actual and modeled neuronal activities, the superpositions of long lasting activations such as bursting are shown to be essential for the neuronal activities to exhibit the $1/f$ PSD. This phenomenon might provide a more basic definition than the conventional polygraphic measures. REMS is conventionally defined as a state where a set of various REMS phenomena is observed, i.e., rapid eye movements, EEG desynchrony, hippocampal theta waves, ponto-geniculo-occipital (PGO) spikes, muscular twitches, autonomic irregularity and muscle atonia. However, not all animals show all of these signs of REMS (Siegel et al., 1996; Siegel, 1997). Such animal species not showing distinct polygraphic characteristics of REMS are frequently encountered in phylogenetic studies of REMS (Siegel et al., 1996). In this case, dynamics of neuronal activities might be useful to detect physiological phenomena of incompletely developed REMS or sleep. If the $1/f$ neuronal dynamics in the extended areas of the brain is a general phenomenon, REMS could be defined as a state where the $1/f$ neuronal dynamics prevails during sleep (Yamamoto et al., 1997).

Physiological substrate of the global neuromodulation

Both serotonergic and noradrenergic neurons show a depression or cessation of firing during REM sleep (McGinty and Harper, 1976; Shima et al., 1986;

Hobson et al., 1975; Jacobs, 1985). The extracellular concentration of serotonin during REM sleep becomes lowest both in the dorsal raphe nucleus (Portas and McCarley, 1994) and in the region of the medial pontine reticular formation (Iwakiri et al., 1993). As for the cholinergic system, the extracellular concentration of acetylcholine becomes minimal during slow wave sleep, and increases during REM sleep, not only in the pontine reticular formation but also in the forebrain structure such as the caudate nucleus (Kodama et al., 1990).

In relation to these experimental findings, following pharmacological experiments were performed on the hippocampal theta and pyramidal neurons and the thalamic ventrobasal complex neurons (Mushiake et al., 1988; Kodama et al., 1989b). Under the administration of p-chlorophenylalanine (PCPA), an inhibitor of serotonin synthesis, the activity of the hippocampal theta and pyramidal neurons as well as thalamic neurons clearly displayed the $1/f$ dynamics (Mushiake et al., 1988; Kodama et al., 1989b), which was accompanied by the similar phenomena characterizing REM sleep, such as rapid eye movements, cortical desynchronization, hippocampal theta waves and PGO activities (Mushiake et al., 1988). Under this condition, administration of 5-Methoxy-N, N-dimethyltryptamine (5MeODMT) nonselective serotonergic agonist, abolished both the polygraphically REMS-like state and the $1/f$ neuronal dynamics, and the white dynamics was obtained (Mushiake et al., 1988; Kodama et al., 1989b). In addition, when atropine sulfate was administered under the condition of the PCPA pretreatment, the $1/f$ dynamics was also abolished, indicating a muscarinic receptor action is necessary for the generation of the 1/f dynamics. Concomitantly, we have preliminarily shown that the nicotinic receptor action is not required for the $1/f$ dynamics of the MRF neuronal activity, because a nicotinic antagonist mecamylamine tended to fail to abolish the $1/f$ neuronal dynamics (Yamamoto et al., 1993).

According to reports concerning the receptor mechanisms, serotonergic influences are excitatory as well as inhibitory (McCormick and Pipe, 1990; Adrien, 1995), of which situation is the same for all other amines. However, in the present simulation study, we assumed each neuron to have an inhibitory postsynaptic input which is common to all neurons of the network, for mimicking the assumed input of an overall biasing from the global neuromodulation system such as the aminergic (serotonergic, noradrenergic, dopaminergic, etc.) and the cholinergic system. The results successfully interpreted the neuronal dynamics-transition, which were suggestive for understanding the concept of the overall effect from the global neuromodulation system. During slow wave sleep, the global neuromodulation from the aminergic and the cholinergic system can be interpreted as inhibitory and disfacilitory, respectively, making the dynamics of the central nervous system white and keeping its activity minimal; therefore the overall biasing effect can be interpreted as inhibitory. Whereas, during REM sleep, overall effect of the global aminergic disinhibition with the cholinergic facilitation can be interpreted as producing the state in which most of the central single neurons are maximally activated with the $1/f$ dynamics. Here, it should be also noticed that the aminergic influence is dependent on the concentration of its ligand, since a higher concentration of serotonin locally diffused generated the $1/f$ dynamics of a thalamic neuronal activities even during slow wave sleep (Nakamura et al., 1997).

Underlying mechanism generating 1/f fluctuations

We have suggested that the structural change of the network attractor associated with inhibitory inputs from the global neuromodulation system could be the underlying mechanism for the neuronal dynamics transition we observed physiologically (Nakao et al., 1990; Nakao et al., 1997). Especially, the metastable properties of the network attractor were found to play a main role for generating the 1/f fluctuations of neuronal activities. In addition, diluting connections in the network was shown to modify the structure of the network attractor so that the dynamics transition of neuronal activities was resulted, which was similar to that induced by modifying the global inhibition (Nakao et al., 1997). This result generalizes the conditions for generating the $1/f$ fluctuations and the dynamics-transition of neuronal activities. Therefore, one of the essential factors is suggested to be a balance between the external inputs such as the global inhibition and the noise, and the inputs from the other interconnected neurons, i.e. network inputs. In other words, when the network inputs are comparable with the external inputs, the metastability of the network attractor distinctly appears: the $1/f$ fluctuations of neuronal activities are shown. In contrast, when the external inputs exceed the network input, the "0" state is highly attractive; the neuronal activities exhibit the flat PSD. Accordingly, mechanisms underlying the actual neuronal dynamics-transition might be anticipated as follows. During SWS, neurons receive either stronger inhibitory inputs from the global neuromodulation system or less input magnitude from interconnected neurons in comparison with REM sleep. In contrast, during REM sleep, either neurons are released from the inhibitory inputs or they receive inputs of comparable magnitude from their interconnected neurons.

Functional implications of REM sleep from the viewpoint of 1/f fluctuations

So far, so many hypotheses on the function of REMS have been proposed, including contributions to development and maintenance of neural circuits in the brain, survival in the severe environment, and higher-order brain function such as learning and memory (Hartmann, 1973; Horne, 1988). However, there have been no theories widely accepted by sleep researchers. Here, in relation to the 1/f fluctuations in neuronal activities during REMS, its possible role in the higher-order brain function is discussed. Crick and Mitchison (1983) postulated that REMS functions so as to remove certain undesirable modes of interaction in neural networks of the cerebral cortex by the nonspecific random excitation (PGO waves), which are known to be generated in the brainstem and to be delivered to the cerebral cortex during REMS. Their idea was concurrent with the "unlearning" algorithm (Hopfield et al., 1983). However, its physiological reality has not yet been known. Concerning the relationship between the *1/f* fluctuations of a single neuronal activity and PGO waves, we have the preliminary physiological result that the *1/f* fluctuations still survived under the abolishment of PGO waves by the administration of mecamylamine (Yamamoto et al., 1993). Nevertheless, the metastable behavior of the network activity appears to be suitable for unlearning, because depth of a well where a metastable state is located is reflected in the corresponding staying time. Appropriate unlearning could be performed by reducing synaptic weights every time when the network stays in the vicinity of a metastable state.

The relationship between the *1/f* fluctuations and dreaming in REMS is more or less to be discussed. Dreaming might be regarded as a random recalling process of memorized patterns without logical context. In this respect, the metastable behavior of the artificial neural network could be analogous to dreaming. Therefore, our model-based approach could provide novel information for investigating the functions of dreaming and REM through the *1/f* fluctuations in neuronal activities. From the cognitive point of view, the computational network models on dreaming have been proposed (Hinton et al., 1995; Antrobus, 1997). In addition, there have been many physiological and psychological studies which suggest an important role of REMS in memory and learning processes such as Wilson and McNaughton (1994) and Smith (1995). The relationship between these ideas and the *1/f* fluctuations deserves to study further.

As for the neuronal dynamics during wakefulness, the power spectrum was different from one neuronal group to another (Yamamoto et al., 1986a; Mushiake et al., 1988; Kodama et al., 1989a). During the quasi-stationary alert state where the cat attended to a couple of small birds, the mesencephalic reticular formation had the *1/f* neuronal dynamics for 0.1-1 Hz and tended to have the white one for lower than approximately 0.1 Hz, and the hippocampal theta neurons had a significant tendency with white dynamics for 0.01-1 Hz (Mushiake et al., 1988; Kodama et al., 1989a). These results must indicate strong influences from the aminergic and cholinergic systems on the global modulation of the brain unit activity during wakefulness, in accordance with the fact that all of these systems are maximally active during wakefulness (McGinty and Harper, 1976; Shima et al., 1986; Hobson et al., 1975; Jacobs, 1985; Portas and McCarley, 1994; Adrien, 1995; Iwakiri, 1993; Kodama et al., 1989b).

Here, we comment on the exceptional neurons which do not show the *1/f* dynamics during REM sleep. Aminergic neurons show no dynamics because of none of firings. Brainstem PGO-on neurons (Sei et al., 1993) and about half of the basal forebrain neurons (Yamamoto et al., 1991) displayed a white dynamics during REM sleep. Moreover, some of the hypothalamic and the state-independent nigral neurons did not have the $1/f$ but white-noise-like dynamics during REM sleep (unpublished observation). Function of these neurons may be modulatory.

Conclusion

Here, we have demonstrated that investigating the dynamics of neuronal activities could provide the novel information concerning REM sleep. The key factors are: $1/f$ fluctuations in neuronal activities, and the metastability of the neural networks. The $1/f$ fluctuation is shown to potentially characterize REM sleep in the cases where the conventional physiological indicators of REMS are not available. On the other hand, the metastability, which is suggested to underlie the $1/f$ fluctuations, could provide the useful framework for understanding functional implications of REM sleep.

Acknowledgments

This study was partly supported by The Toyota Foundation (1987–1990); by Grant-in Aid for Scientific Research (07459003) from the Ministry of Education, Science, Sports, and Culture, Japan (1995–1997); and by Special Coordination Funds for Promoting Science and Technology from the Science and Technology Agency, Japan (1996–1998); and by the Cooperative Research Project Program of the Research Institute of Electrical Communication, Tohoku University (1996–1998). The authors thank all of previous co-workers who studied on the central single neuronal activities.

References

Adrien, J., The serotonergic system and sleep-wakefulness regulation, (In: A. Kales (Ed.), *Handbook of Experimental Pharmacology 116, Pharmacology of Sleep,* Springer-Verlag, Berlin, 1995, pp. 91–116.

Amit, D.J., Modeling Brain Function. Cambridge University Press, Cambridge, 1989.

Antrobus, J., Dream theory 1997: toward a computational neurocognitive model, Sleep Res. Soc. Bulletin, 3 (1997) 5–10.

Crick, F. and Mitchison, G. The function of dream sleep, Nature, 304 (1983) 111–114.

Grüneis, F. and Baiter, H.-J. More detailed explication of a number fluctuation model generating $1/f$ pattern, Physica, 136A (1986) 432–452.

Grüneis, F., Nakao, M., Yamamoto, M., Musha, T. and Nakahama, H. An interpretation of 1/f fluctuations in neuronal spike trains during dream sleep, Biol. Cybern., 60 (1989) 161–169.

Grüneis, F., Nakao, M., Mizutani, Y., Yamamoto, M., Meesmann, M. and Musha, T. Further study on 1/f fluctuations observes in central single neurons during REM sleep, Biol. Cybern., 68 (1993) 193–198.

Hartmann, E.L. The functions of sleep. Yale University Press, New Haven, 1973.

Hinton, G.E., Dayan. P., Frey, B.J. and Neal, R.M. The "wake-sleep" algorithm for unsupervised neural networks, Sciences, 268 (1995) 1158–1161.

Hobson, J.A., McCarley, R.W. and Wyzinski, P.W. Sleep cycle oscillation: reciprocal discharge by two brain stem neuronal groups, Science, 189 (1975) 55–58.

Hopfield, J.J., Feinstein, D.I. and Palmer, R.G. "Unlearning" has a stabilizing effect in collective memories, Nature, 304 (1983) 158–159.

Horne, J., Why we sleep. Oxford University Press, Oxford, 1988.

Iwakiri, H., Matsuyama, K. and Mori, S. Extracellular levels of serotonin in the medial pontine reticular formation in relation to sleep-wake cycle in cats: a microdialysis study, Neurosci. Res., 18 (1993) 157–170.

Jacobs, B.L., Over view of the activity of brain monoaminergic neurons across the sleep-wake cycle. In: A.Wauquier, J.M. Monti, J.M. Gaillard and M. Radulovcacki (Eds.), *Sleep: Neurotransmitters and Neuromodulators,* Raven Press, New York, 1985, pp. 1–14.

Kodama, T., Mushiake, H., Shima, K., Nakahama, H. and Yamamoto, M. Slow fluctuations of single unit activities of hippocampal and thalamic neurons in cats. I. Relation to natural sleep and alert states, Brain Res., 487 (1989a) 26–34.

Kodama, T., Mushiake, H., Shima, K. Hayashi, T. and Yamamoto, M. Slow fluctuations of single unit activities of hippocampal and thalamic neurons in cat. II. Role of serotonin on the stability of neuronal activities, Brain Res., **487** (1989b) 35–44.

Kodama, T., Takahashi, Y. and Honda, Y. Enhancement of acetylcholine release during paradoxical sleep in the dorsal tegmental field of the cat brain stem, Neurosci. Lett., 114 (1990) 277–282.

McCarley, R.W., Greene, R.W., Rainnie, D. and Portas, C.M. Brainstem neuromodulation and REM sleep, Neurosci., 7 (1995) 341–354.

McCormick, D.A. and Pipe, H.C. Noradrenergic and serotonergic modulation of a hyperpolarization-activated cation current in thalamic relay neurons, J. Physiol., London 431 (1990) 292–318.

McGinty, D. and Harper, R.M. Dorsal rahe neurons: depression of firing during sleep in cats, Brain Res., 101 (1976) 569–575.

Mushiake, H., Kodama, T., Shima, K., Yamamoto, M. and Nakahama, H. Fluctuations in spontaneous discharge of hippocampal theta cells during sleep-waking states and PCPA-induced insomnia, J. Neurophysiol., 60 (1988) 925–939.

Nakamura, K., Takahashi, K., Yamamoto, M., Kodama, T., Nakao, M., Katayama, N. and Endo, K. Effects of serotonin on fluctuations in single neuronal activities in the region of the lateral geniculate nucleus of a cat, Abstracts 20th Ann. Meeting Jap. Neurosci. Soc., (1997) S248.

Nakao, M., Takahashi, T., Mizutani, Y. and Yamamoto, M. Simulation study on dynamics transition in neuronal activity during sleep cycle by using asynchronous and symmetry neural network model, Biol. Cybern., 63 (1990) 243–250.

Nakao, M., Honda, I., Musila, M. and Yamamoto, M. Metastable associative network models of dream sleep, Neural Networks, 10 (1997) 1289–1302.

Portas, C.M. and McCarley, R.W. Behavioral state-related changes of extracellular serotonin concentration in the dorsal raphe nucleus: a microdialysis study in the freely moving cat, Brain Res., 648 (1994) 306–312.

Sei, H., Sakai, K., Yamamoto, M. and Jouvet, M. Spectral analyses of PGO-on neurons during paradoxical sleep in freely moving cats, Brain Res., **612** (1993) 351–353.

Shima, K., Nakahama, H. and Yamamoto, M. Firing properties of two types of nucleus raphe dorsalis during the sleep-waking cycle and their responses to sensory stimuli, Brain Res., 399 (1986) 317–326.

Siegel, J.M., Manger, P.R., Nienhuis, R., Fahringer, H.M. and Pettigrew, J.D. The echidna tachyglossus acueatus combines REM and non-REM aspects in a single sleep sate: Implications for the evolution of sleep, J. Neurosci., 16 (1996) 3500–3506.

Siegel, J.M., Sleep in monotremes; Implications for the evolution of REM sleep, In: O. Hayaishi and S. Inoue (Eds.), *Sleep and Sleep Disorders: From Molecule to Behavior,* Academic Press, Tokyo, 1997, 113–128.

Smith, C., Sleep states and memory processes, Behavioral Brain Res., 69 (1995) 137–145.

Takahashi, K., Nakamura, K., Yamamoto, M., Kodama, T., Nakao, M., Katayama, N. and Endo, K. Dynamics-transition of central single neuronal activities during sleep, Abstracts 20th Ann. Meeting Jap. Neurosci. Soc., (1997) S249.

Takahashi, T., Nakao, M., Grüneis, F., Mizutani, Y. and Yamamoto, M. Higher-order spectra of cluster point processes generating 1/f fluctuations, Interdescipl. Inform Sci., in Press.

Yamamoto, M., Nakahama, H., Shima, K., Kodama, T. and Mushiake, H. Markov-dependency and spectral analysis on spike-counts mesencephalic reticular neurons during sleep and attentive states, Brain Res., 366 (1986a) 279–298.

Yamamoto, M., Nakahama, H., Shima, K., Aya, K., Kodama, T., Mushiake, H. and Inase, M. Neuronal activities during paradoxical sleep, Adv. Neurolog. Sci., 30 (1986b) 1010–1022.

Yamamoto, M., McGinty, D., Nakao, M., Szymusiak, R. and Mizutani, Y. Dynamics of single neuronal activities in cat's basal forebrain during sleep states, Jap. J. Psychiat. Neurol., 45 (1991) 971.

Yamamoto, M., Arai, H., Takahashi, T., Sasaki, N., Nakao, M., Mizutani, Y. and Kodama, T. Pharmacological basis of 1/f fluctuations of neuronal activities during REM sleep, Sleep Res., 22 (1993) 458.

Yamamoto, M., 1/f fluctuations observed in single central neurons during REM sleep, In T. Musha and Y. Sawada (Eds.), *Physics of Living State,* Ohmsha, Tokyo, 1994a, pp. 211–222.

Yamamoto, M., Nakao, M., Mizutani, Y., Takahashi, T., Watanabe, K., Arai, H. and Sasaki, N. Pharmacological and model-based interpretation of neuronal dynamics transitions during sleep-waking cycle, Method of Inform. Med., **33** (1994b) 125–128.

Yamamoto, M., Nakao, M., Kodama, T., Hanzawa, A., Nakamura, K. and Katayama, N. Dynamics transition of cortical neuronal activity between REM sleep and slow wave sleep, Abst. 4th IBRO Cong. Neurosci., (1995) D 11.2, 404.

Yamamoto, M., Nakao, M. and Kodama, T. A possible mechanism of dynamics-transition of central single neuronal activity during sleep, (In: O. Hayaishi and S. Inoué (Eds.), *Sleep and Sleep Disorders: From Molecule to Behavior,* Academic Press, Tokyo, 1997, 81–95.

Wilson, M.A. and McNaughton, B.L. Reactivation of hippocampal ensemble memories during sleep, Science, 265 (1994), 676–679.

Author Index